Veterans at Risk

The Health Effects of Mustard Gas and Lewisite

Constance M. Pechura and David P. Rall, Editors

Committee to Survey the Health Effects of Mustard Gas and Lewisite
Division of Health Promotion and Disease Prevention
Institute of Medicine

National Academy Press
Washington, D.C. 1993

NATIONAL ACADEMY PRESS 2101 Constitution Avenue, N.W. Washington, D.C. 20418

NOTICE: The project that is the subject of this report was approved by the Governing Board of the National Research Council, whose members are drawn from the councils of the National Academy of Sciences, the National Academy of Engineering, and the Institute of Medicine. The members of the committee responsible for this report were chosen for their special competences and with regard for the appropriate balance.

This report has been reviewed by a group other than the authors according to procedures approved by a Report Review Committee consisting of members of the National Academy of Sciences, the National Academy of Engineering, and the Institute of Medicine.

The Institute of Medicine was chartered in 1970 by the National Academy of Sciences to enlist distinguished members of the appropriate professions in the examination of policy matters pertaining to the health of the public. In this, the Institute acts under both the Academy's 1863 congressional charter responsibility to be an advisor to the federal government and its own initiative in identifying issues of medical care, research, and education.

The work on which this publication is based was performed pursuant to Contract No. V101(93) P-1326 with the Department of Veterans Affairs. Funds for this contract were provided by the Department of Veterans Affairs.

Library of Congress Cataloging-in-Publication Data

Institute of Medicine (U.S.). Committee to Survey the Health Effects of Mustard Gas and Lewisite.

Veterans at Risk : the health effects of mustard gas and Lewisite / Constance M. Pechura and David P. Rall, editors.

p. cm.

"Committee to Survey the Health Effects of Mustard Gas and Lewisite. Division of Health Promotion and Disease Prevention. Institute of Medicine, National Academy of Sciences."

Includes bibliographical references and index.

ISBN 0-309-04832-X

1. Mustard gas—Toxicology. 2. Lewisite (Poison gas)—Toxicology. I. Pechura, Constance M. II. Rall, David P. III. Title.

RA1247.M8158 1993

615.9'1—dc20 92-40735

CIP

Printed in the United States of America.

The serpent has been a symbol of long life, healing, and knowledge among almost all cultures and religions since the beginning of recorded history. The image adopted as a logotype by the Institute of Medicine is based on a relief carving from ancient Greece, now held by the Staatlichemuseen in Berlin.

Committee To Survey the Health Effects of Mustard Gas and Lewisite

DAVID P. RALL *(Chair),** Director (Retired),National Institute of Environmental Health Sciences, Washington, D.C.

O. MICHAEL COLVIN, Professor of Oncology and Medicine, Johns Hopkins University School of Medicine, Baltimore, Maryland.

ELLEN EISEN, Associate Professor,Department of Work Environment, College of Engineering, University of Massachusetts, Lowell.

WILLIAM EDWARD HALPERIN, Associate Director for Surveillance,Division of Surveillance, Hazard Evaluations and Field Studies, National Institute for Occupational Safety and Health, Cincinnati, Ohio.

CHARLES H. HOBBS, Assistant Director,Inhalation Toxicology Research Institute, Lovelace Biomedical and Environmental Research Institute, Albuquerque, New Mexico.

DAVID G. HOEL,* Director, Biometry and Risk Assessment Division, National Institute of Environmental Health Sciences, Research Triangle Park, North Carolina.

KARL KELSEY, Associate Professor of Occupational Medicine and Radiobiology, Harvard School of Public Health, Occupational Health Program, Boston, Massachusetts.

CHARLES J. McDONALD, Professor and Director,Division of Dermatology, Brown University, Physician in Charge, Division of Dermatology, Roger Williams Medical Center and Rhode Island Hospital, Providence.

JAMES MALCOLM MELIUS, Director,Division of Occupational Health and Environmental Epidemiology, State of New York Department of Health, Albany.

JOHN A. MONTGOMERY, Distinguished Scientist,Southern Research Institute, Birmingham, Alabama.

WILLIAM NICHOLSON, Professor of Community Medicine, Mount Sinai School of Medicine, New York, New York.

ROSWELL ROBERT PFISTER, Past Chairman,Department of Ophthalmology, University of Alabama, Director, Brookwood Eye Research Lab, Birmingham, Alabama.

MARGARET SINGER, Emeritus Adjunct Professor, Department of Psychology, University of California, Berkeley.

BAILUS WALKER,* Dean, College of Public Health, University of Oklahoma, Health Sciences Center, Oklahoma City.

ANNETTA P. WATSON, Research Staff Member, Health and Safety Research Division, Oak Ridge National Laboratory, Oak Ridge, Tennessee.

Former Members

VINCENT MARCHESI,** Director, Boyer Center for Molecular Medicine, Professor of Pathology, Biology, and Cell Biology, Yale University School of Medicine, New Haven, Connecticut.

LINDA ROSENSTOCK, Director of Occupational Medicine and Associate Professor, Department of Medicine and Environmental Health, University of Washington, Seattle.

Institute of Medicine Staff

Constance M. Pechura, Study Director
Catharyn T. Liverman, Research Associate
Jennifer Hope Streit, Project Assistant
Gerri Kennedo, Project Assistant
Gary B. Ellis, Director, Division of Health Promotion and Disease Prevention

* IOM member
** IOM and NAS member

Preface

So vivid were the memories of the first use of "mustard gas" (sulfur mustard) by the Germans in World War I that the United States government began to prepare for chemical warfare even before the Japanese attacked Pearl Harbor in 1941. This work was also spurred by the fury of war in Europe and reports of Japanese use of sulfur mustard against the Chinese. The U.S. preparations included the establishment of war-related research programs organized by President Roosevelt under the White House Office of Scientific Research and Development (OSRD). Two groups under the OSRD became involved in secret testing programs concerned with mustard agents (sulfur and nitrogen mustard) and Lewisite:

- The Committee on Medical Research
  This group studied protective ointments and other treatments through the National Research Council's Committee on Treatment of Gas Casualties.
- The National Defense Research Committee
  This group studied protective clothing and gas masks through military units such as the Chemical Warfare Service.

These testing programs involved the use of close to 60,000 military personnel as human experimental subjects. It was this use of human subjects more than 50 years ago that provided the impetus for the study reported in this volume. The initiation of this study in 1991 was finally prompted by long-delayed official admissions that human subjects had been used and the recognition that these subjects may have suffered adverse, long-term, health consequences as a result of their exposure to mustard agents or Lewisite.

The committee, convened to produce this report by the Institute of Medicine in response to a request by the Department of Veterans Affairs, was comprised of experts in the fields of toxicology, epidemiology, occupational and environmental medicine, ophthalmology, dermatology, oncology, chemistry, and psychology. Its task was to survey the medical and scientific literature on mustard agents and Lewisite, assess the strength of association between exposure to these agents and the development of specific diseases, identify the gaps in the literature, and recommend strategies and approaches to deal with any gaps found. To accomplish this task, the committee met four times, examined nearly 2,000 scientific and medical reports in English and a number of foreign languages, and considered input from 13 military and civilian experts and over 250 affected veterans, including public testimony from 20 veterans. Although this task may have seemed straightforward in the beginning of the study, closer examination of the literature and the World War II (WWII) experimental protocols presented numerous scientific and ethical challenges.

The major scientific challenges were the meager literature on long-term health effects of exposure to these agents and the lack of quantitative exposure data for the veterans who served as human test subjects. The vast majority of the scientific and medical literature was concerned with the short-term, acute effects of mustard agents and Lewisite, because the research priorities of most countries had been placed on treatment of battlefield injuries and the fact that most investigations of mustard agents and Lewisite have been conducted throughout this century under the control of military establishments. Particularly distressing was the essential lack of information regarding the toxicology of Lewisite. Assessing the long-term health effects of mustard agents and Lewisite thus required the committee to integrate many types of data, from studies using laboratory animals to single human case studies, and to examine and compare closely the known biological mechanisms of injury from these agents with agents with similar properties for which more data were available.

The lack of exposure data for the WWII human subjects caused the committee to attempt to gather as much information as possible about the experimental protocols, the equipment used, and any injuries from official reports of the testing programs. The committee found that an atmosphere of lingering secrecy still existed in the Department of Defense regarding some of the testing programs. Reports of the specific experimental protocols were not always easy to obtain; in some cases, reports were not available or were obtained as the study was almost complete. Fortunately, enough information was gathered to allow reasonable estimates of the exposures to human subjects, who were repeatedly exposed to mustard agents and Lewisite in gas chamber tests or under so-called field conditions.

As the full scope of the WWII testing protocols was revealed, compelling ethical questions emerged. At times, it seemed as if every new discovery only posed more questions. As the study progressed, the bits and pieces of information finally coalesced into a picture of abuse and neglect that was impossible for the committee to ignore. One of the first discoveries was that the end point of all the WWII mustard agent and Lewisite experiments was tissue injury—from mild skin burns to severe, and widespread, skin burns that took more than a month to heal. The chamber and field tests were actually called "man-break" tests.

Both veteran self-reports and official documents revealed that some subjects suffered damaging injuries to the lungs and upper respiratory system from inhalation of the agents. Committee analysis of expected gas mask efficiencies further showed that projected normal mask leakage under the hot, humid conditions of the gas chambers would have, in some cases, resulted in exposure levels as high as those reported on World War I battlefields.

The first response of many of the committee members to these discoveries was to try to understand the actions of the investigators in historical context—it was a war and the experiments were conducted before the Nuremberg Code of 1947 established formal principles to govern the proper treatment of human subjects. However, examination of the treatment and care of WWII chemical warfare production workers, and the conduct of later military experiments with human subjects from 1950 to 1975, demonstrated a well-ingrained pattern of abuse and neglect. Although the human subjects were called "volunteers," it was clear from the official reports that recruitment of the WWII human subjects, as well as many of those in later experiments, was accomplished through lies and half-truths.

Most appalling was the fact that no formal long-term follow-up medical care or monitoring was provided for any of the WWII human subjects, other exposed military personnel, or chemical warfare production workers, despite knowledge available by 1933 that mustard agents and Lewisite could produce long-term debilitating health problems, particularly in those people suffering severe burns and inhalation injuries. There was not even adequate short-term follow-up of the human subjects by the Department of Defense. Subjects in the chamber tests were sworn to secrecy and simply released on leave at the conclusion of the experiments. Some of these men still had blisters or evidence of skin burns upon release, but were not given any instructions about how to obtain knowledgeable medical care if they had needed it.

Although the experiments began in a wartime climate of urgency and secrecy, it was clearly a mistake in this case to continue the secrecy after the conclusion of the war. Follow-up of the exposed human subjects could have provided a wealth of information on the effects of these war gases and could have served as a basis for legitimate disability claims by injured subjects. By the end of the war, the use of nitrogen mustard as a chemotherapeutic agent (developed as part of the WWII testing program) clearly showed the serious health effects that the previous "volunteers" might be expected to experience.

In the face of the abuses uncovered, the committee members nevertheless sought to maintain an appropriate balance of their scientific responsibilities in assessing the available literature and their ethical responsibilities as physicians and scientists. In this effort, the committee members were guided by their stated task and their own individual judgments of the scientific and historical information examined. Thus, the committee believes that the findings and recommendations contained in this report are entirely justified by the scientific, medical, and historical evidence examined. There are, however, specific statements the committee wishes to offer as commentary on its findings.

First, the committee believes that each veteran who served as a human subject in the WWII experiments deserves honor for his sacrifice. [1] These men risked their health and safety to help develop better means of protection against chemical warfare. Yet, in most cases, their participation in these experiments was not even acknowledged in their service records and was, in fact, officially denied for decades. Further, these men were ordered to keep their participation secret. They did so for nearly 50 years, in some cases despite serious, disabling diseases that they believed were caused by their exposures. There can be no question that some veterans, who served our country with honor and at great personal cost were mistreated twice—first, in the secret testing and second, by the official denials that lasted for decades. They deserve recognition.

Second, the committee believes that any future military research with human subjects should be conducted according to publicly established ethical principles similar to those that apply to civilian research. The Department of Defense should consider including civilian medical experts in reviews of all proposed military research protocols involving human subjects. As was shown in the examination and evaluation by the Department of the Army Inspector General's report of the military drug and chemical testing programs from 1950 to 1975 (see Appendix F), a climate of secrecy provides a permissive environment for the neglect of

[1] According to all available reports, all the human subjects were males.

established rules of conduct. Such neglect should never be allowed to occur when human experimentation is involved.

Beyond the immense personal costs of the mistakes and failures of the United States government during and after the WWII testing programs, there are societal costs as well. The lack of available biological data concerning these chemical warfare agents also slowed the important fields of toxicology and cancer chemotherapy. Much would have been gained by careful observation after the end of WWII; instead much was lost.

The primary reason to identify and follow-up veterans exposed to mustard agents or Lewisite is to provide needed medical care. In addition, follow-up of these individuals now may also benefit our understanding of carcinogenesis. For example, recent advances in molecular biology have linked some chemical exposures in laboratory animals to specific changes in tumor cells; for example, activated oncogenes with unusual mutations or suppressor genes (and/or their protein products), or chromosome damage. In addition, it is well known that nitrogen mustard cancer chemotherapy can result in second tumors, which show unusual genetic changes. Therefore, study of any sulfur mustard-associated tumors should be explored, because the results could shed light on laboratory animal and human responses to carcinogens.

The committee wishes to acknowledge that this study could not have been done without the assistance of a number of people, many of whom are listed in the acknowledgments section of this report. Before this report was completed, the report draft was reviewed by experts in appropriate fields under the rules of the National Research Council's Report Review Committee. These individuals provided helpful commentary on the draft manuscript and the committee greatly appreciates the care and expertise that the reviewers brought to their task.

The work of the committee's Institute of Medicine staff deserves the highest praise. The committee is especially grateful for the thoughtful input, advice, and support given by Gary Ellis, the Director of the Division of Health Promotion and Disease Prevention. Thanks are also extended to Jennifer Streit, the study's project assistant responsible for planning travel and other meeting arrangements, who also translated some of the French papers requested by the committee. The massive job of finding, organizing, and procuring the hundreds of scientific papers and technical reports was accomplished with great skill by Catharyn Liverman, the study's research associate and medical librarian. The committee is truly indebted to Ms. Liverman—she always knew where something was, kept a thousand details straight, and did a wonderful job tracking down obscure references. Finally, the committee wishes to recognize the major contributions of the study director, Constance

Pechura. She knew and understood the literature, she worked tirelessly to obtain information from reluctant sources, and she organized the study plan, the meetings, the special presentations, and this final report. She clearly foresaw the major problems that this committee faced as it moved from the safe, but complex, problems of risk assessment to the thornier issue of human ethics.

This Preface is somewhat unusual in that it is signed by the entire committee, rather than by the chairman alone. However, the report itself is unusual because it tells a story about veterans involved in a long-secret wartime research program in the United States—a story that the committee and its staff hope will never have to be told again.

David P. Rall, Chairman
O. Michael Colvin
Ellen Eisen
William Halperin
Charles H. Hobbs
David G. Hoel
Karl Kelsey
Charles J. McDonald
James M. Melius
John A. Montgomery
William Nicholson
Roswell R. Pfister
Margaret Singer
Bailus Walker
Annetta P. Watson

Acknowledgments

The committee wishes to acknowledge the valuable help and assistance it received from a number of individuals. First, the committee wishes to thank and honor the more than 250 veterans who participated in the public hearing process. The veterans' openness about their health problems and their understanding of the committee's role was of great value (see Appendix G). The committee is also very grateful for the input of the 13 civilian and military experts, who gave their time to make important presentations before the committee (see Appendix A).

Maria Lloyd and James Gately from the Naval Research Laboratory were tremendously helpful in sending technical reports of the World War II testing projects from their institution. In addition, they have worked diligently to provide documentation of participation to affected veterans. The assistance of Marjorie Ciarlante of the National Archives and Janice Goldblum of the National Academy of Sciences' Office of Archives and Information Services was also of great help in locating and obtaining historically relevant materials.

Susan Mather, Han Kang, and Robert Allen of the Department of Veterans Affairs aided the committee in many ways, but most especially by the determination of the feasibility of identifying the veterans who had served as subjects in the chamber and field tests (see Appendix E). Richard Patterson from the Disabled American Veterans and Richard Christian from the American Legion were valuable resources to the committee and their assistance in publicizing the public hearing process is greatly appreciated. The committee also wishes to thank Sanford Leffingwell of the Centers for Disease Control for sending an early bibliography and helping the committee to locate helpful toxicological reports. Jeffery K. Smart, Historian at the U.S. Army Chemical and Biological Defense Agency, aided the committee by sending technical reports and photographs from work done at the Edgewood Arsenal in Maryland.

The committee appreciates the able editorial work of Paul Phelps, Florence Poillon, and Barbara Bodling. The committee is also indebted to the efforts of the Institute of Medicine's (IOM) managing editor, Michael Edington, and the sponsoring editor from the National Academy Press, Sally Stanfield. Laura Baird, the IOM librarian, provided helpful support and input in the literature search, and Susan Turner-Lowe, from the Office of News and Public Information, provided expert advice on the public hearing and interactions with the press. Gerri Kennedo, a project assistant who stepped in to help produce some of the final drafts of the report, deserves much thanks for her very competent handling of difficult assignments.

Contents

BOXES

FIGURES

TABLES

Veterans at Risk

Executive Summary

BACKGROUND

World War II (WWII) has been called "the unfought chemical war." Both sides had produced millions of tons of chemical weapons and had made massive preparations for their use, yet the weapons were never used. These preparations included the establishment of secret research programs to develop better weapons and better methods of protecting against these weapons. In the United States, some of this research was focused on the development of protective clothing and skin ointments, which could prevent or lessen the severe blistering effects of mustard agents (sulfur and nitrogen mustard) and Lewisite (an arsenic-containing agent).

By the time the war ended, over 60,000 U.S. servicemen had been used as human subjects in this chemical defense research program. At least 4,000 of these subjects had participated in tests conducted with high concentrations of mustard agents or Lewisite in gas chambers or in field exercises over contaminated ground areas. The human subjects had experienced a wide range of exposures to mustard agents or Lewisite, from mild (a drop of agent on the arm in "patch" tests) to quite severe (repeated gas chamber trials, sometimes without protective clothing). All of the men in the chamber and field tests, and some of the men in the patch tests, were told at the time that they should never reveal the nature of the experiments. Almost to a man, they kept this secret for the next 40 or more years.

Public attention was drawn to these experiments when some of the WWII human subjects began to seek compensation from the Depart ment of Veterans Affairs (VA) for health problems that they believed were caused by their exposures to mustard agents or Lewisite. Two factors complicated resolution of these cases. First, there were often no records or documentation available of an individual's participation in the testing programs. Second, there was a great deal of uncertainty about which health problems were in fact the result of mustard agent or Lewisite exposure.

In June 1991 the VA announced guidelines for the handling of these cases. These guidelines included the loosening of normal requirements for documenting the individual's participation in the experiments and the identification of seven diseases that the VA would consider to be caused by mustard agents or Lewisite. These seven are asthma, chronic bronchitis, emphysema, chronic laryngitis, corneal opacities, chronic conjunctivitis, and keratitis (of the eye). In addition, the VA requested that the Institute of Medicine convene a committee to survey the scientific and medical literature in order to assess the strength of association between exposure to these agents and the development of specific diseases. The committee was also charged with identifying the gaps in the literature and making recommendations relevant to closing those gaps. This report details the committee's findings and recommendations.

Between October 1991 and August 1992, almost 2,000 scientific papers, technical reports, and other documents were reviewed by the committee. The experimental protocols used in the WWII testing programs were examined to assess the potential dose levels experienced by the experimental subjects. In addition, the committee consulted with a variety of outside experts and sought information from the affected veterans themselves, through a public hearing process that resulted in written or oral statements from over 260 veterans regarding their exposures to these agents and subsequent health problems.

The committee found large gaps in the literature pertaining to the long-term health effects of exposure to mustard agents and Lewisite. For many diseases, very little or no work had been done in the eight decades following the first use of sulfur mustard in World War I. Almost all of the work in the United States had been conducted or funded by chemical defense sections of the military and was concerned only with the acute effects of these agents and not with their long-term effects. As a result, the committee depended heavily on occupational studies of chemical weapons production workers in other countries, on what could be found on battlefield casualties, and on what was known about the effects of nitrogen mustard derivatives that have been used since WWII as cancer chemotherapy agents. In addition, the committee carefully considered the basic scientific data available regarding the biological mechanisms of tissue damage from mustard agents and Lewisite.

Special attention was directed at estimating the dose levels to which the experimental human subjects had been exposed in gas chambers or field exercises. In these experiments, subjects wore varying amounts of the protective clothing being tested, as well as gas masks. In the chamber tests, human subjects were required to enter gas chambers repeatedly for an hour or more per trial, until, after a number of trials, their skin showed evidence of chemical burns (erythema)—an indication that the agents were penetrating the protective clothing. In the field tests, the agents were dropped over large tracts of land, and human subjects, wearing clothing being tested, were sent into those areas for varying amounts of time. Penetration of the agents through the clothing was assessed in these tests in the same manner as in the chamber tests.

GENERAL CONCLUSIONS

The committee reached the following conclusions on the basis of its analysis of the experimental protocols:

- **The lack of follow-up health assessments of the human subjects in the WWII gas chamber and field tests severely diminished the amount and quality of information that could be applied in the assessment of long-term health consequences of exposure to mustard agents and Lewisite.**
- **The levels of exposure to mustard agents or Lewisite experienced by the human subjects may have been much higher than inferred in the summaries of the gas chamber and field tests.**

The lack of follow-up of these subjects particularly dismayed the committee for a number of reasons. For example, the end point of the chamber and field tests was tissue injury, but it was already known by 1933 that certain long-term health problems resulted from sulfur mustard exposure. Further, it was documented that numerous subjects suffered severe injuries that required up to a month of treatment. Finally, the exposure levels were sufficiently high that even the most efficient gas mask would have leaked enough mustard agent or Lewisite to cause inhalation and eye injuries.

- **The committee was additionally dismayed that there were no epidemiological studies done of mustard agent-exposed, U.S. chemical weapons production workers, war gas handlers and trainers, or combat casualties from WWII.**

Tens of thousands of people (military and civilian) worked in U.S. arsenals that produced mustard agents, Lewisite, and other chemicals. Exposure levels in these facilities were often quite high, as evidenced by the number of injuries reported and by the poor safety record of the

Chemical Warfare Service during the peak years of production. Many other servicemen were trained to handle the gases or were assigned to jobs that put them in contact with mustard agents or Lewisite. A German bombing attack on the harbor of Bari, Italy, released sulfur mustard from a damaged American ship into the water and atmosphere, resulting in thousands of injuries and hundreds of deaths. Yet no follow-up studies were done with any of these groups; the committee had to rely instead on occupational studies from Japan and Great Britain for data on World War II production workers and their long-term health problems.

SPECIFIC FINDINGS

The following is a summary of the major conclusions reached by the committee regarding the association of exposure to mustard agents or Lewisite and the development of specific diseases in different organ systems. Much more is known about mustard agents than is known about Lewisite. Thus, the following summary pertains to mustard agents, except when Lewisite is indicated.

The findings generally fall into one of three categories. In some cases, the data examined were found to indicate a *causal* relationship between exposure and a particular disease. For a few diseases, the data were *suggestive* but not completely clear. Finally, there were many diseases for which very little or no data existed regarding the possible contributions of exposure to mustard agents or Lewisite. This means that many diseases in this category may (or may not) be caused by mustard agents or Lewisite, but no study has been done. It is important to emphasize that *no condition evaluated could be removed from consideration as a health consequence of exposure to these agents.* Thus, for many diseases there remains significant doubt.

The evidence found indicated a causal relationship between exposure and the following health conditions:

- Respiratory cancers

  —Nasopharyngeal
  —Laryngeal
  —Lung

- Skin cancer
- Pigmentation abnormalities of the skin
- Chronic skin ulceration and scar formation
- Leukemia (typically acute nonlymphocytic type, nitrogen mustard)
- Chronic respiratory diseases (also Lewisite)

  —Asthma

—Chronic bronchitis
—Emphysema
—Chronic obstructive pulmonary disease
—Chronic laryngitis

- Recurrent corneal ulcerative disease (Includes corneal opacities; acute severe injuries to eye from Lewisite will also persist.)
- Delayed recurrent keratitis of the eye
- Chronic conjunctivitis
- Bone marrow depression and (resulting) immunosuppression (An acute effect that may result in greater susceptibility to serious infections with secondary permanent damage to vital organ systems.)
- Psychological disorders

—Mood disorders
—Anxiety disorders (including post-traumatic stress disorder)
—Other traumatic stress disorder responses (These may result from traumatic or stressful features of the exposure experience, not a toxic effect of the agents themselves.)

- Sexual dysfunction (Scrotal and penile scarring may prevent or inhibit normal sexual performance or activity.)

The evidence found suggested a causal relationship between exposure and the following health conditions:

- Leukemia (acute nonlymphocytic type, sulfur mustard)
- Reproductive dysfunction (genotoxicity, mutagenicity, etc.; mustard agents)

There was insufficient evidence found to demonstrate a causal relationship between exposure and the following health conditions:

- Gastrointestinal diseases
- Hematologic diseases
- Neurological diseases
- Reproductive dysfunction (Lewisite)
- Cardiovascular diseases (Except for those that may result from serious infections shortly following exposure—heart disease resulting from rheumatic fever, for example.)

RECOMMENDATIONS

There are large gaps in all areas of the knowledge base about the long-term health risks associated with exposure to mustard agents and Lewisite. For example, very little is known about the long-term effects on specific organ systems from studies in animals. The data from human studies lack precise information about the exposure levels in occupational settings. After consideration of these gaps in light of the commit tee's findings regarding the probable long-term health effects of exposure to these agents, as well as the likely exposure levels to the human subjects involved, the committee formulated the following recommendations.

> The committee recommends that the Department of Veterans Affairs (VA) institute a program to identify each human subject in the WWII testing programs (chamber and field tests, and to the degree possible, patch tests), so that these individuals can be notified of their exposures and the likely health risks associated with those exposures. Further, all subjects so identified, if still living, should be medically evaluated and followed by the VA as to their health status in the future. These individuals should also, if they request it, be treated by the VA for any exposure-related health problems discovered. Morbidity and mortality studies should be performed by the VA, comparing chamber, field, and patch test cohorts to appropriate control groups, in order to resolve some of the remaining questions about the health risks associated with exposure to these agents.

The only way to answer some of the key remaining questions is to establish a base of knowledge based on human exposures. There is precedent in the later identification and follow-up of veterans exposed to chemicals, including hallucinogenic drugs, in other military testing programs.

The committee is well aware that a half century has now passed and that many of those who might have benefited from a broader understanding of the toxicity and carcinogenicity of mustard agents and Lewisite are already dead. Nevertheless, their surviving family members deserve to know about the testing programs, the exposures, and the potential results of those exposures. For those veterans still living, diseases such as skin and lung cancer may still appear, and full knowledge of their likely cause might well save their lives.

In the case of the human subjects of the WWII testing programs, it is reasonable to assume that secrecy, uncertainty, and fear may have resulted in adverse psychological effects for the veterans and their families.

> The committee recommends that careful attention be paid by health care providers to the special problems and concerns of the affected veterans and their families. This attention may include the convening of a special task force of experts in stress disorders and risk perception to aid the VA, further than this

committee is able, in the establishment of comprehensive guidelines for handling of these cases.

These recommendations are not meant to ignore the fact that thousands, probably tens of thousands, of other military and civilian personnel were exposed to mustard agents and Lewisite in occupational and training settings, and in combat in the Bari harbor disaster. Some of these exposures will have resulted in one or more of the exposure-related health problems identified in this report; and, in fact, some military personnel who served in the Chemical Warfare Service have qualified for service-connected disability as a result of such exposures. However, many more military personnel were exposed to significant levels of mustard agents or Lewisite than is obvious from service records.

The committee additionally recommends that the Department of Defense (DoD) should use all means at its disposal, including public channels, to identify former chemical warfare production workers (military or civilian) and individuals exposed to mustard agents or Lewisite from gas handling, training, the Bari harbor disaster, or other circumstances. Records of former military personnel could be turned over to the VA for notification, inclusion in morbidity and mortality studies, and health status evaluation. Records of the civilian personnel should be used by the DoD to advise former workers as to their health risks and options for seeking appropriate compensation for any illnesses that resulted from their exposures.

This committee discovered that an atmosphere of secrecy still exists to some extent regarding the WWII testing programs. Although many documents pertaining to the WWII testing programs were declassified shortly after the war ended, others were not. Of those declassified, many remained "restricted" to the present day and, therefore, not released to the public. As a result, the committee often had great difficulty obtaining information. For example, only one of the three major chamber test locations, the Naval Research Laboratory, freely shared technical reports and detailed summaries with the committee from the beginning of the study. For other locations, such information arrived only as the study was in its final stages, despite months of requests and inquiries to a variety of offices. The committee is certain that other relevant information exists that was never obtained. It is also clear that there may be many exposed veterans and workers who took an oath of secrecy during WWII and remain true to that oath even today. Even as this report was going to press, veterans were still contacting the committee for information, having just heard about the study and thinking it might now be permissible to reveal their experiences. This continuing secrecy, in the committee's view, has impeded well-informed health care for thousands of people.

> The committee recommends that the VA and DoD publicly announce and widely advertise that personnel exposed to mustard agents or Lewisite during their service are released from any oath of secrecy taken at the time. In addition, professional educational materials should be prepared by the VA or DoD, or both, and made available for physicians who may be treating affected individuals. These materials should incorporate the latest information regarding the long-term health effects of exposure to mustard agents and Lewisite.

There is no doubt that the long-term health consequences of exposure to mustard agents or Lewisite can be serious and, in some cases, devastating. This report has demonstrated that complete knowledge of these long-term consequences has been and still is sorely lacking, resulting in great costs to some of those exposed in WWII. The lack of knowledge, however, has ongoing ramifications as nations will probably continue to use these chemical weapons in battle or begin to grapple with their disposal. Thus, accidental and deliberate human exposures to mustard agents and Lewisite can only be expected to continue in the foreseeable future.

1

Introduction

The whole distressing story of mustard gas poisoning teems with medical interest. But while its lessons are worth taking to heart, we may well hope that the disease itself need only appear in the future text-books of medicine as a curiosity and the relic of a savage age destined never to return.

The Lancet, March 22, 1919

Modern chemical warfare emerged in 1915, when the German army used chlorine for the first time in a large-scale offensive against the Allies during World War I (WWI). The incident rapidly increased activities on both sides of the conflict toward development of protection against chemical attacks and more effective chemical weapons. As improved gas masks became more effective against inhaled poisons, Germany looked to its stockpile of poison gases to find one that would have damaging effects through absorption into the skin and other mechanisms. Thus, in July 1917, in a field outside of Ypres, Belgium, the blistering agent sulfur mustard, called mustard gas because it had a mustardor garlic-like odor, was used for the first time. Sulfur mustard was destined to cause almost 400,000 casualties during the war, many more than any other chemical agent (Gilchrist, 1928; Prentiss, 1937). Its effectiveness made it a chemical weapon of choice from 1917 to the present day, as evidenced by its use by Iraq against Iran in 1987 (Medema, 1986; United Nations Security Council, 1988).

Although many observers hoped that the end of WWI would be the end of chemical warfare, this was not to be. Mustard agents were used during the next two decades by the British in the Middle East, the

French in Morocco, and the Russians in Central Asia. Strong evidence has been presented that it was used by the Italians in Abyssinia and the Japanese in China. In the beginning of World War II (WWII), the specter of chemical warfare was again raised by reports of the use of mustard by the Germans and the Poles against each other in 1939 (Medema, 1986; Stockholm International Peace Research Institute, 1971; Tanaka, 1988).

Thus, as engagement of U.S. forces in the European conflict began to look inevitable, preparations against the accompanying threat of chemical attack were given a high priority. These preparations included the establishment of secret research programs to develop better protective clothing for the troops and to investigate ointments and salves that might be effective in reducing the blistering effects of sulfur and nitrogen mustard (collectively termed mustard agents) and Lewisite, an organic arsenic-containing compound that shared some properties with mustard agents (Stewart, 1948).

The researchers involved in these programs, frustrated by the lack of adequate animal models of human injury from these chemicals, decided in 1942 that human subjects were needed for experiments with protective clothing materials and ointments (National Research Council, 1942; Office of Scientific Research and Development, 1946). At several centers around the country, top secret research using military personnel as human subjects was begun. The specific protocols, outlined in more detail in the next chapter, included putting drops of sulfur mustard or Lewisite on subjects' arms, either with or without test ointments, and clothing human subjects in suits impregnated with chemicals designed to retard vapor penetration. These subjects repeatedly entered gas chambers filled with vapors of mustard agents or Lewisite until their skin reddened, an indication that the protection of the clothing was failing.

The fears of chemical attacks in the course of WWII were never realized. Military and civilian personnel were injured by sulfur mustard in Bari, Italy, following a German air raid that destroyed a U.S. ship carrying sulfur mustard bombs (Infield, 1976; Perera and Thomas, 1986). The only other individuals injured by mustard agents and Lewisite in WWII were among those who served as subjects in the experiments or who worked with the agents as part of their military, or sometimes civilian, assignments.

In the early 1980s, reports of such injuries and ensuing health problems began to emerge. This seemingly long delay was due to the highly classified nature of the experiments with chemical warfare agents during WWII. After keeping silent for four decades, a few of the human subjects, believing their exposure to these agents was the cause of numerous debilitating diseases, attempted to seek compensation. As time passed and more information became public, claims by individuals to the Department of Veterans Affairs (VA) for service-connected disability for a variety of diseases and injuries increased. However, untangling the specifics of these claims presented major problems. For example, it was common for the periods of time spent as a volunteer in the gas experiments to be unaccounted for in official service records. Individual claimants therefore had great difficulty providing documentation of their participation in the testing programs. In addition, the assessment of the long-term health effects of sulfur mustard and Lewisite proved difficult. The vast majority of scientific and medical literature dealt solely with acute, short-term effects of these agents.

Pressure from individuals, the press, and Congress on the VA to resolve these issues increased over the intervening years and, on June 11, 1991, guidelines were announced by VA Secretary Edward J. Derwinski for compensation of the veterans who had been subjects in the mustard and Lewisite testing programs. These guidelines loosened the normal restrictions on documentation for those veterans who participated in the testing programs and identified the diseases that the VA would consider to be residual effects of sulfur mustard or Lewisite. The diseases identified for compensation were laryngitis, chronic bronchitis, emphysema, corneal opacities, chronic conjunctivitis, and keratitis. Veterans filing claims believed that there were still other diseases caused by their exposure to these agents.

In addition to the changes in guidelines, Secretary Derwinski also requested the Institute of Medicine (IOM) to assemble a committee to survey the health effects of mustard agents and Lewisite. This request asked that special attention be paid to the long-term health effects of these agents and for the assessment to be based, to the degree possible within the time allowed and budget provided, upon published scientific literature dating back to 1917. This report results from the deliberations of the IOM Committee to Survey the Health Effects of Mustard Gas and Lewisite, whose specific charges were to

- survey the published literature on the long-term health effects of mustard agents and Lewisite;
- summarize the strength of association between exposure to these agents and specific diseases;
- identify the gaps in knowledge regarding the contribution of exposure to these agents and disease; and
- generate recommendations aimed toward decreasing the gaps in knowledge that may be found.

The committee met four times in the course of its work to consider as much of the scientific literature pertaining to mustard agents and Lewisite as possible. It was expected from the beginning that the scientific and medical literature would not be complete and that, for some diseases, no literature would exist. Thus, in order to determine the constellation of health problems that exposed individuals might have, the committee held a public hearing and collected written and oral statements from over 250 individuals regarding the details of their exposure and their health problems. This information was used to identify those diseases for which gaps in the medical and scientific literature might be of particular consequence. In addition, by holding scientific workshop sessions and by soliciting input from experts with technical knowledge applicable to the issues, the committee gathered information pertaining to a variety of questions about the molecular mechanisms of mustard and Lewisite toxicity, the specific protocols used and dose levels achieved in the WWII testing programs, the psychological health effects of chemical warfare environments and exercises, and other related topics (Appendix A).

This report focuses on the published scientific and medical literature that pertains to the contribution of mustard agent or Lewisite exposure to the development of disease. It includes a bibliography of most of the published literature in English and foreign languages, as well as government technical reports and material from military archives. All served as the basis of the committee's deliberations. A historical analysis of mustard agent and Lewisite research programs, conducted by the military, is included. This analysis was accomplished in part to determine the possible exposure levels the human subjects received in the WWII testing programs. Also included is a summary of the information obtained through the public hearing process (Chapter 4), consultation with outside experts (Appendix A), and analysis of documents relating to military testing programs that used human subjects (Chapter 3). Further, the report assesses the strength of association between specific diseases and exposure to these agents, identifies the gaps in the present knowledge base, and highlights those gaps that warrant special attention. Finally, the report details the committee's final recommendations.

In order to cover this wide range of topics, the report necessarily includes varied levels of information from complex scientific analysis to descriptive sections regarding the public hearing findings and historical analysis. Efforts have been made to make the scientific sections accessible to as broad an audience as possible.

It is hoped that this report will be useful to policymakers and legislators responsible for programs dealing with the health of veterans in the United States. The report may also be of interest to military and civilian planners who face important issues relating to potential future use of chemical warfare agents and to the destruction of current stockpiles of chemical weapons. Further, for these groups and for the affected veterans and their families, the committee hopes that this report is successful in explaining and clarifying the multiple factors that interact to determine the relationships between exposure to toxic agents and adverse health effects. Finally, the committee hopes that the report will stimulate research to fill the major gaps that still exist in the scientific and medical literature pertaining to the long-term health effects of these agents.

REFERENCES

Gilchrist HL. 1928. A Comparative Study of World War Casualties from Gas and Other Weapons. Washington, DC: U.S. Government Printing Office.

Infield GB. 1976. Disaster at Bari. London: New English Library.

Lancet. 1919. Mustard gas: its brief but inglorious career. 1:471-472.

Medema J. 1986. Mustard gas: the science of H. Nuclear, Biological, and Chemical Defense and Technology International 1:66-71.

National Research Council. Division of Medical Sciences. Committeeon Treatment of Gas Casualties . 1942. Bulletin of the Committee on Treatment of War Gas Casualties. Minutes, March 18, 1942. Available at the National Research Council Archives, Washington, DC.

Office of Scientific Research and Development. 1946. National Defense Research Committee. Chemical Warfare Agents, and Related Chemical Problems. 2 vols. Summary Technical Report of Division 9, NDRC. Washington, DC: NDRC. AD-234 249.

Perera J, Thomas A. 1986. Britain's victims of mustard gas disaster. New Scientist 109:26-27.

Prentiss AM. 1937. Vesicant agents. In: Chemicals in Warfare: A Treatise on Chemical Warfare. 1st ed. New York: McGraw-Hill. 177-300.

Stewart I. 1948. Organizing Scientific Research for War. Science in World War II: Office of Scientific Research and Development. Boston: Little, Brown.

Stockholm International Peace Research Institute. 1971. The Problem of Chemical and Biological Warfare: A Study of the Historical, Technical, Military, Legal, and Political Aspects of Chemical and Biological Warfare and Possible Disarmament Measures. Vol. 1, The Rise of Chemical and Biological Weapons. Stockholm: Almqvist &Wiksell.

Tanaka Y. 1988. Poison gas: the story Japan would like to forget. Bulletin of the Atomic Scientists 44:10-19.

United Nations Security Council. 1988. Report of the mission dispatched by the SecretaryGeneral to investigate allegations of the use of chemical weapons in the conflict between the Islamic Republic of Iran and Iraq. April 25, 1988. S/19823 and S/19823/ Addendum 1. New York: United Nations.

2

Methods of Literature Collection and Survey

The primary task of this study was a survey of the scientific and medical literature on the health effects of mustard gas and Lewisite published from 1917 to the present. Early on in the course of this study, methods to identify, collect, and disseminate the literature were discussed and decided upon.

ONLINE DATABASES

The initial emphasis was a comprehensive search of relevant online databases. These computerized databases offer the most effective means of identifying international scientific literature and, in general, cover the time span from 1965 to the present. Databases were accessed through Dialog, a commercial database vendor, and through the National Library of Medicine's (NLM's) Medical Literature Analysis and Retrieval System (MEDLARS).

To maximize retrieval, the search strategy incorporated synonymous terms for mustard gas and Lewisite. Thus, the databases were searched by using the following terms: mustard gas, Yperite, sulfur mustard, schwefellost, yellow cross, dichlorodiethyl sulfide, Lewisite, and chlorovinylarsine dichloride. Enhanced accuracy in online searching was gained by using Chemical Abstracts Service (CAS) Registry Numbers, which uniquely identify each individual chemical. The final search strategy combined the CAS Registry Numbers (505-60-2 for mustard gas and 541-25-3 for Lewisite) and the synonymous terms for each chemical. Individual searches were customized to reflect the structure of each database. For applicable databases, searching was done on the standard ized terminology and alphanumeric designators for each chemical found in NLM's Medical Subject Headings (MeSH) and the MeSH tree structures.

Although there is subject and content overlap, each database serves a unique function, has a distinct subject emphasis, and indexes literature not available elsewhere. For example, the two prominent medical databases, NLM's Medline and Excerpta Medica's EMBASE, have only an approximate 36 percent content overlap. To serve the comprehensive goals of this study, it was decided to search all relevant databases in their entirety. A total of 46 online databases were searched, covering biomedical, toxicological, chemical, and regulatory information. As shown in Table 2-1, the majority of these databases were bibliographic, providing citations to scientific literature. Factual databases, Table 2-2, were also searched to provide toxicological and chemical information.

OTHER SOURCES

Online databases were developed in the mid-1960s, and few offer retrospective coverage. Identifying the literature published prior to this time required the use of a variety of sources. The volumes of *Index Medicus* covering the years 1917-1965 were an important bibliographic source. Reference lists of major review articles and books were also examined for relevant citations; several provided extensive reference lists (Goldman and Dacre, 1989; Gray, 1989; Papirmeister et al., 1991; Smith and Dunn, 1991; Somani and Babu, 1989). Document collection of published literature involved accessing the collections of the National Library of Medicine, the National Institutes of Health Library, the Himmelfarb Health Sciences Library of The George Washington University, and the National Research Council Library, as well as the use of interlibrary loans.

In conjunction with World War II (WWII) research on chemical warfare agents, scientists collected and reviewed the scientific literature and compiled bibliographies, which recorded the pre-WWII literature as well as ongoing military research. These include the bibliography for the Office of Scientific Research and Development, National Defense Research Committee's *Summary Technical Report of Division 9: Chemical Warfare Agents and Related Chemical Problems;* chapter reference lists in the three volumes of the National Research Council's *Fasciculus on Chemical Warfare Medicine;* and an unpublished bibliography compiled by the National Research Council's Committee on Treatment of Gas Casualties, entitled *Bibliography of the Medical Aspects of Chemical Warfare: Published Literature.*

Literature identification was an ongoing process throughout the study and, in addition to the above sources, input was received from veterans, interested persons, committee members, and speakers at committee meetings. All retrieved citations were reviewed to determine whether the citation was relevant to the study, and if relevant whether

TABLE 2-1 Bibliographic Databases Searched

Database	No. of Citations
Toxline	635
CA Search (Chemical Abstracts)	477
BIOSIS Previews	300
EMBASE	285
Toxlit and Toxlit65	266
Medline	240
Occupational Safety and Health (NIOSH)	172
NTIS (National Technical Information Service)	156
Cancerlit	152
Scisearch	129
Environmental Mutagen Information Center Backfile (Toxnet)	119
Life Sciences Collection	60
Pascal	38
Conference Papers Index	21
SSIE Current Research	19
Analytical Abstracts Online	16
World Translations Index	15
Chemical Industry Notes	12
Dissertation Abstracts Online	11
Federal Register	10
Pollution Abstracts	9
Compendex Plus	8
Environmental Bibliography	8
REMARC	8
Chemical Regulations and Guidelines Systems	7
AGRICOLA	7
Chemical Safety Newsbase	6
International Pharmaceutical Abstracts	5
Enviroline	5
LC MARC Books	5
Developmental and Reproductive Toxicology (Toxnet)	5
Environmental Teratology Information Center Backfile (Toxnet)	5
Books in Print	3
U.S. Copyrights	3
Legal Resource Index	3
Congressional Record	3
Biotechnology Abstracts	2
Psycinfo	2
British Books in Print	1
Nursing and Allied Health	1
Federal Research in Progress	1

to obtain the full paper or to first obtain an abstract. Citations were entered into the study's bibliographic database, which at the conclusion of the study contained 2,124 references to abstracts, journal articles, books, military and civilian reports, dissertations, and conference proceedings relating to the health effects of mustard gas and Lewisite.

TABLE 2-2 Factual Databases Searched

Chemical Carcinogenesis Research Information System (CCRIS)
Development and Reproductive Toxicology (DART)
Genetic Toxicology (GENETOX)
Hazardous Substances Data Bank (HSDB)

SUPPLEMENTAL WORLD WAR II MILITARY REPORTS

In addition to the published scientific literature, essential supplemental information was made available to the committee through military and technical reports and also through archival research. Retrieving technical and military documents involved searching archival records of the Office of Scientific Research and Development (OSRD), requesting access to military documents, and ordering technical reports from the National Technical Information Service (NTIS).

WWII civilian research was coordinated through the OSRD, and two of its divisions conducted chemical warfare agent research, the National Defense Research Committee (NDRC) and the Committee on Medical Research (CMR). Within NDRC it was Division 9 (Chemistry) that was responsible for overseeing this research, and Division 9 records are stored in Series 29, Record Group 227, Civil Reference Branch, National Archives and Records Administration, Washington, D.C. The records of the CMR's Committee on Treatment of Gas Casualties, the subcommittee involved in chemical warfare agent research, are housed in the National Research Council Archives and are also available in Series 29, Record Group 227 at the National Archives.

Military reports were also obtained to supplement the journal literature. Reports obtained from the Naval Research Laboratory (Washington, D.C.) provided background information on WWII test conditions and protocols. Chemical Warfare Service administrative records and correspondence were retrieved from Record Group 175, Suitland Reference Branch of the National Archives, Suitland, Maryland. Freedom of Information Act (FOIA) requests were sent to the U.S. Army Chemical Research Development and Engineering Command (Aberdeen Proving Ground, Maryland), U.S. Army Dugway Proving Ground (Dugway, Utah), and Naval Training Center (Great Lakes, Illinois). The Medical Research Institute of Chemical Defense (Aberdeen Proving Ground, Maryland) was also contacted. Reports on field tests involving sulfur mustard were sent by Dugway Proving Ground. The committee was informed that the archives at the Great Lakes Naval Training Center did not have documents on chemical warfare agent testing. Aberdeen Proving Ground provided information on three publicly releasable documents and reviewed on a case-by-case basis other FOIA requests for military reports. British documents on WWII chemical warfare agent research were sent by the Chemical and Biological Defence Establishment, Ministry of Defence, Great Britain. The Australian Commonwealth Department of Veterans' Affairs provided summary information on the WWII Australian chemical warfare agent testing program.

TABLE 2-3 Foreign Language Citations

Database Statistics	
2,124	Total records in database
458	Foreign language records
French	136
German	128
Japanese	67
Italian	34
Russian	21
Polish	14
Danish	11
Dutch	9
Swedish	7
Czech	7
Hungarian	6
Romanian	5
Spanish	3
Bulgarian	3
Portuguese	3
Hebrew	2
Serbo-Croatian	2
Acquisition and Translation Statistics	
250	Foreign language documents acquired
97	English abstracts available or summary section translated
60	Entire articles translated

LITERATURE TRANSLATION, DISSEMINATION, AND ANALYSIS

As the documents were collected, one of the first considerations was dealing with the significant portion of this literature (22 percent) published in a foreign language. English summaries and translated titles were reviewed, and papers with pertinent original research not available in English were professionally translated (Table 2-3). Given the large number of documents on this topic, it was necessary early on in the study to determine a method that would most effectively organize the material and serve as a tool in the dissemination of the literature to the committee. A list of index terminology was developed and subsequently revised at the first committee meeting. Each paper was indexed, and updated subject bibliographies were distributed to the committee throughout the study to reflect new acquisitions, allowing committee members to request copies of the papers they needed for their information and analysis.

In assessing the associations between exposure and specific health conditions, the committee generally followed the guidelines proposed by Hill (1971). These guidelines include six considerations that can be brought to bear on judgments of causality. Strength of association reflects the relative risk or odds of an association. A dose-response relationship can reinforce the judgment of causality when the strength of association increases with increases in exposure. Further, associations need to be temporally correct; the effect occurs in a reasonable or expected time period following exposure. Consistency and specificity of associations are also important considerations. A consistent association is one that is found in a variety of studies. If, however, a particular health condition is reliably predicted by a given exposure, then specificity of an association is demonstrated. Finally, for an association to be judged causal it must be biologically plausible or explainable by known biological mechanisms.

Each of the considerations listed above is based on certain assumptions and requires varying amounts and types of information. In other words, the application of Hill's guidelines can be difficult when assessing literature that is incomplete or sparse. The assumptions made and the difficulties encountered in assessing the literature regarding the health effects of mustard agents and Lewisite are discussed in this report in the chapters dealing with specific organ systems and health conditions.

The complete bibliography at the end of this report is divided into three sections: published literature, military reports, and technical reports. Availability information is also provided at the end of the bibliography. A separate subject bibliography has been compiled and is available through the NTIS under the title *Health Effects of Mustard Gas and Lewisite: Subject Bibliography.*

REFERENCES

Goldman M, Dacre J. 1989. Lewisite: its chemistry, toxicology, and biological effects. Reviews of Environmental Contamination and Toxicology 110:75-115.

Gray PJ. 1989. A Literature Review on the Mechanism of Action of Sulphur and Nitrogen

Mustard. Maribyrong, Australia: DSTO Materials Research Laboratory. AD-A125-070.

Hill AB. 1971. Principles of Medical Statistics. 9th ed. New York: Oxford University Press.

National Research Council. Division of Medical Sciences. Committee on Treatment of Gas Casualties. 1945. Bibliography of the Medical Aspects of Chemical Warfare: Published Literature. Unpublished, available at the National Research Council Archives, Washington, DC.

National Research Council. Division of Medical Sciences. Committee on Treatment of Gas Casualties. 1945. Fasciculus on Chemical Warfare Medicine. 3 vols. Washington, DC: Prepared for the Committee on Medical Research of the Office of Scientific Research and Development. Available at the National Research Council Archives, Washington, DC.

Office of Scientific Research and Development. National Defense Research Committee. 1946. Summary Technical Report of Division 9, NDRC. Washington, DC: NDRC. AD-234 249.

Papirmeister B, Feister AJ, Robinson SI, Ford RD. 1991. Medical Defense Against Mustard Gas: Toxic Mechanisms and Pharmacological Implications. Boca Raton, FL: CRC Press.

Smith WJ, Dunn MA. 1991. Medical defense against blistering chemical warfare agents. Archives of Dermatology 127:1207-1213.

Somani SM, Babu SR. 1989. Toxicodynamics of sulfur mustard. International Journal of Clinical Pharmacology, Therapy and Toxicology 27:419-435.

3

History and Analysis of Mustard Agentand Lewisite Research Programsin the United States

This chapter begins with an introduction that briefly describes sulfur mustard and Lewisite and their effects, accompanied by an overview of their development. This is followed by a description of the organization of chemical warfare research during World War I (WWI) and the postwar period of 1919 to 1940, including the development of Lewisite and nitrogen mustard. The major focus of this chapter, however, is to describe the research programs and protocols relating to mustard agents and Lewisite, initiated just prior to World War II (WWII) and continued throughout the war.

This committee also investigated as many protocols and supporting military documents as it could obtain for use in estimating the possible exposure levels experienced by the men who participated in the mustard and Lewisite tests. These estimates were intended to put into context the concentrations of vesicant used in animal and other types of experiments, which the committee was also charged to survey. As these protocols were investigated, it became apparent to the committee that the full body of knowledge available to the wartime scientists, especially information relevant to the long-term health outcomes of exposure to these agents, was not applied in the conduct of the human experimentation. Thus, this chapter begins to address compelling questions that emerged through the course of this study regarding the appropriate use of the existing scientific and medical literature in WWII testing programs, the lack of medical follow-up of human research subjects, and the probable exposure levels experienced by these subjects.

Finally, the chapter overviews the research programs since the end of WWII, including the continuing investigations concerning the mechanisms of toxicity of these agents. Description of the chemical stockpile disposal program is also included. The chapter concludes with an outline of some of the conclusions drawn by the committee from analysis of the historical records and calculations of exposure levels.

INTRODUCTION

Sulfur Mustard

Sulfur mustard ($C_4H_8Cl_2S$) is one of a class of chemical warfare agen *ts* known as vesicants because of their ability to form vesicles, or blisters, on exposed skin (see Figure 3-1). During WWI, exposed troops described the odor of this agent as a stench like mustard or garlic, hence its common name. Table 3-1 summarizes some characteristics of mustard agents and Lewisite. First noted for its toxic properties by dye chemists in the late 1880s, sulfur mustard has been referred to by a number of synonyms: S-mustard, to distinguish it from nitrogen mustard; "Lost" or "S-Lost," from the names of two chemists who suggested it be used as a war gas (Lommel and Steinkopf); "yellow cross," for the identifying mark on WWI shells containing sulfur mustard; or Yperite, after the site of its first use in 1917. Although commonly and inaccurately referred to as mustard gas, the agent is a liquid at room temperature.

Sulfur mustard produces skin blisters and damage to the eyes and respiratory tract, and it can be lethal at sufficiently high doses. It is a cellular poison and mutagen and a recognized human carcinogen. Battlefield use of sulfur mustard decreases the opponent's ability to fight by producing chemical burns on tissues that come into contact with either vapors or liquid droplets and aerosols. Exposed skin surfaces, eyes, and the linings of both the respiratory and the gastrointestinal tracts are all at risk, and the risks increase dramatically under hot, humid conditions.

From a military standpoint, one of sulfur mustard's most useful properties is its persistence. Droplets of this agent released in an explosion can deposit on numerous surfaces, evaporating slowly and posing a risk from inhalation as well as contact with the skin. Indeed, this very set of conditions was observed in WWI after mustard shelling (Haber, 1986). One reason for this persistence is the characteristic freezing temperature of sulfur mustard (13°C to 15°C). Droplets or bulk quantities would thus be expected to remain where initially deposited during cool or winter weather, under forest canopies, or under overgrown vegetation. Under certain conditions, bulk quantities of mustard agent spilled or splashed onto the soil would not degrade for months.

The exact date of the first sulfur mustard synthesis is somewhat unclear, but the first report may have been by Despretz in 1822. An 1860 report by Neimann describes a delayed-effect vesicant oil as a reaction product of ethylene on a mixture of sulfur chlorides. At that time, this

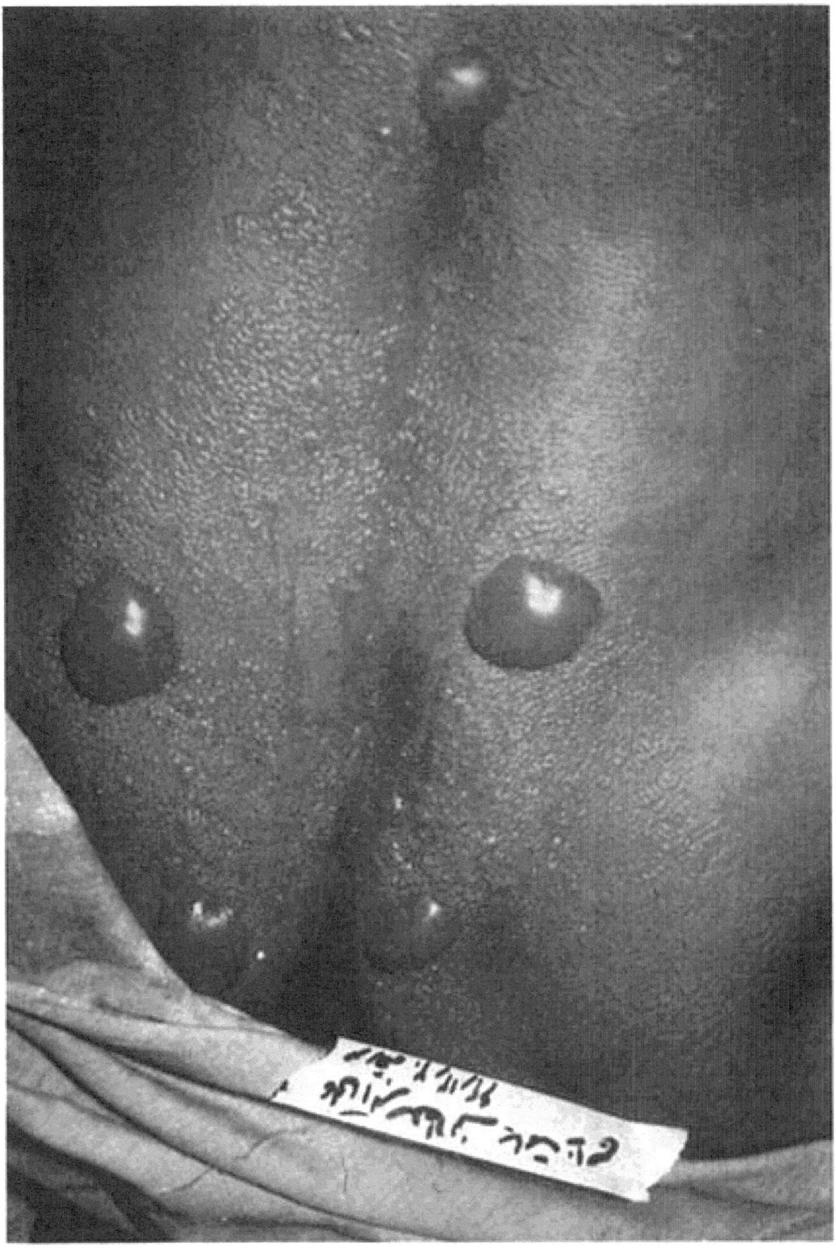

FIGURE 3-1 Vesicle formation on an Iranian patient, 16 hours after battlefield exposure to sulfur mustard. Reprinted from Willems, 1989, with permission from Annales Medicinae militaris Belgicae.

TABLE 3-1 Chemical and Physical Data

	Sulfur Mustard	Lewisite	Nitrogen Mustard
Chemical Abstracts Registry Number	505-60-2	541-25-3	51-75-2
Chemical formula	C_4H8Cl_2S	C_2H_2AsCl3	$C_5H_{11}Cl2N$
Chemical structure	CH_2-CH_2Cl S CH2-CH2Cl	H $AsCl_2$ Cl H	CH2-CH2Cl H3C-N CH2-CH2Cl
Synonyms	1,1'-Thiobis(2-chloroethane); 2,2'-dichloroethyl sulfide; bis (2-chloroethyl) sulfide; ,'-dichloroethyl sulfide; mustard gas; Schwefel-Lost; S-Lost; Yperite; yellow cross; Senfgas; Kampstoff "Lost"; dichlorodiethyl sulfide	Chlorovinyldichloroarsine; 2-chlorovinyldichloroarsine; chlorovinylarsine dichloride; dichloro(2-chlorovinyl)arsine	Mechlorethamine; chlormethine; 2-chloro-N-(2-chloroethyl)-Nmethylethanamine; Stickstofflost; di(chloroethyl)methylamine
Abbreviations	H (Levinstein mustard), HD (distilled mustard), HT (impure mixture)	L	HN2; related compounds include HN1, ethylbis(chloroethyl)amine; and HN3, tris(-chloroethyl)amine
Melting point	13C-14C	0.1C	-60C
Boiling point	215C-217C at 760 mm Hg	190C at 760 mm Hg (decomposes)	87C at 18 Hg
Molecular weight	159.08	207.32	156.1
Solubility	Very sparingly soluble in water; soluble in oily solvents; high lipid solubility	Insoluble in water, soluble in ordinary organic solvents	Very slightly soluble in water
Appearance and odor	Colorless when pure, normally yellow to brown oily liquid, slight garlic-type odor	Liquid, faint odor of geranium	Liquid, faint fishy odor

SOURCES: Budavari, 1989; Hazardous Substances Databank, 1991; IARC, 1975; Somani, 1992; U.S. Army CRDEC, 1988, 1990.

product was identified as a compound $[(C_2H_4)_2S_2Cl_2]$ different from sulfur mustard; however, the observed severe skin blistering, latent period of several hours, and subsequent slow healing are all typical of skin exposure to sulfur mustard. At about the same time, Guthrie (1859, 1860) published investigations describing yet another variant compound (thought to be $C_4H_4S_2Cl_2$), also produced from sulfur chloride in reaction with ethylene. The odor was "pungent," resembling that of "oil of mustard." Guthrie noted destruction of the epidermis when the thin skin between the fingers and around the eyes was exposed to the "vapour" of this compound. When the liquid was allowed to remain on the skin, blister formation was observed. Finally, in 1886, a process to produce significant quantities of pure sulfur mustard was described by Meyer using sodium sulfide, ethylene chlorohydrin, and hydrochloric gas (Jackson, 1936; Meyer, 1886; Prentiss, 1937; West, 1919). This process was the one eventually used by the German war factories to fill the shells fired at Ypres (Haber, 1986).

Lewisite

Lewisite ($C_2H_2AsCl_3$) is a vesicant that contains organic arsenic. During WWI, a U.S. chemical warfare research laboratory investigating arsenic compounds as potential war gases developed the potent vesicant, subsequently named "Lewisite" after the research group director. Purified Lewisite is a colorless, oily liquid at room temperature with a faint "geranium-like" odor. More volatile than sulfur mustard, this agent can be used as a vapor over large distances and has been mixed with sulfur mustard to achieve greater effectiveness in combat. With a freezing point between -18°C and 0°C, Lewisite is effective over a wider temperature range than sulfur mustard.

Lewisite is also a cellular poison, but works via a different mechanism than sulfur mustard. It is readily absorbed through the skin and respiratory tract, but moist tissues are particularly vulnerable and eyes exhibit the greatest sensitivity (Trammell, 1992; Watson and Griffin, 1992). In contrast to sulfur mustard, Lewisite exposure is characterized by immediate onset of pain. The agent is lethal at sufficient doses, produces chromosomal aberrations in some mammalian cellular assays, and is a systemic poison when absorbed into the bloodstream. Finally, some evidence suggests that Lewisite might be a carcinogen (Centers for Disease Control, 1988).

The development of Lewisite as a war gas was made by W. Lee Lewis in 1918, while working at the Chemical Laboratory of the Catholic University of America in Washington, D.C. (Lewis and Perkins, 1923). The thrust of the work in this laboratory during WWI was the evaluation of substituted arsines (arsenic-containing chemicals) as potential chem ical warfare agents. Lewis had noticed a paragraph in a 1904 student dissertation by J.A. Nieuwland that documented the formation of an "extremely poisonous" substance after a reaction of arsenic chloride with dry acetylene in the presence of aluminum chloride (cited in Lewis and Perkins, 1923). The toxicity that had caused Nieuwland to stop further work on the reaction spurred Lewis to investigate the substance more fully. In addition, Lewis and his group worked out safer and more efficient production methods and elaborated plans for large-scale production (Lewis and Perkins, 1923; Lewis and Steigler, 1925). A production plant was eventually constructed in Willoughby, Ohio, and approximately 150 tons of Lewisite were in transit to Europe when the Armistice was signed in November 1918. The vessel was sunk at sea (Spiers, 1986; Tarbell and Tarbell, 1981; Trammell, 1992), and all experimental work with Lewisite in the U.S. Chemical Warfare Service abruptly ceased until WWII (Gates et al., 1946).

RESEARCH PROGRAMS OF WORLD WAR I AND THE POSTWAR PERIOD

As outlined above, prior to the actual use of sulfur mustard as a war gas in 1917, the substance was little more than an interesting compound produced, along with hundreds of other compounds, by the emerging science of industrial chemistry in the last half of the nineteenth century. Thus, tragically, the combat casualties of WWI became the first large group of experimental subjects in studies of the medical effects of sulfur mustard. Organized research into chemical warfare agents began in earnest in Britain and France after the German chlorine gas attack in 1915. In the United States, it was 1917 before a formally organized chemical warfare research program was established. The history of the program has been documented by various authors and summarized by Cochrane (1946) in a classified report released to the public in 1991. The program began with an offer from the Bureau of Mines to the National Research Council (NRC) to mobilize the bureau's unique and specialized laboratories toward the investigation of poison gases.[1] With the U.S. declaration of war against Germany in 1917, the NRC Committee on Noxious Gases was formed to administer the research programs concerning poison gases, including sulfur mustard and later Lewisite.

In the United States and Europe, much of the research was focused on methods of mass production of sulfur mustard, development of other vesicants and war gases, and development of better gas masks and other

[1] The National Research Council was in 1917 and is today part of the National Academy of Sciences. The NRC was directly involved in defense research programs during both World War I and World War II. A description of this involvement is included in Appendix C.

equipment to protect troops from chemical attack. The overall research program was divided into sections, each of which was responsible for specific types of research, ranging from gas production processes to treatment of gas casualties. As the program matured, the various organizational structures were modified. The details of these modifications are not presented here because they are largely irrelevant to consideration of the health effects of mustard agents and Lewisite. One modification, however, may have set the stage for how research into these substances' medical effects has been conducted and directed ever since.

In 1918, a presidential order moved the research program from essentially civilian control under NRC to military control under the War Department. This move gave birth to the Chemical Warfare Service (CWS), which, to the present day, is responsible for the majority of research concerning chemical and biological warfare agents. As time went on this administrative change altered the direction of almost all investigations into the toxicology of vesicants and other chemical agents. Mainstream biomedical science is "hypothesis driven": when interesting results are obtained by an investigator, either that investigator or other groups begin further research to better understand what has been discovered, even if the interesting results are not directly relevant to the original questions being asked. In addition, the results of most biomedical research are published in "open literature," critically reviewed by outside experts and available to all.

In contrast, most military research is "applications driven": priorities are determined on the basis of military needs (e.g., treatment of acute injuries, development of protective clothing), and results not directly relevant to the original questions are seldom pursued. Such research is commonly classified and is published only for other military groups. The tight controls and restrictions on military research can result in a "stunted" body of literature that presents major limitations to later assessments in areas that were never pursued —in this case, the long-term health effects caused by exposure to chemical agents in general, and mustard agents and Lewisite in particular.

Researchers in the medical aspects of chemical warfare began their work in 1917 with few of the guideposts that are normally available from previous studies. The only literature available on the medical effects of sulfur mustard and Lewisite was that produced by the English and French, who had only a small head start with their research programs. Nevertheless, perusal of the significant papers published after the war from these groups reveals that multiple lines of investigation were quickly initiated and pursued. Some of the work done by the medical research groups examined the mechanisms of absorption of mustard agents into human skin, the effects of various ointments and antidotes on the severity of blisters and skin damage, and the pathological changes in the respiratory tract following inhalation of sulfur mustard vapor. Cochrane's history lists 22 papers and books published by 1920, which stood as the distillation of all significant investigations completed in the United States during WWI (Cochrane, 1946).

With the end of the war, the research program came to an end and CWS decreased drastically in size. The many research and production locations were also completely consolidated at the Edgewood Arsenal in Maryland. After a short period of relative uncertainty about its continued existence, CWS became part of the U.S. Army by an amendment to the National Defense Act on July 1, 1920.

Two later books included comments on the long-term effects of warfare gases. In 1925, Vedder published *Medical Aspects of Chemical Warfare,* in which he discounted any significant long-term sequelae of gassing and introduced the concept that "neurasthenic" conditions (an archaic term describing lassitude, decreased energy, and impaired functioning that, in Vedder's work, seemed to be used as a synonym for "psychosomatic") were the underlying factors in most veterans' claims of disability. Colonel Harry L. Gilchrist, in contrast, published an extensive comparative study on WWI casualties in 1928 that included detailed clinical descriptions of men who had been gassed in combat and a carefully researched chapter on the probable residual health effects of various gases (Gilchrist, 1928a,b). Gilchrist's work was a major contribution to knowledge about vesicant toxicity. Based on clinical examination of human gas victims and some animal experiments, as well as later follow-up studies of WWI veterans, Gilchrist found that the long-term effects of sulfur mustard were mainly respiratory, including emphysema, chronic asthma, and chronic bronchitis. Chronic conjunctivitis and corneal opacities were also described later by Gilchrist and Philip B. Matz (1933a,b; also see Appendix B for excerpt from Gilchrist and Matz, 1933b).

The period from 1920 until 1936 also saw the establishment of a Medical Research Division within the CWS. This group continued toxicological studies of chemical warfare agents, including sulfur and nitrogen mustards and Lewisite; investigated the lethal and sublethal concentrations of the agents; and renewed investigations into protective ointments. In addition, the group formalized what was known regarding treatment of gas casualties and attempted to examine the residual effects of exposure to various chemical warfare agents. Its work involved both animal and human experimentation. The experiments with human subjects, however, used only a few subjects mostly drawn from personnel, including the scientists themselves, working at Edgewood Arsenal.

For mustard agents and Lewisite, no major breakthroughs were made by the research efforts between 1920 and 1936. According to Cochrane, the lack of progress was traceable to a variety of factors. One of these factors was funding, often in short supply in a peacetime environment. Also, to appease public concern about poison gas production, significant efforts were spent trying to find peacetime uses for these agents. Another factor was the constant struggles and competition between different branches of the military, different departments within the CWS, and different scientific disciplines. Cochrane reports that the medical researchers, in direct competition with apparently more productive chemists, were especially vulnerable to funding shifts.

Thus, by the dawn of WWII, what was known about mustard agents and Lewisite (or many other agents, for that matter) was not organized into a cohesive body of literature. The clinical picture of the acute effects of exposure and some of the mechanisms of toxicity were well known (Gilchrist, 1928a,b; Vedder, 1925). There were clear guidelines for treatment of casualties, but the treatments were solely palliative. No effective ointments had been developed and nothing was available to prevent skin and lung damage. Even less was known about the long-term effects. So unorganized was the scientific base concerning vesicants that, when the 1941 version of the training manual for treatment of gas casualties (TM 8-285) was prepared for use in treating expected casualties in WWII, it did not include the carefully documented long-term effects of exposure reported by Gilchrist in 1928 and by Gilchrist and Matz in 1933.

This omission, not explained in Cochrane's history or elsewhere, was surprising to this committee. It is improbable that CWS did not know of Gilchrist's work, because it had been published in "open, nonclassified" literature, including one journal. In retrospect, we know that such an omission may well have unfavorably influenced the treatment and long-term follow-up of gassed soldiers in WWII, had such casualties occurred. In terms of the WWII testing programs with human subjects, this omission—coupled with an apparent disregard for the long-term effects of gas exposure—may have contributed to the absence of follow-up of these human subjects, despite the fact that the end points of many of the experiments were skin injury and burns (also see Chapter 4).

TESTING PROGRAMS AND CHEMICAL WARFARE PRODUCTION IN WORLD WAR II

As the war in Europe eroded U.S. neutrality, preparations began to revitalize and expand the activities of the Chemical Warfare Service. In order to obtain a greater base of scientific expertise, the War Department again came to the National Research Council for help (see Appendix C). In 1941, the research effort was reorganized and subsumed under a newly established Office of Scientific Research and Development

(OSRD), eventually comprised of two working branches, the Committee on Medical Research (CMR) and the National Defense Research Committee (NDRC), and an Advisory Council. NRC's Committee on the Treatment of Gas Casualties (CTGC), experts in the fields of medicine and biological sciences, began working closely with the CMR to aid the effort by administering and supervising government grants to researchers and universities for a wide range of research regarding chemical warfare agents (see Figure 3-2 for OSRD organizational chart).

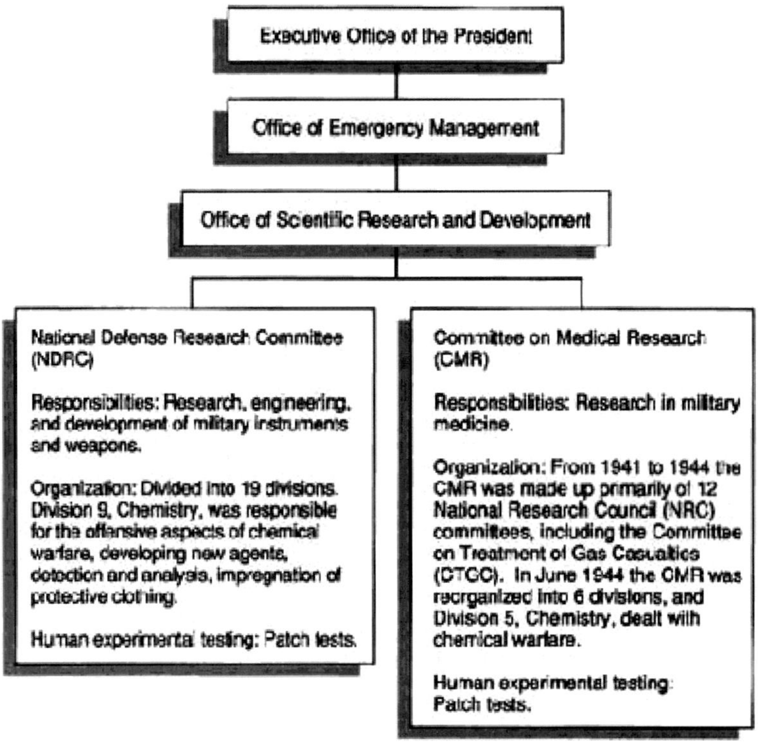

FIGURE 3-2 Organization of World War II civilian scientific research and testing programs. Chamber and field tests were conducted by the Chemical Warfare Service and the Navy Department, Office of Research and Inventions. Civilian researchers from the NDRC and CMR worked in close communication with the military. SOURCE: OSRD, 1946; Stewart, 1948.

One of the first assignments given to the CTGC was to review the scientific literature on the physiological effects of sulfur mustard and Lewisite, and on the methods tested for protection against injury and treatment of gas burns. The focus of the literature review was on acute toxicity data, but included longer-term effects. The first reviews were distributed among the various NDRC and CMR groups in August 1943 (Smith, 1943). There is no evidence that these reviews resulted in any changes or modifications of the protocols for the treatment or follow-up of human subjects in the experimental programs. On the subject of long-term effects of exposure, only Vedder's work, and not Gilchrist's, is mentioned in this review.

The focus of the following section is on the experiments conducted in the United States using human subjects; consideration of the animal experiments is included in later chapters, which survey the scientific literature. Similar experiments were also conducted in Great Britain, Canada, Australia, and the Soviet Union, as well as in Germany and Japan. The details of these experiments and these countries' chemical weapons programs are not included here. However, numerous references that provide information regarding these programs were examined by the committee, are included in the bibliography, and are specifically cited in this report where appropriate. One of the richest sources for information on the gas chamber experiments was reports released from the Naval Research Laboratory (NRL). Other primary source information regarding gas chamber experiments in locations other than the NRL was obtained by the committee from Edgewood Arsenal in Aberdeen, Maryland.

CWS carried out three basic types of experiments with human subjects. According to Cochrane, these testing programs involved the use of approximately 60,000 human subjects. *Patch, or drop, tests* were the most common and were used to assess the efficacy of a multitude of protective or decontamination ointments, treatments for mustard agent and Lewisite burns, effects of multiple exposures on sensitivity, and the effects of physical exercise on the severity of chemical burns. In addition, drop application of liquid mustard agents was commonly used in basic training to raise single blisters to impress upon the trainees the toxicity of these agents and the need for immediate responses to any orders to don gas masks. *Chamber tests* of various types were conducted to test the effectiveness of protective clothing, all of which had been impregnated with chemicals to retard vapor penetration. Finally, *field tests* involved the contamination of large or small areas of land with sulfur mustard or Lewisite. Human subjects were used in field tests to test protective clothing, to monitor the effects of the agents on animals in the test sites, and to take measurements of agent concentrations in soil and water samples. Table 3-2 summarizes the known major locations of these tests and the types of experiments done in each location.

Many veterans who were subjects in the chamber tests have obtained detailed records of their exposures from the Naval Research Laboratory. These reports often employ an outdated scientific notation. Table 3-3 summarizes the various notations used in NRL reports, along with other sources, to express concentrations, and compares these with modern notation. In addition, this table illustrates the difference between atmospheric concentration and the concept of cumulative exposure, in which

TABLE 3-2 Known Gas Testing Facilities and Test Typesa

Location	Type of Experiments
Edgewood Arsenal, Maryland	Chamber and patch tests Small-scale field tests Gas production
Bainbridge, Maryland	Chamber tests
Dugway Proving Ground, Utah	Large-scale field tests
Camp Sibert, Alabama	Chamber and patch tests (1943-1944 only)
Naval Research Laboratory, Virginia	Chamber and patch tests
Great Lakes Naval Training Center, Illinois	Chamber and patch tests
Camp Lejeune, North Carolina	Chamber tests
San Jose Island, Canal Zone	Large-scale field tests Chamber tests
Bushnell, Florida	Large-scale field tests
Other Allied Tests	
Finschhafen, New Guinea	Large-scale field tests
Innisfail, Australia	Large-scale field tests Chamber tests
Porton Down, England	Chamber tests (reported) Patch tests (reported)

[a] This represents only a partial list of locations, especially for patch tests because patch exposures were a frequent part of training at Chemical Warfare Schools. In addition, British testing reports are still classified and not. available

SOURCES: Cochrane, 1946; Gillis, 1985; Office of Scientific Research and Development, 1946.

TABLE 3-3 Concentration Versus Cumulative Exposure Level: Explanation of Notations in NRL Reports and Modern Literature

Concentration	
y/L[a]	Used in NRL reports to signify micrograms (μg, also called gamma) per liter (1, according to modern notation) or μg/l
mg/m^3	milligrams per cubic meter, equivalent to μg/l
ppm	parts per million, a volume to volume measurement that, at 25°C at sea level, is equal to 6.5 mg/m3 of sulfur mustard
Cumulative Exposure	
CT	Used in NRL reports to signify concentration (C) multiplied by time (T); equal to mg·min/m3, modern notation uses t to signify time (Ct)

[a] The use of L to signify liters is confusing because L is also the abbreviation for Lewisite. However, used with micrograms in the NRL reports, L signified liters.

SOURCE: Taylor et al., 1943.

both concentration and time of exposure are noted. For purposes of this report, the more modern notation is used and cumulative exposures are expressed in Ct (equivalent to mg min/m^3) to avoid repetition, because the vast majority of chamber tests were conducted in 60-minute trials. To place the concentrations that follow into additional context, Table 3-4 summarizes the concentrations required to produce specific physiological effects.

For further comparison, it may be useful to consider two additional calculations. First, according to the International Agency for Research on Cancer (IARC, 1975), the average and maximum atmospheric concentrations of sulfur mustard in combat zones were estimated to be approximately 20 and 33 mg/m^3. To determine exposure levels, however, one must consider the duration of exposure to a given atmospheric concentration. The exposure threshold for death from respiratory damage has been estimated to be between 1,000 and 1,500 mg·min/m^3 (Ct, see Table 3-4). Thus, fatal exposures on the battlefield in WWI must have lasted between 50 and 75 minutes (the product of 50 minutes and 20 mg/m^3 would equal a Ct of 1,000), if the estimated atmospheric concentrations were sustained, or longer if the concentrations dropped substantially.

Recent Centers for Disease Control (CDC) recommendations for safe levels of exposure to mustard agents and Lewisite provide a second useful frame of reference. Responding to the mandated destruction of all unitary lethal and chemical munitions, CDC published chemical agent control limits for atmospheric exposures to chemical munitions in the *Federal Register* in 1988. For general population exposures, the limits are 0.0001 and 0.003 mg/m^3 for sulfur mustard and Lewisite, respectively. For workers directly involved in munitions removal and destruction, the limits (averaged over 8 hours) are 0.003 mg/m^3 for both agents (CDC, 1988).

It is important to remember, when considering these comparisons, that the battlefield and estimated safe occupational levels of sulfur mustard and Lewisite refer to *atmospheric concentrations,* rather than *"dose" level.* However, it is the concentration and cumulative exposure to an unprotected target tissue (e.g., the eye, skin, or breathing passages) that determines the dose received and thus the damage to tissues from these agents. The presence of protective clothing and/or a gas mask reduces considerably the amount of agent reaching such a target tissue. Sections of this chapter to follow contain estimates and analysis of the probable cumulative exposures achieved in the chamber and field tests, as well as occupational situations.

Patch or Drop Tests

Information on the specific protocols used in patch tests was obtained from a variety of sources, including archived materials from the NRC

TABLE 3–4 Summary of Major Biological End Points Characterizing Sulfur Mustard Exposure to Humans (at 16°C–27°C unless otherwise noted)

	Effects End Points				
Exposure Route	Threshold[a]	Injured (nondisabling)	Incapacitation[b]	Death or Permanent Injury	References
Eye/vapor ($mg{\cdot}min/m^3$)	2 (at ≥32°C) 12	50–100	200 (ICt_{50})[c]	>800	Gates and Moore, 1946; McNamara et al., 1975; Papirmeister et al., 1991; Project Coordination Staff, 1946; Stepanov and Popov, 1962 Urbanetti, 1988; U.S. Army, 1974; U.S. Army CRDEC, 1990; U.S. Army and Air Force, 1975
Respiratory tract/vapor ($mg{\cdot}min/m^3$)	12–70	<100	200 (ICt_{50})	1,000 1,500 (LCt_{50})[d]	Ganas, 1969; McNamara et al., 1975; Project Coordination Staff, 1946; Robinson, 1967; Sidell, 1990; Stepanov and Popov, 1962; Stroykov, 1970; U.S. Army, 1974; U.S. Army CRDEC, 1990
Skin/vapor ($mg{\cdot}min/m^3$)	5 (maximum safe Ct) 50 (at 32°C) 100–200	100–300	1,000 (ICt_{50} at 32°C, low humidity) 2,000 (ICt_{50} at 21°C–27°C, humid)	10,000 (LCt_{50})	Gates and Moore, 1946; McNamara et al., 1975; Papirmeister et al., 1991; Project Coordination Staff, 1946; Sidell and Hurst, 1992; U.S. Army, 1974; U.S. Army CRDEC, 1990
Skin/liquid	10 μg 10–20 $\mu g/cm^2$ 32 mg/man (HD) 3.5–4 mg/man (HT)	No reported estimates	770 mg/man 3,000–7,000 mg/man	100 mg/kg (LD_{50}) 64 mg/kg (reported human lethal dose)	Papirmeister et al., 1991; Robinson, 1967; Sidell, 1990; U.S. Army, 1974; U.S. Army CRDEC, 1990; WHO, 1970

[a]Threshold is considered first indication of nondisabling signs and symptoms (eye reddening, erythema, and blisters).
[b]Incapacitation is considered inability to perform designated duties.
[c]The Ct expected to incapacitate 50 percent of those exposed.
[d]The Ct expected to cause death in 50 percent of those exposed.
SOURCE: Adapted from Papirmeister et al., 1991, Watson and Griffin, 1992.

Committee on Treatment of Gas Casualties, the Summary Technical Report of the National Defense Research Committee (OSRD, 1946, declassified in 1960), and from the *Fasciculus on Chemical Warfare Medicine,* Volume III: *Skin and Systemic Poisons* (NRC, 1945, declassification date unknown). It should also be noted that some subjects participated in tests in which only the protective ointments were applied to test skin sensitivity to the ointments themselves (many of the ointments were found to be highly irritating and corrosive to the skin). Analysis of the amounts of vesicant used was difficult, however, due to great variability in reporting of concentrations and cumulative exposures in individual experiments.

There were three types of delivery systems for patch testing. One type was called "Edgewood Rods," which were stainless steel rods with tips of varying diameters that were dipped into liquid sulfur or nitrogen mustard, or Lewisite, and then touched to the skin of a subject, usually on the forearm. A second type, "drod," was constructed from a small syringe that could deliver a measured amount of liquid to the skin. Various types of "vapor cups" were also used. The most common was the Edgewood Vapor Cup, a small glass cup similar to a beaker in which a section of filter paper saturated with liquid vesicant was placed. The cup was placed on the skin, allowing the vapor to rise from the filter paper and contact the skin. The cumulative exposures achieved in the vapor cups have been estimated to be 40,000 to 78,000 Ct. In some experiments, vapor cups were left on the skin for 15 minutes; in others, the cups were applied every 5 minutes for up to 3 hours and 40 minutes; in yet others, the cups were left on for more than an hour. Liquid patch tests, employing rods or drods, were more common than vapor cup tests and exhibited a wide variability in cumulative exposures.

Most of these experiments involved the application of liquid vesicant either before or after some test ointment. Most often, there were two or more sites on the forearm to which the vesicants were applied, thus providing for control sites at which no ointments were applied, and the liquids were allowed to remain on the skin for up to 2 minutes. The amounts used in these types of patch tests ranged from 0.15 to 7 mg for mustard agents and 1.4 to 7 mg for Lewisite. In some experiments, concentrations were expressed in micrograms. In still other experiments of this type, concentrations were also expressed as dilutions, ranging from 1:100 to 1:50,000 sulfur mustard, or Lewisite, to solvent. To further complicate analysis, a number of different solvents were used, including benzene, alcohol, paraffin oil, and chloroform.

Chamber Tests

Because the chamber tests were largely designed for the technical development of protective clothing, these tests were conducted by CWS and NRL in communication with the NDRC, rather than the CMR and CTGC. Thus, the sources used for information on concentrations and protocols for chamber tests included the Summary Technical Report of the NDRC and NRL technical reports. However, these sources are not exhaustive, and the details of chamber tests in locations such as Edgewood Arsenal and Great Lakes Naval Training Center were not made available to the committee for evaluation. Further, only the NRL has maintained accurate records of the individuals who participated in the tests (close to 2,500 men). Lacking similar information from the other locations, the total number of individuals involved in chamber tests is unknown. The vast majority of those participating in chamber tests were Caucasian men. A small number of African American and Japanese American soldiers were recruited for tests to determine possible differential skin effects of sulfur mustard on members of these races.

Similar to patch tests, there were a variety of types of chamber tests. For some chamber tests, the major questions were how long, under what conditions of temperature, and under what concentrations of gas would chloramide- or activated carbon-impregnated clothing afford protection of personnel against chemical attack with vesicants. The vesicant used most often was sulfur mustard, but nitrogen mustard and, probably, Lewisite were also used. These tests were called "man-break" tests. The common procedure was to equip men with gas masks[2] and clothe them in the impregnated suits (see Figure 3-3). The men would then enter the gas chamber (Figure 3-4) and remain there for periods from 60 minutes to 4 hours. The interiors of the chambers were most often maintained at 90°F and 65 percent relative humidity, because investigators were specifically interested in the durability of protective clothing under tropical conditions. Following the period in the chamber, the men wore their gas masks for an additional 5 minutes and remained in the suits for additional periods of time, ranging from 4 to 24 hours[3] (Taylor et al., 1943; see Appendix D for excerpts and the Military Reports section, U.S. Navy, of the Bibliography for a complete listing of NRL reports examined). Twenty-four hours after each chamber trial, the men were examined for reddening of the skin (erythema), evidence that the vapor had penetrated the

[2] The gas masks used were Navy Mark III or Mark IV diaphragm-type masks, designed to facilitate speaking during mask use. These masks were eventually removed from use by the military, because the diaphragms were leaky (for a fuller discussion, see the Conclusions and Further Analysis section of this chapter).

[3] It is not clear whether sulfur mustard, which is very persistent and evaporates slowly, was still present on the surface of the suits and, thus, a possible source of further contamination by inhalation or contact (fuller discussion is included at the end of this chapter).

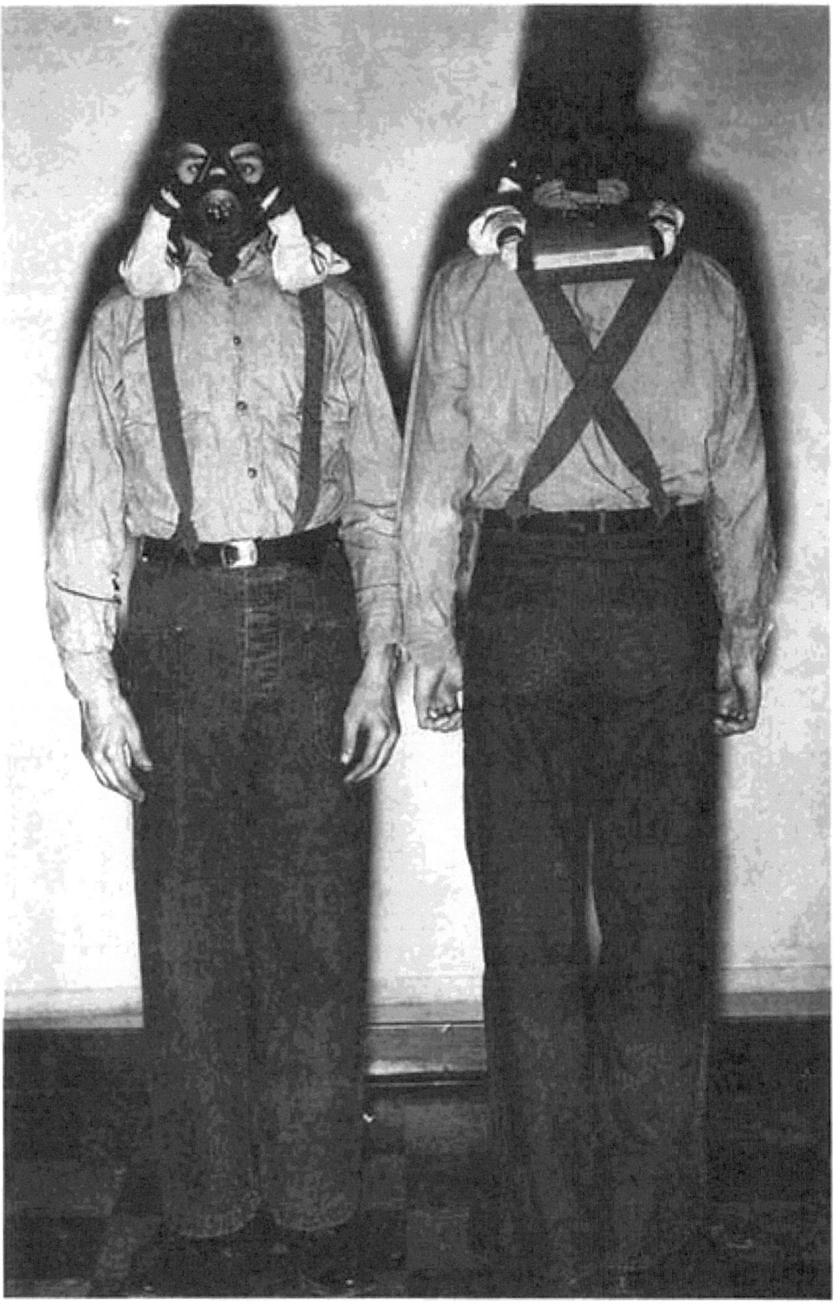

FIGURE 3-3 U.S. Naval personnel dressed for World War II sulfur mustard experiments. Gas masks shown are Navy diaphragmtype masks. SOURCE: Heinen et al., 1945. Photograph provided by Naval Research Laboratory.

FIGURE 3-4 (A) Naval Research Laboratory (NRL) gas chamber. Interior dimensions of this chamber were 10 ft. by 15 ft., with a 12 ft. high ceiling. The chamber was designed to fit ten men and allow room for moderate exercise. (B) Photograph of the inside of a similar gas chamber used at Edgewood Arsenal for World War II chamber tests (the numbers refer to specific equipment). SOURCE: Taylor et al., 1943. Photographs provided by Naval Research Laboratory (A) and U.S. Army Chemical and Biological Defense Agency (B).

suits and burned the skin. The men were required to repeat the procedure and enter the chambers either every day or every other day until they developed moderate to intense erythema. The number of trials tolerated depended largely on the vesicant concentration in the chambers. For example, a technical report by Taylor and his colleagues documents a specific set of 60-minute trials in which, at a cumulative sulfur mustard exposure level of 600 Ct per trial, the average number of trials tolerated was 5.3, representing a low of 4 trials and a high of 14 trials (Taylor et al., 1945). In trials with cumulative exposures of 9,600 Ct per trial, the average number of trials tolerated was estimated at two, because all the men developed some erythema after the first trial.

The anatomical locations and intensity of erythema are reported in Taylor's paper for each individual. The majority of men experienced intense erythema that was widespread over their bodies, especially in moist areas of skin folds, such as behind the knee and under the arms, in large areas of the chest and shoulders, and on their arms and legs. Little involvement of areas such as the scrotum and buttocks was reported for this particular set of trials, possibly because the men wore an extra layer of impregnated undergarments.

In another set of trials reported by Heinen and his colleagues (1945), multiple cumulative exposure levels ranged from 50 to 700 Ct. In these experiments, the subjects were engaged in different levels of physical activity before, during, and after the chamber trials. No data, however, are included regarding the tests in which activity was performed inside the chambers. Significantly, subjects in this set of trials were not completely dressed in protective clothing. Most were dressed in standard issue attire but wore carbon-impregnated suspenders. It was thought that the suspenders would protect a strip of skin that could then be compared with skin areas that were unprotected. In addition, only a few men were given impregnated underwear. Results from this set of trials are documented with photographs that show burns to the genital areas of many of the men. In one series, of 24 men participating, 13 experienced crusted lesions to the scrotum that were characterized as severe, and 8 experienced severe lesions to the penis. These lesions took up to one month to heal, according to the report.

In general, these two reports are representative of many of the chamber tests conducted at NRL (Washington, D.C.), Edgewood Arsenal (Maryland), Great Lakes Naval Training Center (Illinois), and, for a short time, at Camp Sibert, Alabama. The concentrations, times of exposures, and types of chemical agents used in other locations may not be similar, however, and full reports of other chamber tests were not made available to this committee. There is evidence that some chamber tests may have been done with higher cumulative exposures, because

Taylor and colleagues (1945) refer to ranges of 3,000 to 11,000 Ct of sulfur mustard used at Edgewood Arsenal.

Chamber tests were also conducted in Panama, North Carolina, and Maryland. These tests were called "wear tests," designed to see how the impregnated clothing stood up to use under possible combat conditions. Thus, the men wore the protective suits during drills and combat simulations, ranging from 1 day of amphibious training to 6 weeks of simulated combat. Following this, the men entered gas chambers in these suits, for 60-minute trials (as in the man-break tests), until erythema developed. The data reported in the Summary Technical Report of the NRDC (OSRD, 1946) show a range of 1.0 to 2.2 hours chamber time before erythema developed in this group of subjects. These data are difficult to compare with other chamber tests, however, because only micrograms of sulfur mustard were reported without the usual accompanying notation regarding volume. If one assumes that the micrograms listed were per liter, then the exposures ranged from 300 to 2,400 Ct in this series of chamber trials. Finally, some men participated in arm chamber tests of protective ointments or clothing materials. In these, the arms of the men were placed in a wind tunnel with cumulative exposures reported to be 1,200 Ct.

Some physiological measurements, including temperature and blood counts, were done on men participating in some of the tests, but these physiological measurements were not generally reported in the technical summaries. For those men who participated in Naval Research Laboratory tests, records of the experimental conditions, as well as any physiological measurements, were kept for each test subject and are available from NRL to individuals through the Freedom of Information Act (also see Chapter 4).

Field Tests

Many field tests with mustard gas were conducted with human subjects, but relatively little information is available. Known field tests were conducted by the United States and Australia in various locations (Freeman, 1991; Gillis, 1985; OSRD, 1946) (Table 3-2). Apparently, 1,000 U.S. servicemen participated in these field tests, a number that is supported by the discovery of a list of 1,000 servicemen recommended for special citation for participation in CWS testing programs (Cochrane, 1946; see also Appendix E). There is also evidence that some U.S. field tests involved human subjects who were not protected by clothing or even gas masks. The Summary Technical Report of the NDRC (OSRD, 1946, Table 8, p. 58) presents data about the exposure levels of mustard gas required to produce injuries in man, based on field tests in varied temperatures and climates in which none of the men wore protective clothing and only some of the men wore gas masks. The cumulative exposures reported for these tests ranged from 50 to 10,000 Ct.

The appendixes to the Report of the Chemical Warfare Service Conference of October 10-13, 1944, obtained from the National Archives, describe various field tests (CWS, 1944). Some of these tests may have contributed data to the summary mentioned above. Appendix VIII of the conference report outlines field tests in which bombing runs dropped from 125 to 550 tons of sulfur mustard over a specified area. Subjects wearing varying levels of protective clothing traversed the area in simulated patrols from 1 to 72 hours following the bombing. Such a protocol required the men to drop to the ground intermittently, thus coming into direct contact with contaminated surfaces. The resultant injuries were classified on the basis of the men's probable fitness for combat. Evidence of accidents during such trials can also be found in CWS documents. For example, one note describes how a group of men involved in a field test removed their gas masks after a rain storm and within two hours experienced ocular pain; three were hospitalized with acute conjunctivitis (Adler, 1944).

Gas Production, Gas Handling, and Chemical Warfare Production

Preparations for chemical warfare before and during WWII involved many additional people in the production, handling, shipping, and training to use and defend against chemical warfare agents. By the end of the war, the four CWS production facilities had produced close to 175 million pounds of ordinary sulfur mustard (H) and over 9 million pounds of distilled, purified sulfur mustard (HD) (Brophy et al., 1959). These production sites were at Edgewood Arsenal in Maryland, Huntsville Arsenal in Alabama, Pine Bluff Arsenal in Arkansas, and Rocky Mountain Arsenal in Colorado. An additional 40 million pounds of Lewisite and 200,000 pounds of nitrogen mustard were produced. Once produced, the agents were shipped to various storage facilities, depots, and proving grounds around the United States and were shipped overseas through ports such as Seattle, New York, New Orleans, and others.

This elaborate network of supply, coupled with the needs for training and chemical weapons testing, required many people from both the military and the civilian sectors. In 1939, CWS listed 803 enlisted men on its personnel rolls; this number grew to over 5,500 by December 1941 and over 61,000 by June 1943. Some 17 percent of military personnel assigned to CWS units were African Americans, a very high percentage when compared to all other units of the War Department. Women from the Women's Army Auxiliary Corps (WAAC) were also assigned to

CWS in jobs ranging from clerks and housekeepers to chemists and toxicologists. Civilian workers numbered 7,000 at the beginning of the war and 28,000 by 1943, of whom 40 percent were female and 45 percent were African Americans. The latter percentage was lower at the Rocky Mountain Arsenal, where there were fewer African Americans available from the surrounding community (Brophy and Fisher, 1959).

Although many of the specific jobs performed by these military and civilian personnel did not involve handling of, or even proximity to, warfare gases, the number of documented injuries was quite high. CWS, in fact, had the worst safety record of any branch of the War Department in both 1942 and 1943, the peak years of production (Brophy and Fisher, 1959). According to these authors, the safety record improved considerably after that, becoming among the best in the War Department by the end of the war. Nevertheless, the dismal safety record meant that many injuries were themselves studied by those involved in the CWS research branches or in studies contracted under NRC's Committee on Treatment of Gas Casualties. For example, many of the eye injuries at Edgewood were referred to and studied at the Johns Hopkins University Medical School under CTGC contracts (Andrus et al., 1948).

One study of these accidental injuries, reported that over 1,000 cases of mustard poisoning, resulting in eye, ear, nose, and throat symptoms, occurred at Edgewood Arsenal over a two-year period (Uhde, 1946). Of these, 790 were eye injuries; these injuries occurred to both males and females. Slow leaks of mustard vapor accounted for close to 80 percent of the problems. An additional 7 percent were from short-term exposures and accidents, such as explosions and mistaken use of real mustard in training exercises designed for simulated gas exposure. While the study did not present adequate information with which to judge the overall severity of injuries, it does report one death from sulfur mustard poisoning during this period. Little information is available from other locations, but Cochrane (1946) noted that during the first two weeks of December 1941, 577 patients were treated for eye and respiratory tract injuries from exposure to chemical warfare agents, especially sulfur mustard. The CWS locations where these injuries occurred were not reported. Finally, there is anecdotal evidence that the atmospheric concentrations of sulfur mustard around manufacturing areas at Edgewood Arsenal exceeded the odor threshold concentrations and thus may have been high enough to cause physiological effects (Howard Skipper, personal communication; see also Appendix A).

It is important to note that CWS personnel were exposed to a variety of toxic materials. For example, in addition to mustard agents, gases such as phosgene (a choking agent), hydrogen cyanide and cyanogen chloride (blood poisoning agents), and chloroacetophenone (tear gas) were also produced at the arsenals. Other personnel were involved in biological warfare research and production, in locations such as Fort Detrick, Maryland, and a civilian plant in Terre Haute, Indiana (Brophy and Fisher, 1959). Chemicals including napalm and white phosphorus were also stored and packed into bombs by CWS personnel. Even the production and testing of gas masks and filter canisters involved the use of toxic chemicals such as asbestos.[4] Finally, many people, including women, were assigned duties in the preparation of impregnated clothing, the most common method of which involved the use of two extremely toxic chemicals, chloroamide and acetylene tetrachloride.

The Bari Harbor Disaster

The only combat casualties from sulfur mustard in WWII were those injured or killed following a German air raid on the harbor of Bari, Italy, on December 2, 1943 (Alexander, 1947; Cochrane, 1946; Gage, 1946; Harris and Paxman, 1982; Infield, 1976; Perera and Thomas, 1986). Under conditions of secrecy, 2,000 bombs, each of which held 60 to 70 pounds of sulfur mustard, had been loaded on the merchant marine ship S.S. *John Harvey* before it had sailed from Baltimore to Bari. During the raid on Bari harbor, the *John Harvey* was sunk and some of its load of mustard bombs was damaged, causing liquid mustard to spill out into water already heavily contaminated with an oily slick from other damaged ships. Men who abandoned their ships for the safety of the water became covered with this oily mixture that provided an ideal solvent for sulfur mustard. The casualties were pulled from the water and sent to medical facilities unaware of what they carried with them on their clothes and skin. Equally unaware were the medical personnel who treated these casualties. Before a day passed, symptoms of mustard poisoning appeared in both the casualties and the medics. This disturbing and puzzling development was further compounded by the arrival of hundreds of civilians for treatment; they had been poisoned by a cloud of sulfur mustard vapor that blew over the city from some of the bombs that had exploded when the ship sank.

As the medical crisis worsened, little information was available about what was causing these symptoms. U.S. military command did not want to reveal to the enemy its preparations to position sulfur mustard in Europe for possible use against German forces. Eventually, however, the secret could not be kept (Harris and Paxman, 1982). The destroyer

[4] Canisters equipped with asbestos filters included the M9A1 and the M10A2 canisters (Brophy et al., 1959). The model numbers of canisters used in gas masks employed in the chamber or field tests are not known, but M9A1 canisters were in common use prior to July 1943 (Brophy et al., 1959).

U.S.S. *Bistera,* well outside the harbor and undamaged by the raid, had pulled 30 men from the water in a rescue effort. By the next day, the officers and crew of the *Bistera* were blinded from the effects of the sulfur mustard carried onto the ship by those rescued. Bari was overloaded with casualties by then, and the *Bistera* and its crew struggled to nearby Taranto for treatment. Soon the U.S. command had no choice but to confirm the cause of these injuries. With the assistance of Colonel Stewart Alexander, a military physician with extensive knowledge of mustard poisoning, better precautions and treatment were begun. By the end of the disaster, over 600 victims of mustard poisoning were treated from the harbor area alone; of these, 83 died (Alexander, 1947). Close to 1,000 civilians from the town also died (Harris and Paxman, 1982). Unfortunately, no long-term medical follow-up of survivors of the Bari harbor disaster has been reported.

Medical Applications of Chemical Warfare Research

As history has repeatedly shown, the experience of medical personnel and researchers in wartime can lead to major innovations in medical treatment practices. Such was the case with chemical warfare research in WWII. Numerous advances were made in the treatment of metal poisoning, development of antibiotics, treatment of burn injuries, and other areas. Many of these advances were reviewed soon after the war in a two-volume summary *Advances in Military medicine* (Andrus et al., 1948). The story of the use of nitrogen mustard as a cancer chemotherapy agent is especially relevant to the present report.

Nitrogen mustards were first synthesized in the 1930s. These compounds were modifications of sulfur mustard and were found to have greater systemic toxicity than sulfur mustard (Gilman and Philips, 1946; OSRD, 1946). Particularly potent was the effect of nitrogen mustard on cells that are actively proliferating, including the lymphoid tissue, bone marrow, and certain cells lining the gastrointestinal tract. During WWII, the Committee on the Treatment of Gas Casualties authorized a contract between the Office of Scientific Research and Development and Yale University (Andrus et al., 1948). Under this contract, Louis C. Goodman headed a group that was responsible for the study of the pharmacologic effects of nitrogen mustards. The group, including Alfred Gilman, Frederick Philips, and Roberta Allen, focused its efforts on the study of the cytotoxic properties of nitrogen mustard. Enlisting the help of anatomist Thomas Dougherty, the group expanded its work to examine the effect of nitrogen mustard on experimental tumor cells in mice. It was found, but not published until later, that systemic administration of nitrogen mustard caused dramatic regression of these mouse tumors. These data formed the experimental basis of the first clinical trials of nitrogen mustard as a cancer chemotherapy agent (Gilman, 1946, 1963; Gilman and Philips, 1946; Rhoads, 1947).

Although the concept of chemotherapy does not seem radical today, in 1942 the idea of injecting poisons into cancer patients, especially poisons marked "compound X" due to their classified status, would have been viewed by most physicians as "the act of a charlatan" (Gilman, 1963). With the help of Gustav Lindskog however, clinical trials were begun in December 1942 with a patient dying of lymphosar-coma for whom all other treatments had failed. The patient's tumors regressed, the outlook brightened, and another patient was begun on the nitrogen mustard therapy. In all, six patients made up this first trial. However, as had happened in the animal studies, the tumors reappeared as the bone marrow recovered, and no long-lasting cure was attained. The challenge remained to establish the regimen of therapy that would kill the cancer cells completely, yet preserve enough of the bone marrow to regenerate needed healthy cells. In addition, there was no reason to assume that all types of cancer cells would be equally affected by nitrogen mustard therapy.

The Yale group dispersed in June 1943, but clinical trials with nitrogen mustard continued in several other locations. By 1948, close to 150 patients in the terminal stages of Hodgkin's disease, lymphosarcoma, or certain leukemias had been treated with this agent (Gilman and Cattell, 1948). The best results were obtained in cases of Hodgkin's disease. Derivatives of nitrogen mustard (hydrochloride forms) are still used today, particularly for treatment of lymphoma, in a regimen that includes an array of other drugs and chemicals administered with the nitrogen mustard (see also Chapter 6).

RESEARCH, USE, AND DISPOSAL OF CHEMICAL WEAPONS AFTER WORLD WAR II

Postwar Research Programs

The vast majority of the post-WWII research concerning mustard agents and Lewisite has been done in animal studies or in model systems, such as skin tissue culture. This research has been aimed toward the development of pretreatments to prevent mustard toxicity or toward improved treatments against acute poisoning. Emphasizing these issues, Papirmeister and colleagues reviewed the literature on sulfur mustard in 1991, including consideration of all such work published up to 1990. For the purposes of the present report, discussion is confined to only those research programs that used human subjects.

Once WWII was over, all of the research programs of the Chemical Warfare Service were scaled down. Very little research was done during the period between 1946 and 1950, and by the time research in chemical and biological weapons was revitalized in the 1950s, military priorities had shifted to agents perceived to pose greater threats than sulfur mustard or Lewisite. For example, improvements to early nerve gases developed in WWII gave new importance to the development of antidotes to nerve agents. Chemicals with intense psychoactive properties, such as lysergic acid diethylamide (the hallucinogen LSD) and phencyclidine (PCP, known on the street as "angel dust") were also of special interest. Most of this research was done at or supervised by personnel from Edgewood Arsenal; it involved approximately 6,700 human subjects between 1950 and 1975. Only a few projects tested sulfur mustard or Lewisite.

Other groups that participated in this research included the Central Intelligence Agency and the Special Operations Division of the Department of the Army (Taylor and Johnson, 1975). As has been documented in numerous government and popular press publications, abuses of human subjects in these research programs began to emerge almost as soon as the projects were begun but were largely covered up until the early 1970s (Harris and Paxman, 1982; Taylor and Johnson, 1975; see also Appendix F). Finally, congressional hearings into these abuses in 1974 and 1975 resulted in fuller disclosures, eventual notification of all subjects as to the nature of their chemical exposures, and compensation of a few families of those who had died while serving as human subjects in these projects (Harris and Paxman, 1982; Taylor and Johnson, 1975).

As part of its effort to rectify the abuses discovered, the Department of the Army asked the National Research Council to assess the likelihood of long-term health consequences of exposure to the chemicals tested and to report on the current health status of the soldiers who participated in the 1950-1975 testing programs. The resulting study was published in three volumes in 1982, 1984, and 1985. The vast majority of these test subjects, however, had been exposed to nerve agents or hallucinogenic drugs. In the 1984 volume, the NRC committee reported that only 150 individuals had been exposed to vesicants. In a section on vesicants, no conclusions were drawn for Lewisite on the basis of scanty information; for sulfur mustard, however, the group concluded:

> Mustard gas is mutagenic in various organisms and test systems. One cannot readily predict the degree of genetic risk that it poses for man, however, because data on its mutagenicity in mammalian germ cells are very limited, and the mutagenic potency of mustards varies considerably among assay systems. Nevertheless, the available evidence suggests that the possibility of mutagenic effects of mustard gas in human germ cells should not be disregarded. The clear muta

genicity of mustard gas in various assays is consistent with its carcinogenic potential.

Mustard gas is not only a vesicant, but also a systemic poison. Its acute effects have been demonstrated in bone marrow, intestinal tract, and respiratory tract. It can cause blindness and permanent skin scarring with a potential for skin tumors. It probably can also cause acute and chronic bronchitis. Other nonmalignant chronic effects have not been adequately documented.

Single exposures, even if severe, as in military service, are not associated with statistically verifiable increases in mortality from tuberculosis and cancer; but repeated small exposures, such as occur in industrial operations, do increase cancer deaths significantly.

The NRC committee's 1985 report summarized the investigations of the current health status of test subjects and concluded that the number of subjects exposed to mustard gas was too small to detect any long-term health effects. Also cited were the only long-term follow-up studies of WWI sulfur mustard casualties. Overall mortality and morbidity data for a sample of men treated for sulfur mustard injuries in American Expeditionary Forces hospitals from August through November 1918 revealed a slightly increased incidence of lung cancer among gassed veterans, but this increase was not sufficiently high for statistical significance (Beebe, 1960). A further study of this cohort 10 years later did not alter these results (Norman, 1975). [5] To the present committee's knowledge, no human subjects have been used in tests of mustard agents or Lewisite in the United States since the 1960s.

Continuing Use of Sulfur Mustard and Other Chemical Weapons in International Conflicts

Military use of sulfur mustard was a topic at the Paris Conference on the Prohibition of Chemical Weapons in January 1939. Due to continued use of these weapons around the world, however, chemical weapons bans remain an ongoing issue of negotiation at the current chemical convention talks in Geneva, Switzerland. The Stockholm International Peace Research Institute (SIPRI) published a book analyzing the historical, technical, military, legal, and political aspects of chemical and biological warfare in 1971 (SIPRI, 1971). This document reports use of sulfur mustard by the Egyptians in Yemen in 1965 and the Iraqis against the Kurds in 1965.

[5] Review and analysis of the Beebe and Norman papers are included in Chapter 6.

Numerous reports of use of other agents, including tear gas, smokes, and herbicides, are also reported. Additional reports have surfaced of use of sulfur mustard by the Vietnamese in Cambodia and Laos between 1976 and 1980 (Medema, 1986). There are more recent reports of use of sulfur mustard and cyanide by Armenians against the Azerbaijanis in the Nakhichevan Autonomous Republic (CBW News, 1992).

Probably the greatest use of sulfur mustard, however, has been in the ongoing conflicts between Iran and Iraq, and many of these incidents have been confirmed (D'Halluin and Roels, 1984; Dunn, 1986a,b; Heyndrickx and Heyndrickx, 1984; Mandl and Freilinger, 1984; Medema, 1986; Physician's for Human Rights, 1989; Requena et al., 1988; United Nations Security Council Reports, 1986, 1987, 1988a,b,c,d). Some of the Iranian casualties were treated in European hospitals and thus could be documented medically. These patients suffered from pulmonary, eye, and skin lesions at similar incidence levels as battlefield casualties from WWI. In WWI, 80 to 90 percent of sulfur mustard casualties suffered skin lesions, 86 percent suffered eye involvement, and 75 percent had pulmonary damage (Sidell and Hurst, 1992). Among the Iranian casualties, 83 percent suffered skin lesions, 92 percent had eye problems, and 95 percent had pulmonary damage (Balali-Mood and Navaeian, 1986).

There are also sketchy data that indicate that some Iranian soldiers may have been exposed to Lewisite. London physicians who examined and treated the lesions of these soldiers reported that the signs exhibited were similar to those associated with Lewisite, rather than sulfur mustard (Perera, 1985). For example, pain occurred very quickly following vapor exposure, and skin lesions showed none of the pigmentation changes characteristic of sulfur mustard exposure. In addition, the victims reported that the agent did not smell like garlic, as does sulfur mustard.

U.S. Chemical Stockpile Disposal Program

The U.S. stockpile of sulfur mustard, currently stored at seven military installations on the continental United States (Aberdeen Proving Ground, Maryland; Anniston Army Depot, Alabama; LexingtonBlue Grass Army Depot, Kentucky; Pine Bluff Arsenal, Arkansas; Pueblo Depot Activity, Colorado; Tooele Army Depot, Utah; Umatilla Depot Activity, Oregon) and one location in the South Pacific (Johnston Island, U.S. Pacific Territory), is under congressional mandate for destruction (Carnes, 1989; Carnes and Watson, 1989). Lewisite is stored in only one location, Tooele Army Depot, in ton containers. Although the locations listed here are the official storage facilities, it is not known on how many former military bases small amounts of agents such as sulfur mustard were left or buried when the bases were deactivated. For example, it was recently discovered that a dumping site, used to dispose of 55-gallon drums of sulfur mustard in the mid-1940s, now lies near a large business complex in Edison, New Jersey (Gallotto, 1992). This site may be one of many located on what was once the Raritan Arsenal, where reports of former soldiers claim that toxic chemicals were poured into pits, along with the emptied drums and shells, treated with lime, and covered over with soil. Such reports are not only relevant to the issue of toxic waste from chemical weapons production in this century; they also point out locations, not apparent in official CWS histories, where military and possibly civilian personnel were exposed to chemical agents during WWII.

The Department of Defense (DoD) Authorization Act of 1986 (P.L. 99-145) directed and authorized the Secretary of Defense to destroy the United States' aging and obsolete stockpile of lethal unitary chemical munitions and bulk agent by September 1994. In response, DoD established the Chemical Stockpile Disposal Program in 1986, but the target completion date has been postponed to 2004. Unitary munitions contain a lethal chemical agent at the time the munition is loaded; in contrast, binary munitions contain agent precursors that mix and react to form lethal agent after the munition is fired. The unitary stockpile includes the vesicant agents sulfur mustard and Lewisite, as well as organophosphate nerve agents. All but approximately 6 percent of the U.S. stockpile of unitary munitions and bulk agent is currently stored in the continental United States as bombs, cartridges, mines, projectiles, spray tanks, and ton containers. Approximately 60 percent of the unitary stockpile tonnage is stored in bulk as ton containers, spray tanks, or similar large containers. The remainder is stored on Johnston Island, including the North Atlantic Treaty Organization's stockpile that was moved in 1990 from a military site near Clausen, Germany.

DoD has tested, considered, and discarded a number of proposed disposal methods in favor of high-temperature incineration (Carnes, 1989; Carnes and Watson, 1989; U.S. Department of the Army, 1988). The first step in this approach involves "reverse assembly" of the munition inside an explosive-containment room, resulting in the separation of agent from any explosive materials and munition hardware or containers. These different fractions are sent to separate incinerators, and materials are incinerated by a specially designed system using four two-staged furnaces (a furnace and an afterburner) for each component (e.g, liquid agent, contaminated metal parts). Temperatures reach between 540°C and 1370°C for the furnaces and approximately 1090°C for the afterburners. Stack gases and incinerator ash are treated in advanced pollution-abatement systems intended to ensure safe handling and eventual disposal in a hazardous waste facility (see Carnes, 1989, for details of emission control systems).

Incineration remains a matter of continuing controversy among environmental groups, citizens who live near some of the proposed incineration sites, and the involved government agencies. The Programmatic Environmental Impact Statement, released by the Department of the Army in 1988, concludes that timely disposal of the stockpile of chemical weapons entails less of a hazard than continued storage.

Technical support and oversight for the Chemical Stockpile Disposal Program (and the companion Chemical Stockpile Emergency Preparedness Program that assists nearby communities in developing emergency preparedness programs) is provided by numerous Army commands and a host of civilian institutions. These include the National Center for Environmental Health of the Centers for Disease Control (U.S. Department of Health and Human Services), Federal Emergency Management Agency, U.S. Environmental Protection Agency, U.S. Department of Agriculture, as well as state and local planning agencies in the 10 affected states, and two national laboratories (Oak Ridge National Laboratory, Oak Ridge, Tennessee; Argonne National Laboratory, Argonne, Illinois). Analyses of vesicant toxicity and long-term health risks from these groups have been considered, along with other information, in generating the present report.

CONCLUSIONS AND FURTHER ANALYSIS

The committee reached two principal conclusions based on its analysis of the chemical warfare testing programs from WWI through 1975. These conclusions relate directly to the estimated level of exposure to mustard agents and Lewisite experienced by the WWII chamber and field test subjects and to the exposures of workers in the Chemical Warfare Service during WWII. In addition, the committee's conclusions are pertinent to the health care concerns of those who have been injured by use of these agents in recent wars and conflicts, or who may be exposed in the future from belligerent use of these agents or through accidental exposure during their disposal.

1. **The lack of follow-up health assessments of the human subjects in WWII gas chamber tests and field tests severely diminished the amount and quality of information that could be applied in the assessment of long-term health consequences of exposure to mustard agents and Lewisite.** Although the reasons underlying the lack of follow-up health assessments are not explicit from the numerous documents and materials considered by this committee, a number of factors may have played a role:

   - There was no unified body of information, based on WWI research and the research done in the period from 1918 to 1939, when research was intensified in the early pre-WWII period. This lack of information seems to have contributed directly to a lack of appreciation for the serious long-term health risks associated with exposure to mustard agents and Lewisite, specifically chronic bronchitis, emphysema, chronic laryngitis, corneal opacities, chronic conjunctivitis, and keratitis.

- Scientific inquiry was controlled by the military establishment, whose primary concern was with acute rather than long-term injury. This control also probably contributed to the paucity of animal or other types of studies, following WWII, aimed toward elucidation of long-term consequences of damage to specific physiological systems. For example, no long-term follow-up was done on workers involved in chemical warfare materials production, despite the high level of injuries that occurred.
- The atmosphere of immediacy caused by the outbreak of war, and the resulting prioritization of expected combat injuries, at least strengthened the focus on acute damage from chemical warfare agents, and at worst dampened any sensitivities that were present regarding the future health of human subjects or chemical warfare production workers.
- Once the war was over, there may also have been ambiguities about which federal department or agency should have had responsibility for follow-up of veterans. Although the former Veterans Administration (VA) had that role traditionally, the VA could not have been expected to know about the testing programs and their possible effects on the health of human subjects without communication from the military.
- Finally, and related to the issue of responsibility for follow-up, the continued secrecy maintained by the military regarding the WWII testing programs also created a barrier to follow-up assessments of exposed individuals. Even during the present study, which follows a five-year period of intensifying public scrutiny of these WWII programs, obtaining certain types of information was not easy and often involved piecing together bits of data from numerous sources. In fact, this committee was commonly required by many DoD and Department of the Army offices to file all requests under the Freedom of Information Act. These requirements were often imposed even on the present Department of Veterans Affairs (VA), when it attempted to aid this committee by making certain requests for information regarding the possible existence of records of individuals who participated in the testing programs (see Appendix E). The most valuable primary source data were received from the Naval Research Laboratory and NRC's Office of Archives and Information Services. This committee is especially grateful to the NRL for its commitment to open its files. The NRL

stands alone among sections of the Department of Defense in the maintenance of files and reports, and the sharing of those files with this committee and with the affected veterans.

2. **The levels of exposure to mustard agents or Lewisite experienced by the test human subjects may have been much higher than inferred in the summaries of the experiments and field tests. As in all chemical exposures, such exposure levels directly relate to the types and severities of exposure-induced injuries and diseases.** One can infer the cumulative exposures to the skin of chamber subjects strictly on the basis that skin damage was the end point of these experiments (see Table 3-4). Therefore, if all other types of exposures were held to zero, these subjects received between 100 and 300 Ct. As has been documented, some of the subjects were hospitalized for as long as "a month or so" (Taylor et al., 1943). Thus, exposures to the skin may have been as high as 1,000-2,000 Ct. Under the hot, humid conditions in the chambers, however, lower exposure levels would have resulted in similar injuries (Papirmeister et al., 1991). The dose to the skin from such exposures would have been as high as those observed under battlefield and occupational conditions. Further, some sulfur mustard would also have been absorbed from the skin into the systemic circulation.

In the chamber experiments, unmasked subjects were required to remain in their protective clothing from 4 to 24 hours following chamber trials, allowing ample opportunity for additional contact and inhalation exposures from contaminated surfaces and clothing. Another factor that probably resulted in some inhalation exposure of subjects in the chamber tests was vomiting during the period subjects were in the chamber. This was reported by at least one of the subjects who spoke at the public hearing; this person reported conjunctivitis and laryngitis following such a vomiting incident on his seventh day of testing (Elmer Hood, public hearing statement; see also Appendix G). Vomiting presumably would result in removal of the mask while in the chamber, with a resulting inhalation exposure of unknown duration at the chamber concentration being tested.

The most important route of additional exposure in the chamber and field tests was probably gas mask leakage. From the information available to the committee, it appears that the vast majority of the human subjects in the chamber and field tests wore full-face gas masks during their exposures. In fact, the documented exposures at the Naval Research Laboratory were delivered at concentrations and for durations that would have caused lethal respiratory effects if the subjects had not been equipped with respiratory protection. Thus, exposure of the respiratory tract and eye to the agent would have depended on the protection factor (PF) afforded by the gas masks. The PF of a full-face respirator (e.g., a gas mask) is calculated as the ratio of the ambient concentration of the contaminant to the concentration inside the mask, which in turn depends on both leakage around the respirator and contaminant penetration of the gas mask canister. A PF of 100 equals a penetration of 1 percent of the contaminant into the mask (Adley and Uhle, 1969). A PF of 50 to 100, based primarily on leakage around facemasks, has been reported for relatively modern (post-WWII) full-face respirators (Hyatt, 1976). Estimates from industrial hygiene research, however, indicate that the level of protection achieved in actual use of a respirator is usually below the stated PF for that respirator (National Institute of Occupational Safety and Health, 1974). Thus, modern respirators are likely to function closer to the lower PF estimate of 50. In practical terms, even if the respirator actually achieved a PF of 100, subjects exposed to a concentration of 100 mg/m^3 of sulfur mustard would be breathing a concentration as high as 1 mg/m^3 inside the mask, corresponding to a cumulative exposure of 60 Ct over a single 60-minute trial. At even lower concentrations—under the odor threshold (0.6 mg/m^3)—the subjects may well have been unaware of any leakage through their masks (see Box 3-1). Information on the breakthrough capacity of the gas mask cartridges used in the WWII chamber tests was not available to the committee, but it is known that prolonged use of cartridges can result in breakthrough of the agent by exceeding the capacity of the absorbent filter material (Stampfer, 1982).

BOX 3-1 ODOR THRESHOLD FOR SULFUR MUSTARD AND LEWISITE: COMPARISON WITH TISSUE DAMAGE THRESHOLDS

Even when enough agent had broken through their gas mask canisters to produce symptoms, chamber test subjects may not have noticed it, at least by odor. The odor threshold for sulfur mustard is reported to be about 0.6 mg/m^3, and the median concentration of Lewisite detectable by odor is reported to be 14 to 23 mg/m^3 (OSRD, 1946). However, both agents have effects on the eye and respiratory tract at lower concentrations (Papirmeister et al., 1991; Urbanetti, 1987). For example, sulfur mustard exposure at a concentration of 0.5 mg/m^3 for 30 minutes (15 Ct) would result in both respiratory and eye symptoms (see Table 3-4). For Lewisite, such irritating effects are reported to be noticeable at concentrations estimated to be as low as 6 to 8 mg/m^3 (Papirmeister et al., 1991; Urbanetti, 1987). Thus, for sulfur mustard exposures at 0.5 mg/m^3, an exposure of only about 25 minutes (12.5 Ct) could be expected to cause eye and respiratory tract symptoms without the subject being aware of the exposure, at least by odor.

In the NRL chamber test reports examined by the present committee, when gas mask types were listed, the masks used were Mark III or Mark

IV Navy diaphragm gas masks. These masks were probably equipped with M9A1 (prior to July 1943) or M9A2 canisters that contained Whetlerite, a copper-, chromium-, and silver-impregnated activated charcoal as the sorbent (Brophy et al., 1959). The PF afforded by these masks for sulfur mustard or Lewisite was not available to the committee. However, an individual involved in this testing reported that the WWII British and U.S. masks were very effective in removing sulfur mustard (Howard Skipper, personal communication; see Appendix A). Yet some chamber tests were conducted at high concentrations. For example, a test conducted at chamber concentration of 100 mg/m^3 for 60 minutes would have resulted in a cumulative, unprotected, exposure of 30,000 Ct over five trials. Even an assumed PF of 1,000 for the gas mask (10 times greater than that estimated for modern full-face respirators) would have resulted in concentrations as high as 0.1 mg/m^3 in each trial, corresponding to a cumulative exposure of 6 Ct just from the inspired air in each trial. This would have been below the odor threshold for sulfur mustard and, over five trials, would have resulted in a cumulative inhalation exposure of 30 Ct, enough to cause signs and symptoms in the eyes and respiratory tract (see Table 3-4). If a more realistic estimate is used, such as a mask with a PF of 100, the per trial exposure would have been 60 Ct. Over five trials then, a subject could have had an inhalation exposure of 300 Ct, more than sufficient to cause an incapacitating injury (see Table 3-4). It is important to remember that any such inhalation exposure would have been *in addition* to any skin exposure through breakdown of the protective clothing.

It is important to note also that the gas masks and clothing used in the NRL tests were worn repeatedly by the subjects. In at least one series of studies, it was reported that the rubber of the gas mask facepieces and connecting tubes absorbed enough sulfur mustard after 12 to 15 exposures to cause conjunctivitis, laryngitis, and erythema of the face (Taylor et al., 1943). Therefore it is clear that some exposure to the respiratory tract occurred from absorption of sulfur mustard on masks. Finally, as mentioned previously, the special diaphragm element in the types of gas masks used in the NRL chamber tests was eventually shown to provide an additional route of mask leakage, independent of the filter capacity (Brophy et al., 1959).

The presence of erythema of the face, conjunctivitis, laryngitis, or bronchitis within 24 to 72 hours following an exposure to sulfur mustard or Lewisite would be clear evidence that a significant inhalation and eye exposure had occurred, even if the subject was wearing a mask during the exposure. Conversely, it would appear that a lack of such symptoms following even a low-level exposure of 5 to 6 days to sulfur mustard would indicate a cumulative exposure (Ct) of less than about 12 Ct (see Table 3-4). However, in terms of the Centers for Disease Control's estimates of permissible exposure levels (CDC, 1988), the exposures actually reaching the breathing zone of chamber subjects (from the above example, 0.1 mg/m^3 sulfur mustard breakthrough with a gas mask rated at 1,000 PF) may have been more than 1,000 times the general population agent control limits (0.0001 mg/m^3 for sulfur mustard), and 33 times the control limits for occupational exposure (0.003 mg/m^3 for sulfur mustard). In reality, some of the subjects in the chamber tests and field trials almost certainly breathed concentrations 10 or more times the 0.1 mg/m^3 level for at least a part of their exposures.

The focus here on chamber and field test subjects is not meant to discount the probable exposure levels experienced by those who were involved in the production or handling of mustard agents and Lewisite. Indeed, as outlined above, the poor safety record of the Chemical Warfare Service during the peak years of production, the high rate of agent-induced injuries, and the anecdotal reports of perceptible odors of sulfur mustard in the manufacturing areas argue that workers and gas handlers were often exposed to levels of mustard agents and Lewisite sufficient to cause short- and long-term health effects. Thus, these individuals should also be considered at risk for any of the adverse health effects this report identifies.

In conclusion, the dose of sulfur mustard to the skin, eye, and respiratory tracts of the human subjects was substantial, especially in the case of the subjects involved in the chamber tests. Doses to the skin were probably equivalent to those received under combat conditions. Consideration of the probable gas mask leakage, additional exposures from contact or vapors from the clothing, accidents, and the documented signs and symptoms in the chamber test records indicate that the doses received by the human subjects were equivalent to those received in occupational exposures and, perhaps, even battlefield exposures.

REFERENCES

Adler FH. 1944. Report of consultant in ophthalmology on 6 cases of H vapor burns occurring at Bushnell Field Installation on April 20, 1944. Memo to Colonel C.P. Rhoads, dated May 1, 1944. Available at the National Archives, Suitland Reference Branch, Suitland, MD. Record Group 175, Group 4B, Folder 319.1.

Adley FE, Uhle RJ. 1969. Protection factors and self-contained compressed-air breathing apparatus. American International Hygiene Association Journal 30:355-359.

Alexander SF. 1947. Medical report of the Bari Harbor mustard casualties. Military Surgeon 101:1-17.

Andrus EC, Bronk DW, Carden GA Jr, Keefer CS, Lockwood JS, Wearn JT, Winternitz MC, eds. 1948. Advances in Military Medicine. Science in World War II: Office of Scientific Research and Development. Boston: Little, Brown and Company.

Balali-Mood M, Navaeian A. 1986. Clinical and paraclinical findings in 233 patients with sulfur mustard poisoning. In: Heyndricks B, ed. Terrorism: Analysis and Detection of

Explosives. Proceedings of the Second World Congress on New Compounds in Biological and Chemical Warfare. Gent: Rijksuniversiteit. 464-473.
Beebe G. 1960. Lung cancer in World War I veterans: possible relation to mustard gas injury and 1918 influenza epidemic. Journal of theNational Cancer Institute 25:12311252.
Brophy LP ,Fisher G. 1959. The Chemical Warfare Service: Organizing for War. United States Army in World War II: The Technical Services. Washington, DC: Office of the Chief of Military History, Department of the Army.
Brophy LP ,Miles WD, Cochrane RC. 1959. The Chemical Warfare Service: From Laboratory to Field. United States Army in World War II: The Technical Services. Washington, DC: Office of the Chief of Military History, Department of the Army.
Budavari S, ed. 1989. The Merck Index. 11th ed. Rahway, NJ: Merck & Co.
Carnes SA. 1989. Disposing of chemical weapons: a desired end in search of an acceptable means. Environmental Professional 11:279-290.
Carnes SA, Watson AP. 1989. Disposing of the U.S. chemical weapons stockpile: an approaching reality. Journal of the American Medical Association 262:653-659.
CBW News, Quarterly Bulletin of Chemical and Biological Weapons Issues. 1992. 9(July): 8.
Centers for Disease Control (CDC). 1988. Final recommendations forprotecting the health and safety against potential adverse effects of long-term exposure to low doses of agents: GA, GB, VX, Mustard Agent (H, HD,T) and Lewisite (L) . Federal Register 53:8504-8507.
Chemical Warfare Service (CWS). 1944. Report of Chemical Warfare Service Conference. October 10-13, 1944. Available at the National Archives, Suitland Reference Branch, Suitland, MD. Record Group 175, Group 4B, Folder 337.
Cochrane RC. [1946]. Medical research in chemical warfare. Available through the U.S. Army Chemical Defense Research, Development and Engineering Center, Aberdeen Proving Ground, MD.
Despretz M. 1822. Chlorine compounds. Ann Chim Phys 21:428. [In French]
D'Halluin F, Roels H. 1984. Autopsy observations in an Iranian soldier exposed to war gases. Archives Belges (Supplement):284-290.
Dunn P. 1986a. The chemical war: journey to Iran. Nuclear, Biological and Chemical Defense and Technology International 1(2):28-37.
Dunn P. 1986b. The chemical war: Iran revisited-1986. Nuclear, Biological and Chemical Defense and Technology International 1(3):32-39.
Freeman K. 1991. The unfought chemical war. Bulletin of the Atomic Scientists 47:30-39.
Gage EL. 1946. Mustard gas (dichloroethyl sulfide) burns: in clinical experiences. West Virginia Medical Journal 42:180-185.
Gallotto AA. 1992. Corps tries to ease Raritan Arsenal cleanup fear. Sunday Star-Ledger, Newark, New Jersey. February 2, 1992.
Ganas P. 1969. New developments in chemical and biological warfare. Forces Aeriennes Francaises 24:449-475.
Gates M, Moore S. 1946. Mustard gas and other sulfur mustards. In: Division 9, National Defense Research Committee. Chemical Warfare Agents, and Related Chemical Problems. Summary Technical Report of Division 9, NDRC. Washington, DC: Office of Scientific Research and Development. 30-58.
Gates M, Williams JW, Zapp JA. 1946. Arsenicals. In:Division 9, National Defense Research Committee, comp. Chemical Warfare Agents, and Related Chemical Problems. Summary Technical Report of Division 9, NDRC. Washington, DC: Office of Scientific Research and Development.
Gilchrist HL. 1928a. A Comparative Study of World War Casualties from Gas and Other Weapons. Washington, DC: U.S. Government Printing Office.

Gilchrist HL. 1928b. Chemical warfare and its medical significance. Military Surgeon 63:477-492.

Gilchrist HL, Matz PB. 1933a. The residual effects of warfare gases. Medical Bulletin (US Veterans Administration) 9:339-390.

Gilchrist HL, Matz PB. 1933b. The Residual Effects of Warfare Gases. Washington, DC: U.S. Government Printing Office.

Gillis RG. 1985. Australian Field Tests with Mustard Gas 1942-1945. Department of Defence, Australia.

Gilman A. 1946. Symposium on advances in pharmacology resulting from war research: therapeutic applications of chemical warfare agents. Federation Proceedings 5:285-292.

Gilman A. 1963. The initial clinical trial of nitrogen mustard. American Journal of Surgery 165:574-578.

Gilman A, Cattell M. 1948. Systemic agents: action and treatment. In: Andrus EC, Bronk DW, Carden GA Jr, Keefer CS, Lockwood JS, Wearn JT, Winternitz MC, eds. Advances in Military Medicine. Science in World War II: Office of Scientific Research and Development. Boston: Little, Brown. 546-564.

Gilman A, Philips FS. 1946. The biological actions and therapeutic applications of the b-chloroethyl amines and sulfides. Science 103:409-415.

Guthrie F. 1859. On some derivatives of the olefines. Quarterly Journal of the Chemical Society 12:109-126.

Guthrie F. 1860. On some derivatives of the olefines. Quarterly Journal of the Chemical Society 13:129-135.

Haber L. 1986. The Poisonous Cloud. Oxford:Clarendon Press. 250-257.

Harris R, Paxman J. 1982. A Higher Form of Killing: The Secret Story of Chemical and Biological Warfare. New York: Hill and Wang.

Heinen JH, Carhart HW, Taylor WH, Stolp BN, Conner JC, Clausen NM. 1945. Chamber Tests with Human Subjects. IX. Basic Tests with H Vapor. NRL P-2579. Washington, DC: Naval Research Laboratory.

Heyndrickx A, Heyndrickx B. 1984. Treatment of Iranian soldiers attacked by chemical and microbiological war gases. Archives Belges (Supplement):157-159.

Hyatt. 1976. Respirator Protection Factors. Los Alamos:Los Alamos Scientific Laboratory of the University of California.

Infield GB. 1976. Disaster at Bari. London: New English Library.

International Agency for Research on Cancer (IARC). 1975. IARC Monographs on the Evaluation of Carcinogenic Risk of Chemicals to Man. Vol. 9, Some Aziridines, N,S- & O-Mustards and Selenium. Lyon: IARC.

Jackson KE. 1936. The history of mustard gas. Journal of the Tennessee Academy of Science 11:98-106.

Lewis WL, Perkins GA. 1923. The β-chlorovinylchloroarsines. Industrial and Engineering Chemistry 15:290.

Lewis WL, Stiegler HW. 1925. The b-chlorovinyl chloroarsines and their derivatives. Journal of the American Chemical Society 47:2546-2556.

Mandl H, Freilinger G. 1984. First report on victims of chemical warfare in the Gulf war treated in Vienna. Archives Belges (Supplement):330-340.

McNamara BP, Owens EJ, Christensen MK, Vocci FJ, Ford DF, Rozimarek H. 1975. Toxicological Basis for Controlling Levels of Mustard in the Environment. Edgewood Arsenal Special Publication EB-SP-74030. Aberdeen Proving Ground, Maryland: U.S. Army Armament Command. Edgewood Arsenal Biomedical Laboratory.

Medema J. 1986. Mustard gas: the science of H. Nuclear, Biological, and Chemical Defense and Technology International 1:66-71.

Meyer V. 1886. Compounds of thiodiglycol. Chemischte Berichte 19:3259-3266. [In German]

National Institute of Occupational Safety and Health. 1974. Research Report. Publication No. 74-104.

National Research Council (NRC). Committee on Toxicology. 1985. Possible Long-Term Health Effects of Short-Term Exposure to Chemical Agents. 3 vols. Washington, DC: National Academy Press.

National Research Council. Division of Medical Sciences. Committee on Treatment of Gas Casualties. 1945. Fasciculus on Chemical Warfare Medicine. Volume III, Skin and Systemic Poisons. Washington, DC: Prepared for the Committee on Medical Research of the Office of Scientific Research and Development.

Niemann A. 1860. The action of brown sulfur chloride on ethylene gas. Annalen der Chemie und Pharmazie 113:288. [In German]

Norman JE Jr. 1975. Lung cancer mortality in World War I veterans with mustard gas injury, 1919-1965. Journal of the National Cancer Institute 54:311-318.

Office of Scientific Research and Development (OSRD). National Defense Research Committee. 1946. Summary Technical Report of Division 9, NDRC. Washington, DC: NDRC. AD-234 249.

Papirmeister B, Feister AJ, Robinson SI, Ford RD. 1991. Medical Defense Against Mustard Gas: Toxic Mechanisms and Pharmacological Implications. Boca Raton, FL: CRC Press.

Perera J. 1985. Lewisite: new gas weapon in the Gulf war. New Scientist 105(1450):8.

Perera J, Thomas A. 1986. Britain's victims of mustard gas disaster. New Scientist 109:26-27.

Physicians for Human Rights. 1989. Winds of Death: Iraq's Use of Poison Gas Against Its Kurdish Population. Somerville, MA: Physicians for Human Rights.

Prentiss AM. 1937. Vesicant agents. In: Chemicals in Warfare: A Treatise on Chemical Warfare. 1st ed. New York: McGraw-Hill. 177-300.

Project Coordination Staff, Chemical Warfare Service. 1946. Technical Aspects of Chemical Warfare in the Field. 2 vols. Washington, DC: Chemical Warfare Service.

Requena L, Requena C, Sanchez M, Jaqueti G, Aguilar A, Sanchez-Yus E, Hernandez Moro B. 1988. Chemical warfare: cutaneous lesions from mustard gas. Journal of the American Academy of Dermatology 19:529-536.

Rhoads CP. 1947. Edward Gamaliel Janeway lecture: the sword and the ploughshare (dichloroethyl sulfide poisoning at Bari, 1943 and work of Chemical Warfare Service, especially on nitrogen mustards or chloroethylamines). Journal of Mt. Sinai Hospital 13:299-309.

Robinson JP. 1967. Chemical warfare. Science Journal 4:33-40.

Sidell FR. 1990. Clinical Notes on Chemical Casualty Care. USAMRICD Technical Memorandum 90-1. Aberdeen Proving Ground, MD: U.S. Army Medical Research Institute of Chemical Defense.

Sidell FR, Hurst CG. 1992. Clinical considerations in mustard gaspoisoning . In: Somani SM, ed. 1992. Chemical Warfare Agents. San Diego: Academic Press.

Smith HW. 1943. Progress Report on Review of the Literature on the Systemic Action of Mustard Gas. OSRD Report 1717. Washington, DC: National Defense Research Committee.

Somani SM. 1992. Chemical Warfare Agents. New York: Academic Press.

Spiers EM. 1986. Chemical Warfare. Urbana, IL: University of Illinois Press.

Stampfer JF. 1982. Respirator canister evaluation for nine selected organic vapors. American International Hygiene Association Journal 43:319-328.

Stepanov AA, Popov VN. 1962. [Chemical Weapons and Principles of Antichemical Defense]. Translated by Joint Publications Research Service. JPRS 15107. Washington, DC: JPRS.

Stewart I. 1948. Organizing Scientific Research for War. Science in World War II: Office of Scientific Research and Development. Boston: Little, Brown.

Stockholm International Peace Research Institute (SIPRI). 1971. TheProblem of Chemical and Biological Warfare: A Study of the Historical, Technical, Military, Legal, and Political Aspects of Chemical and Biological Warfare and Possible Disarmament Measures . Vol. 1, The rise of chemical and biological weapons. Stockholm: Almqvist & Wiksell.

Stroykov, KN. 1970. Medical Aid for Toxic Agent Victims. Moscow: Meditsina. [In Russian]. As cited in Papirmeister B, Feister AJ, Robinson SI, Ford RD. 1991. Medical Defense Against Mustard Gas: Toxic Mechanisms and Pharmacological Implications. Boca Raton, FL: CRC Press.

Tarbell DS, Tarbell AT. 1981. Roger Adams: Scientist and Statesman. Washington, DC: American Chemical Society.

Taylor JR, Johnson WN. 1975. Research Report Concerning the Use of Volunteers in Chemical Agent Research. DAIG-IN 21-75. Washington, D.C.: Department of the Army, Office of the Inspector General and Auditor General.

Taylor WH, Carhart HW, Daily LE. 1943. Chamber Tests with Human Subjects. I. Design and Operation of Chamber. II. Initial Tests of Navy Issue Protective Clothing Against Vapor. NRL P-2208. Washington, DC: Naval Research Laboratory.

Taylor WH, Carhart HW, Heinen JH. 1945. Chamber Tests with Human Subjects. VII. The Effect of Concentration of H Vapor and Time of Exposure on the Protection Afforded by CC-2 Impregnated Clothing. NRL P-2528. Washington, DC: Naval Research Laboratory.

Trammell GL. 1992. Toxicodynamics of organoarsenic chemical warfare agents. In: Somani SM, ed. Chemical Warfare Agents. San Diego: Academic Press.

Uhde GJ. 1946. Mustard gas (dichloroethyl sulfide) burns of human eyes in World War II. American Journal of Ophthalmology 29:929.

United Nations, Security Council. 1986. Report of the mission dispatched by the Secretary-General to investigate allegations of the use of chemical weapons in the conflict between the Islamic Republic of Iran and Iraq. March 12, 1986. S/17911, S/17911/ Addendum 1, and S/17911/Addendum 2. New York: United Nations.

United Nations, Security Council. 1987. Report of the mission dispatched by the SecretaryGeneral to investigate allegations of the use of chemical weapons in the conflict between the Islamic Republic of Iran and Iraq. May 8, 1987. S/18852 and S/18852/ Addendum 1. New York: United Nations.

United Nations, Security Council. 1988a. Report of the mission dispatched by the Secretary-General to investigate allegations of the use of chemical weapons in the conflict between the Islamic Republic of Iran and Iraq. April 25, 1988. S/19823 and S/19823/Addendum 1. New York: United Nations.

United Nations, Security Council. 1988b. Report of the mission dispatched by the Secretary-General to investigate allegations of the use of chemical weapons in the conflict between the Islamic Republic of Iran and Iraq. July 20, 1988. S/20060 and S/20060/Addendum 1. New York: United Nations.

United Nations, Security Council. 1988c. Report of the mission dispatched by the Secretary-General to investigate allegations of the use of chemical weapons in the conflict between the Islamic Republic of Iran and Iraq. July 25, 1988. S/20063 and S/20063/Addendum 1. New York: United Nations.

United Nations, Security Council. 1988d. Report of the mission dispatched by the Secretary-General to investigate allegations of the use of chemical weapons in the conflict between the Islamic Republic of Iran and Iraq. August 19, 1988. S/20134. New York: United Nations.

Urbanetti JS. 1987. In: Chan P, ed. Proceedings of the Vesicant Workshop, February 1987. Aberdeen Proving Ground, MD: U.S. Army Medical Research Institute. AD-A188 222.

Urbanetti JS. 1988. Battlefield chemical injury. In: Loke J, ed. Pathophysiology and Treatment of Inhalation Injuries. New York: Marcel Dekker.

U.S. Army. 1974. Chemical agent data sheets, Vol. 1. Technical Report EO-SR-74001. Edgewood Arsenal, MD.

U.S. Army. 1988. Final Programmatic Environmental Impact Statement for the Chemical Stockpile Disposal Program. Aberdeen Proving Ground,MD .

U.S. Army and U.S. Air Force. 1975. Military chemistry and chemical compounds. Field Manual No. FM3-9, Regulation No. AFR 355-7.

U.S. Army Chemical Research, Development, and Engineering Center (CRDEC). 1988. Lewisite. Material safety data sheet. Aberdeen Proving Ground, MD: U.S. Army Chemical Research, Development and Engineering Center.

U.S. Army Chemical Research, Development, and Engineering Center (CRDEC). 1990. HD and THD. Material safety data sheet. HCSDS No. 20058A. Aberdeen Proving Ground, MD: U.S. Army Chemical Research, Development and Engineering Center.

Vedder EB. 1925. The Medical Aspects of Chemical Warfare. Baltimore: Williams & Wilkins.

Watson AP, Griffin GD. 1992. Toxicity of vesicant agents scheduled for destruction by the Chemical Stockpile Disposal Program. Environmental Health Perspectives 98:in press.

West CJ. 1919. History of mustard gas. Science 49:412-417.

Willems JL. 1989. Clinical management of mustard gas casualties. Annales Medicinae Militaris Belgicae 3:S1-61.

World Health Organization. 1970. Health Aspects of Chemical and Biological Weapons: Report of a WHO Group of Consultants. Geneva: WHO.

4

Findings from the Public Hearing Process

The original purpose of the public hearing process associated with the present study was twofold:

1. to determine what the experimental conditions were for the veterans involved in the testing programs; and
2. to identify the constellation of diseases from which the veterans were suffering.

In terms of testing conditions, it was not clear at the beginning of the study whether the committee would be able to obtain authoritative documentation regarding the experimental protocols employed in the World War II (WWII) testing programs. The committee felt that estimates of conditions, such as vapor concentrations in the chambers, would be helpful in relating the information from the scientific literature to the possible exposure levels experienced by the veterans. The committee therefore sought descriptions of the veterans' symptoms in the period shortly following their exposure, the duration of the gas chamber tests, and the number of times they were in the gas chamber, for the purpose of estimating the possible exposure. Fortunately, as detailed in Chapter 3, the committee was able to obtain numerous official reports concerning the tests and, as the study progressed, knowledge of many of the experimental protocols grew.

The second purpose of identifying the diseases took on a special importance as it became obvious that there were great gaps in the scientific and medical literature. Particularly regarding the long-term health consequences of exposure to mustard agents and Lewisite, it seemed as if the gaps outnumbered the answers. As a result, the set of health problems reported to the committee by the veterans became extremely valuable in highlighting those gaps that were important to consider.

Input to the public hearing process was solicited in a number of ways, beginning in late January 1992. The Disabled American Veterans (DAV) and the American Legion generously provided space for hearing announcements in their respective official publications (see Appendix G). The committee was also provided with a list of affected veterans who had contacted the offices of U.S. Congressman Porter Goss (Florida), and letters of invitation were sent to each person on that list. At the committee's request, the Department of Veterans Affairs (VA) also sent hearing announcements to each individual who had a claim pending for injuries from exposure to mustard agents or Lewisite. Finally, some input was received as a result of discussions among veterans and scattered stories in the press.

In order to allow the greatest flexibility, veterans who were unable to come to the public hearing were offered two alternative methods of giving statements: oral statements taken by the study staff over the telephone, and written statements in letter form. Each letter received from a veteran was acknowledged with a letter from the study director. In some cases, additional information was needed and requested, such as current health problems or more detail regarding the veteran's exposure. There were a number of cases in which the study staff imparted useful information to the veterans, both over the telephone and in letters. For example, of all the testing program locations, only the Naval Research Laboratory (NRL) had maintained records of individuals who had been human subjects during WWII. In a number of cases, the study staff was able to inform veterans about how to obtain their records from the NRL.[1] In other cases, the study staff informed veterans about how to arrange for a local DAV representative to assist them in gathering information and filing claims with the VA.

Prior to the hearing, summaries of each telephone call and copies of each letter received were sent to the committee members. Twenty veterans appeared in person to present statements at the hearing, held in Washington, D.C., on April 15, 1992 (Appendix G). Each veteran had five minutes to make his presentation, and ample time for committee questions was allowed. In addition, speakers were also given the

[1] These NRL records included copies of actual data sheets for the experiment in which the individual was involved. Data sheets contained information about the gas concentration, the number of trials, any blood counts and temperature measurements, as well as the location and severity of erythema for each person. In some cases, such records provided the only documentation that an individual had been in the tests.

opportunity to supplement their statements with written documentation.

After the hearing, each veteran who spoke at the hearing received a letter of thanks from the chairman of the committee. Other veterans who participated through the mail and telephone received a memorandum from the study director informing them that the hearing had taken place and outlining the committee's next steps.

The press coverage generated by the public hearing elicited additional input from veterans who were previously unaware of the hearing or the committee's activities. Statements from these veterans were accepted and incorporated into the committee's deliberations until the end of August 1992. Thus, input from veterans was accepted by the committee during approximately seven months of the study.

RESULTS AND FINDINGS

Types of Veteran Exposures

A total of 257 individuals, including veterans and surviving spouses or relatives, reported a variety of types of exposures to the committee. Although some types of exposures were expected, other types had not been foreseen by the committee. For example, as expected, many men reported their experiences in gas chamber tests, such as those conducted at the NRL, and others reported having participated in patch tests. In addition, a number of men who had participated in field tests contacted the committee. The largest additional group of veterans consisted of those who had been trained to handle toxic gases as part of their military assignments, often as part of military units organized under the Chemical Warfare Service (CWS). As outlined in Chapter 3, such men performed many types of functions, including loading gas bombs and decontamination of test sites and equipment.

The committee also heard from a very small group of veterans who had been exposed to sulfur mustard in combat. These individuals were either World War I veterans or veterans who had been injured in the Bari harbor bombing in WWII. Another large group of veterans who were exposed to the chemical warfare agents during various training exercises also contacted the committee. Experiences among this group were quite heterogeneous, ranging from drops of sulfur mustard applied to the skin to reports of use of sulfur mustard (as opposed to tear gas) in gas mask training exercises. One woman who had been an Army pilot during WWII reported having transported a variety of toxic chemicals from place to place within the United States, in addition to being exposed to Lewisite during flight training.

It should be emphasized that the vast majority of veterans contacting the committee served during WWII. However, a number of men had served during later time periods, including the wars in Korea and Vietnam, and had participated in sulfur mustard and Lewisite testing programs at Edgewood Arsenal in the 1950s and 1960s. In addition, some veterans had experienced exposures to multiple types of chemicals or radiation. This was especially true for former CWS personnel, who were trained with and exposed to a multitude of chemical agents, including phosgene. Finally, some of the men who had participated in the chemical warfare testing programs also participated in later atomic bomb or drug tests (see also Chapter 3 and Appendix F).

Health Problems Reported

The many different types of health problems reported by the veterans exposed to mustard agents and Lewisite (Appendix G) are summarized here. However, no analysis of the frequency of specific health problems was carried out, nor was any of the information reported by the veterans compared with data from unexposed populations. Such analysis was not possible due to the manner in which the information was gathered, nor was it appropriate to the task of this committee. In addition, it should be emphasized that disease and health condition categories were based on veteran self-reports and, thus, did not always fall into strict medical diagnostic categories. Nevertheless, consideration of the health problems reported did aid the committee in identifying important gaps in the knowledge base about the health effects of these warfare agents.

Various types of cancer were reported by these veterans. Most frequent were skin cancers, followed by lung or laryngeal cancer, bladder cancer, and prostate cancer. Tumors or polyps, not identified as cancer, were reported most often in the skin, larynx, and intestines. Among nonmalignant diseases, by far the most frequent problems reported were pulmonary and respiratory diseases, including asthma, chronic bronchitis, emphysema, laryngitis, sinusitis, and other respiratory problems, including repeated bouts of pneumonia and chronic respiratory infections. Skin problems were also common and included scars, repeated and varied types of irritations, and chronic rashes. Among eye diseases, chronic conjunctivitis and corneal opacities were reported, as well as cataracts, glaucoma, and other problems.

Cardiovascular problems ranging from heart attacks to strokes and high blood pressure were commonly reported. Gastrointestinal difficulties included difficulty swallowing, esophageal and laryngeal strictures (narrowing), chronic nausea, stomach ulcers, and Crohn's disease (chronic inflammation and scarring of the small intestine, often leading to obstruction). In addition, benign prostate disease was reported. There were reports of diabetes, allergies, liver and kidney diseases, blood and lymphatic diseases, and disturbances in immune function. There were also frequent reports of arthritis and bone disease, headaches, and muscle spasms. Finally, a few reported hair and tooth loss.

Quite frequent were reports of neurological disease and psychological difficulties. Neurological problems included multiple sclerosis and amyotrophic lateral sclerosis (degenerative diseases of the central nervous system), abnormal sensory disturbances, Alzheimer's disease, paralysis and weakness, and chronic pain, among others. Among those reporting psychological difficulties, some had been diagnosed with post-traumatic stress disorder, others experienced chronic depression or anxiety, and still others described themselves as being very nervous or tense. It should be noted that psychological problems and any sexual problems, such as those resulting from genital scars left by gas burns to the scrotum and penis, were often extremely difficult for the men to discuss, especially on the telephone or in person, and may well have been underreported.

Personal Aspects of Veteran Reports

Beyond the facts of their exposure and subsequent health problems, there were certain aspects of the veterans' experiences that are compelling and drew the attention of the committee. For example, men who participated in the chamber tests commonly reported that they had originally volunteered to "test summer clothing" in exchange for extra leave time before being sent overseas. It was not until they arrived at the test location that they were told about the gas chamber tests and, even then, many were not told to what agent they would be exposed. Those who became sick during the "man-break" tests reported being threatened with court martial if they did not continue the test and reenter the gas chambers. Some even reported that they saw other subjects collapse in the chamber and that they never saw these men again after they were removed from the chamber. In all such cases, the men reported that they had assumed the person had died. Other men recalled that the chamber door could not be opened from the inside and that this frightened them by making them feel trapped.

In the majority of statements, experiences were related in such precise detail that a supportable conclusion could be drawn that many of these men experienced intense fear about what was happening to them during the tests. All the men in the chamber tests vividly recalled being told that they would be sent to prison if they ever revealed their participation in these tests. Some were even shown pictures of Fort Leavenworth to reinforce the threat of prosecution. They further reported that this possibility prevented them from telling their wives, parents, family doctors, or anyone else about what had happened to them. Some of the men told the committee that they had been asked directly by physicians if they had been exposed to sulfur mustard; these men typically denied any exposure in such circumstances. Finally, in a few cases, wives reported that their husbands revealed their participation in these tests only on their deathbeds.

Most of the men involved in field tests had experiences similar to those in the chamber tests regarding the instructions about secrecy. However, except for individuals who were injured or witnessed a severe injury or death of a comrade, this group did not uniformly report the level of intense fear during the tests as that reported by the chamber test group. Most often the men involved in field tests were drawn from CWS units, such as the 94th and 95th Medical Gas Treatment Battalions, some of whom participated in field tests in Florida. Such subjects often had additional training in toxic gases in advance of their participation in tests, and this training may have better prepared them for the experience.

Veterans whose exposures occurred during their training or work as part of the CWS also reported varied levels of fear. Again, the most intense feelings of fear were reported by those who had been involved in some kind of accident, such as one veteran who described being severely injured by an explosion of mustard gas shells during a drill that resulted in the deaths of two other soldiers. Two important additional factors were reported by those who routinely worked with chemical warfare agents. One common factor was that many of the men were very young (often as young as 17 years of age), had little formal education, and were afraid of the chemicals. The second common factor was that protective measures, including impregnated clothing and even gas masks on occasion, were not always used or available.

The least amount of fear was reported by veterans who participated in patch tests, a few of whom said that they only remembered the incident because of the faint scars on their arms. Some of these veterans reported no health problems that they attributed to their exposure to the agents.

The final but significant personal difficulty reported, especially by those who participated in chamber tests, was how frustrating it had been to be ill and not be able to file a disability claim, often because there was no proof or record of the tests and no one knew or believed that they had happened. Even among those working in CWS units, there was great variability in the handling of cases after separation from the military. For example, some men were discharged with full disability due to sulfur mustard or other chemical injuries, while others with similar health problems were not. Some also reported that their military records did not include certain assignments and time periods; others had complete military records and numerous citations for their work with chemical agents.

TREATMENT OF HUMAN SUBJECTS

In the course of this study, the committee examined many government documents and technical summaries of experiments with mustard agents and Lewisite that involved the use of human subjects. Although information regarding the treatment of human subjects was scarce, it was possible to piece together a general picture. Certain aspects of these reports were striking and, coupled with very precise and matching statements of many veterans, were impossible for the committee to ignore. A brief description of these aspects is included here to corroborate the statements of the veterans. It is also presented to make manifest all of the information and challenges that faced this committee and to offer additional background for some of the directions taken by the committee during the study.

Detailed descriptions, or copies of official instructions, of how human subjects were recruited are lacking, but are outlined in Cochrane's (1946) historical description of the research done under the CWS (declassified in 1991) and other papers. One report, "Chamber Tests with Human Subjects," includes a short section that describes the treatment of subjects in the initial chamber tests at the NRL (Taylor et al., 1943; see also Appendix D). This section details how the men were induced to participate by offers of extra leave and a "change of scenery." It further states that the men should not be told too much in the beginning, but that after a few times in the chamber they can be told "almost anything without affecting their morale."

In contrast, Cochrane states that tests at Camp Sibert in Alabama had to be halted due to a lack of willing subjects. The official explanation was that the commanding officers actively discouraged men from becoming subjects, because they did not want to have to replace them. Cochrane, who was present at Camp Sibert and notes in the text that men were sometimes burned more than necessary, writes that the "apathy may have been due to the look of the scars on the men returned to the training companies after the tests." The NRL report provides additional evidence for severe injuries during the testing programs: in praising the morale of the subjects, it describes how men sent to the hospital and incapacitated for a month were "not upset and even volunteered for further trials" (Taylor et al., 1943; see also Appendix D). However, morale may not have been uniformly high at NRL, because the same report also gives instructions for dealing with uncooperative individuals. Such subjects were to be given "a short, explanatory talk and, if necessary, a slight verbal 'dressing down.'" Finally, although no man was to be sent into the chamber without the Medical Officer's approval, "malingerers and psychoneurotics" were handled by "minimizing their symptoms and sending them into the chamber."

In a section on physical examination and requirements, warnings are given against mistaking certain physical ailments caused by "physical unfitness" for gas manifestations. The ailments listed are conjunctivitis, laryngitis, nausea, and shock, all of which were well-known symptoms of mustard agent or Lewisite poisoning. However, no instructions for distinguishing these symptoms as real gas manifestations are given in this summary report, or anywhere else the committee could find. Evidence that such distinctions were possibly made inappropriately was provided by an individual veteran's NRL record sent to the committee. This record shows a notation of mild laryngitis after the third time in the chamber; no such notation was recorded prior to the first chamber trial. Yet, the person was sent back into the chamber two more times and the final record notes "severe laryngitis." As noted in Chapters 3 and 7, occurrence of this symptom would have been a clear indication of cumulative inhalation exposure beyond 100 Ct.

Physical examinations, according to the NRL summary, included a complete blood count, urinalysis, and body temperature. Blood counts were to be redone after a cumulative Ct of 4,800. Yet cursory examination of the very few individual records submitted to the committee indicates that blood counts may not have been repeated in all such cases.

This short summary is not exhaustive of all the information available, nor does it cover field tests and other procedures. It is an attempt to portray the atmosphere in which the experiments were done and to describe the attitudes brought to these experiments by the military research establishment. Some would argue that this description has no place in a report of a committee charged to survey scientific literature. Further, it was a war, a worldwide emergency that understandably required certain goals to take precedence over others, possibly to the detriment of sound medical research practices concerning individual well-being. In fact, the authors of the NRL summary state their belief that the men coming through their program benefited from their experience and were better prepared than most to confront the realities of gas warfare.

It may also be fair to argue that no formalized set of rules, carrying the weight of law, existed in 1942 to govern the treatment of human subjects. However, a Department of the Army Inspector General's report in 1975 documented how these patterns of neglect of human subjects, established during WWII, continued through the 1950s and 1960s, well beyond the immediacy of wartime concerns (Taylor and Johnson, 1975; see also Chapter 3 and Appendix F). These patterns even continued after the formal set of rules, the Nuremberg Code of 1947, was developed and adopted (see Appendix F).

The committee believes that the individual investigators involved in the WWII testing programs were convinced of the likelihood of great numbers of gas casualties and that they believed their work to be necessary to save lives. Yet, exposure level and injury are key questions in the determination of risk for the affected individuals, and these questions cannot be separated from the consideration of scientific data gleaned from other studies, especially those done with animals. The reliability of the military's official human exposure data is another key question, and these data are undermined by the demonstrated patterns of inconsistency in the reporting of injuries to the subjects, their severity, and their cause.

CONCLUSIONS AND ACTIONS TAKEN

There is no doubt that some veterans who were involved in the chemical warfare testing programs and other circumstances of exposure to mustard agents and Lewisite have been dealing with serious and debilitating diseases for decades. This burden has been further compounded by the secrecy oath taken by the veterans and faithfully kept for nearly 50 years, only to experience the denials of government agencies and their representatives that such tests and activities ever occurred. The committee understands the anger of veterans who believe they have been victims of injustice and neglect. In addition, the committee is greatly impressed by the level of patriotism exhibited by these individuals; almost to a man, they obeyed their orders. Finally, the committee is indebted to the veterans for helping to identify key gaps in the scientific and medical literature. Special attention was given after the public hearing to reviewing again those areas of the literature that were especially lacking in substantive information yet represented the only work relating to certain diseases reported by the veterans.

Another action taken in partial response to the findings of the public hearing process was the addition of a clinical psychologist to the committee. This added expertise facilitated the review of information available regarding the psychological effects of chemical and biological warfare environments and environmental toxins. Thus, the possible psychological health effects of exposure to mustard agents and Lewisite, and of the circumstances of exposure, were treated by the committee with care and importance equal to that of the physiological health effects.

Recognizing the difficulties the veterans had experienced in communicating with various agencies over the years, the committee also requested input from an expert in risk communication. The resulting presentation by Professor Peter Sandman of Rutgers University offered the committee a strong base of understanding regarding risk perception and how such perceptions can be lessened or made worse (Appendix A). Of special value was a better understanding of how the study of risk communication could aid the committee in framing conclusions about exposure and disease in ways, and in language, that would be least likely to increase the perception of risk already felt by the veterans.

Finally, the committee sought input from a bioethicist regarding the conduct of the WWII experiments. The primary motivation for this request was the committee's wish to inform itself about the ethical and legal issues of informed consent and to explore what its responsibilities may be from a bioethical viewpoint, as physicians and scientists confronted with unanticipated and disturbing information about these testing programs. Professor Jay Katz from Yale University met with the committee in June 1992 to outline the history and development of the Nuremberg Code of 1947 and its ethical and legal ramifications, especially as they might apply to the issues in the present study (Appendix A; see also Appendix F). In addition, Dr. Katz commented about the way the experiments were conducted, the secrecy of the experiments, and the lack of medical follow-up of the human subjects, and urged the committee to take a strong stand on these issues. His presentation was followed by a letter further explicating his view that the committee would miss an important and needed opportunity if it simply completed an isolated survey of the scientific and medical literature, without comment on the experiments themselves. This letter is included in Appendix H.

The committee has drawn valuable information and guidance from the presentations described above. This report, its contents and its recommendations, reflects long and careful consideration of all the issues and suggestions, much discussion, and a final consensus. The inclusion of information about how the experiments were conducted and the medical treatment afforded to the human subjects is based on what the committee believes to be justified scientific, as well as humanitarian and ethical grounds.

REFERENCES

Cochrane RC. 1946. Medical research in chemical warfare. Available through the U.S. Army Chemical Defense Research, Development and Engineering Center, Aberdeen Proving Ground, MD.

Taylor JR, Johnson WN. 1975. Research Report Concerning the Use of Volunteers in Chemical Agent Research. DAIG-IN 21-75. Washington,DC : Department of the Army, Office of the Inspector General and Auditor General.

Taylor WH, Carhart HW, Daily LE. 1943. Chamber Tests with Human Subjects. I. Design and Operation of Chamber. II. Initial Tests of Navy Issue Protective Clothing Against H Vapor. NRL 2208. Washington, DC: Naval Research Laboratory.

5

Chemistry of Sulfur Mustard and Lewisite

SULFUR MUSTARD

This chapter reviews the important chemical reactions of sulfur mustard and Lewisite. It is included for those readers interested in these chemical reactions and their concomitant biological effects. Thus, the chapter may be of interest only to those readers with a background in chemistry. To ensure brevity, all sources used in the preparation of this chapter are listed at the end of the chapter and are not specifically cited in the text.

The synthesis and chemistry of sulfur mustard, or mustard gas, have been studied and reviewed extensively. Chemical and physical data regarding sulfur mustard were presented in Table 3-1. Meyer first prepared pure sulfur mustard by the reaction of thiodiglycol with phosphorus trichloride (5-2). Thiodiglycol was prepared by the reaction of 2-chloroethanol with potassium sulfide (5-1):

$$2ClCH_2CH_2OH + K_2S \rightarrow S(CH_2CH_2OH)_2 + 2KCl \qquad (5\text{-}1)$$

$$3S(CH_2CH_2OH)_2 + 2PCl_3 \rightarrow 3S(CH_2CH_2Cl)_2 + 2P(OH)_3 \qquad (5\text{-}2)$$

Concentrated hydrochloride, thionyl chloride, and phosgene have all been used in place of phosphorus trichloride.

Sulfur mustard was produced for use in warfare by what is known as the Levinstein process, the reaction of ethylene with sulfur dichloride. The fundamental reactions are the addition of sulfur dichloride to ethylene to form 2-chloroethylsulfenyl chloride (5-3) and the addition of that compound to a second molecule of ethylene (5-4):

$$SCl_2 + CH_2{=}CH_2 \rightarrow ClCH_2CH_2SCl \qquad (5\text{-}3)$$

$$ClCH_2CH_2SCl + CH_2{=}CH_2 \rightarrow S(CH_2CH_2Cl)_2 \qquad (5\text{-}4)$$

So-called Levinstein mustard gas as manufactured on a large-scale contains 69.3 percent sulfur mustard new and 71.5 percent after aging. To this day, no one knows exactly what is in this material, but physiological tests have disclosed no appreciable difference between it and the highly purified material used in chemical studies. Sulfur mustard is a heavy, somewhat oily liquid that is clear or straw colored when pure but dark when crude. Its molecular weight is 159.08, boiling point 215°C-217°C, freezing point 14.45°C, specific gravity 1.27. It is sparingly soluble in water but very soluble in organic solvents, animal oils, and fats. It is stable for weeks at room temperature, slowly hydrolyzed by water, and destroyed by strong oxidizing agents.

Studies on the mechanism and kinetics of the hydrolysis of sulfur mustard have shown that the first step in this reaction is the formation of a transient cyclic sulfonium cation, which then reacts quickly with water to form 2-chloroethyl-2-hydroxysulfide and a hydrogen ion. The reaction sequence is repeated to give dithioglycol (5-5):

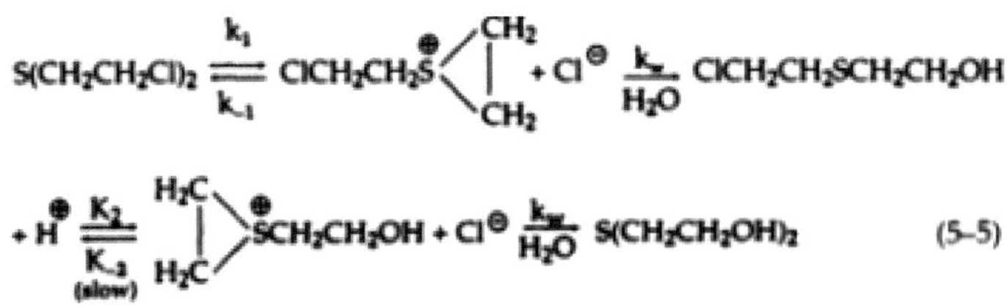

(5-5)

To ensure pure first-order kinetics, sulfur mustard is predissolved in a polar organic solvent, and its concentration is kept low in solution so that the rate of the reverse reactions become negligible compared to k_W. The overall reaction—the formation of dithioglycol and 2 HCl—can be described as a quasi-monomolecular process with first-order kinetics. The rate constant for the hydrolysis of sulfur mustard, as determined by acid production, is markedly dependent on temperature and the presence of chloride ion, which retards the hydrolysis rate without altering the reaction products. The retardation of hydrolysis by added chloride is consistent with the reversibility of the activation step to cyclic sulfonium ion. The rate of hydrolysis is not pH dependent and is not altered by metal ions.

At greater substrate concentrations in the absence of an organic solvent, however, the reaction is more complex, since both dissolution and reaction take place simultaneously and the initial product from the reaction with water accumulates in the aqueous phase and reacts with the sulfonium cation to form a dimeric sulfonium cation (5-6):

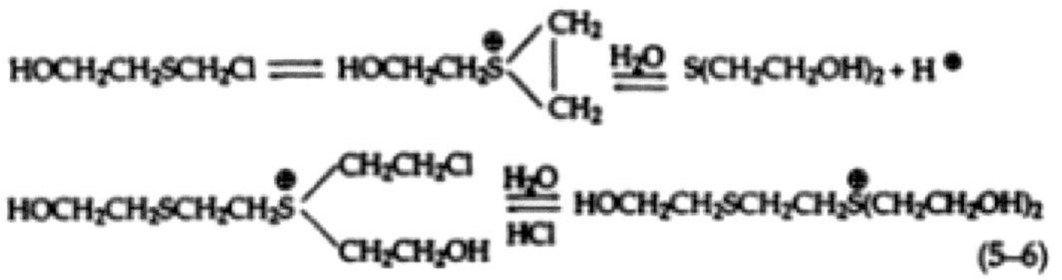

(5-6)

This secondary reaction may occur via a transient dithiane disulfonium ion intermediate (5-7):

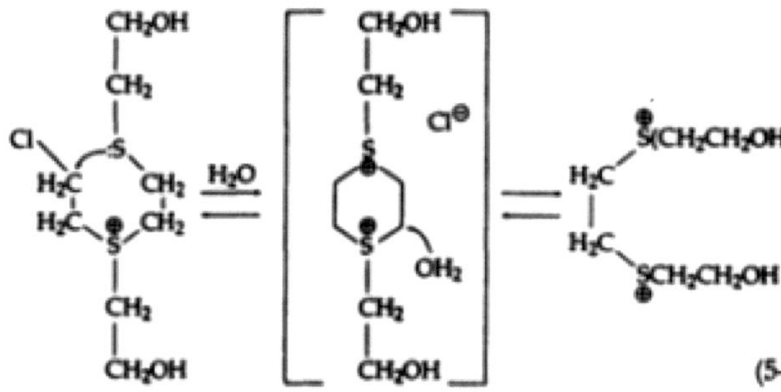

(5-7)

It should be emphasized that all the sulfonium salts (5-6), especially the 2-chloroethyl compound, possess noteworthy toxicity. This toxicity may be due to the decomposition of the sulfonium salts under physiological conditions to form alkylating moieties. The conversion of these sulfonium salts to reactive species is considerably slower than for sulfur mustard. The chemical reactions of the sulfonium salts have been studied in detail, but it is not known whether they are actually formed *in vivo* . It is certainly possible that such toxic products might be formed on moist areas of the skin, which is consistent with the high susceptibility of these regions to the vesicant action of sulfur mustard. The physiological effects and toxicities of the sulfonium salts need to be investigated, since the proposed mechanism of the cytotoxicity of sulfur mustard is based on the simplified S_N1 hydrolysis and is not fully understood.

The relative affinities of nucleophiles are quantitatively described by their competition factors, which compare the rate of constants for bimolecular reactions of cyclic ethylene sulfonium ion with a given nucleophile (K_a) and water (K_o), respectively (5-8):

$$F_a = K_a/K_o\ [H_2O] \qquad (5\text{-}8)$$

The dimensions of Fa are 1/concentration, so the reciprocal of Fa is the concentration of nucleophile that must be present in water so that it reacts with 50 percent of the sulfur mustard. An extensive list of competition factors for sulfur mustard was compiled during World War II. It should be emphasized, however, that despite the large differences in affinities of some nucleophiles, the overall rates of reaction of sulfur mustard are approximately equal. This is consistent with the proposed reaction mechanism, in which the rate-limiting step in the reaction of sulfur mustard in aqueous media is the formation of the cyclic sulfonium intermediate.

In addition to the potential contribution of sulfonium salts to the biologic activity of sulfur mustard, the oxidized forms of sulfur mustard may also be of importance. The reactions of the sulfoxide [$OS(CH_2CH_2Cl)_2$] are much slower than those of the sulfone [$O_2S(CH_2\ CH_2Cl)_2$], leading to a detoxification mechanism (oxidation of sulfur mustard to its sulfoxide). The sulfone, on the other hand, is quite reactive via the elimination of HCl to form the divinylsulfone to which nucleophiles add (5-9):

$$S(CH_2CH_2Cl)_2 + 2R_2NH \rightarrow S(CH_2CH_2NR_2)_2 \rightarrow R_2NCH_2CH_2SCH{=}CH_2 + R_2NH \qquad (5\text{–}13)$$

The sulfone is particularly important, since conjugates of it have been identified in the urine of rats dosed intravenously with sulfur mustard.

Reaction of Sulfur Mustard with Various Nucleophiles

Sulfur mustard reacts with sodium salts of alcohols (R; ethanol, methanol, etc.) to give ethers, but the yields are only fair (5-10):

$$O_2S(CH_2CH_2Cl)_2 \rightarrow O_2S(CH{=}CH_2)_2 + 2HCl$$

$$\overset{X^{\ominus}}{\rightleftarrows} O_2S(\overset{\ominus}{C}HCH_2X)_2 \rightleftarrows O_2S(CH_2CH_2X)_2 \qquad (5\text{–}9)$$

With the corresponding sulfur compounds, almost quantitative yields are obtained (5-11):

$$S(CH_2CH_2Cl)_2 + 2NaOR \rightarrow S(CH_2CH_2OR)_2 + 2NaCl \qquad (5\text{–}10)$$

The formation of the dimethyl derivative, which is harmless and can be distilled, has been used to characterize sulfur mustard.

With salts of organic acids, esters of thiodiglycol are produced (5-12):

$$S(CH_2CH_2Cl)_2 + 2NaSR \rightarrow S(CH_2CH_2SR)_2 + 2NaCl \qquad (5\text{–}11)$$

Sulfur mustard reacts readily with secondary amines, but one amine group of the product may be eliminated (5-13):

$$S(CH_2CH_2C1)_2 + 2RCO_2{}^{\oplus}M^{\ominus} \rightarrow S(CH_2CH_2O_2CR)_2 + 2MCl \qquad (5\text{–}12)$$

With ammonia and primary amines, a thiomorpholine is formed (5-14):

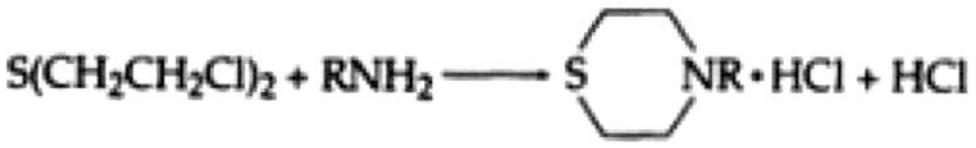

(5-14)

Two molecules of amine may react with one of sulfur mustard (5-15):

$$S(CH_2CH_2Cl)_2 + H_2NCH_2CO_2Et \rightarrow S(CH_2CH_2NHCH_2CO_2Et)_2 + 2HCl \quad (5\text{-}15)$$

Tertiary amines form quaternary ammonium salts (5-16):

$$S(CH_2CH_2Cl)_2 + 2R_3N \rightarrow S(CH_2CH_2{}^{\oplus}NR_3)_2 + 2Cl^{\ominus} \quad (5\text{-}16)$$

When heated with a concentrated aqueous solution of thiourea, sulfur mustard gives the isothiouronium salt, which is decomposed by aqueous NaOH. Acidification produces the mercaptan in high yield

(5-17):

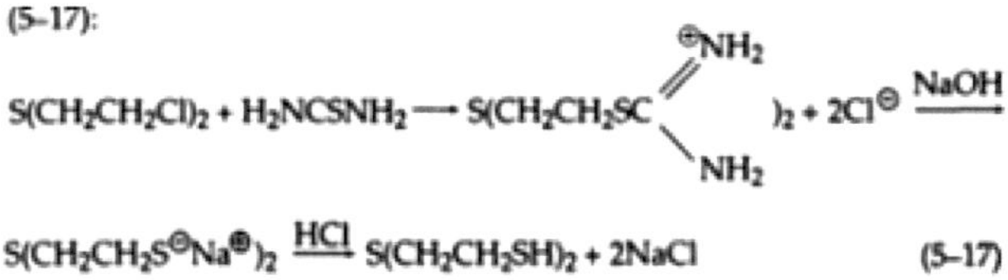

(5-17)

Reactions of Biologic Importance

As is obvious from the chemistry described above, sulfur mustard can react with a number of important functional groups of the large variety of compounds present in cells and tissues. The reactive groups that are of greatest interest are the sulfhydryl group; the phosphate and pyrophosphate ions; organic phosphates such as nucleotides and phospholipids; aromatic nitrogen atoms such as in nicotinamide, adenine, cytosine, and histidine; the carboxyl groups of amino acids and of intermediates of glucose metabolism; the sulfides such as methionine and thiodiglycol; and the amino groups of amino acids, peptides, purines, and pyrimidines. It should be noted, however, that at physiologic pH, most amines are present predominantly in the protonated form rather than as the free base, diminishing the probability of extensive reaction with sulfur mustard.

Evidence that the cytotoxicity of sulfur mustard is due to the alkylation of DNA was first obtained in the late 1940s from studies with bacteria, DNA-containing bacterial viruses, and transforming DNA. The later discovery that the sensitivity of bacterial and mammalian cells is critically dependent on the cell's capacity for repairing sulfur mustard-induced DNA damage strongly supports the DNA target hypothesis.

The relevance of DNA damage and repair to the vesicant action of sulfur mustard is supported by the observation that inhibitors of DNA repair significantly exacerbate skin injury.

Sulfur mustard at neutral pH alkylates purines, pyrimidines, nucleosides, and nucleotides, preferentially at N-7 of guanine and N-1 of adenine. Reactions with 0-6 and N-2 of guanine and N-6 of adenine have also been reported. The following products have been isolated from the reaction of sulfur mustard with DNA (5-18):

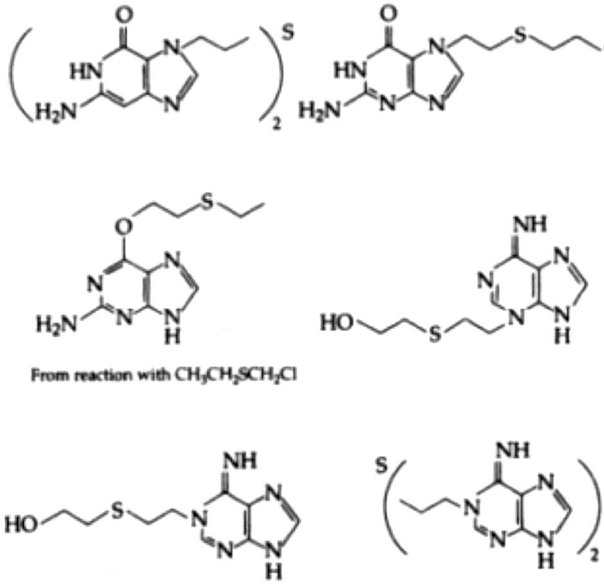

(5-18)

Sulfur mustard, because of its bifunctional nature, is more cytotoxic than is its monofunctional analogue. The molecular basis for this greater toxicity is the ability of sulfur mustard to form interstrand cross-links between guanines of the double helix, which prevents strand separation during replication. In addition, 7-alkylguanines and 3-alkyladenines of DNA are unstable and are released spontaneously from sulfur mustard-treated DNA at physiologic pH and temperature by cleavage of the N-9 glycosyl bond to give an apurinic site. Opening of the imidazole ring of this alkylated purine may also occur under physiologic conditions.

Although sulfur mustard also reacts with RNA, proteins, and phospholipids, the consensus of opinion has been for some time that it is the alkylation of DNA that is by far the most important of its actions. The interstrand DNA cross-link produced by bifunctional mustard com pounds is probably the lesion that produces lethality at the lowest frequency of occurrence and at the lowest concentration of the agent. However, cell death from this lesion is delayed for a number of hours, until the cell replicates its DNA or undergoes division. At higher cellular exposures, mechanisms other than DNA cross-linking become important and produce more rapid cell death. The acute damage to the cornea, mucous membranes, and skin seen with sulfur mustard is probably generated by one or more of these other mechanisms.

One mechanism that may be involved in acute damage is nicotinamide adenine dinucleotide (NAD) depletion. The nuclear enzyme poly-(adenosine diphosphoribose) polymerase is activated by DNA strand breaks, such as those produced by sulfur and nitrogen mustards. The enzyme cleaves NAD between nicotinamide and adenine diphosphoribose (ADP) and joins the ADP molecules into polymers of ADP-ribose, which are then linked to nuclear proteins, including the enzyme itself. This process can rapidly deplete cellular pools of NAD, which is required for ATP synthesis. The subsequent depletion of ATP rapidly produces loss of energy-dependent functions in the cell and results in cell death.

Other potential mechanisms of rapid cell death are related to the rapid inactivation of sulfhydryl peptides, especially glutathione, and proteins. These sulfhydryl compounds are critical to maintaining the appropriate oxidation-reduction state of cellular components. In particular, enzymes that maintain calcium homeostasis are sulfhydryl dependent, and sulfhydryl depletion may lead to elevated cellular calcium levels and cell death. Glutathione is also thought to be critical in reducing reactive oxygen species in the cell and preventing lipid peroxidation and loss of membrane integrity.

The toxicities of sulfur mustard to specific organs and tissues are described in detail in subsequent sections of this report. Essentially all of the data on the effects of sulfur mustard on humans are derived from either gas exposure or topical application to the skin. Because of the extensive use of nitrogen mustards in cancer chemotherapy, there is an extensive body of literature on these compounds in man after systemic administration, with doses and clinical follow-up. Since the fundamental mechanisms of interaction of sulfur mustard and nitrogen mustards with biological molecules are very similar, it should be useful to consider the major effects of nitrogen mustards, especially the long-term effects, in trying to ascertain the long-term clinical effects of sulfur mustard. The acute effects of nitrogen mustard are initially nausea and vomiting, followed in a few days by hematopoietic depression. At higher doses, neurotoxicity and damage to the gastrointestinal epithelium are seen. The major delayed effect of nitrogen mustards has been carcinogenesis, especially the development of myelocytic leukemia, although an in crease in other types of tumors now seems certain. Another long-term effect of nitrogen mustard treatment is pulmonary fibrosis, produced by damage to the pneumocytes.

Certainly, hematopoietic depression is seen with sulfur mustard exposure in man, although (except in massive exposures) the degree and frequency do not seem to be as intense or frequent as with the nitrogen mustards. This difference is likely due to the more direct exposure to the bone marrow of the nitrogen mustards when given by systemic exposure. This same rationale probably explains why acute leukemia has not been recognized as a consequence of sulfur mustard exposure. However, the increased incidence of solid tumors seen with nitrogen mustard would support the conclusion that exposure of the lungs and skin to sulfur mustard produces a carcinogenic effect on these tissues. Similarly, the delayed pulmonary toxicity seen in a small percentage of patients treated with nitrogen mustards would suggest that long-term damage to the lungs would be expected with intense exposure of the lungs to sulfur mustard.

LEWISITE

The preparation of Lewisite (L-1) by the original procedure is complicated and dangerous. It involves the reaction of acetylene with arsenic trichloride, by using aluminum chloride as a catalyst. The reaction yields three principal products (5-19):

$$\mathrm{ClCH{=}CHAsCl_2} \underset{}{\overset{\text{(fast)}}{\rightleftarrows}} \mathrm{ClCH{=}CHAs(OH)_2} + 2\mathrm{HCl} \underset{\text{(slow)}}{\rightleftarrows} \mathrm{ClCH{=}CHAsO} + \mathrm{H_2O} \underset{\text{(slow)}}{\rightleftarrows} \mathrm{(ClCH{=}CHAsO)}_n \qquad (5\text{–}20)$$

The optimum yield of Lewisite is about 20 percent, obtained along with L-2, L-3, tar, and an explosive material. Acetylene reacts with $AsCl_3$ in hydrochloric acid solution, with mercuric chloride as a catalyst, to give Lewisite in 80 to 85 percent yield (based on $AsCl_3$). Cuprous chloride and ethanolamine hydrochloride used together, however, constitute the best catalyst for the reaction.

The hydrolysis of Lewisite by water involves the following equilibria (5-20):

$$\mathrm{HC{\equiv}CH} + \mathrm{AsCl_3} \rightarrow \underset{\text{L-1}}{\mathrm{ClCH{=}CHAsCl_2}} + \underset{\text{L-2}}{\mathrm{(ClCH{=}CH)_2AsCl}} + \underset{\text{L-3}}{\mathrm{(ClCH{=}CH)_3As}} \qquad (5\text{–}19)$$

The only substance isolated is polymeric 2-chlorovinylarsinoxide, a white insoluble powder.

The cold aqueous media of pH $\leq$ 0.5 Lewisite decomposes as follows (5-21):

$$\mathrm{ClCH{=}CHAsCl_2 + R_2NCSS^{\ominus}Na^{\oplus} \rightarrow ClCH{=}CHAs(SSCNR_2)_2} \quad (5\text{-}25)$$

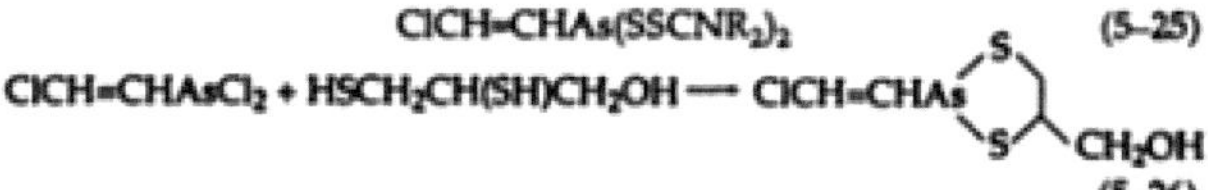

(5-26)

The vesicant character of arsenicals such as Lewisite is not a property of the $AsCl_2$ group exclusively, since carefully prepared solutions of corresponding oxide or dihydroxide are equally vesicant. Lewisite reacts with sodium alkoxides to give derivatives that are volatile, vesicant liquids (5-22) that hydrolyze irreversibly on contact with water:

$$\mathrm{ClCH{=}CHAs(OH)_2 + OH^{\ominus} \rightarrow Cl^{\ominus} + HC{\equiv}CH + As(OH)_3} \quad (5\text{-}21)$$

Reaction with sodium mercaptides gives the analogous thioethers (5-23), which are only slightly soluble in water and in general are hydrolyzed reversibly, giving toxic and sometimes vesicant solutions, although the equilibrium generally favors thioether formation (5-24):

$$\mathrm{ClCH{=}CHAsCl_2 + RO^{\ominus}Na^{\oplus} \rightarrow ClCH{=}CHAs(OR)_2 + 2NaCl} \quad (5\text{-}22)$$

Aqueous and alcoholic solutions of sodium dialkyldithiocarbamates react readily with Lewisite to give crystalline, sharp-melting solids that are useful for its characterization (5-25). These dithiocarbamates are much more stable than the simple thioethers. However, hydrolysis of cyclic thioethers, such as the reaction product of Lewisite and BAL (British Anti-Lewisite) (5-26) is negligible.

$$\mathrm{ClCH{=}CHAsCl_2 + RS^{\ominus}Na^{\oplus} \rightarrow ClCH{=}CHAs(SR)_2 + 2NaCl} \quad (5\text{-}23)$$

$$\mathrm{ClCH{=}CHAs(SR) + 2H_2O = ClCH{=}CHAs(OH)_2 + 2RSH} \quad (5\text{-}24)$$

Alkali hydrolyzes all of these compounds with the evolution of acetylene (5-21). Hydrogen peroxide causes decomposition of the ethers and thioethers in neutral or acid solution, giving free arsenic acids.

Little information is available in the literature concerning the reactions of Lewisite with biologically important molecules, although it is reasonable to assume that, as with sulfur mustard, DNA is a major target.

REFERENCES

Annals of the New York Academy of Sciences. 1958. Comparative clinical and biological effects of alkylating agents. 68:657-1266.

Annals of the New York Academy of Sciences. 1969. Biological effects of alkylating agents. 163:589-1029.

Berger NA. 1985. Poly(ADP-ribose) in the cellular response to DNA damage. Radiation Research 101:4-15.

Colvin M, Chabner BA. 1990. Alkylating agents. In: Chabner BA, Collins JM, eds. Cancer Chemotherapy: Principles and Practice. Philadelphia: J.B. Lippincott.

Gates M, Williams JW, Zapp JA. 1946. Arsenicals. In: Division 9, National Defense Research Committee, comp. Chemical Warfare Agents, and Related Chemical Problems. Summary Technical Report of Division 9, NDRC. Washington, DC: Office of Scientific Research and Development.

Grunicke H, Putzer H, Scheidl F, Wolff-Schreiner E, Grunewald K. 1982. Inhibition of tumor growth by alkylation of the plasma membrane. Biosci Rep 2:601-604.

Jarman GN. 1959. Chemical Corps experience in the manufacture of Lewisite in metal-organic compounds. In: Advances in Chemistry. Vol. 23. Washington, DC: American Chemical Society.

Papirmeister B, Feister AJ, Robinson SI, Ford RD. 1991. Medical Defense Against Mustard Gas: Toxic Mechanisms and Pharmacological Implications. Boca Raton, FL: CRC Press.

Reid EE. 1958. Mustard gas. In: Organic Chemistry of Bivalent Sulphur. Vol. 2. New York: Chemical Publishing. 237-451.

Ross WCJ. 1962. Biological Alkylating Agents: Fundamental Chemistry and Design of Compounds for Selective Toxicity. Butterworths: London.

Waters WA, Williams JH. 1950. Hydrolyses and derivatives of some vesicant arsencials. Journal of the Chemical Society (London) 18-22.

6

Relationship of Mustard Agent and Lewisite Exposure to Carcinogenesis

There is substantial evidence that some chemical exposures can cause cancer in human tissues. Not all of the mechanisms by which chemicals cause cancer have been delineated, but it is clear that agents that change the genetic memory of the cells—their DNA—are prime candidates for causing cancer. Clearly, not all chemicals are capable of causing cancer, and not all chemicals that cause cancer in experimental animals have been demonstrated to do so in studies of humans. In some cases, a large enough group has not been studied; in others, too little time has elapsed after exposure for the expression of cancer, or the exposure was not at a high enough level to discern the added cancers against the background level of cancer that occurs naturally or due to some other exposure, such as cigarette smoking or occupational exposure to asbestos.

There are several bodies of human data upon which to form a judgment of whether an agent is a carcinogen. The first involves biologic mechanisms: essentially, if a chemical acts in a fashion parallel to a known human carcinogen, it is evidence for a conclusion that the chemical is itself a carcinogen. The second involves evidence of an adverse effect in human cells grown and exposed in the laboratory. The third line of evidence is epidemiologic studies in human populations exposed to the agent in question for some other reason, often because the exposed individuals were involved in its manufacture. Human epidemiologic information about the carcinogenicity of war gas comes from epidemiologic studies of workers exposed in its manufacture, soldiers exposed on the battlefield, and patients exposed to the agents when used for therapeutic purposes—ironically, usually to fight cancer.

In considering the potential carcinogenicity of a chemical, it is important to acknowledge a dose-response relationship—that is, the more exposure, the more effect. This is important for two reasons:

- If a study is small enough, or if the level of exposure in the study is low enough, the study may not detect a chemical-cancer association, even if it is present.
- Even if a chemical is a carcinogen at a high level of exposure, it may only rarely cause cancer at low levels of exposure.

Two further questions must therefore be addressed in order to make a contribution to the well-being of surviving experimental subjects exposed to chemical war agents. The first is whether the specific agents are likely to be carcinogens. The second is whether, at the level of exposure experienced by these subjects, the added risk that they carry is small or large.

ACUTE EFFECTS AND BIOLOGICAL MECHANISMS

Sulfur Mustard

Sulfur mustard has been produced primarily for its acute toxic effects. Concern has been raised, however, about the long-term health effects of exposure to sulfur mustard in humans. Because the various sulfur mustards are known to be animal carcinogens, much of this concern has centered around their potential carcinogenicity to humans. One part of the process of assessing the carcinogenic risk involves examination of the biologic fate of this compound, its potential genotoxicity, and its ability to induce mutations in living systems.

Biologic Fate and Mechanisms of Action

After absorption, sulfur mustard undergoes intermolecular cyclization to form an ethylene episulfonium ion intermediate, liberating a free chloride anion. This process is facilitated by heat and by water, a likely explanation for the vulnerability of the warm and moist regions of the body (mucous membranes, eyes, respiratory tract, etc.) to the acute toxic effects of this compound (Somani and Babu, 1989; Ward and Seider, 1984). Cyclization can occur on both ends of the molecule. The cyclic intermediate reacts rapidly with a variety of nucleophiles, according to the affinity of neighboring compounds for the reaction. In pure aqueous media, most sulfur mustard is hydrolyzed to thiodiglycol and hydrochloric acid.

Boursnell and colleagues (1946) have shown that ^{35}S-labeled sulfur mustard diffused rapidly throughout the body after intravenous (IV) injection in experiments employing rabbits. Activity was retained chiefly in the liver, lungs, and kidneys, with approximately 20 percent of the ^{35}S activity being excreted in 12 hours. In rodents the majority of IV-injected sulfur mustard was excreted in the urine within 72 hours (Davison et al., 1961). The urinary metabolites included thiodiglycol and its conjugate (15 percent), glutamine-bis(b-chloroethylsulfide) conjugates (45 percent), glutamine-bis(b-chloroethylsulfone) conjugates (7 percent), and bis(b-chloroethysulfone) and conjugate (8 percent), with minute amounts of cysteine conjugates. These findings are comparable to subsequent work in rodents after intraperitoneal injection (Roberts and Warwick, 1963).

Nucleic Acid and Protein Conjugation

The reactive cyclic intermediate, the sulfonium ion, reacts avidly with proteins and nucleic acids, producing alkylation products (see Chapter 5). The alkylation of DNA by sulfur mustards has been studied by many investigators (Ball and Roberts, 1972; Boursnell et al., 1946; Davison et al., 1957, 1961; Gross et al., 1981; Habraken and Ludlum, 1989; Kohn et al., 1965; Lawley and Brookes, 1965; Ludlum et al., 1986; Meier et al., 1984; Papirmeister and Davison, 1964; Papirmeister et al., 1969, 1970, 1984a,b; Price et al., 1968; Roberts et al., 1971; Walker, 1971; Wheeler, 1962).

The sulfur mustards can be bifunctional, in that some ion intermediates covalently bind adjacent strands of DNA (a DNA cross-link). This interstrand link has been the subject of much of the study of the genotoxic effect of these agents. DNA cross-links induced by these mustards were shown by Wheeler (1962) to be extremely lethal to cells. Several workers also studied the cell cycle-specific toxicity of this bifunctional agent (Ludlum et al., 1978; Mauro and Elkind, 1968; Roberts et al., 1968, 1986). They have shown that cells in late G_1 phase or S phase of the cell cycle are particularly sensitive to the biologic effects of alkylation.

In addition, the repair of DNA lesions induced by sulfur mustards has been studied in many systems, including those employing cells known to be naturally deficient in certain repair enzymes (Ball and Roberts, 1970; Fox and Fox, 1973; Gilbert et al., 1975; Lawley and Brookes, 1968; Murnane and Byfield, 1981; Plant and Roberts, 1971; Reid and Walker, 1966, 1969; Roberts and Kotsaki-Kovatsi, 1986; Roberts et al., 1986; Savage and Breckon, 1981; Walker, 1966; Walker and Reid, 1971; Walker and Smith, 1969). As expected, DNA repair-deficient cells generally are much more sensitive to the DNA cross-linking, and the cells die at significantly lower doses.

Recent work has specifically shown that ring nitrogens on DNA are the primary sites of attack. Among the products identified are N-7 alkylguanine and N-3 alkyladenine. Intrastrand and opposite cross-links have been identified at N-7 guanine (Ball and Roberts, 1972; Ludlum et al., 1986; Walker, 1971; Wheeler, 1962). Sulfur mustards also alkylate the 0-6 position of guanine in DNA (Habraken and Ludlum, 1989; Ludlum et al., 1984, 1986). Interestingly, the well-understood DNA repair enzyme O-6-alkylguanine-DNA alkyltransferase does not act to repair these O-6 lesions (Ludlum et al., 1986). Therefore, the 0-6 alkyl product in DNA may be the major mutagenic lesion produced by the sulfur mustards.

Cytogenetic and Mutagenic Effects

Sulfur mustard induces chromosome aberrations—gross structure breaks visible under light—in a variety of cell systems. In fact, sulfur mustard was the first compound reported to induce chromosome abnormalities in the fruit fly *Drosophila melanogaster* (Auerbach, 1943). Subsequent work has shown that this type of genetic damage is dose related and that the spectrum of genetic change is similar to that of X-rays, in that it is cell cycle-specific (Nasrat, 1954; Nasrat et al., 1954; Sobels and van Steenis, 1957).

Subsequent studies have generalized the data to demonstrate that sulfur mustard induces chromosome aberrations in *Vicia faba* (fava bean) and marsupial lymphocytes (Popa, 1969; Scott and Bigger, 1972). Further, when cell lines have been studied, cytogenetic (chromosome) sensitivity to sulfur mustard has paralleled that of X-rays (Scott et al., 1974). Interestingly, fishermen exposed to sulfur mustard through netting of leaky barrels of mustard agents dumped at sea after World War II (WWII) have been found to have elevated sister chromatid exchange (SCE) frequencies in their peripheral blood lymphocytes (Wulf et al., 1985). SCEs represent breakage and rejoining of chromosomes. DNA alkylation is well known to be associated, in other circumstances, with induction of SCEs in human cells.

Sulfur mustards are also mutagens, inducing heritable alterations in dividing cells. They have been shown to induce mutations in *Drosophila* (Auerbach, 1951; Auerbach and Robson, 1946, 1947; Auerbach and Sonbati, 1960; Fahmy and Fahmy, 1972; Lee, 1975; Luening, 1951; Sobels, 1962; Sonbati and Auerbach, 1960); in L5178Y mouse lymphoma cells (Capizzi et al., 1973); in the red bread mold *Neurospora crassa* (Auerbach and Moser, 1950; Jensen et al., 1950; Stevens and Mylroie, 1950); and in the bacteria *Salmonella* (Ashby et al., 1991). The potency of sulfur mustard in most of these systems is comparable to X-rays. The compound clearly induces somatic mutations in exposed cells in a dose-related fashion. One study has also demonstrated that occupational exposure to sulfur mustard and Lewisite (manufactured in com bination) induces mutations *in vivo* in human lymphocytes at the hypoxanthine phosphoribosyltransferase *(hprt)* enzyme gene locus (Yanagida et al., 1988).

Summary

Mustard agents are mono- and bifunctional DNA-alkylating agents that are extremely cytotoxic at low doses. They alkylate RNA and proteins and produce DNA lesions, which may be repaired only at low doses. Sulfur mustards also alkylate the 0-6 position of guanine; this lesion is likely primarily responsible for the mutagenic consequence of cellular exposure. The sulfur mustards are genotoxic in a wide variety of cells, producing chromosome aberrations and gene mutations *in vitro* in a dose-related fashion. They also induce SCE and *hprt* mutations *in vivo* in the peripheral blood lymphocytes of individuals exposed at low doses.

Lewisite

There are limited data in the literature on the genetic toxicology of Lewisite. Although data on many types of arsenical compounds have demonstrated significant genotoxic potential, data on Lewisite are incomplete.

Biologic Fate and Mechanism of Action

Lewisite undergoes a complex hydrolysis involving several reversible reactions. Lewisite oxide (Cl-CH=CH-AsO) is approximately 1 percent soluble in water and 2 percent soluble in a saline solution. It is slightly more soluble at an alkaline pH. At higher pH, Lewisite oxide is cleaved by hydroxyl ions to yield arsenite and acetylene.

Lewisite has labile chlorine atoms, trivalent arsenic, and multiple bonds of carbon atoms. It is a very reactive compound. It will undergo rapid nucleophilic substitution by water. In the presence of hydrogen sulfide it forms 2-chlorovinylarsine, an extremely irritating compound. Lewisite also reacts rapidly with mercaptans to form alkylarsine.

Lewisite penetrates skin rapidly on contact. It binds avidly to proteins and thiols, and the mechanism of its local and systemic toxicity is likely mediated through this binding. It is concentrated in the thiol-containing tissues throughout the body (e.g., skin and hair). The toxicity of Lewisite is reversed by administration of the dithiol compound BAL, British Anti-Lewisite, or other thiol-containing compounds. The precise chemical nature of any of the genetic lesions (DNA-based lesions) induced by cellular exposure to Lewisite appears to be unknown.

Cytogenetic and Mutagenic Effects

Little *in vitro* genetic toxicology testing appears to have been done on Lewisite. Jostes and colleagues (1989) have completed the most extensive study in Chinese hamster ovary (CHO) cells. They found that Lewisite was cytotoxic after a one-hour exposure to micromolar amounts. Cell survival experiments yielded a D_{37} of 0.6 μM with an extrapolation number of 2.5. Interestingly, at the dose ranges of 0.125 to 2.0 μM no consistent significant induction of mutations at the *hprt* gene locus was observed. At doses of 0.25 to 1 μM, Jostes and colleagues noted no significant induction of SCEs in CHO cells, although the dose-response trend was toward a positive response. Lewisite did significantly induce chromosomal aberrations at doses of 0.5 and 0.75 μM, with a definite positive dose-response. Stewart and colleagues (1989) tested the mutagenicity of Lewisite in the Ames *Salmonella* assay. Four *Salmonella* strains were tested with and without S9 microsomal activation. This compound (S9) is used in this bioassay system to activate the agent of interest to its biologic intermediates. These intermediates are often the bioactive species and are the compounds of real interest. The strain most sensitive to killing was TA 102. No mutagenic response was observed in any strain with or without S9 activation.

No other data evaluating the genotoxicity of Lewisite appear to exist, but there have been studies of the genotoxicity of other arsenicals. Jacobson-Kram and Montalba (1985) have shown that inorganic arsenic induces both chromosome aberrations and SCEs in mammalian cells. Arsenite enhances ultraviolet light (UV) mutagenicity in bacteria (Ross-man, 1981) and viral transformation in mammalian cells (Casto et al., 1979). Arsenite synergistically enhances cis-platinum (a DNA-alkylating agent) and UV-plus-psoralen induced chromosome aberrations (Lee et al., 1986a,b). Using sulfur dioxide and arsenite, Beckman and Nordenson (1986) noted no enhanced induction of SCEs. Recent work has also shown that arsenic will induce gene amplification (an increase in the number of copies of an actively transcribed gene) in mouse cells in culture (Barrett et al., 1989; Lee et al., 1988). Arsenic exposure *in vivo* also has been reported to induce chromosome aberrations in human lymphocytes (Nordenson et al., 1978).

Arsenite reacts avidly with protein sulfhydryl groups. Relatively recent work has shown that arsenite is highly selective in reacting with small, closely spaced dithiol groups in proteins (Joshi and Hughes, 1981; Knowles and Benson, 1983). Dexamethasone binding to glucocorticoid receptors is inhibited by arsenite via a putative mechanism involving the formation of a stable dithioarsenite complex with a single dithiol group within the binding domain of the receptor (Lopez et al., 1990). Arsenite also blocks DNA binding by the receptor, presumably via a similar mechanism within the DNA-binding domain of the relevant protein (Simons et al., 1990).

Recent work of Wiencke and Yager (1992) hypothesized that arsenite might interact with DNA repair proteins that are known to contain so-called zinc fingers. These zinc fingers, which contain closely spaced dithiols, are likely important in gene regulation. Proteins that contain zinc fingers include the UVRA protein (Husain et al., 1986), polyadenosine diphosphoribose polymerase (Cherney et al., 1987; Uchida et al., 1987), the RAD-18 protein (Jones et al., 1988), and the XPAC protein (Tanaka et al., 1990). All of these are proteins that are thought to be central to DNA repair of genetic lesions. Wiencke and Yager showed that arsenite alone induced both SCEs and chromosome aberrations. However, when the DNA cross-linking agent diepoxybutane (DEB) was added to human lymphocyte culture in the presence of arsenite, the induction of chromosome aberrations was synergistically enhanced. The induction of SCEs was only additively increased. Interestingly, the synergistic enhancement of chromosome aberrations was most pronounced in individual lymphocytes previously known to be relatively more sensitive to the clastogenic action of DEB (Wiencke et al., 1991). Wiencke and Yager concluded that the specific co-clastogenic effects of arsenite were mediated by its interference with DNA repair activities. All of this work may indicate that arsenicals could interact with DNA-alkylating agents when they are given concomitantly. Although there is no direct evidence that the genetic effects of sulfur mustard exposure are enhanced by concomitant Lewisite exposure, it remains a possibility.

Summary

In contrast to mustard agents, the genetic toxicology of Lewisite has been poorly studied. Its hydrolysis has been examined and arsenite is one significant product likely produced in man after exposure. Lewisite itself clearly induces chromosome aberration in one type of cellular assay. It appears not to be mutagenic in *Salmonella.* Arsenicals in general and arsenite have been shown to be clastogenic and to induce SCE in human and other mammalian cells. Arsenite synergistically enhances the clastogenic action of alkylating agents, perhaps through binding to DNA repair proteins.

EVIDENCE OF LONG-TERM HEALTH EFFECTS

Animal Studies

This section reviews the results of the published experimental carcinogenesis studies of sulfur mustard and nitrogen mustard. The latter are included because there is a dearth of experiments on sulfur mustard. Further, the similarity of action of nitrogen mustard to sulfur mustard provides information that is useful in assessing the types and sites of malignancy that may occur from exposure to sulfur mustard.

In what follows, in accordance with historical usage, we use the term sulfur mustard, although the terms HS, HD, or H were often used in past experimental literature. HD and HS refer to distilled sulfur mustard (approximately 96 percent pure), while H refers to an impure preparation known as Levinstein mustard. Experiments have been conducted with two nitrogen mustards, HN2 [methylbis(b-chloroethyl)amine or the hydrochloride] and HN3 [tris(b-chloroethyl)amine]. By far the most data are available for HN2, which is a common chemotherapeutic agent, usually in combination with other chemicals (see Medical and Therapeutic Exposure section of this chapter). The designations HN2 and HN3 are used as appropriate.

Early Studies in Mice and Rats

One of the earliest reports of the carcinogenic effect of the nitrogen mustards is an interim report describing preliminary results on the intravenous, intraperitoneal, and subcutaneous injections of HN2 and HN3 into Swiss mice and albino rats (Griffin et al., 1950). At the time of the report, tumors had developed in 15 to 20 percent of treated animals, whereas no tumors had appeared in controls. The tumors included fibrosarcomas, lymphosarcomas, and adenocarcinomas (no sites or numbers were given), and administration of a single dose proved as effective in inducing tumors as multiple injections.

Another early study on the carcinogenicity of mustard compounds was conducted at the Chester Beatty Research Institute in London (Boyland and Horning, 1949). Aqueous solutions of HN2 or HN3 were administered by subcutaneous injection weekly for 50 weeks to two groups of 20 stock mice each. The weekly dose of each substance was 1 mg/kg body weight. This dose was toxic and led to a high early mortality; only 10 animals administered HN2, and 4 animals administered HN3, survived 150 days. These 14 survived from 284 to 580 days. At autopsy, among the 10 injected with HN2, there were 3 lung carcinomas, 1 lung adenoma, and 1 with "early bronchogenic tumors." One animal had a liver lymphosarcoma and another had a uterine fibromyoma. Three were free of tumors. Among the 4 administered HN3, there were 2 lung carcinomas and 1 lung adenoma. One of the animals with carcinoma also had a spindle-celled sarcoma at the site of injection. Substantial other non- or premalignant pathology was described in all of the treated animals. Of 40 untreated mice sacrificed between 14 and 18 months of age, 6 had lung adenomas and 2 hepatomas. The number of tumors at any site and the number of lung tumors in animals administered HN2 were significantly greater than expected at the 95 percent confidence level. Boyland and Horning noted a particular excess of lung tumors and lymphosarcomas compared to the number expected.

Heston's Studies in Mice

From 1949 through 1953, W. E. Heston and his colleagues conducted a series of studies in which sulfur mustard and nitrogen mustard were administered to strain A mice and the occurrence of pulmonary tumors was studied (Heston, 1949, 1950, 1953a,b; Heston et al., 1953). Strain A mice are highly inbred and have an extremely high genetic susceptibility for development of pulmonary adenomas; the incidence of spontaneous pulmonary tumors is about 50 percent in animals 12 months of age and 90 percent at 18 months of age. Thus, in several of Heston's experiments all of the animals developed adenomas, and comparisons can be made only between the total number of tumor nodules found in different exposure groups. The administered amounts of nitrogen and sulfur mustards were highly toxic, and a significant number of early deaths of animals occurred.

The results of the first two studies (Heston, 1949, 1950) are summarized in Table 6-1. In Experiment I, 100 percent of the experimental animals that survived acute mortality had tumors at sacrifice (13 to 16 weeks), compared to 13 percent of controls (the remaining 8 experimental and 8 control mice were followed longer, but no data were provided on their outcome). The mean number of lung nodules in the experimental group was 3.48, compared with 0.13 in the control group. The 9 animals receiving 4 injections had 5.11 nodules on average, the 13 receiving 3 injections had 2.62 nodules, and the lone animal with 2 injections had 2 nodules. The administered doses of nitrogen mustard clearly increased the number of nodules and number of strain A mice affected.

In Experiment II, all experimental animals received four injections and follow-up was continued for 10 months, rather than 3 or 4. The full dosing scheme led to several animals dying prior to completion of the experiment, but the results are in agreement with those of Experiment I, with a greater number of nodules being associated with the longer follow-up period. In Experiment III, a single larger injection of HN2 resulted in higher short-term mortality, but the incidence of tumor nodules in animals that survived 10 months was similar to the group that received the full dose in four injections.

Experiments IV and V followed a similar methodology to examine the effects of intravenous injection of sulfur mustard. Experiment IV saw a high early mortality: 9 males and 4 females of 30 animals injected died shortly after completion of the injections. The results for the animals surviving 16 weeks showed somewhat fewer mice with lung nodules and a lower mean number of nodules compared to Experiment I with HN2. Further, an unusually high number of controls developed adenomas for which no cause could be identified. When Experiment IV was repeated as Experiment V, by using a lower concentration of sulfur mustard, fewer mice developed lung nodules (68 vs. 93 percent) and the mean number of lung nodules was substantially less (1.09 vs. 2.60). Obviously, there is uncertainty about the actual dose to the animals in the sulfur mustard experiments, particularly in Experiment IV, where the administered dose may have exceeded 1 mg/kg body weight.

TABLE 6-1 Pulmonary Tumors in Strain A Mice Injected Intravenously with Nitrogen Mustard and Sulfur Mustard

Experiment	Follow-up Duration	No. of Mice	% with Nodules	Mean No. of Nodules
Nitrogen mustard				
Experiment I: two to four intravenous injections of 1 mg/kg HN2 (14 received 4 injections; 22 received 3; and one, 2) at 2-day intervals	13–16 wks	29	100	3.48
Control	13–16 wks	30	13	0.17
Experiment II: Four intravenous injections of 1 mg/kg HN2 at 2-day intervals	10 mos	20	100	9.60
Control	10 mos	32	63	0.81
Experiment III: One intravenous injection of 4 mg/kg HN2	10 mos	9	100	7.56
Control	10 mos	31	58	0.94
Sulfur mustard				
Experiment IV: Four intravenous injections of 0.65 mg/kg at 2-day intervals	16 wks	15	93	2.60
Control	16 wks	28	61	0.93
Experiment V: Four intravenous injections of 0.65 mg/kg at 2-day intervals	16 wks	47	69	1.09
Control	16 wks	46	13	0.13

SOURCE: Heston, 1949, 1950.

A later study by Heston and colleagues (1953) compared the separate and combined effects of intravenous HN2 and X-irradiation (Table 6-2). With only two injections of 1 mg/kg HN2, some animals were tumor free after 11 months of follow-up. Those receiving the radiation had fewer tumors than controls; those receiving both radiation and HN2 had a tumor frequency similar to controls, but higher than those receiving radiation.

TABLE 6-2 Pulmonary Tumors in Strain A Mice Injected Intravenously with Nitrogen Mustard and/or Exposed to X-Radiation

Experiment	Follow-up Duration	No. of Mice	% with Nodules	Mean No. of Nodules
Control mice	11 mos	64	39	0.7
900 R of X-radiation	11 mos	55	24	0.6
Two injections of 1 mg/kg bw of nitrogen mustard	11 mos	59	93	4.2
900 R of X-radiation plus 2 injections of 1 mg/kg bw of nitrogen musard	11 mos	55	38	1.1

SOURCE: Heston et al., 1953.

The carcinogenicity of sulfur mustard and HN2 was further studied by Heston (1953a), using a variety of subcutaneous injections of sulfur mustard and HN2 in strain A, C3H, or C3Hf mice. (C3Hf female mice are less susceptible to mammary tumors than C3H mice, but C3Hf male mice are more likely to develop hepatomas.) The results, summarized in Table 6-3 for C3H and C3Hf mice and Table 6-4 for strain A mice, show that tumors, including sarcomas, were commonly found at the site of injection: among C3H and C3Hf mice, 7 of 87 injected with sulfur mustard and 12 of 73 injected with HN2 had injection site tumors. No tumors appear at the site of injection of 16 mice administered olive oil, nor did any tumors occur in noninjected controls in the region chosen for injection. Of the tumors at sites other than the site of injection, only pulmonary nodules from HN2 injections were significantly different from control animals: 34 of 73 injected animals developed pulmonary tumors, compared with 14 of 74 control animals. For all C3H and C3Hf animals, tumors appeared in 8.4 percent injected with sulfur mustard, compared with 15.8 percent of controls. In the strain A mice only 1 tumor was observed in the injection site and there were no differences between the control and experimental group in number of tumors in other sites (Table 6-4).

Heston conducted another experiment in which strain A mice were exposed to sulfur mustard by inhalation. It is likely that the concentration of sulfur mustard in the experimental chamber reached relatively high levels during the 15-minute exposure: of a total of 80 mice exposed, 13 died in the 4 months prior to follow-up. Control and experimental mice were sacrificed periodically from 4 to 11 months after exposure. At

TABLE 6–3 Tumors from Subcutaneous Injection of Sulfur and Nitrogen Mustard into C3H Mice and Sulfur Mustard into C3Hf Mice

No. of Injections	Sex	No.	Average Months Survival	Tumors at Site of Injection	Tumors at Other Sites
				0.05 cc 0.05% soln. of sulfur mustard in olive oil into C3H mice	
6	M	8	14.3	1 Sarcoma	2 Pulmonary tumors
6	F	8	8.9	0	8 Mammary tumors 1 Hepatoma
6	M	24	13.1	1 Rhabdomyosarcoma	7 Hepatomas
				0.25 mg HN2 in 0.25 cc water into C3H mice	
8	M (8)	7	10.6	1 Papilloma 1 Squamous cell carcinoma	3 Pulmonary tumors
8	F	8	8.6	1 Papilloma	2 Pulmonary tumors 7 Mammary tumors
6	M (24)	21	17	3 Sarcomas 2 Neurogenic sarcomas 1 Hemangioendothelioma	8 Pulmonary tumors 9 Hepatoma 2 Liver hemangioendothelioma
				C3H control mice	
	M	8	14		1 Pulmonary tumor 2 Hepatomas
		8	7.4		7 Mammary tumors
	M (8)	6	14.3		1 Pulmonary tumor
	F	8	11.1		2 Pulmonary tumors 7 Mammary tumors
	M (24)	21	15.5		5 Pulmonary tumors 9 Hepatomas
				0.05 cc 0.05% soln. of sulfur mustard in olive oil into C3Hf mice	
6	F (10)	9	15.4	2 Sarcomas 1 Mammary tumor	1 Pulmonary tumor 1 Mammary tumor
6	M (40)	38	16.8	2 Sarcomas	5 Pulmonary tumors 22 Hepatomas
				0.25 mg HN2 in 0.25 cc distilled water into C3Hf mice	
6	M (41)	37	16.1	2 Sarcomas 1 Papilloma	21 Pulmonary tumors 17 Hepatomas 1 Reticulum cell sarcoma 1 Lymphocytic leukemia 1 Liver hemangioendothelioma
				C3Hf control mice	
	M (40)	39	16.3		6 Pulmonary tumors 18 Hepatomas 2 Lymphoytic leukemias 1 Skin squamous carcinoma

NOTE: Table does not include data from one high-dose group, due to the high mortality reported. Numbers in parentheses reflect number of animals initially injected.

SOURCE: Heston, 1953a.

TABLE 6-4 Tumors from Subcutaneous Injection of Sulfur Mustard into Strain A Mice

No. of Injections	Sex	No.	Average Months Survival	Tumors at Site of Injection	Tumors at Other Sites
		0.05 cc 0.05% soln. in olive oil into strain A mice			
5	M	(16) 14	6.7	1 Sarcoma	6 Pulmonary tumors 1 Lymphocytic leukemia
5	F	(14) 12	11.2		7 Pulmonary tumors 1 Mammary tumor
		Strain A control mice injected with 0.05 cc olive oil			
5	M	16	11.7		4 Pulmonary tumors 2 Hepatomas 1 Myoepithelioma
5	F	14	14.1		11 Pulmonary tumors 2 Lymphocytic leukemias 1 Myoepithelioma

SOURCE: Heston, 1953a.

the conclusion of the experiment, 33 of 67 experimental mice had developed pulmonary tumors, compared with 21 of 77 control mice. Additionally there were 3 lymphocytic leukemias in the experimental group and none in controls.

Later Studies of Mustard Agents and Other Alkylating Agents

Several studies conducted at Edgewood Arsenal examined the toxic and carcinogenic effects on rats of chamber (whole body) exposure to sulfur mustard (McNamara et al., 1975). Experimental animals were exposed to sulfur mustard at concentrations of 0.001 mg/m^3 (continuously) or 0.1 mg/m^3 (6.5 hours per day, 5 days per week) for periods of from 1 to 52 weeks. The results of these studies, summarized in Tables 6-5 and 6-6, demonstrated that sulfur mustard readily produced skin malignancies in rats, but no excess tumors at other sites. Table 6-7 summarizes the data on all tumors observed in the rats. Data on other species were equivocal: experiments on guinea pigs, rabbits, and dogs were not of sufficient duration for a valid carcinogenicity study. No tumors were observed in any animal other than rats.

Conklin and colleagues (1965) compared the carcinogenicity of HN2 with X-radiation, triethylenemelamine, and 1,4-di(methanesulfonoxy)butane (Myleran). The data in Table 6-8 show that HN2 produced a statistically significant increase in thymic lymphomas and lung tumors, and a nonsignificant increase in ovarian tumors, compared with controls. The effect of X-radiation was somewhat greater than HN2 in producing thymic lymphomas, much greater in producing myeloid leukemias, and much less than HN2 in producing lung tumors.

TABLE 6-5 Number of Rats Developing Tumors Following Exposures to HD (Toxicity Study)

		Incidence of Tumors		
Duration of Exposure (months)	Postexposure (days)	Controls	0.001 mg/m³	0.1 mg/m³
2		0/5	0/10	0/10
3		0/10	0/10	0/10
4				0/10
6		0/10	0/10	
8		0/10	1/10	1/10
12		0/10	1/10[a]	0/10
	70			4/4 (4[b], 1[a])
	90	1/8[c]	0/9	0/6
	180	2/11 (1[a],1[d])	1/20[e]	6/19 (4[f],1[g],1[h])
Rats with tumors		3/64	2/79	10/79
Rats with agentrelated tumors[i]		0/64	0/79	9/79

[a] Subcutaneous fibroma.
[b] Skin, squamous cell carcinoma (agent related).
[c] Uterus, squamous cell carcinoma.
[d] Pulmonary adenoma.
[e] Skin, papilloma.
[f] Skin, squamous cell carcinoma (probably agent related).
[g] Skin, basal cell carcinoma (probably agent related).
[h] Thyroid, adenoma.
[i] As designated by McNamara et al.
SOURCE: McNamara et al., 1975.

Bioassays of 29 alkylating compounds, including nitrogen mustard, conducted at the National Cancer Institute showed similar results (Shimkin et al., 1966). Each assay consisted of 12 thrice-weekly intraperitoneal injections of the various compounds; at least four dose levels were used for each compound and follow-up continued for 39 weeks. Although negative results were obtained for many compounds, a strong positive effect was seen with nitrogen mustard: total doses of HN2 between 0.02 and 17.5 μM/kg produced pulmonary tumors in 30 to 95 percent of the experimental animals. Compared with other alkylating agents, in fact, HN2 is second only to uracil mustard in tumorigenicity among the 29 compounds. These estimates are supported by the work of Abell and colleagues (1965), in which a wider range of doses was used.

Epstein (1984) investigated the separate and combined effects of HN2 administration and UVB radiation on tumor formation on the skin of Uscd (Hr) strain, albino hairless mice. Twice-weekly applications of 0.1 ml of HN2 to the skin for 52 weeks produced tumors in 34 percent of the mice. An additional thrice-weekly exposure to 1.98 x 10^2 mJ/cc^2 of UVB energy produced an 88 percent incidence of skin tumors; UVB energy alone produced tumors in 76 percent of mice. Twice-weekly application of HN2 with UVB produced more tumors than once-weekly applications of HN2 with UVB, and there was no tumorigenic activity of the alcohol vehicle among control animals when combined with UVB.

TABLE 6-6 Number of Rats Developing Tumors Following Exposures to HD (Carcinogenicity Study)

		Incidence of Tumors		
Duration of Exposure (weeks)	Postexposure (months)	Controls	0.001 mg/m^3	0.1 mg/m^3
1	13		0/1	
	15			1/1[a]
	21		1/4[b]	0/4
2	20		0/5	1/5[c]
4	16		0/1	0/1
	20		0/4	0/5
8	15	0/4	0/2	0/4
	17		0/1	
	18		1/1[d]	
12	12		0/2	4/5 (3[f],1[g])
	17		1/3e	
26	14		0/4	3/4[f]
	18			1/1[f]
39	11		1/3e	4/4 (4[f], 1[h])
52	2			1/1[f]
	4			1/1[h]
	6			1/1[f]
	7			0/1
	10	1/22[e]	1/17	3/14 (3[e], 2[f], 1[i])
	17	1/1[e]		1/1[e]
	18			4/4[f]
Rats with tumors	2/27	5/48	25/57	
Rats with agentrelated tumors [j]	0/27	0/48	22/57	

[a] Subcutaneous lipoma.
[b] Auxillary lipoma.
[c] Subcutaneous fibroma.
[d] Astrocytoma.
[e] *Skin,* fibroma.
[f] *Skin,* squamous cell carcinoma (agent related).
[g] Skin, basal cell carcinoma (agent related).
[h] *Skin,* trichoepithelioma (agent related).
[i] Skin, keratoacanthoma (agent related).
[j] As designated by McNamara et al.
SOURCE: McNamara et al., 1975.

A final experiment on the carcinogenicity of HN2 was a skin painting experiment by Zackheim and Smuckler (1980). They applied topical HN2 in concentrations of 0.3 mg/0.2 ml 95 percent ethanol, once per week for 20 weeks, or 0.1 mg/0.2 ml 95 percent ethanol, three times per week for 23 or 26 weeks. Just over 27 percent of the mice in the low-dose group (23 weeks) developed squamous cell carcinoma of the skin; incidence of papillomas in other experimental groups was 21 to 33 percent. The authors concluded that HN2 was a potent skin carcinogen.

TABLE 6-7 Tumors Observed in Rat Carcinogenicity and Toxicity Studies (Includes All Animals and Only Those Observed for 12 or More Months)

			Exposed			
	Controls		0.001 mg/m^3		0.1 mg/m^3	
Study	All	12+ mo.	All	12+ mo.	All	12+ mo.
Carcinogenicity	2/27	2/23	5/48	1/17	25/57	11/23
Toxicity	3/64	3/29	2/79	2/39	10/79	10/39
Both studies	5/91	5/52	7/127	3/56	35/136	21/62

SOURCE: McNamara et al., 1975.

TABLE 6-8 Percentage of Female RF Mice with Neoplasms from Exposure to HN2 or X-Rays

	Percentage of Animals with Neoplasm		
Type of Tumor	2.4 mg/kg HN2 x 4 (115 mice)	300 R X-Rays x 4 (110 mice)	Controls (113 mice)
Thymic lymphoma	21	33	10
Myeloid leukemia	2	27	4
Other leukemia	34	20	37
Lung tumors	68	13	15
Ovarian tumors	26	39	20

SOURCE: Conklin et al., 1965.

Summary of Animal Studies

Sulfur mustard has produced a significant increase in pulmonary tumors as a result of intravenous injection in highly susceptible strain A mice. A 15-minute inhalation exposure to a very high but unknown vapor concentration of sulfur mustard produced a marginally significant increase of pulmonary tumors, again in strain A mice. Sprague-Dawley-Wistar rats developed a substantial increase in skin malignancies from long-term chamber exposure to 0.1 mg/m^3 of sulfur mustard. Such increased malignancies were not observed in chamber exposures with A/J mice, guinea pigs, rabbits, and dogs, although follow-up periods were inadequate. Subcutaneous injections totaling about 6 mg/kg of sulfur mustard produced sarcomas and other malignancies at injection sites in C3H, C3Hf, and strain A mice, but did not produce an increase of malignancies at other sites.

Nitrogen mustard, particularly HN2, was more widely tested than sulfur mustard and was found to be a potent tumorigen and carcinogen. It produced an increase in lung tumor nodules in strain A mice from both intravenous and intraperitoneal injections. Subcutaneous exposures produced injection site tumors and pulmonary tumors in selected strains of mice. Its tumorigenic potency appeared to be similar to or somewhat greater than sulfur mustard; HN2 was one of the most potent alkylating agents tested in the strain A bioassay program. HN2 increases the action of both X and UV radiation. Excess lymphosarcomas were reported in two studies of this type, and thymic lymphomas in a third. However, the numbers of animals used in these studies were small.

Human Studies

Epidemiologic evidence about the carcinogenicity of mustard agent and Lewisite comes from three sources: (1) studies of soldiers who were exposed to these agents on the battlefield; (2) studies of workers who manufactured these agents; and (3) studies of the unintended adverse effects on patients exposed to these agents for medical therapy. This review describes key studies from each of these arenas and characterizes their attributes, including study design, comparison population, case ascertainment and definition, control of other (confounding) important variables, and estimation of exposure to mustard and Lewisite.

Occupational Exposure

Japanese Studies. A number of studies have followed workers at a weapons factory on Okuno-jima, an island in the Inland Sea and in the south of Hiroshima Prefecture. The factory manufactured various chemical agents for use in World War II, and many investigators believe that these workers were exposed to multiple poisonous agents other than sulfur mustard and Lewisite, including HCN, diphenyl cyanarsine, choroacetopheonone, and phosgene. The studies reflect use of various methodologies in epidemiology ranging from case reports and case series to formal cohort mortality studies. The weight of the evidence firmly demonstrates the carcinogenicity of sulfur mustard in the occupational setting, which implies repeated exposures but does not allow comment on the level of exposure.

Yamada (1963) reported a case of bronchogenic carcinoma in a former worker in the Hiroshima plant. Other cases are included in this report, representing a pioneering effort to quantify the degree of excess cancer among former workers. Its epidemiologic value is surpassed by subsequent work by Wada on the same population (see below), but Yamada does provide a qualitative description of exposure not included in subsequent reports: "Each working room was so filled with the leaked gas shrubs around the room became withered." Workers apparently wore gasproof masks and rubber clothes, although this personal protective gear was not provided in "sufficient amount." Wada and colleagues (1962) added a description of the differential exposure to various jobs at the plant, information that was used in subsequent epidemiologic studies of this population.

Nishimoto and colleagues (1983) extended this analysis by comparing mortality experience with the likelihood of exposure to sulfur mustard and Lewisite. Three cohorts were defined: (1) those involved in gas production; (2) those in laboratory work, inspection, incineration, etc; and (3) those in the production of tear gas and other jobs not exposed to mustard, Lewisite, etc. There is almost no discussion of methods of epidemiologic analysis used. The authors reported that only the first two groups had excesses of lung cancer compared with local rates. Inspection of one table of mortality ratios, however, suggests that all three groups were in substantial excess, although the third group did not reach statistical significance. Due to the limits of the methods used and the brevity of the report, this study adds little weight to the evidence that there was a differential response by degree of exposure in the workers at this plant.

A subsequent report by Nishimoto and colleagues (1988) further clarified the Japanese experience in two ways. First, it indicates, albeit indirectly, that workers in the two groups with lung cancer excess (cohorts 1 and 2) were not exposed to phosgene; phosgene-exposed workers appear to be included in cohort 3, "the production of other gases." Secondly, this paper reports 11 cases of cancer of the larynx among the 674 workers exposed to sulfur mustard and only 1 case among the group not exposed to sulfur mustard. Although no formal estimates are made of the number of expected cases of laryngeal cancer in the first group, the number of cases clearly suggests an excess.

Kurozumi and colleagues (1977) described four cases of respiratory tract malignancy among former workers in the same Hiroshima plant studied by Wada. The case reports provide no additional information in evaluating the carcinogenicity of sulfur mustard. However, in evaluation of the dose-response for sulfur mustard, one case is reported to have worked in the factory for only six months; her duties and extent of exposure are not described.

Wada and colleagues (1968) conducted a cohort mortality study of 2,620 individuals who worked in the plant between 1929 and 1945, including 495 who reported that they manufactured sulfur mustard. The study is unusual in that the cohort of exposed workers was not determined from a plant employment roster, but by a series of indirect methods that included community surveys, interviews with former workers, etc. It is not clear from their report whether the cause of death was ascertained only from the death certificate or multiple methods. However, 33 deaths due to cancer of the "respiratory passages" were identified that occurred between 1952 and 1967 as compared with 0.9 expected. Among plant workers not exposed to sulfur mustard, there were 3 deaths from respiratory cancer compared to 1.8 expected.

With only one exception, the cases of respiratory and lung cancer in this study were located in the central, major airways, rather than in peripheral regions of the lung. None were adenocarcinomas. No comparison distributions for these findings were provided.

The authors did not provide information describing the level of exposure of the workers. It is unclear whether exposures were substantially low and not associated with acute symptoms, or high enough to cause dermatologic or pulmonary symptoms (see Chapters 7 and 9). Length of employment, which is frequently used as a surrogate for exposure in epidemiologic studies, was provided for workers with cancer but not for the entire population at risk. This prevented the authors from estimating any dose-response relationships. Other potential problems with this study include (1) the possibility of differential selection of cohort members if they had a malignancy; (2) a potential bias toward ascribing employment in mustard production among cases of cancer; and (3) the possibility of more extensive searching for a history of mustard exposure among the cases of respiratory malignancy. However, none of these potential methodologic problems are likely to explain the extraordinarily elevated mortality ratio for cancer of the respiratory tract or the unusual anatomical distribution of malignancies. Taken as a whole, this paper provides substantial evidence that occupational exposure to sulfur mustard is associated with cancer. It does not, however, provide information by which to judge the extent of risk among those exposed to low levels over a short duration.

Inada and colleagues (1978) provide a case series among former workers at the Japanese plant of Bowens' disease, a dyskeratotic hyperplastic growth of the epidermis that is thought to be precancerous. The case series draws attention to the possible association of this disease entity with exposure, but does not provide sufficient information to evaluate whether the occurrence of this number of cases is unusual and hence possibly causally related to exposure.

British Studies. A second series of studies examined exposure of workers in the manufacture of war gas in World War II in the United

Kingdom. Manning and colleagues (1981) conducted a cohort mortality study among 511 former employees of a plant that manufactured sulfur mustard, alone and in combination with other toxic gases, from 1939 to 1945. The most prominent finding was a statistically significant excess of cancer of the larynx, based upon 2 cases of cancer of the larynx and 1 of the trachea; the relative risk was 7.5. Examination of cancer incidence, rather than cancer mortality, provides a parallel conclusion: 7 workers developed cancer of the larynx, compared with 0.75 expected, yielding a relative risk of 9.3. The relative risk for lung cancer was elevated, albeit not significantly, compared to the general population of England and Wales (21 observed versus 13.43 expected; relative risk equal to 1.6), although the relative risk is lessened to 1.2 when compared with the experience of the local population, which has higher lung cancer rates normally. The cohort also had a significant excess of mortality due to pneumonia (relative risk 2.0; 14 observed versus 6.86 expected). This study had three major deficits:

- No analysis is provided by length of employment in this manufacturing facility, the usual surrogate for exposure.
- No description or estimate of the exposure is provided by which to gauge the magnitude of exposure.
- No information is available on other exposures that are associated with cancer of the larynx, such as cigarette smoking.

Given these limitations, this study still provides substantial indication that there is a relationship between employment in this facility, which the authors equate with exposure to sulfur mustard, and cancer of the larynx.

Easton and colleagues (1988) conducted an expanded study, including follow-up study of the Manning group's cohort, and provided more qualitative information on exposure. For example, it was documented that gas had escaped into the air of the facility and that acute respiratory and dermatologic symptoms had occurred. Consistent with the prior and smaller Manning study, significant excesses for cancers of the pharynx (standardized mortality ratio [SMR] = 549) and lung (SMR = 145) were observed when compared to national rates for cancer. New findings included significant excesses of cancer of the esophagus and stomach and an excess for nonmalignant respiratory disease. Again, the excesses were lower when compared to local rates. A significant contribution of this study is the added use of duration of employment as a surrogate for exposure. Dose-response relationships were observed for cancers of the lung, pharynx, larynx, but not for stomach or esophagus. The findings of a dose-response relationship with malignancies ob served in other studies adds weight to the evidence of the carcinogenicity of sulfur mustard.

Battlefield Exposure

Jackson and Adams (1973) conducted a case control study to explore the causes of aggressive and disfiguring cases of basal cell carcinoma: 33 cases of "horrifying" basal cell carcinoma were compared with 435 other cases of basal carcinoma. Of interest to this discussion, 2 of the 33 cases had histories of mustard vapor burns. One case, reported in some detail, developed a basal cell carcinoma in the scar on his neck left by sulfur mustard in World War I (WWI). This study does not provide an estimate of the risk to veterans with mustard agent burns of subsequently developing dermatologic cancers. The study nevertheless reveals a possible etiologic association between mustard agent burns and aggressive basal cell carcinoma.

Case and Lea (1955) studied a group of 1,267 men who were receiving pensions for the effects of mustard agent poisoning resulting from WWI combat exposures. The mortality experience of this group was compared with expected mortality for males in England and Wales and with the mortality experience of 1,421 other WWI pensioners disabled by bronchitis and another 1,114 pensioners with limb amputations, but without bronchitis, who had not been exposed to sulfur mustard. The results showed that both the cohort exposed to sulfur mustard and the cohort with bronchitis, but not the amputees, had significant excesses of mortality for "all causes" and for "cancer of the lung and pleura." The authors did not feel that these results support the hypothesis that sulfur mustard acts as a direct carcinogen. The major limitation of this study was the lack of information on cigarette smoking. Pension records showed that 81 percent of the exposed cohort were disabled by bronchitis or other pulmonary conditions, but there is no way to determine if their bronchitis is mustard-associated, smoking-associated, or both. It is possible that members of this cohort, like U.S. veterans who were exposed to sulfur mustard (Norman, 1975), smoked less than their peers. Thus, this study provided weak evidence that mustard exposure is not associated with an excess of cancer of the lung.

Beebe (1960) conducted a similar study of U.S. veterans of WWI that was designed to compare the mortality and morbidity of three groups: (1) soldiers hospitalized for exposure to mustard vapor; (2) soldiers with pneumonia; and (3) soldiers requiring amputation. Unlike the British study, all members of the gas cohort had to have had medical evidence of eye or skin burns from mustard agents. Smoking data were also available on a sample of all three cohorts and documented a deficit of smoking in the gas cohort. The sulfur mustard cohort had 1.3 percent die of lung cancer compared with 1.0 percent of the wounded control, which while elevated was not significant. Compared with the mortality patterns of the general population, however, the gas cohort had 39 observed deaths from respiratory cancer compared with 26.6 expected. This statistically significant excess is even more impressive if one considers that the gas cohort had a deficit of smokers. Significant lung cancer excess was not observed for the pneumonia or the wounded cohorts. Beebe concluded that, since there was no substantial difference in lung cancer mortality between sociologically similar groups of soldiers, yet there was a substantial difference with the general population, this study is "quite suggestive" but "unproved."

Norman (1975) followed Beebe's cohorts for an additional 10 years. The mortality ratios for the mustard-exposed, pneumonia, and wounded populations were 0.99, 0.93, and 0.92, respectively, which may reflect an excess in that soldiers must pass a screening examination before induction. The mortality ratio of the gas-exposed cohort compared to wounded controls was 1.3, with 95 percent confidence intervals extending from 0.9 to 1.9. With "a more sensitive test," this difference in lung cancer deaths in the mustard-exposed cohort was even higher. A further case control analysis, for which little methodological detail was provided, indicated that mustard exposure added little excess risk compared with cigarette smoking. The authors recognized that this comparison might not accurately reflect the cumulative amount of cigarette smoking in each group; indeed, they provided data that smokers who were gassed quit substantially earlier than those in the other two groups. This study provides some evidence supporting the conclusion that a battlefield exposure associated with hospitalization is later associated with an excess mortality from lung cancer. This study also demonstrates the degree of information that would be essential to disentangle unequivocally the effect of mustard exposure from cigarette smoking.

Medical Therapeutic Exposure

In the early 1970s, reports began appearing of acute nonlymphocytic leukemia (ANL) in patients with multiple myeloma and lymphoma who had been treated with nitrogen mustards (Kyle et al., 1970; Rosner and Grunwald, 1975). Initially, it was thought that this occurrence might represent a consequence of the late natural history of these diseases. However, ANL was also found to occur in other types of cancer patients who had been treated with nitrogen mustards and other alkylating agents (Einhorn, 1978; Fisher et al., 1985; Greene et al., 1982). In particular, patients with ovarian cancer treated with melphalan and chlorambucil developed this complication. The reason why the initial observations of this complication were in multiple myeloma, lymphoma, and ovarian cancer patients is undoubtedly that these were the first patients to receive prolonged therapy with nitrogen mustards who survived long enough to develop this complication.

As experience with cancer chemotherapy has increased, it has become obvious that this complication occurs in approximately 3 to 5 percent of all patients treated with therapeutic courses of nitrogen mustards and other alkylating agents (Tucker et al., 1988), and in some groups of patients treated with prolonged, intensive courses, this rate has been as high as 30 percent (Einhorn, 1978). The acute leukemia that is seen after alkylating agent therapy is very malignant and responds poorly to conventional therapy. The peak time of onset of this leukemia has been reported to be between three and nine years after the original treatment (Blayney et al., 1987). As larger numbers of cancer patients have been treated with alkylating agents, the evidence has become very strong that there is also an increase in the rate of solid tumors in these patients (Tucker et al., 1988). In one study of patients who had been treated for Hodgkin's disease, the patients' 10-year actuarial risk of ANL was estimated to be 5.9 percent, for lymphoma 3.5 percent, and for solid tumors 5.8 percent (Koletsky et al., 1986). Similar risks have been described in other studies, including children treated for cancer.

Therapeutic nitrogen mustards are administered systemically (except for a small experience with topical application to the skin, see Chapter 9) and are given repeatedly for periods of weeks to months. They are less reactive and have a different systemic pharmacology than sulfur mustard (Colvin and Chabner, 1990). Therefore, it is difficult to make quantitative extrapolations to the carcinogenicity of sulfur mustard and to which tumors sulfur mustard would be expected to produce. For example, acute leukemia has not been reported as a late consequence of sulfur mustard exposure. This might well be because those patients with sufficient skin or inhalation exposure to deliver a leukemic dose to the bone marrow would have succumbed to pulmonary and other complications. However, because sulfur mustards and nitrogen mustards have similar effects on DNA, the clinical experience with nitrogen mustards supports the evidence that sulfur mustard is carcinogenic in man.

SUMMARY

Gaps in the Literature

There are three major gaps in the epidemiologic literature on occupational exposure. The first gap concerns the limited cohorts of workers that have been studied, Japanese and British. Other countries (including the United States) manufactured war gases; hence other cohorts of workers could be identified and studied, which would contribute additional information.

The second gap concerns the limited information available on exposure levels. Historical records vary considerably in their detail. Exposure measurements, if any, were usually done to determine levels for particularly troublesome parts of the manufacturing process or were collected for specific areas of the plant, which limits their value in characterizing particular jobs.

The third gap concerns assessment of risk. The available epidemiologic studies focus on determining whether exposure is associated with a increased rate of malignancy in the exposed workers. This endeavor is complicated, and its conclusions—whether or not war gases are carcinogenic—are not the most appropriate questions to be addressed by this committee. Even if one concludes that the studies demonstrate a relationship between exposure and human cancer, the question remains: What would be the likely rate of cancer in the chamber and field test volunteers, or those who worked with these agents, exposed at levels different from the battlefield or occupational situations in Japan or Great Britain? This question requires a quantitative risk assessment, which cannot be developed from the data available in the literature.

Conclusions

Mustard agents are well known to be monofunctional and bifunctional DNA alkylating agents. They are extremely cytotoxic at low doses. They alkylate RNA and proteins and produce DNA lesions, which may be repaired only at low doses. The sulfur mustards also alkylate the 0-6 position of guanine. DNA alkylation is likely primarily responsible for the mutagenic consequence of cellular exposure. The sulfur mustards induce a wide variety of genetic lesions in many types of mammalian cells *in vitro* in a dose-related fashion. They also induce genetic damage *in vivo* in peripheral blood lymphocytes at low doses.

In contrast to mustard agents, the genetic toxicology of Lewisite has been poorly studied. Lewisite induces chromosome aberration in one type of cellular assay. It appears not to be mutagenic in *Salmonella*.

Sulfur mustard produces a variety of cancers through different exposure routes. It produced skin malignancies in chamber exposure in rats. Intravenous injection produced a significant increase in pulmonary tumors in highly susceptible strain A mice. Subcutaneous injections produced sarcomas and other tumors at the injection site in C3H, C3Hf, and strain A mice, but did not produce an increase of tumors at other sites.

Nitrogen mustard, particularly HN2, was more widely tested and

found to be a carcinogen. It produced pulmonary tumors from both intravenous and intraperitoneal injections in strain A mice. Subcutaneous exposures produced injection site tumors and pulmonary tumors in selected strains of mice. Its carcinogenic potency appeared to be similar to sulfur mustard, and it was one of the most potent carcinogens amongst the alkylating agents tested in the strain A bioassay program.

Evidence indicates that occupational exposure to sulfur mustard is associated with respiratory tract cancer. The battlefield experience is somewhat more equivocal, although the lung cancer excess is suggestive of an association.

Evidence from therapeutic use of nitrogen mustard clearly indicates a causal association with skin cancer (see Chapter 9) and leukemia. An excess of skin cancer or leukemia was not evident in the occupational or battlefield studies.

The weight of the evidence—cellular, epidemiological, and toxicologic—indicates a causal association between sulfur mustard exposure and the occurrence of excess respiratory cancer, and skin cancer, and possibly leukemia. Inadequate exposure information limits accurate estimation of the cancer excesses that may be expected. The evidence is insufficient to indicate a causal relationship for Lewisite carcinogenesis.

Based on the foregoing, the committee concludes that human subjects of the WWII chamber tests are probably at increased risk of respiratory tract and skin cancer. This conclusion is based upon estimates of exposure to sulfur mustard that occurred among the subjects of the chamber tests (see Chapter 3), which approximated the battlefield exposure of surviving WWI soldiers. Studies of WWI gassing victims demonstrate a suggestive excess of cancer of the respiratory tract. Limitations on information on exposure of study subjects and WWI gassing victims limit the precision of this risk projection.

REFERENCES

Abell CW, Falk HL, Shimkin MB. 1965. Uracil mustard: a potent inducer of lung tumors in mice. Science 147:1443-1445.

Ashby J, Tinwell H, Callander RD, Clare N. 1991. Genetic activity of the human carcinogen sulfur mustard toward *Salmonella* and the mouse bone marrow. Mutation Research 257:307-312.

Auerbach C. 1943. Chemically induced mutations as rearrangements. Drosophila Information Service 17:48-50.

Auerbach C. 1951. The effect of oxygen concentration on the mutagenic action of mustard gas. Kurze Mitteilungen 15:341-342.

Auerbach C, Moser H. 1950. Production of mutations by monochloro-mustards. Nature 166:1019-1020.

Auerbach C, Robson JM. 1946. Chemical production of mutations. Nature 157:302.

Auerbach C, Robson JM. 1947. Production of mutations by chemical substances. Proceedings of the Royal Society of Edinburgh. Section B: Biology 62:271-283.

Auerbach C, Sonbati EM. 1960. Sensitivity of the *Drosophila* testis to the mutagenic action of mustard gas. Zeitschrift fur Vererbungslehre 91:237-252.

Ball CR, Roberts JJ. 1970. DNA repair after mustard gas alkylation by sensitive and resistant Yoshida sarcoma cells *in vitro*. Chemico-Biological Interactions 2:321-329.

Ball CR, Roberts JJ. 1972. Estimation of interstrand DNA cross-linking resulting from mustard gas alkylation of HeLa cells. Chemico-Biological Interactions 4:297-303.

Barrett JC, Lamb PW, Wang TC, Lee TC. 1989. Mechanisms of arsenic-induced cell transformation. Biological Trace Element Research 21:421-429.

Beckman L, Nordenson I. 1986. Interaction between some common genotoxic agents. Human Heredity 36:397-401.

Beebe G. 1960. Lung cancer in World War I veterans: possible relation to mustard gas injury and 1918 influenza epidemic. Journal of the National Cancer Institute 25:1231-1252.

Blayney DW, Longo DL, Young RC, Greene MH, Hubbard SM, Postal MC, Duffey PL, DeVita VT Jr. 1987. Decreasing risk of leukemia with prolonged followup after chemotherapy and radiotherapy for Hodgkin's disease. New England Journal of Medicine 316:710-714.

Boursnell JC, Cohen JA, Dixon M, Francis GE, Greville GD, Needham DM, Wormall A. 1946. Studies on mustard gas (b,b-dichlorodiethylsulphide) and some related compounds. 5. The fate of injected mustard gas (containing radioactive sulphur) in the animal body. Biochemical Journal 40:756-764.

Boyland E, Horning ES. 1949. The induction of tumours with nitrogen mustards. British Journal of Cancer 3:118-123.

Capizzi RL, Smith WJ, Field R, Papirmeister B. 1973. A host mediated assay for chemical mutagens using the L5178Y/Asn murine leukemia (abstract). Mutation Research 21:6.

Case RM, Lea AJ. 1955. Mustard gas poisoning, chronic bronchitis, and lung cancer: an investigation into the possibility that poisoning by mustard gas in the 1914-1918 war might be a factor in the production of neoplasia. British Journal of Preventive and Social Medicine 9:62-72.

Casto BC, Meyers J, DiPaolo JA. 1979. Enhancement of viral transformation for evaluation of the carcinogenic and mutagenic potential of inorganic metal salts. Cancer Research 39:193-198.

Cherney BW, McBride OW, Chen C, Alkhatib H, Bhatia K, Hensley P, Smulson ME. 1987. cDNA sequence, protein structure, and chromosomal location of the human gene for poly(ADP-ribose) polymerase. Proceedings of the National Academy of Sciences (USA) 84:8370-8374.

Colvin M, Chabner BA. 1990. Alkylating agents. In: Chabner BA, Collins JM, eds. Cancer Chemotherapy: Principles and Practice. Philadelphia: J.B. Lippincott.

Conklin JW, Upton AC, Christenberry KW. 1965. Further observations on late somatic effects of radiomimetic chemicals and X-rays in mice. Cancer Research 25:20-28.

Davison C, Rozman RS, Bliss L, Smith PK. 1957. Studies on the metabolic fate of bis(2-chloroethyl) sulfide (mustard gas) in the mouse and human. Proceedings of the American Association for Cancer Research 2:195.

Davison C, Rozman RS, Smith PK. 1961. Metabolism of bis-b-chloroethyl sulfide (sulfur mustard gas). Biochemical Pharmacology 7:65-74.

Easton D, Peto J, Doll R. 1988. Cancers of the respiratory tract in mustard gas workers. British Journal of Industrial Medicine 45:652-659.

Einhorn N. 1978. Acute leukemia after chemotherapy. Cancer 41:444.

Epstein J. 1984. Nitrogen mustard (mechlorethamine) and UVB photocarcinogenesis: a dose-response effect. Journal of Investigative Dermatology 83:320-322.

Fahmy OG, Fahmy MJ. 1972. Mutagenic selectivity for the RNA-forming genes in relation to the carcinogenicity of alkylating agents and polycyclic aromatics. Cancer Research 32:550-557.

Fisher B, Rockette H, Fisher ER, Wickerham DL, Redmond C, Brown A. 1985. Leukemia in breast cancer patients following adjuvant chemotherapy or postoperative radiation: the NSABP experience. Journal of Clinical Oncology 3:1640-1658.

Fox BW, Fox M. 1973. The nature of the resistance to methylene dimethane sulphonate in the Yoshida sarcoma (abstract). British Journal of Cancer 28:81.

Gilbert RM, Rowland S, Davison CL, Papirmeister B. 1975. Involvement of separate pathways in the repair of mutational and lethal lesions induced by a monofunctional sulfur mustard. Mutation Research 28:257-275.

Greene MH, Boice JD Jr, Greer BE, Blessing JA, Dembo AJ. 1982. Acute nonlymphocytic leukemia after therapy with alkylating agents for ovarian cancer. New England Journal of Medicine 307:1416-1421.

Griffin AC, Brandt EL, Tatum EL. 1950. Nitrogen mustards as cancer-inducing agents. Journal of the American Medical Association 144:571.

Gross CL, Watson CV, Petrali J, Papirmeister B. 1981. Effect of sulfur mustards on lysosomes from rat liver *in vitro*. Toxicology and Applied Pharmacology 61:147-151.

Habraken Y, Ludlum D. 1989. Release of chloroethyl ethyl sulfide-modified DNA bases by bacterial 3-methyladenine-DNA glycosylases I and II. Carcinogenesis 10:489-492.

Heston WE. 1949. Induction of pulmonary tumors in strain A mice with methyl-bis(ßchloroethlyl) amine hydrochloride. Journal of the National Cancer Institute 10:125-130.

Heston WE. 1950. Carcinogenic action of the mustards. Journal of the National Cancer Institute 11:415-423.

Heston WE. 1953a. Occurrence of tumors in mice injected subcutaneously with sulfur mustard and nitrogen mustard. Journal of the National Cancer Institute 14:131-140.

Heston WE. 1953b. Pulmonary tumors in Strain A mice exposed to mustard gas. Proceedings of the Society for Experimental Biology and Medicine 82:457-460.

Heston WE, Lorenz E, Deringer MK. 1953. Occurrence of pulmonary tumors in strain A mice following total body x-radiation and injection of nitrogen mustard. Cancer Research 13:573-577.

Husain I, van Houten B, Thomas DC, Sancar A. 1986. Sequences of *E. coli* uvrA gene and protein reveal two potential ATP binding sites. Journal of Biological Chemistry 261:4895-4901.

Inada S, Hiragun K, Seo K, Yamura T. 1978. Multiple Bowen's diseaseobserved in former workers of a poison gas factory in Japan with special reference to mustard gas exposure . Journal of Dermatology 5:49-60.

Jackson R, Adams RH. 1973. Horrifying basal cell carcinoma: a study of 33 cases and a comparison with 435 non-horror cases and a report on four metastatic cases. Journal of Surgical Oncology 5:431-463.

Jacobson-Kram D, Montalba D. 1985. The Reproductive Effects Assessment Group's report on the mutagenicity of inorganic arsenic. Environmental Mutagenesis 7:787-804.

Jensen KA, Kirk I, Westergaard M. 1950. Mutagenic activity of some mustard gas compounds. Nature 166:1020-1021.

Jones JS, Weber S, Prakash L. 1988. The *Saccharomyces cerevisiae* RAD18 gene involves a protein that contains potential zinc finger domains for nucleic acid binding and a putative nucleotide binding sequence. Nucleic Acids Research 16:7119-7131.

Joshi S, Hughes JB. 1981. Inhibition of coupling factor B activity by cadmium ion, arsenite-2, 3-dimercaptopropanol, and phenylarsine oxide, and preferential reaction by dithiols. Journal of Biological Chemistry 256:11112-11116.

Jostes RF, Rausch RJ, Miller BM, Sasser LB, Dacre JC. 1989. Genotoxicity of Lewisite in Chinese hamster ovary cells (abstract). Toxicologist. 232.

Knowles FC, Benson AA. 1983. The biochemistry of arsenic. Trends in Biomedical Sciences 8:178.

Kohn KW, Steibigel NH, Spears CL. 1965. Cross-linking and repair of DNA in sensitive and resistant strains of *E. coli* treated with nitrogen mustard. Proceedings of the National Academy of Sciences (USA) 53:1154-1160.

Koletsky AJ, Bertino JR, Farber LR, Prosnitz LR, Kapp DS, Fischer D, Portlock CS. 1986. Second neoplasms in patients with Hodgkin's disease following combined modality therapy: the Yale experience. Journal of Clinical Oncology 4:311-317.

Kurozumi S, Harada Y, Sugimoto Y, Sasaki H. 1977. Airway malignancy in poisonous gas workers. Journal of Laryngology and Otology 91:217-225.

Kyle RA, Pierce RV, Bayrd ED. 1970. Multiple myeloma and acute myelomonocytic leukemia. New England Journal of Medicine 283:1121.

Lawley PD, Brookes P. 1965. Molecular mechanism of the cytotoxic action of difunctional alkylating agents and of resistance to this action. Nature 206:480-483.

Lawley PD, Brookes P. 1968. Cytotoxicity of alkylating agents towards sensitive and resistant strains of *Escherichia coli* in relation to extent and mode of alkylation of cellular macromolecules and repair of alkylation lesions in deoxyribonucleic acids. Biochemical Journal 109:433-447.

Lee TC, Lee KC, Tzeng YJ, Huang RY, Jan KY. 1986a. Sodium arsenite potentiates the clastogenicity and mutagenicity of DNA cross-linking agents. Environmental Mutagenesis 9:119-128.

Lee TC, Tzeng YJ, Chang WJ, Lin YC, Jan KY. 1986b. Posttreatments with sodium arsenite during G_2 enhance the frequency of chromosomal aberrations induced by S-dependent clastogens. Mutation Research 163:263-269.

Lee TC, Tanaka N, Lamb PW, Gilmer TM, Barrett JC. 1988. Inductionof gene amplification by arsenic . Science 241:79-81.

Lee WR. 1975. Comparison of the mutagenic effects of chemicals and ionizing radiation using *Drosophila melanogaster* test systems. In: Nygaard OF, Adler HI, Sinclair WK, eds. Radiation Research: Biomedical, Chemical, and Physical Perspectives. New York: Academic Press. 976-983.

Lopez S, Miyashita Y, Simons SS. 1990. Structural based selective interaction with arsenite with steroid receptors. Journal of Biological Chemistry 265:16039-16042.

Ludlum DB, Metha JR, Steiner RF, DeWitt J. 1978. Physical properties of poly(3,N4ethenocytidylic acid). Biophysical Chemistry 7:339-346.

Ludlum DB, Tong WP, Mehta JR, Kirk MC, Papirmeister B. 1984. Formation of O^6 ethylthioethyldeoxyguanosine from the reaction of chloroethylethyl sulfide with deoxyguanosine. Cancer Research 44:5698-5701.

Ludlum DB, Kent S, Mehta JR. 1986. Formation of O^6-ethylthioethylguaninein DNA by reaction with the sulfur mustard, chloroethyl sulfide, and its apparent lack of repair by O^6-alkylguanine DNA alkyltransferase . Carcinogenesis 7:1203-1206.

Luening KG. 1951. Mustard gas and gynandromorph production in *Drosophila melanogaster.* Hereditas 37:488-500.

Manning KP, Skegg DCG, Stell PM, Doll R. 1981. Cancer of the larynx and other occupational hazards of mustard gas workers. Clinical Otolaryngology and Allied Sciences 6:165-170.

Mauro F, Elkind MM. 1968. Differences in survival variations during the growth cycle of cultured Chinese hamster cells treated with sulfur mustard and x-rays. Cancer Research 28:1150-1155.

McNamara BP, Owens EJ, Christensen MK, Vocci FJ, Ford DF, Rozimarek H. 1975. Toxicological Basis for Controlling Levels of Mustard in the Environment. Edgewood Arsenal Special Publication EB-SP-74030. Aberdeen Proving Ground, Maryland: U.S. Army Armament Command. Edgewood Arsenal Biomedical Laboratory.

Meier HL, Gross CL, Papirmeister B. 1984. The use of human models for validating the biochemical mechanism of mustard-induced injury and for developing and evaluating therapeutic regimens to prevent mustard gas incapacitation.In: Proceedings of the Army Science Conference. West Point, NY.

Murnane JP, Byfield JE. 1981. Irrepairable DNA cross-links and mammalian cell lethality with bifunctional alkylating agents. Chemico-Biological Interactions 38:75-76.

Nasrat GE. 1954. Some cytological observations on the delayed effect of mustard gas. Nature 174:968-969.

Nasrat GE, Kaplan WD, Auerbach C. 1954. A quantitative study of mustard gas induced chromosome breaks and rearrangements in *Drosophila melanogaster*. Zeitschrift fur Induktive Abstammungs-und Vererburgslehre 86:249-262.

Nishimoto Y, Yamakido M, Shigenobu T, Onari K, Yukutake M. 1983. Long-term observation of poison gas workers with special reference to respiratory cancers. Sangyo Ika Daigaku Zasshi 5(Suppl):89-94.

Nishimoto Y, Yamakido M, Ishioka S, Shigenobu T, Yukutake M. 1988. Epidemiological studies of lung cancer in Japanese mustard gas workers. International Symposium of the Princess Takamatsu Cancer Research Fund. Vol. 18, Unusual Occurrences as Clues to Cancer Etiology. 95-101.

Nordenson I, Beckman G, Beckman L, Nordstrom S. 1978. Occupational and environmental risks in and around a smelter in northern Sweden. II. Chromosomal aberrations in workers exposed to arsenic. Hereditas 88:47-50.

Norman JE Jr. 1975. Lung cancer mortality in World War I veterans with mustard gas injury, 1919-1965. Journal of the National Cancer Institute 54:311-318.

Papirmeister B, Davison CL. 1964. Elimination of sulfur mustard-induced products from DNA of *Escherichia coli*. Biochemical and Biophysical Research Communications 17:608617.

Papirmeister B, Westling AW, Schroer J. 1969. Mustard: The Relevance of DNA Damage to the Development of the Skin Lesion. EATR-4294. Edgewood Arsenal, MD: Medical Research Laboratory. DTIC 688-866.

Papirmeister B, Dorsey JK, Davison CL, Gross CL. 1970. Sensitization of DNA to endonuclease by adenine alkylation and its biological significance (abstract). Federation Proceedings 29:726A.

Papirmeister B, Gross CL, Petrali JP, Hixson CJ. 1984a. Pathology produced by sulfur mustard in human skin grafts on athymic nude mice. 1. Gross and light microscopic changes. Journal of Toxicology—Cutaneousand Ocular Toxicology 3:371-392.

Papirmeister B, Gross CL, Petrali JP, Meier HL. 1984b. Pathology produced by sulfur mustard in human skin grafts on athymic nude mice. 2. Ultrastructural changes. Journal of Toxicology —Cutaneous and Ocular Toxicology 3:393-408.

Plant JE, Roberts JJ. 1971. Extension of the pre-DNA synthetic phase of the cell cycle as a consequence of DNA alkylation in Chinese hamster cells: a possible mechanism of DNA repair. Chemico-Biological Interactions 3:343-351.

Popa NE. 1969. Dynamics and spectrum of structural chromosome mutations of *Vicia faba* L. var— minor under the effect of mustard gas. Tsitologiya i Genetika 3:136-141. [In Russian]

Price CC, Gaucher GM, Konero P, Shibakowa R, Sowa JR, Yamaguchi M. 1968. Relative reactivities for monofunctional nitrogen mustard alkylation of nucleic acid components. Acta Biochimica et Biophysica Hungarica 166:327-359.

Reid BD, Walker IG. 1966. Resistance to sulfur mustard: a comparison of some properties of strain L cells and a resistant subline. Cancer Research 26:1801-1805.

Reid BD, Walker IG. 1969. The response of mammalian cells to alkylating agents. II. On the mechanism of the removal of sulfur mustard-induced cross links. Biochimica et Biophysica Acta 179:179-188.

Roberts JJ, Kotsaki-Kovatsi VP. 1986. Potentiation of sulfur mustard or cisplatin-induced toxicity by caffeine in Chinese hamster cells correlates with formation of DNA double-strand breaks during replication on a damaged template. Mutation Research 165:207-220.

Roberts JJ, Warwick GP. 1963. Studies of the mode of action of alkylating agents. 6. The metabolism of bis-2-chloroethylsulphide (mustard gas) and related compounds. Biochemical Pharmacology 12:1329-1334.

Roberts JJ, Brent TP, Crathorn AR. 1968. Mechanism of the cytotoxic action of alkylating agents on mammalian cells. In: Campbell PN, ed. Interaction of Drugs and Subcellular Components in Animal Cells. Boston: Little, Brown. 5-27.

Roberts JJ, Brent TP, Crathorn AR. 1971. Evidence for the inactivation and repair of the mammalian DNA template after alkylation by mustard gas and half mustard gas. European Journal of Cancer 7:515-524.

Roberts JJ, Friedlos F, Scott D, Ormerod MG, Rawlings CJ. 1986. The unique sensitivity of Walker rat tumour cells to difunctional agents is associated with a failure to recover from inhibition of DNA synthesis and increased chromosome damage. Mutation Research 166:169-181.

Rosner F, Grunwald H. 1975. Multiple myeloma terminating in acute leukemia. American Journal of medicinae 58:339.

Rossman TG. 1981. Enhancement of UV-mutagenesis by low concentrations of arsenite in *E. coli.* Mutation Research 91:207-211.

Savage JRK, Breckon G. 1981. Differential effects of sulfur mustard on S phase cells of primary fibroblast cultures from Syrian hamsters *Mesocricetus-auratus.* Mutation Research 84:375-387.

Scott D, Bigger TRL. 1972. The induction of chromosomal aberrations by sulphur mustard in marsupial lymphocytes. Chromosomes Today 3:162-176.

Scott D, Fox M, Fox BW. 1974. The relationship between chromosomal aberrations, survival, and DNA repair in tumour cell lines of differential sensitivity to x-rays and sulphur mustard. Mutation Research 22:207-221.

Shimkin MB, Weisburger JH, Weisburger EK. 1966. Bioassay of 29 alkylating chemicals by the pulmonary-tumor response in strain A mice. Journal of the National Cancer Institute 36:915-935.

Simons SS, Chakraborti KK, Cavanaugh AH. 1990. Arsenite and cadmium(II) as probes of glucocorticoid receptor structure and function. Journal of Biological Chemistry 265:1938-1945.

Sobels FH. 1962. Rates of forward and reverse mutation in *Drosophila* after exposure to mustard gas and x-rays. Genetica 33:31-44.

Sobels FH, Van Steenis H. 1957. Chemical induction of crossing-over in *Drosophila* males. Nature 179:29-31.

Somani SM, Babu SR. 1989. Toxicodynamics of sulfur mustard. International Journal of Clinical Pharmacology, Therapy and Toxicology. 27:419-435.

Sonbati EM, Auerbach C. 1960. The brood pattern for intragenic and intergenic changes after mustard gas treatment of *Drosophila* males. Zeitschrift fur Vererbungslehre 91:253-258.

Stevens CM, Mylroie A. 1950. Mutagenic activity of b-chloroalkyl amines and sulphides. Nature 166:1019.

Stewart DL, Sass EJ, Fritz LK, Sasser LB. 1989. Toxicology studies on Lewisite and sulfur mustard agents: mutagenicity of Lewisite in the *Salmonella* histidine reversion assay. Richland, WA: Pacific Northwest Laboratory. AD-A213-146.

Tanaka K, Miura N, Satokata I, Miyamoto I, Yoshida MC, Satoh Y, Kondo S, Yasui A, Okayama H, Okada Y.1990. Analysis of a human DNA excision repair gene involved in group A xeroderma pigmentosum and containing a zinc-finger domain. Nature 348:73-76.

Tucker MA, Coleman CN, Cox RS. 1988. Risk of second cancers after treatment for Hodgkin's disease. New England Journal of Medicine 318:76-81.

Uchida K, Morita T, Sato T, Ogura T, Yamashita R, Noguchi S, Suzuki H, Nyunoya H, Miwa M, Sugimura T. 1987. Nucleotide sequence of afull-length cDNA for human fibroblast poly (ADP-ribose) polymerase . Biochemical and Biophysical Research Communications 148:617-622.

Wada S, Nishimoto Y, Miyanishi M, Katsuta S, Nishiki M. 1962. Review of Okuno-jima poison gas factory regarding occupational environment. Hiroshima Journal of Medical Sciences 11:75-80.

Wada S, Nishimoto Y, Miyanishi M, Kambe S, Miller RW. 1968. Mustard gas as a cause of respiratory neoplasia in man. Lancet 2:1161-1163.

Walker IG. 1966. Sulfur mustard: reaction with L-cells treated with 5-fluorodeoxyuridine. Science 151:99-101.

Walker IG. 1971. Intrastrand bifunctional alkylation of DNA in mammalian cells treated with mustard gas. Canadian Journal of Biochemistry 49:332-336.

Walker IG, Reid BD. 1971. Some properties of substrains of L-cells with a decreased sensitivity to bis(2-chloroethyl)sulfide. Cancer Research 31:510-515.

Walker IG, Smith JF. 1969. Protection of L-cells by thiols and againstthe toxicity of sulfur mustard . Canadian Journal of Physiology and Pharmacology 47:143-151.

Ward JR, Seider RP. 1984. Activation energy for the hydrolysis of bis(2-chloroethyl) sulfide. Thermochimica Acta 81:343-348.

Wheeler GP. 1962. Studies related to the mechanisms of action of cytotoxic alkylating agents: A review. Cancer Research 22:651-688.

Wiencke JK, Yager JW. 1992. Specificity of arsenite in potentiating cytogenetic damage induced by the DNA cross-linking agent diepoxybutane. Environmental and Molecular Mutagenesis 19:195-200.

Wiencke JK, Christiani DC, Kelsey KT. 1991. Bimodal distribution of sensitivity to SCE induction by diepoxybutane in human lymphocytes. I. Correlation with chromosomal aberrations. Mutation Research 248:17-26.

Wulf HC, Aasted A, Darre E, Niebuhr E. 1985. Sister chromatid exchanges in fishermen exposed to leaking mustard gas shells (letter). Lancet 1:690-691.

Yamada A. 1963. On the late injuries following occupational inhalation of mustard gas, with special references to carcinoma of the respiratory tract. Acta Pathologica Japonica 13:131-155.

Yanagida J, Hozawa S, Ishioka S, Maeda H, Takahashi K, Oyama T, Takahashi M, Hakoda M, Akiyama M, Yamakido M. 1988. Somatic mutation in peripheral lymphocytes of former workers at the Okuno-jima poison gas factory. Japanese Journal of Cancer Research 79:1276-1283.

Zackheim HS, Smuckler EA. 1980. Tumorigenic effect of topical mechlorethamine, BCNU and CCNU in mice. Experientia 36:1211-1212.

7

Nonmalignant Respiratory Effects of Mustard Agents and Lewisite

It is well documented that inhalation of sulfur mustard or Lewisite causes acute damage to the respiratory tract. In fact, pulmonary injury is the principal cause of mortality in the first few days to weeks after exposure to sufficiently high concentrations of sulfur mustard (Hosseini et al., 1989; Papirmeister et al., 1991; Willems, 1989).

The primary question at issue in this report, however, is whether repeated daily exposures to these agents, at the concentrations used in the World War II (WWII) testing programs, are associated with long-term respiratory effects. If a risk of chronic effects is judged to exist among these human subjects, a secondary question arises as to whether an acute response is a necessary precursor to the development of long-term respiratory effects.

There are few directly relevant studies available for answering these questions and, as discussed in Chapter 3, the long-term health status of the subjects has not been followed. Nevertheless, some data do exist on the chronic health effects among populations exposed to sulfur and nitrogen mustards in far different contexts. To make inferences to the WWII testing program experiments, however, would require knowing the exposure levels present in each of the contexts. Thus, the conclusions reached in Chapter 3 regarding the possible exposure levels of human subjects in the WWII testing programs become key aspects of this assessment.

This chapter deals with chronic, nonmalignant respiratory effects. It is based primarily on a review of the epidemiologic and clinical literature on chronic effects of exposures incurred in the manufacture of mustard agents, in combat, and in medical therapeutic situations. Finally, in an attempt to understand the relationship between the acute and chronic responses to these agents, other respiratory hazards known to cause both acute and chronic effects are also reviewed.

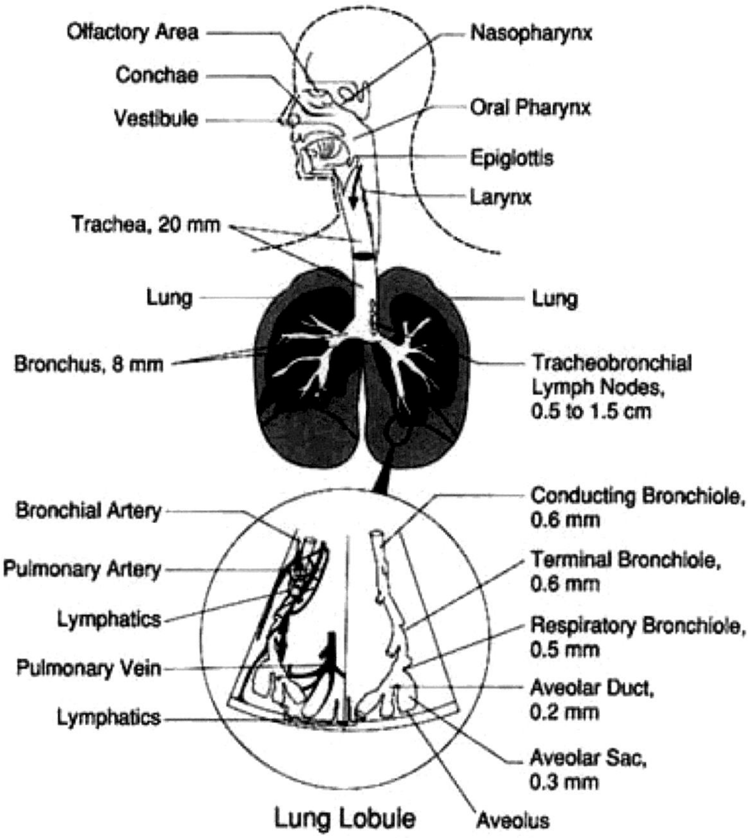

FIGURE 7-1 Structure of the respiratory tract. SOURCE: National Research Council, 1979.

PHYSIOLOGY AND ANATOMY

The basic anatomy of the respiratory tract is illustrated in Figure 7-1. The respiratory tract is lined with epithelial cells of different types, depending on their location and functions. A stratified squamous epithelium lines the nasal vestibule, followed by pseudostratified ciliated and ciliated columnar to cuboidal epithelium that lines the remainder of the nose, trachea, bronchi, and bronchioles (Figure 7-2). These cells are interspersed with nonciliated goblet and Clara cells. The cells are coated with a thin layer of mucus secreted by the goblet cells and

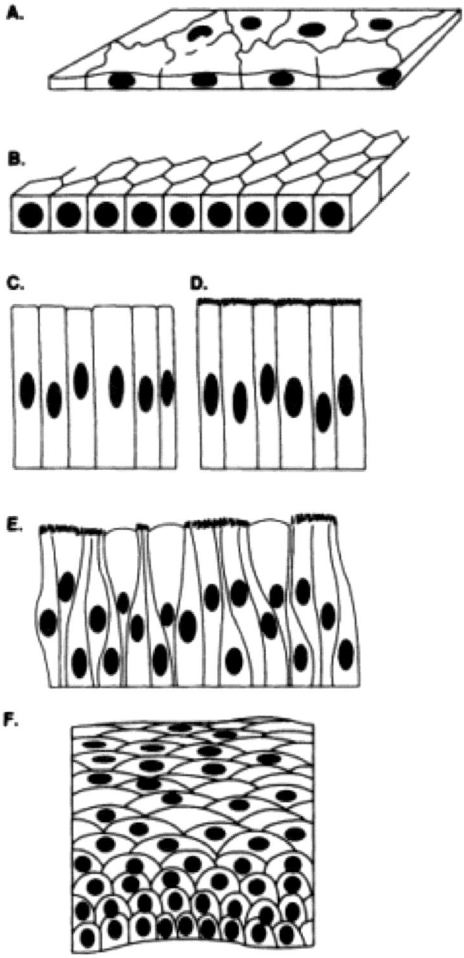

FIGURE 7-2 Illustration of the various types of cellular arrangements of epithelium that line internal body passages, including the respiratory system. Specialized cells, such as goblet cells, that secrete mucus into the passages are often interspersed with the epithelial cells. Clara cells take the place of goblet cells in certain parts of the respiratory system. A: simple squamous; B: simple cuboidal; C: simple columnar; D: ciliated columnar; E: pseudostratified ciliated columnar with goblet cells; and F: stratified squamous. SOURCE: Adapted from Stedman's Medical Dictionary, 1976, with permission from Williams & Wilkins.

mucous glands. The ciliated cells and the mucus serve as an escalator to move materials from the deep lung to the oral cavity where they are swallowed and excreted.

The conducting airways extend through the respiratory bronchioles, which terminate into the acini, consisting of alveolar ducts, alveolar sacs, alveoli and associated blood vessels, lymphatic tissues, supportive tissues, and nerve endings. The alveoli are the primary site of gas exchange with the blood, and are lined by two specialized types of epithelial cells (Type I and Type II alveolar epithelial cells). The total alveolar surface area in the adult human is about 100 m^2 during deep inspiration (Menzel and Amdur, 1986; Phalen and Prasad, 1988).

The most important type of inhalation exposure of the respiratory tract to sulfur mustard or Lewisite, including the chamber tests and field trials, has been to vapors of these compounds (Papirmeister et al., 1991). Inhaled vapors rapidly contact airway surfaces by the process of molecular diffusion (Lippmann, 1992). Surface uptake of the vapor then depends on the chemical properties of the inhaled compound. Highly reactive compounds with characteristics similar to sulfur mustard are generally removed higher in the respiratory tract and, thus, cause most of their damage in nasal, laryngeal and bronchial regions of the respiratory tract (Dahl, 1990; Dahl et al., 1991; Lippmann, 1992).

ACUTE EFFECTS AND BIOLOGICAL MECHANISMS

The inhalation of sulfur mustard or Lewisite causes acute damage to the respiratory tract, but the symptoms of exposure are not immediate and develop over a period of several days. Pulmonary injury is in fact the principal cause of mortality in the first few days to weeks after exposure to sulfur mustard (Hosseini et al., 1989; Papirmeister et al., 1991; Willems, 1989). The signs and symptoms of respiratory tract damage following inhalation of various levels of sulfur mustard are summarized in Table 7-1.

Damage to the respiratory tract involves acute edema (swelling), inflammation, and destruction of the airway epithelial lining. Depending on the dose, the destruction may be mild to severe. Severe damage includes destruction of the epithelium with subsequent formation of pseudomembranes (such as those formed in diphtheria infections), which may slough and obstruct the airway, resulting in death. In most cases, the injury is most severe in the larynx, trachea, and bronchi, with small bronchi less affected than large bronchi. The basement membranes that underlie the epithelium are edematous and are infiltrated by white blood cells. In some cases, presumably with high exposures, damage extends into the deeper alveolar regions, resulting in generalized edema of the lung. Finally, allergic hypersensitivity reactions to inhaled sulfur

TABLE 7-1 Effects of Acute Sulfur Mustard Vapor Exposure on the Human Respiratory Tract (at 16°C-27°C unless otherwise noted)

Estimated Exposure	Effects	Latency	Reference
12-70 mg·min/m^3	Hoarseness; nasal mucosa irritation; nondisabling; recovery in approximately 2 weeks	12 hours to 2 days	Project Coordination Staff, 1946
100-500 mgmin/m^3	Pulmonary effects evident (very little data)	Not available	Sidell, 1990
200 mg·min/m^3	Median incapacitating Ct (ICt_{50}). Upper airway effects of sneezing, lacrimation, rhinorrhea, sore throat, hoarseness, nosebleed, from which recovery usually occurs after two weeks. Lower airway effects of tracheobronchitis, hacking cough, rapid breathing, and pseudomembrane formation. Pulmonary edema, bronchial pneumonia, and other secondary infections may develop after 36-48 hours. Prolonged recovery (1-2 months), particularly after infections.	4 to 6 hours	Ganas, 1969; Stepanov and Popov, 1962
1,000-1,500 mg·min/m^3	1,500 mg·min/m^3 is estimated median lethal Ct (LCt_{50}). Injuries as for ICt_{50} above, progressing to edema in pharynx and tracheobronchial tree; death due to severe edema, secondary infection or necrotic bronchopneumonia. Incapacitating for survivors; recovery over period of several mnths.	24 to 48 hours	Robinson, 1967; Sidell, 1990; Stepanov and Popov, 1962; U.S. Army, 1974; U.S. Army CRDEC, 1990; U.S. Army and U.S. Air Force, 1975; WHO, 1970
150 mg/m^3 (10 min) 1,500 mg·min/m^3	Reported human lethal concentration; no temperature data	Not available	Back et al., 1972
70 mg/m^3 (30 min) 2,100 mg·min/m^3	Reported human lethal concentration; no temperature data	Not available	Inada et al., 1972

SOURCE: Adapted from Papirmeister et al., 1991; Watson and Griffin, 1992.

mustard vapors have not been well studied, but could be involved in cases of chronic exposures to sulfur mustard (Morgenstern et al., 1947; Papirmeister et al., 1991).

Infection of the respiratory tract resulting in bronchopneumonia is a common complication of respiratory tract injury from inhaled sulfur mustard. It appears that most deaths following inhalation of sulfur mustard result, at least in part, from pulmonary infection, often complicated by septicemia (Hosseini et al., 1989; Keshmiri, 1989; Papirmeister et al., 1991; Willems, 1989). Immunosuppression from systemic absorption of sulfur mustard probably plays a major role in the pathogenesis of these infections, although its importance relative to the local pulmonary injury is not well understood. Pulmonary infection and septicemia, despite modern antibiotic therapy, were still significant causes of death in Iranians taken to Europe for treatment (Willems, 1989).

The above descriptions of the acute effects come primarily from humans who died following World War I (WWI) gas attacks and from animals that died as a result of experimentally induced lesions (Papirmeister et al., 1991). There is little contemporary information regarding the pathogenesis of the respiratory lesions or few data from people or animals exposed to nonlethal concentrations of sulfur mustard vapor. There are even fewer studies in which the histopathology of the recovery process has been studied in animals exposed to sulfur mustard. However, two studies conducted during WWI suggest that low-level exposure or survivable exposures in dogs and rabbits may produce scar tissue following small ulcerations in the trachea and larynx, causing contractions of these areas (Warthin and Weller, 1919; Winternitz, 1919). Nevertheless, the descriptions of the more severe lesions of the respiratory tract of animals exposed to sulfur mustard vapor and those in humans appear to be quite similar in type and location (Papirmeister et al., 1991). Thus, further studies in animals conducted to follow the repair of the acute respiratory tract lesions from nonlethal, inhaled sulfur mustard vapor would be of value in determining the persistence and course of such lesions in people. However, it has been difficult to induce a chronic bronchitis that persists following cessation of exposure in animals (Greene et al., 1984).

The acute effects of inhaled Lewisite vapors are similar to those discussed above for inhaled sulfur mustard vapor. A major exception is that the irritating effect of Lewisite is immediately detectable by the exposed person. Data on the histopathological lesions in the respiratory tract from inhaled Lewisite are available only from animals, but the reported lesions are generally very similar to those discussed above for sulfur mustard. However, at high concentrations, Lewisite-induced pulmonary edema appears to be more prominent than with sulfur mustard. Hemoconcentration, presumably due to this pulmonary edema, has also been reported to be a prominent finding following inhalation of Lewisite (Urbanetti, 1988).

EVIDENCE OF LONG-TERM HEALTH EFFECTS

Animal Studies

Inhaled sulfur mustard vapor produces destruction of the epithelium of the respiratory tract. Apparently, much of the vapor is removed higher in the respiratory tract: the nasal, laryngeal, and tracheobronchial regions appear to be the most severely affected (Papirmeister et al., 1991). Many of the animal studies of toxicity from inhaled sulfur mustard were conducted during WWI. In rabbits, inhalation of sulfur mustard produced damage that was particularly prominent in the upper respiratory tract, including the nasal passages, pharynx, larynx, trachea, and large bronchi (Warthin and Weller, 1919). The damage increased with increasing exposure concentrations. Low levels of exposure caused congestion of these areas without hemorrhage. Degeneration of mucous cells was observed in the pharynx and larynx. The highest levels of exposure caused necrosis of the epithelium, infiltration of white blood cells, and the formation of diphtheritic-like pseudomembranes.

Experiments in dogs showed necrosis of the epithelium of the upper respiratory passages with pseudomembrane formation. This usually extended to the bronchioles. Animals that died from two to ten days following exposure had evidence of necrotizing pneumonia. Animals sacrificed at later times showed localized ulceration or constriction of the trachea (Winternitz, 1919).

Lung damage also occurred following intravenous injection of sulfur mustard into animals (Office of Scientific Research and Development, 1946). Intravenous injection of a solution of sulfur mustard in either propylene glycol or thiodiglycol was reported to cause diffuse pulmonary congestion and edema, but pure sulfur mustard given rapidly caused more serious necrotic and hemorrhagic lesions of the lung. However, these authors attributed this lung damage from injected sulfur mustard to localization of particulate sulfur mustard in the pulmonary capillaries, because pulmonary injury was not observed with other parenteral routes of administration.

Sulfur mustard is absorbed through the skin into the systemic circulation. Following intravenous administration of radiolabeled sulfur mustard in rabbits, the level of radioactivity in tissue was highest in kidney, followed by lung and then liver (Boursnell et al., 1946). Thus, it is conceivable that some sulfur mustard exposure of respiratory tract tissue with subsequent biological effects could occur following systemic absorption from skin exposure.

Human Studies

The clinical and epidemiologic literature on health effects of sulfur mustard covers three types of exposure situations: (1) chronic occupational exposures incurred in the manufacture of the mustard agent; (2) acute combat exposures; and (3) medical use as antitumor drugs.

Occupational Exposure

There is some relatively recent epidemiologic literature on occupational exposure to mustard agents in British and Japanese munitions factories. Several studies suggest that workers who were chronically exposed to mustard agents developed chronic nonmalignant respiratory effects (Easton et al., 1988; Manning et al., 1981; Nishimoto et al., 1970). In the British population, chronic respiratory disease has been reported to occur even among workers with only a few years of employment (Easton et al., 1988).

In a cohort mortality study of 511 employees at a manufacturing plant in England, a significant excess of deaths due to pneumonia was found (Manning et al., 1981). In a more extensive examination of this same plant, Easton examined the mortality patterns of an enlarged cohort of 3,500 workers and reported standardized mortality ratios (SMRs, measures of relative risk) for specific causes of death (Easton et al., 1988). Findings included statistically significant excesses of nonmalignant respiratory disease (SMR = 143), including subcategories of influenza and pneumonia (SMR = 143), bronchitis (SMR = 159), and asthma (SMR = 151). These excesses were present even among those with less than three years of employment at the plant, and were not related to duration of employment. The finding of excess mortality due to nonmalignant respiratory disease observed shortly after initial exposure is consistent with the follow-up studies of combat survivors, in which bronchitis and emphysema were found to be present within months of the acute exposure (see below).

Workers exposed to sulfur mustard and Lewisite in a Japanese production plant were surveyed for respiratory morbidity 25 years after production had ceased (Nishimoto et al., 1970). The survey included chronic symptoms and pulmonary function assessments, that is, forced vital capacity (FVC) and forced expiratory volume in 1 second (FEV1). The study group of 1,403 represented 62 percent of those identified from among the 5,000 total former employees of the plant. Based on an internal comparison, more highly exposed workers reported more chronic bronchitis and had slightly lower FEV1/FVC than either a less-exposed or unexposed clerical group of coworkers. Compared to groups of unexposed patients at a Chicago respiratory clinic and "ordinary" patients at a Japanese clinic, the group exposed to mustard agent and Lewisite had more chronic bronchitis, chronic cough, and reduced FEV1/FVC. Among nonsmokers, almost half of those previously exposed to mustard agents or Lewisite reported persistent productive cough, compared to 30 percent in the unexposed group. The former poison gas workers showed a more bronchitic type of airway obstructive pattern than did the ordinary patients with chronic obstructive pulmonary disease (COPD); however, they did not have less anatomical evidence of emphysema.[1] Based on this literature, the Agency for Toxic Substances and Disease Registry (ATSDR) concluded in the recent *Toxicological Profile for Mustard "Gas"* (1991) that, "when the vapors are breathed for a prolonged period, other respiratory diseases, such as chronic bronchitis . . . can eventually occur."

None of the epidemiologic studies of occupational exposures provided estimates of exposure level. It has been reported that worker protection was inadequate in the Japanese factories and that employees were exposed to significant levels of sulfur mustard, and to a lesser extent to Lewisite (Wada et al., 1962). Conditions may have been better, but still relatively poor, in British munitions factories (Haber, 1986). It is difficult to judge the safety conditions, however. For example, as discussed in Chapter 3, all sulfur mustard and Lewisite production in the United States was accomplished at military bases controlled by the Chemical Warfare Service. This group's safety record was the worst in the military during the peak years of production, with hundreds and probably thousands[2] of documented injuries resulting from sulfur mustard and Lewisite (Brophy and Fisher, 1959; Cochrane, 1946). Given this high incidence of injury, it is surprising that follow-up studies of chemical warfare production workers have not been conducted on the U.S. worker populations.

Battlefield Exposure

The risks associated with acute combat exposures are probably more relevant for predicting the likely long-term effects of the experimental chamber exposures. Chamber exposures were delivered over a period of

[1] Bronchitis is an inflammation of the mucous membrane lining the respiratory passages. Emphysema is a disease in which the air spaces of the lungs are widened and their walls or linings are destroyed. Although symptoms of these diseases may be similar and the diseases can commonly occur together, they are different anatomically. Both emphysema and bronchitis can be called chronic obstructive pulmonary diseases (COPDs).

[2] One thousand injuries of this type were reported for a two-year period just at Edgewood Arsenal.

days to weeks and were designed to simulate combat exposures in terms of concentration, humidity, and temperature.

World War I. In a 1922 clinical study of 83 pensioners with recognized disability due to gas poisoning, the principal symptom was shortness of breath (Sandall, 1922). Persistent cough, expectoration, and chest tightness were also frequent. Sandall reported that, on physical examination, 26 percent showed signs of "emphysema," while another 20 percent had some definite signs of bronchitis.

A similar clinical study published in the same year involved 166 sanatorium patients, who gave a definite history of having been gassed during the war and hospitalized for at least 20 days (Hankins and Klotz, 1922). Of these young adults, 25 percent gave a history of influenza or pneumonia, and all reported having never regained their health since the gassing. Shortness of breath, cough, and expectoration were common, as were asthmatic symptoms. Based on X-ray findings, the clinical picture was one of chronic peribronchitis, resulting from a permanent thickening of the bronchial mucous membrane.

Berghoff reported clinical data based on 2,000 U.S. servicemen who had been gassed with chlorine or sulfur mustard and were examined in the course of their related discharge (Berghoff, 1919). Of those exposed to sulfur mustard, most had been exposed in an explosive attack three to four months prior to discharge. Of the total group, 30 percent were diagnosed to have bronchitis, characterized by prolonged expiration and coarse moist rales. Another 22 percent had characteristics associated with emphysema, including a rigid chest, limited diaphragm movement, and impaired expansion.

Gilchrist and Matz (1933) selected 89 living and 53 deceased cases from among 1,016 U.S. servicemen who had been gassed with sulfur mustard during World War I. The 89 living cases were examined clinically and roentgenologically eight to ten years after exposure to sulfur mustard. The basis for selection of cases was the availability of full and well-documented histories of exposure to sulfur mustard, based on military records, and the severity of the effects, also from information in military records. These investigators concluded that 27 of the 89 men who were examined nine to ten years after gassing had evidence of anatomic or symptomatic disease attributable to their exposure to sulfur mustard. The residual effects on the respiratory tract they noted were chronic bronchitis, emphysema, and bronchial asthma.

There is also evidence that British soldiers exposed to mustard gas during combat in WWI had a higher incidence of chronic bronchitis than the general population (Case and Lea, 1955). One study, designed to examine for lung cancer, identified a group of 1,267 war pensioners who had suffered from sulfur mustard poisoning; 80 percent of the group had chronic bronchitis. Although this study group may not have been representative of the entire cohort of veterans with exposure to sulfur mustard, the findings strongly suggest an elevated risk of chronic bronchitis associated with combat exposure.

Iran-Iraq War. There are several more recent reports of chronic respiratory effects among people with combat exposure in the Iran-Iraq war. Hosseini and colleagues described 61 victims of sulfur mustard injury, between the ages of 15 and 30, seen two to four weeks following exposure (Hosseini et al., 1989). Twenty-one of these patients were followed for 15 months with spirometry and X-rays. The symptomatic profile of the patients indicated that all 61 had cough, 75 percent had sputum, and 62 percent experienced shortness of breath. An improvement in symptoms during follow-up was observed for one subject. Only 20 percent appeared to have "normal" patterns of pulmonary function, defined as predicted FVC and FEV1/FVC values greater than 80 percent. The FVC was less than this for 71 percent of those followed, and the FEV1/FVC was below this figure in 52 percent of the cases. The authors were unable to explain this observed inconsistent pattern of abnormalities. In another report of survivors of the Iran-Iraq conflict, Somani and Babu (1989) described delayed effects, present two years after the exposure, which included chronic bronchitis and recurrent pneumonia.

In summary, the literature on the longer-term respiratory effects of acute combat exposures suggests the presence of significant chronic lung disease of both an obstructive and a restrictive nature. Yet, most of these studies were based on individuals who had sustained an initial acute injury with documented symptomatology. There is no direct examination of whether such chronic effects can occur in the absence of an observable acute response.

Medical Therapeutic Exposure

A variety of chemicals, including antitumor drugs with similarities to sulfur mustard, are known to damage lung tissue following systemic administration. Table 7-2 is a listing of antitumor drugs for which there is good evidence of pulmonary toxicity, particularly pulmonary fibrosis, following their administration in patients (Muggia, 1983; Rosenow et al., 1985; Smith and Walker, 1990; Weiss and Muggia, 1980). Although five of these drugs are alkylating agents (as is sulfur mustard), the alkylating agents nitrogen mustard and thio TEPA have not been reported to produce pulmonary fibrosis (Weiss and Muggia, 1980). Thus, the question of pulmonary effects following absorption of sulfur mustard, or for that matter for Lewisite, from systemic absorption following an exposure to skin, cannot be answered with available information.

TABLE 7-2 Antitumor Drugs That Produce Pulmonary Fibrosis

Classical Alkylating Agents	Antibiotics	Nitrosoureas	Miscellaneous
Busulfan	Bleomycin	Carmustine	Cytosine arabinoside
Chlorambucil	Mitomycin	Semustine	Methotrexate
Cyclophosphamide	Zinostatin	Chlorozotocin	6-Mercaptopurine
Melphalan		Lomustine	Procarbazine
Uracil mustard			VM-26 Vinblastine
			Vincristine

SOURCES: Muggia, 1983; Rosenow et al., 1985; Smith and Walker, 1990.

RELATIONSHIP BETWEEN ACUTE AND CHRONIC EFFECTS CAUSED BY EXPOSURES TO OTHER RESPIRATORY HAZARDS

There is clear evidence from descriptions of severe overexposure to sulfur mustard that exposure can cause both acute and chronic respiratory disease. However, there is little information available on the effects of lower levels of acute exposure, and apparently no information on the relatively brief exposures to levels that typify the experience in the chamber and some field studies.

Even in the absence of follow-up studies that directly address the question, however, it can still be pursued by examining indirect evidence from studies of compounds that might behave similarly. To evaluate the respiratory health risk associated with repeated brief overexposures at nonlethal levels, the committee requested an expert review of indirect evidence focusing on the link between acute and chronic respiratory responses to similar agents (see Appendix J). The review included exposures to irritant gases (chlorine, sulfur dioxide, and combustion products), materials of plant origin (cotton), chemicals (isocyanates), and inorganic dusts (silica, beryllium, and asbestos).

Sources of similar exposures include accidental overexposures to irritant gases, which have occurred in the workplace as well as the general community. Becklake and colleagues have reviewed the literature on acute and chronic effects associated with massive accidental exposures to sulfur dioxide, chloride, and oxides of nitrogen (Becklake et al., 1988). In 16 studies of sulfur dioxide, persistent abnormalities were consistently reported, including airflow limitation, with symptoms lasting up to 10 years following a single overexposure (Haskonen et al., 1979). For oxides of nitrogen, persistent abnormalities were less consis tent; for overexposure to chlorine, abnormalities were found to be rarely persistent. In a community study based on a population exposure to chlorine, subjects exposed during the derailment of a tank car were followed up six years after the accident. The rate of annual decline in FEV1 was found to be unrelated to the initial clinical status of the subjects (Jones et al., 1986).

Becklake concludes that none of the many case reports, nor the one community study, provides evidence to suggest that the pathophysiologic process induced by the exposure to an irritant gas was progressive. Persistent pulmonary damage did not appear to be determined by the degree of acute response at the time of the exposure, or by the extent of initial recovery that occurred during the several months following the exposure.

Exposures to cotton dust or components of the dust have been noted to cause both acute and chronic effects. The acute effects are represented by both a disabling respiratory syndrome (byssinosis) and a cross-shift decrement in FEV1. Chronic effects include both accelerated annual decrements in FEV1 and chronic bronchitis. The acute syndrome does not necessarily remit on removal from exposure, and there is a growing body of evidence to suggest a relationship between the acute effects and chronic airflow limitation. There is evidence that short-term exposures result in acute effects (Castellan et al., 1987; Martin and Higgins, 1976), but no evidence for whether such exposures, with or without an acute response, ultimately lead to chronic effects.

Isocyanates have been noted to cause several acute conditions, including chemical bronchitis and allergic bronchoconstriction (Axford et al., 1976; Brooks et al., 1985) and large dose-related cross-shift losses in FEV1 (Peters et al., 1968). The chronic respiratory effects caused by isocyanates include accelerated loss in pulmonary function over several years, suggesting the development of chronic airway limitation, and irreversible asthma (Peters and Wegman, 1975). There is also evidence that the presence of acute effects predicts both the development of chronic asthma and an accelerated rate of loss in lung function in nonasthmatics. Finally, there is evidence that short-term exposures to isocyanates can cause acute responses that are irreversible and progressive, but there is no evidence as to whether such short-term exposures without acute response result in irreversible respiratory effects.

Beryllium exposures have also been associated with both acute and chronic pulmonary disease, both of which have been shown to be disabling (AMA Archives of Industrial Health, 1959). Relevant to the present question, beryllium exposures, in one instance as brief as one week and in several instances occurring for less than ten weeks, have resulted in disease certified for inclusion in the registry of cases of beryllium disease established at the Massachusetts General Hospital (Hall et al., 1959).

Silica exposures are generally considered important only for chronic disease production. However, the condition of acute silicosis is caused by relatively short-term, high-intensity exposures to very fine silica particulate (Jones et al., 1975). Exposures as brief as six months have been associated with acute silicosis. Yet even in the absence of such severe overexposures and acute clinical disease, there is evidence that low-level exposures, not associated with clinical symptoms or chest X-ray abnormality, can result in irreversible pathology (Craighead and Vallyathan, 1980).

Similarly, asbestos exposures are noted for causing chronic respiratory illness, as well as cancer. Brief high exposures are not recognized to cause respiratory complaints. However, relatively short-term exposures of less than four years have been associated with increased respiratory symptoms and decreased vital capacity (Rodriques-Roisin et al., 1986).

It should be noted that beryllium, asbestos, and silica are relatively insoluble particles that remain in the lungs for months to years following inhalation. Thus even a brief exposure results in a chronic exposure of lung tissue to the agent.

Based on this review, a set of answers has been suggested for the following key questions regarding sulfur mustard and Lewisite exposure and chronic nonmalignant pulmonary disease. However, it should be kept in mind that there is a striking absence of knowledge about the early stages of environmentally related pulmonary diseases and almost no knowledge on the natural (longitudinal) evolution of the clinical conditions. Thus, there is little to guide us towards the appropriate disease model or towards identification of the relevant pattern(s) of exposure for disease etiology.

Does the occurrence of an acute pulmonary reaction following exposure to toxic chemicals identify an individual at risk for long-term respiratory sequelae? Ample evidence has been provided for all of the agents reviewed that a significant portion of individuals who react acutely to short, high exposures (and even some to short, relatively low exposures) go on to develop a variety of long-term respiratory effects. The agents differ in the probability of long-term adverse outcomes, but none appear to be free of this risk. The literature reviewed does not allow identification of a minimum magnitude of acute response necessary in order for there to be long-term sequelae.

If acute pulmonary reactions can identify individuals at risk for long-term sequelae of chemical exposures, can the probability or degree of damage be predicted from the magnitude of the acute

response? This question has not been adequately studied. An association with dose is found for acute as well as chronic responses, which provides support for the association, but not for a relationship between the *magnitude* of the acute response and the *magnitude* of the chronic response. If the disease model invoked requires the acute exposure to cause acute irreversible damage, then one might reasonably expect the magnitude of acute response to predict the magnitude of the chronic response. However, if another disease model is invoked, one in which the acute exposure results in or leads to an alteration in individual risk factors, then it is quite likely that the magnitude of the acute and chronic responses would be unrelated.

What assurance is there that the absence of an acute pulmonary reaction identifies chemically exposed individuals who will not develop long-term sequelae? This may be the most important question to ask and, unfortunately, the one for which no direct evidence could be found. The indirect evidence, however, would suggest the need to invoke an unusual disease model, one in which changes in individual risk factors could be excluded from the list of possible mechanisms, so that the absence of an acute reaction would eliminate the possibility of any chronic effects related to a short-term or acute exposure.

SUMMARY

Gaps

There are few directly relevant data for evaluating the risk of chronic nonmalignant respiratory disease associated with the specific exposure conditions present in the WWII chamber and field studies. The range of exposure concentrations that were actually inhaled by the subjects can be only crudely estimated, but they may have reached the levels experienced by those exposed in combat or production facilities. Without precise knowledge about the exposures, the only way to estimate the true risk would be to identify the cohort of test subjects and follow them for the occurrence of respiratory morbidity. To date, the entire cohort of subjects has not been identified, nor has any subset of them been followed for long-term health effects. The absence of a follow-up study of the experimental subjects from the WWII testing programs is the single largest gap in the literature relevant to the assessment of chronic nonmalignant respiratory disease risk in individuals experiencing short-term exposures to sulfur mustard or Lewisite.

Conclusions

The evidence indicates a causal relation between exposure to sufficient concentrations of sulfur mustard (and presumably nitrogen mustard and Lewisite) and chronic nonreversible respiratory effects in humans. It is well known that pulmonary injury is the principal cause of mortality in the first few days to weeks after intense exposure to sulfur mustard. In addition, the evidence is consistent with a causal relationship between occupational exposure to these agents and chronic obstructive lung diseases, including chronic bronchitis, asthma, and pneumonia. Finally, evidence from studies of combat survivors of gas attacks is consistent with a causal relationship between acute overexposure and bronchitis and emphysema. Further, given the likely exposure levels outlined in Chapter 3, **the evidence is consistent with a causal relation between the concentrations of sulfur mustard (and Lewisite) used in the WWII experiments and chronic nonreversible lung diseases.**

Indirect evidence, based on a review of the relationships between acute and chronic effects caused by other substances, suggests that the likelihood of long-term respiratory effects may not necessarily be linked to the presence of an acute respiratory response. Review of the evidence does not support a minimum magnitude of acute response necessary in order for there to be long-term sequelae. Further, if the disease model requires the acute exposure to cause acute irreversible damage, then the magnitude of acute response might well predict the magnitude of the chronic response. However, if acute exposure led to an alteration in individual risk factors, then it is possible that the magnitude of the acute and chronic responses would be unrelated. Finally, indirect evidence suggests that only an unusual disease model would exclude the possible mechanism of change in individual risk factors so that the absence of an acute reaction would eliminate the possibility of a chronic effect related to an acute exposure. Thus, there is insufficient evidence to conclude that long-term respiratory responses occur only in cases where an earlier acute response has been documented.

REFERENCES

Agency for Toxic Substances and Disease Registry (ATSDR). 1991. Draft, Toxicological Profile for Mustard "Gas." U.S. Department of Health and Human Services.

AMA Archives of Industrial Health. 1959. Beryllium disease and its control. Conference held at Massachusetts Institute of Technology, Sept. 30-Oct. 1, 1958. AMA Archives of Industrial Health 19.

Axford AT, McKerrow CB, Jones AP, Le Quesne PM. 1976. Accidental exposure to isocyanate fumes in a group of firemen. British Journal of Industrial Medicine 33:65-71.

Back KC, Thomas AA, MacEwen JD. 1972. Reclassification of material listed as transportation health hazards.Office of Hazardous Materials of the Assistant Secretary for Safety and Consumer Affairs, Department of Transportation. TSA-2072-3. PB-214 270/1.

Becklake MR, Bourbeau J, Menzies R, Ernst P. 1988. The relationship between acute and chronic airway responses to occupational exposure. In: Current Pulmonology. Chicago: Year Book Medical Publishers.

Berghoff RS. 1919. The more common gases: their effect on the respiratory tract. Observation on two thousand cases. Archives of Internal Medicine 24:678-684.

Boursnell JC, Francis GE, Wormall A. 1946. Studies on mustard gas (b,b-dichlorodiethylsulphide) and some related compounds. 6. The fate of injected b,b-dichlordiethylsulphone and b,b-dichlorodiethylsulphoxide (containing radioactive sulphur) in the animal body. Biochemical Journal 40:765-768.

Brooks SM, Weiss MA, Bernstein IL. 1985. Reactive airways dysfunction syndrome (RADS). Chest 88:376-384.

Brophy LP, Fisher G. 1959. The Chemical Warfare Service: Organizing for War. United States Army in World War II: The Technical Services. Washington, DC: Office of the Chief of Military History, Department of the Army.

Case RM, Lea AJ. 1955. Mustard gas poisoning, chronic bronchitis, and lung cancer: an investigation into the possibility that poisoning by mustard gas in the 1914-1918 war might be a factor in the production of neoplasia. British Journal of Preventive and Social Medicine 9:62-72.

Castellan RM, Olenchock SA, Kinsley KB, Hankinson JL. 1987. Inhaled endotoxin and decreased spirometric values. New England Journal of Medicine 317:605-610.

Cochrane RC. [1946]. Medical research in chemical warfare. Available through the U.S. Army Chemical Defense Research, Development and Engineering Center, Aberdeen Proving Ground, MD.

Craighead JE, Vallyathan NV. 1980. Cryptic pulmonary lesions in workers occupationally exposed to dust containing silica. Journal of the American Medical Association 244:1939-1941.

Dahl AR. 1990. Contemporary issues in toxicology. Toxicology and Applied Pharmacology 103:185-197.

Dahl AR, Schlesinger RB, Heck HD, Medinsky MA, Lucier GW. 1991. Comparative dosimetry of inhaled materials: differences among animal species and extrapolation to man. Fundamental and Applied Toxicology 16:1-13.

Easton D, Peto J, Doll R. 1988. Cancers of the respiratory tract in mustard gas workers. British Journal of Industrial Medicine 45:652-659.

Ganas P. 1969. New developments in chemical and biological warfare. Forces Aeriennes Francaises 24:449-475.

Gilchrist HL, Matz PB. 1933. The Residual Effects of Warfare Gases. Washington, DC: U.S. Government Printing Office.

Greene SA, Wolff RK, Hahn FF, Henderson RF, Mauderly JL, Lundgren DL. 1984. Sulfur dioxide induced chronic bronchitis in beagle dogs. Journal of Toxicology and Environmental Health 13:945-958

Haber L. 1986. The Poisonous Cloud. Oxford: Clarendon Press. 250-257 . Hall TC, Wood CH, Stoeckle JD, Tepper LB. 1959. Case data from the beryllium registry. AMA Archives of Industrial Health 19:100-103.

Hankins JL, Klotz WC. 1922. Permanent pulmonary effects of gas inwarfare . American Review of Tuberculosis 6:571-574.

Haskonen H, Nordman H, Korhonen O. 1979. Long-term effects of exposure to sulfur dioxide: lung function four years after a pyrite dust explosion. American Review of Respiratory Diseases 119:555-560.

Hosseini K, Moradi A, Mansouri A, Vessal K. 1989. Pulmonary manifestations of mustard gas injury: a review of 61 cases. Iranian Journal of Medical Sciences 14:20-26.

Inada S, Hiragun K, Seo K, Yamura T. 1978. Multiple Bowen's disease observed in former workers of a poison gas factory in Japan with special reference to mustard gas exposure. Journal of Dermatology 5:49-60.

Jones RN, Weill H, Ziskind M. 1975. Pulmonary function in sandblasters' silicosis. Bulletin de Physio-Pathologie Respiratoire 11:589-595.

Jones RN, Hughes JM, Glindmeyer H, Weill H. 1986. Lung function after acute chlorine exposure. American Review of Respiratory Disease 134:1190-1195.

Keshmiri M. 1989. Pulmonary cause of death from chemical warfare agents: the Halabche experience. Iranian Journal of Medical Sciences 14:10-19.

Lippman M, ed. 1992. Environmental Toxicants, Human Exposures and Their Health Effects. New York: Van Nostrand Reinhold.

Manning KP, Skegg DCG, Stell PM, Doll R. 1981. Cancer of the larynx and other occupational hazards of mustard gas workers. Clinical Otolaryngology and Allied Sciences 6:165-170.

Martin CVF, Higgins JE. 1976. Byssinosis and other respiratory ailments. Journal of Occupational Medicine 16:455-462.

Menzel DB, Amdur MO. 1986. Toxicology responses of the respiratory system. In: Klaassen CD, Amdur MO, Doull J, eds. Casarett and Doull's Toxicology. 3rd ed. New York: Macmillan.

Morgenstern P, Koss FR, Alexander WW. 1947. Residual mustard gas bronchitis: effects of prolonged exposure to low concentrations of mustard gas. Annals of Internal Medicine 26:27-40.

Muggia FM. 1983. Pulmonary toxicity of anti-tumor agents. Cancer Treatment Reviews 10:221-243.

National Research Council. 1979. Airborne Particles. Baltimore: University Park Press.

Nishimoto Y, Burrows B, Miyanishi M, Katsuta S, Shigenobu T, Kettel LJ. 1970. Chronic obstructive lung disease in Japanese poison gas workers. American Review of Respiratory Disease 102:173-179.

Office of Scientific Research and Development. National Defense Research Committee. 1946. Summary Technical Report of Division 9, NDRC. Chemical Warfare Agents and Related Chemical Problems. Parts I-VI. Washington, DC: NDRC. AD-234 249.

Papirmeister B, Feister AJ, Robinson SI, Ford RD. 1991. Medical Defense Against Mustard Gas: Toxic Mechanisms and Pharmacological Implications. Boca Raton, FL: CRC Press.

Peters JM, Wegman DH. 1975. Epidemiology of toluene diisocyanate (TDI) induced respiratory disease. Environmental Health Perspectives 11:97-100.

Peters JM, Murphy RLH, Pagnotto LD, Van Ganse WF. 1968. Acute respiratory effects in workers exposed to low levels of toluene diisocyanate (TDI). Archives of Environmental Health 16:642-647.

Phalen RF, Prasad SB. 1988. Morphology of the respiratory tract. In: McClellan RO, Henderson RF, eds. Concepts in Inhalation Toxicology. New York: Hemisphere.

Project Coordination Staff, Chemical Warfare Service. 1946. Technical Aspects of Chemical Warfare in the Field. 2 vols. Washington, DC: Chemical Warfare Service.

Robinson JP. 1967. Chemical warfare. Science Journal 4:33-40.

Rodriques-Roisin R, Picado C, Roca J, Arrigo S, Agusti-Vidal A. 1986. Early lung function changes after short heavy exposure to chrysotile asbestos in non-smoking women. Bulletin Europeen de Physiopathologie Respiratoire 22:225-229.

Rosenow EC, Wilson WR, Cockerill FR. 1985. Pulmonary disease in theimmunocompromised host . Mayo Clinic Proceedings 60:473-487.

Sandall TE. 1922. The later effects of gas poisoning. Lancet 2:857-859.

Sidell FR. 1990. Clinical Notes on Chemical Casualty Care. USAMRICD Technical Memorandum 90-1. Aberdeen Proving Ground, MD: U.S. Army Medical Research Institute of Chemical Defense.

Smith G, Walker J. 1990. The histopathology of pulmonary reactions to drugs. Clinics in Chest Medicine 11:95-117.

Somani SM, Babu SR. 1989. Toxicodynamics of sulfur mustard. International Journal of Clinical Pharmacology, Therapy and Toxicology. 27:419-435.

Stedman's Medical Dictionary. 1976. 23rd ed. Baltimore: Williams & Wilkins.

Stepanov AA, Popov VN. 1962. [Chemical Weapons and Principles of Antichemical Defense]. Translated by Joint Publications Research Service. JPRS 15107. Washington, DC: JPRS.

Urbanetti JS. 1988. Battlefield chemical injury. In: Loke J, ed. Pathophysiology and Treatment of Inhalation Injuries. New York: Marcel Dekker. 281-348.

U.S. Army. 1974. Chemical agent data sheets, Vol. 1. Technical Report EO-SR-74001. Edgewood Arsenal, Maryland.

U.S. Army Chemical Research, Development, and Engineering Center (CRDEC). 1990. HD and THD. Material Safety Data Sheet. HCSDS No. 20058A. Aberdeen Proving Ground, MD: U.S. Army Chemical Research, Development and Engineering Center.

U.S. Army and U.S. Air Force. 1975. Military chemistry and chemical compounds. Field Manual No. FM3-9, Regulation No. AFR 355-7.

Wada S, Nishimoto Y, Miyanishi M, Katsuta S, Nishiki M. 1962. Review of Okuno-jima poison gas factory regarding occupational environment. Hiroshima Journal of Medical Sciences 11:75-80.

Warthin AS, Weller CV. 1919. The lesions of the respiratory and gastrointestinal tract produced by mustard gas (dichlorethyl sulphide). Journal of Laboratory and Clinical Medicine 4:229-264.

Weiss RB, Muggia FM. 1980. Cytotoxic drug-induced pulmonary disease: update 1980. American Journal of medicinae 68:259-266.

Willems JL. 1989. Clinical management of mustard gas casualties. Annales Medicine Militaris Belgicae 3:S1-61.

Winternitz MC. 1919. Anatomical changes in the respiratory tract initiated by irritating gases. Military Surgeon 44:476-493.

World Health Organization (WHO). 1970. Health Aspects of Chemical and Biological Weapons: Report of a WHO Group of Consultants. Geneva: WHO.

8

Ocular Effects of Mustard Agents and Lewisite

The external eye is uniquely vulnerable to injury by mustard compounds and Lewisite. Over 90 percent of patients exposed to sulfur mustard in World War I (WWI) sustained injury to the conjunctiva; about 10 percent of these injuries were severe, involving the cornea as well. Animal studies suggest that these acute injuries can be followed by recurrent or delayed symptoms, including ulceration, inflammation, and corneal erosion. Many WWI veterans in fact developed such symptoms some 8 to 25 years after their initial injuries. As with pulmonary injuries (Chapter 7), however, there have been no long-term studies of the ocular status of World War II human subjects.

PHYSIOLOGY AND ANATOMY OF THE EYE

The housing of the eye is a fibrous envelope composed of clear cornea over the front one-sixth and opaque sclera surrounding the rest (Figure 8-1). The sclera is covered by the conjunctiva, a transparent mucous membrane that also covers the inner surface of the eyelids. Both the cornea and the conjunctiva are covered with squamous epithelium, six to seven and three to four layers thick, respectively. A three-layered tear film (lipid, aqueous, and mucoprotein) bathes the surface cells constantly, providing the eye with lubrication and protection. The lids also close to protect the eye, and in doing so they periodically replenish the tear film lost from evaporation and lacrimal drainage.

The cornea itself is a layered structure. Underlying the corneal epithelium is a specialized region of collagen tissue. The rest of the corneal collagen is organized into lamellae, or layers, that are very regularly oriented and contribute significantly to corneal clarity. The inner surface of the cornea is lined with a single layer of endothelium, a metabolically active cellular layer that pumps electrolytes, water, and metabolites out of the cornea. The corneal epithelium is derived from a specific group of stem cells, encircling the cornea, that multiply and ultimately form mature corneal cells. A different population of stem cells, in the conjunctival fornices, give rise to the conjunctival epithelium.

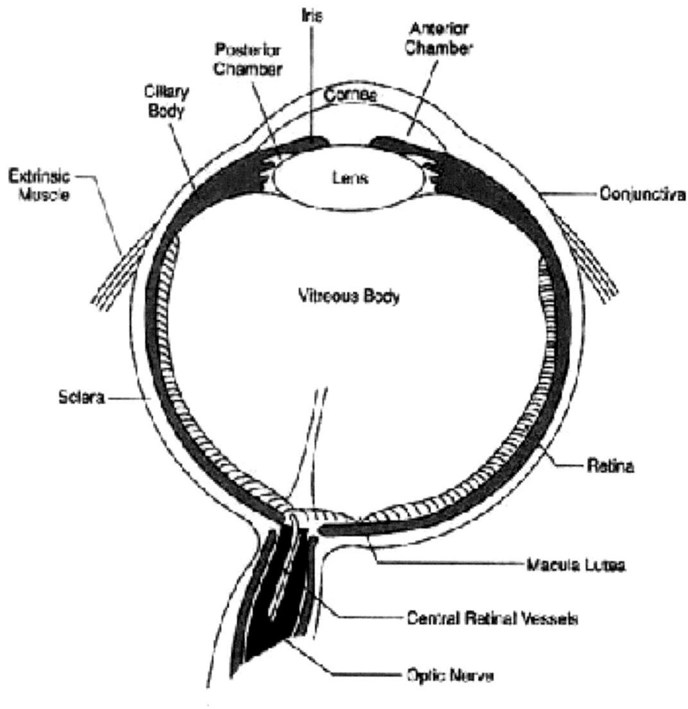

FIGURE 8-1 Anatomy of the eye.

Corneal nutrition is supplied almost exclusively by the aqueous humor circulating behind the cornea in the front of the eye. This allows the cornea to obtain its nutrition in the absence of blood vessels, an obvious aid to clarity. When blood vessels do enter the cornea (in disease or injury), scarring occurs, with a loss of transparency and consequently loss of vision.

ACUTE EFFECTS AND BIOLOGICAL MECHANISMS

Dose-Related Considerations

Because of the constant presence of a tear film over its surface and mucous membranes, the eye is more sensitive to sulfur mustard than any other organ of the body. The fact that fluid is present in the eye at all times probably accounts for the fact that varying conditions of humidity do not influence the degree of injury as compared, for example, to skin. Table 8-1 summarizes dose-related effects of sulfur mustard on eyes at 16°C-7°C. The degree of visual loss depends on the concentration and exposure time to sulfur mustard. Concentrations of less than 50 to 100 mg·min/m^3 cause simple conjunctivitis (inflammation of the conjunctiva) that, at most, can disable an individual for one to two weeks (Papirmeister et al., 1991). When doses exceed 200 mg·min/m^3, however, corneal edema or swelling occurs. At even higher doses, severe corneal damage takes place, with significant loss of vision (Papirmeister et al., 1991).

Ocular injury with sulfur mustard liquid increases the risk of perforation of the cornea and thus is far more damaging to the cornea than vapor-induced injury. Some permanent loss of vision is also more likely as the severity of such exposure is increased.

The time between injury and the appearance of clinical symptoms varies, depending on the severity and duration of exposure. The less severe the injury, the longer is this latency period. With conjunctival injury alone, the latent period is 4 to 12 hours. After more severe exposures in which the cornea is damaged, this latent period may be decreased to as little as 1 to 3 hours. The latency period after liquid sulfur mustard injury is less than 1 hour (Papirmeister et al., 1991).

The exquisite sensitivity of the eye, compared to respiratory tract and skin, to sulfur mustard is evident in the data previously presented in Table 3-4. Symptoms appear in the eye before most other tissues. As the concentration of sulfur mustard increases, however, the injury to the eye parallels that of the respiratory tract. Unlike sulfur mustard, Lewisite exposure causes immediate pain and blepharospasm (spasm of eyelid muscle), especially when the agent is aerosolized (Adler and Leopold, 1945). Eye morbidity from liquid and vapor Lewisite is summarized in detail in Table 8-2.

Clinical Observations

The acute clinical course of severe sulfur mustard injuries of the eye has been well described in the rabbit (Figure 8-2). Hughes (1942) described five stages following severe damage to the cornea in rabbit and man:

TABLE 8–1 Effects of Acute Sulfur Mustard Exposure on the Human Eye at (16°C–27°C unless otherwise noted)

Estimated Exposure	Effects	Latency	Reference
Vapor (Ct of mg·min/m³)			
2 (≥32°C)	Threshold for conjunctivitis and reddening; nondisabling; maximum safe Ct for eyes	Not available	McNamara et al., 1975; U.S. Army CRDEC, 1990
≤12	Threshold for conjunctivitis and reddening; nondisabling	Several hours to several days	McNamara et al., 1975; Project Coordination Staff, 1946
12	Marginal effects	Not available	U.S. Army, 1974
30 (60 min)	Conjunctivitis	Not available	Reed, 1920
50–100	Conjunctivitis, sensation of grittiness under eyelids, tearing, light sensitivity; nondisabling; healing of injury 2–7 days (14 days if severe)	4–12 hours	Gates and Moore, 1946; McNamara et al., 1975; Project Coordination Staff, 1946
70	Reddening, no incapacitation	Not available	U.S. Army, 1974
90	Reddening, mild incapacitation	Not available	U.S. Army, 1974
200	Median incapacitating Ct (ICt_{50}); corneal edema and clouding; eyelid edema, photophobia, severe blepharospasm (temporary blindness); healing of injury over several weeks; some hypersensitivity for months	3–12 hours	Project Coordination Staff, 1946; Urbanetti, 1988; U.S. Army, 1974; U.S. Army and U.S. Air Force, 1975
400–800	Corneal damage with possible ulceration and secondary infection; incapacitating; eye damage may be permanent; possible prolonged hospitalization (several months)	1–4 hours	Karnofsky and Nolen, 1944
>800	Severe corneal damage with possible permanent loss of vision; possible systemic effects; incapacitating; possible prolonged hospitalization	1–3 hours	Papirmeister et al., 1991; Stepanov and Popov, 1962
Liquid			
Unknown (droplets or liquid splash)	Severe pain; edema; corneal damage; possible corneal perforation; incapacitating, possible permanent eye damage or loss (rare); hospitalization required	<1 hour	Maumenee and Scholz, 1948; Papirmeister et al., 1991

SOURCE: Adapted from Papirmeister et al., 1991; Watson and Griffin, 1992.

TABLE 8–2 Effects of Acute Lewisite Exposure on the Mammalian Eye

Estimated Exposure	Species	Effects	Exposure time (min)	References
Vapor (mg/m³)				
0.8	Man	No irritation (estimate)	Not reported	A.D. Little, 1986
1.0	Rabbit	Permanent lesions (minimum effective dose)	30	Gates et al., 1946
1.0	Man	Lesions	30	U.S. Army, 1974
1.5	Man	Severe (estimate)	Not reported	Gates et al., 1946
2	Man	Threshold of irritation	Not reported	A.D. Little, 1986
10	Man	Inflammation	15	Ottinger, 1973
10–30	Man	Irritation	1	A.D. Little, 1986
20	Dog	Permanent lesions (minimum effective dose)	30	Gates et al., 1946
Saturated	Rabbit	Perforation	22.5	Hughes, 1942
Vapor (mg·min/m³)				
<300	Man	Median incapacitating Ct (ICt_{50})	Not reported	U.S. Army CRDEC, 1990
Liquid (mg)				
0.005	Rabbit	Mild	Not reported	Wallen et al., 1943
0.01–0.02	Rabbit	Permanent damage	Not reported	Wallen et al., 1943
0.1	Rabbit	75% perforated	Not reported	Wallen et al., 1943
0.1	Man	Severe (estimated)	Not reported	Gates et al., 1946
12	Rabbit	Destruction	Not reported	Irwin, 1954

SOURCE: Adapted from Solana, unpublished.

- Immediate damage to the corneal epithelium, with edematous clouding and necrosis of the stroma.
- After five hours, dense infiltration of polymorphonuclear neutrophils at the sclerocorneal junction, extending into the corneal stroma.
- Clinical improvement of the opacity after five to seven days, with diminished edema of the stroma in less severe injuries.
- Progressive vascularization of the cornea extending in from the limbal vessels. This process may continue for several weeks.
- Persistent ulceration of the cornea for weeks or recurrent ulceration after a latent period of years.

Over 90 percent of patients exposed to sulfur mustard in World War I sustained conjunctival injury. Corneal changes were apparent in a much smaller group and were likely due to vapor, as opposed to liquid, exposure. All those exposed, however, had photophobia and blepharospasm. Hughes (1945b) divides these patients into three categories:

- Class I: about 75 percent had mild symptoms without corneal involvement.
- Class II: about 15 percent were moderately affected, the corneal

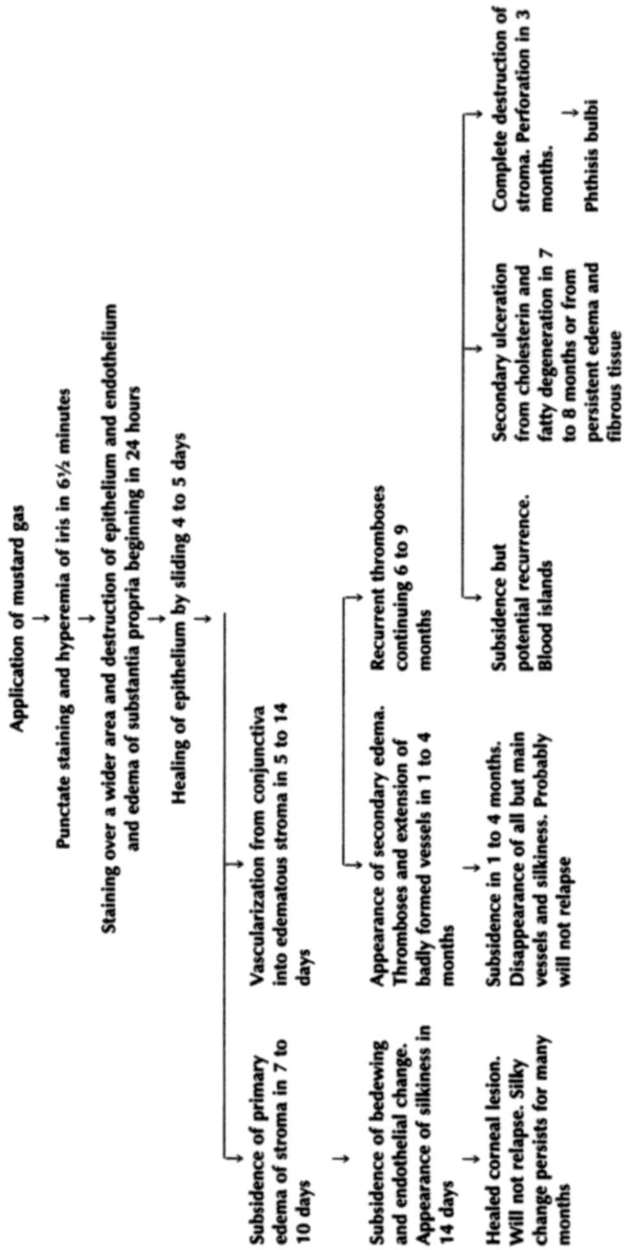

FIGURE 8-2 Course of mustard gas lesions of rabbit cornea. SOURCE: Mann and Pullinger, 1944.

epithelium was mildly injured, but would return to normal in one or two days.

- Class III: about 10 percent had severely affected eyes, both the cornea and the conjunctiva being involved. Class III was subdivided into IIIa (mild corneal changes with a prognosis of return to active duty in six weeks to three months) and IIIb (severe corneal changes resulted in disability of more than three months). It is from members of the latter 10 percent of Class III that "delayed keratitis" developed 8 to 25 years later (see below).

Experience with Lewisite eye injuries is much less extensive. However, Lewisite can produce some changes similar to those of sulfur mustard (Goldman and Dacre, 1989; Mann et al., 1946; also see Table 8-3). Although severe visual loss occurs with Lewisite exposure, no long-term ocular effects were reported by Mann and colleagues (1946). However, animals in the Mann study were only followed for 30 days following exposure.

Physiology and Histopathology of Injury

The corneal epithelium continues to appear viable and respond in a normal manner for hours after exposure to sulfur mustard, even if the epithelium is separated from the underlying layers of the cornea. However, even very low doses of sulfur and nitrogen mustard cause cessation of mitotic activity in the corneal epithelium. Exposed cells in mitosis complete their cell division normally. If exposed to the poison before the onset of mitosis, however, the mitosis is either greatly prolonged or completely suppressed (Friedenwald, 1945). This is consistent with the effects of these agents on all rapidly proliferating cells.

Mann and Pirie (as cited in Friedenwald, 1945) found that corneal collagen reacts with sulfur mustard and, in the presence of an excess of sulfur mustard, more molecules of sulfur mustard are bound by the protein than the number of sulfhydryl groups present in the material before exposure. Further they reported that collagen was abnormally resistant to attack by pepsin after reacting with the agent. This suggests that a specific physical or chemical reaction occurred between the collagen and the sulfur mustard, namely, a denaturation of collagen, potentially making it more vulnerable to degradative enzymes.

The histological changes taking place after sulfur mustard injury of the eye have been summarized by Scholz (1945). Thirty to 60 minutes after exposure, the first change noted was edema of the basal epithelial cells of the cornea. At one to two hours, the basal nuclei relocated toward the central portion of the cell. Between two and 12 hours, the goblet cells had lost their mucus and were sloughed off, followed shortly by loss of the conjunctival and corneal epithelial cells. Edema of the stroma developed as a consequence of endothelial cell damage and loss. From 12 to 24 hours after exposure, the conjunctiva was edematous, the endothelium of the blood vessels was lost, and an infiltrate accumulated, composed primarily of neutrophils.

TABLE 8–3 Characteristics of Sulfur Mustard and Lewisite Ocular Lesions

Types	Sulfur Mustard	Lewisite
Onset of ocular action	No initial reaction; symptoms do not appear for some hours.	Immediate and painful reaction.
Pupillary reaction	Not affected.	Immediate strong miotic action.
Vascularization	Never occurs unless limbus is damaged.	Independent of the site of the primary lesion; occurs when a sufficient dose reaches the cornea or limbus.
Vascular lesions	Do not all perforate; tend to be chronic, to relapse, and to show intracorneal hemorrhages from newly formed vessels; vessels have peculiar and characteristic form.	Not all vascular lesions perforate; there are no relapses and no recurrent hemorrhages.
Cholesterin and other lipoid scars	Follow some vascular lesions, and subsequently these tend to break down (delayed mustard gas keratitis).	Do not occur, and there is no late breakdown due to them.
Perforation and loss of an eye	Caused by relatively large doses (0.005 cc), and even then are long delayed.Perforation never occurs as a primary lesion before the stage of vascularization.	Caused by relatively small doses (0.001 cc); perforation may occur within a few days without vascularization, or later after the entry of blood vessels.
Edema	Edema of the conjunctiva and cornea is present, but not excessive.	Edema of the lids and conjunctiva is immediate and severe. Edema of the cornea is extreme in all but the smallest doses.
Iris and ciliary body	Relatively little involvement. No late effect on pigment.	Early and severe involvement, followed by gradual depigmentation and shrinkage of the iris stroma.
Vessel formation	Characteristic vessels form in cornea and conjunctiva.	Corneal vessels do not show the characteristic varicosities of mustard vessels.

SOURCE: Adapted from Mann et al., 1946.

The conjunctival epithelium began to regenerate two days after injury. If the corneal and limbal epithelium had been lost, conjunctival epithelium was observed to cross the limbus to resurface the cornea.

Conjunctival epithelium thickened in one to two weeks postinjury, but the corneal epithelial layer remained very thin, often with "skip" areas referred to as defects. When such defects were long-lasting, necrotic ulcers, with or without bacterial infection, often supervened. Depending on the severity of the original injury, a scarring, or "hazing," of the corneal stroma was noted.

Lewisite injuries cause necrosis of the deep corneal layers with loss of nuclei of all keratocytes and loss of normal staining characteristics in collagen fibers (Adler et al., 1947). Neutrophils invade the stroma at 24 hours but are replaced by round, chronic inflammatory cells at the end of 10 days. Corneal vascularization is not visible until 5 days after the injury, becoming intense at 14 days. Topical British Anti-Lewisite treatment diminished the severity of these injuries only if given within two to five minutes of injury (Hughes, 1946).

EVIDENCE OF LONG-TERM HEALTH EFFECTS

Animal Data and Cellular Bioassay

Long-term studies of the effect of sulfur mustard were conducted by Mann and Pullinger (1944). Over a period of 18 months they intermittently examined 138 rabbits injured by sulfur mustard (Figure 8-2). Based on the lifetime of a rabbit as one-tenth that of man, an observation period of 18 months would be sufficient time to develop delayed keratitis in this experimental animal. The results showed that, similar to the human condition, migration of fatty or cholesterin deposits to the surface of the eye could occur after 7 to 8 months and cause recurrent secondary ulceration. In these models of severe burns, it can be concluded that delayed and recurrent keratitis is demonstrable and reproducible.

Comparison of sulfur mustard injuries to those sustained with Lewisite is summarized in Table 8-3. A summary of the toxicology of Lewisite liquid and vapor is separately tabulated in Table 8-2. Many of the rabbits in the Lewisite study were watched for more than one year; no secondary lipoid degeneration or cholesterin deposition was noted, nor any long-term effect, in particular (Mann et al., 1946).

Human Studies

Occupational Exposure

A large number of injuries occurred in the production of sulfur mustard. A summary of the toxic accidents in all three mustard gas production factories in England showed that 10.4 percent of 939 eye casualties showed improvement of the cornea, and only 1 percent had severe Class IIIb corneal lesions (Hughes, 1945a). A report of 1,097 patients treated at Edgewood Arsenal during the 17 months preceding March 1, 1943, showed that 91 percent of the injuries were due to mustard agent vapor, and of that number 78 percent had eye burns (Uhde, 1946). Eighty of these patients were exposed to a sudden break in the shell filling line, which released large quantities of mustard agent vapor. The remaining 93 percent were exposed to slow leaks that were not detected by smell. Unfortunately, there have been no long-term studies of these patients to determine their ocular status many years after exposure.

Battlefield Exposure

The British reported many thousands of eye casualties during World War I. Sulfur mustard was responsible for 77 percent of all gas injuries in WWI; of these, 75 percent were relatively mild conjunctival irritations, forcing hospital care for an average of two weeks before return to active duty (Gilchrist and Matz, 1933; Hughes, 1942). Another 15 percent were described as intermediate, with incapacitation for four to six weeks. Finally, 10 percent were severe, requiring hospitalization or rehabilitation for a four- to six-month period before stabilizing (Hughes, 1942). A total of 51 British soldiers were reported as blinded, and there were 180 vision-related pensions (Phillips, 1940). Phillips (as cited in Hughes, 1945a) also reported 80 patients with late recurrent ulceration of the cornea following exposure to sulfur mustard in WWI. He stated that there were a total of 300 reported cases of delayed keratitis as of 1939.

Two French reports describe only a handful of ocular lesions. Teulieres (as cited in Hughes, 1942) reported on 1,500 mustard gas casualties, of which only 23 patients sustained ocular lesions severe enough to necessitate observation. Of these, 3 patients showed ulceration of the cornea and 1 developed inflammation of the entire eye (panophthalmitis). In a parallel report of 1,800 sulfur mustard casualties, Beauvieux (1920) found only 2 patients in whom severe corneal ulceration developed; both cases ultimately recovered useful vision. Beauvieux also examined the retinal blood vessels and condition of the retina in 120 cases of severe generalized sulfur mustard lesions and noted venous dilatations in 34 percent and hyperemia of the optic disk in 23 percent.

If all of these reports are combined, up to 90 percent of casualties would be expected to have ocular involvement, with symptoms peaking 6 to 12 hours after exposure. However, 90 percent of these cases would have no significant corneal involvement. Common symptoms included gritty sensation, conjunctivitis, chemosis (edema of the conjunctiva causing swelling around rim of cornea), lid edema, blepharospasm, photophobia, blurred vision, and tearing. The 10 percent of patients with corneal involvement exhibited corneal edema, keratitis, ocular pain, temporary blindness, tissue necrosis, iridocyclitis (inflammation of iris and ciliary body), glaucoma, vascularization, and, rarely, ulceration or perforation of the eye.

Anatomical investigations by Ashoff and colleagues, cited in Oswald (1920), have shown that sulfur mustard poisoning probably causes blood clots in the precapillary arterioles of the eye. This might explain a report of progressive narrowing of the retinal blood vessels in a 34-year-old man who had sustained a relatively mild exposure to sulfur mustard in 1917 that nevertheless resulted in bilateral blindness (Oswald, 1920).

Chronic Course

Many patients continue to have recurrent corneal erosions and inflammatory keratitis for an indefinite number of years after the serious corneal injury. Approximately 243 cases of late recurrent ulceration of the cornea have been reported following severe sulfur mustard burns in WWI (Scholz and Woods, 1945).

In the acute stage the limbal region frequently presents a marbled appearance in which porcelain-like areas of ischemia (decreased or blocked blood flow) are surrounded by blood vessels of irregular diameter. Later, the vascularized scars of the cornea often contain deposits of cholesterin, calcium, and fat. There were reports of a sudden increase in these symptoms and findings some 8 to 25 years after the initial injury. This information first appeared in the U.S. literature in a 1947 article by Scholz and Woods, but many such cases had already been reported in the British, French, and German literature (Genet, 1925; Heckford, 1937; Moore and Heckford, 1929; Proceedings of the Royal Society of Medicine, 1940; Rohrschneider, 1937; Sourdille, 1936; Weill, 1939).

Phillips, who collected 70 cases, called attention to the delayed keratitis due to sulfur mustard exposure (Proceedings of the Royal Society of Medicine, 1940). He noted that, after the initial early and intermediate hospitalization from 4 to 6 months, these patients were often symptom free for 10 to 13 years, whereupon delayed keratitis developed, characterized by photophobia, lacrimation, and failing vision. Superficial ulcerations also occurred in these patients, and the sensitivity of the cornea to touch and other stimuli was reduced. Marbling of the cornea appeared in the acute stage of these cases along with additional linear marks referred to as "skate marks on fresh ice." Also found were pale triangular patches on either side of the cornea, indicating an absence of small conjunctival and episcleral blood vessels that left large areas of the sclera bare. In each of these cases there was a history of a severe corneal and perilimbal burn that involved an extensive amount of the perilimbal conjunctiva. This corresponds to Class III injuries in which there is (a) damage to the limbal blood supply, associated with moderate corneal edema, superficial collections of inflammatory cells, with or without superficial corneal vascularization; or (b) ischemic necrosis of the limbal regions, marked and persistent corneal edema, deep corneal inflammatory cells, and deep blood vessel ingrowth into the cornea (vascularization).

The problem of delayed keratitis was neither trivial nor infrequent in the severe injuries. Mann's studies of 84 cases in 1944 showed that an inflammatory keratitis (inflammation of the cornea) developed intermittently in the most severely injured veterans for a period of 17 years after initial exposure (Mann, 1944); however, during the next 7- to 8-year period, a sudden substantial increase in these numbers was observed.

The use of Lewisite as a wartime gas was much less common. A combination of sulfur mustard and Lewisite was reported to have been used in the Japanese attack on Ichang in 1941 (SIPRI, 1971). No medical reports are available to document the short- or long-term effects to the health of survivors. No large-scale reports of pure Lewisite eye injury to humans are noted in the literature.

Medical Therapeutic Exposure

Sulfur mustard has not been used to treat eye disease. However, other alkylating agents, such as cyclophosphamide and chlorambucil, have been shown to be effective in arresting a number of eye diseases, including peripheral ulcerative keratitis, Mooren's corneal ulcer, some ulcerative and nonulcerative scleral diseases, and recalcitrant uveitis (inflammation of the iris, ciliary body, and choroid). Each of these diseases is regarded as, or has been proven to be, an autoimmune process. There are no reports of secondary cancers developing as a consequence of immunosuppression from treatment of eye disease.

SUMMARY

Gaps in Knowledge

Limbal Vascular Damage

The whitened appearance of perilimbal ocular tissues after sulfur mustard injury suggested to early investigators that destruction of the limbal vasculature resulted in the disastrous effects on the cornea and the late development of recurrent keratitis. This hypothesis was based on the presumption that the cornea derived its entire nutritional supply from these vessels. Since then it has been discovered that corneal nutrition is principally derived from the aqueous humor and probably receives only a small facultative supply from perilimbal vessels. It is unknown whether sulfur mustard can damage blood vessels in the ciliary body—the only way that nutrition of the cornea could be influenced.

Ocular Epithelial Damage

The most trying and difficult problems associated with mustard gas injuries are the corneal complications that can occur in the early, intermediate, but especially the late phases after exposure. Recent research suggests that the loss of ocular epithelium is a key factor in persistent epithelial defects of the cornea, giving rise to new disease entities called "ocular surface disease" and "stem cell disease." Even when the cornea itself has not been damaged, loss of stem cells can result in persistent epithelial defects that will encourage inflammatory cell invasion, vascularization, and scarring. When corneal scarring interferes with vision, corneal transplantation is hazardous in the absence or with a deficiency of stem cells.

The concept of corneal stem cell injury and destruction has become of central importance in a variety of diseases of the eye, especially severe alkali and acid injuries. Frequently in these cases all corneal, limbal, and extensive conjunctival epithelium is lost.

No clinical trial has been initiated to study this problem, but numerous individual reports and clinical observations have shown that small conjunctival transplants from the limbus of an unaffected eye to the injured eye can eliminate these recurrent erosions and inflammatory propensities and restore the injured eye to full function (Kenyon and Tseng, 1989). When the injury is bilateral it is possible that conjunctival stem cell transplants from a donor eye would survive in these injured eyes to create a stable epithelial surface. If the corneal problem is one of constant epithelial breakdown, then stem cell transplant should stabilize the epithelial surface and thereby improve vision. In those patients who show severe tear film instability, irremediable eyelid malappositions, recurrent corneal transplant failures, and severe distortions of the anterior segment, a keratoprosthesis (artificial cornea) is an option to provide some vision when live corneal tissue consistently fails. Patients with severe corneal opacities also benefit from modern techniques of corneal transplantation and postoperative management.

Protection of the Corneal Epithelium with a Contact Lens

Mechanical protection of the corneal epithelium has been shown to reduce the irritation and improve the vision in a large series of cases of recurrent keratitis. The large and cumbersome scleral lenses were difficult to make, could only be worn for short periods by most individuals, and gave only a limited but welcome response. The development of soft contact lenses has dramatically improved our ability to protect the corneal epithelium in those patients when it is fragile (Pfister, 1986). Research to study the effect of soft contact lenses on the chronic relapsing keratitis after sulfur mustard injuries might provide an updated approach to mechanical protective devices.

Sulfur Mustard Compared to Alkali Injury: The Stromal Component

Sulfur mustard injuries bear distinct similarities to alkali or very strong acid injuries of the eye. If these two injuries present similar problems, then treatments that have been initiated for alkali burns of the eye might well be effective in mustard gas or Lewisite injuries. An animal model of the alkali-injured eye has yielded significant data showing that sodium citrate and sodium ascorbate, when applied topically to the eye, significantly reduce the incidence of corneal ulceration and perforation (Pfister and Paterson, 1980; Pfister et al., 1981, 1982). A national clinical trial on the treatment of human alkali-injured eyes with ascorbate and citrate is currently in progress. Other chelators that inactivate metalloproteases, such as collagenases, are also available. These metalloprotease inhibitors interfere with the enzymes leading to corneal ulceration. Although there is no clear-cut evidence available, it seems likely that this treatment could be quite effective in sulfur mustard injuries of the eye.

CONCLUSIONS

Acute, severe injury of the eye with sulfur mustard might result in recurrent corneal ulcerative disease for the remainder of the patient's life, with a maximum incidence occurring 15 to 20 years after the injury. **Based on extensive data, there is a causal relationship between severe exposure to sulfur mustard and the development of delayed recurrent keratitis.**

There is a causal relationship between exposure to sulfur mustard and the development of prolonged, intractable conjunctivitis.

There is evidence, in laboratory animals, to indicate that no causal relation is present between exposure to Lewisite and any long-term ocular disease process. However, any corneal scarring or vasculariza tion occurring soon after the injury will persist. In spite of the arsenical nature of Lewisite, no association has been noted between Lewisite and the development of neoplasia of the eye.

REFERENCES

Adler FH, Leopold IH. 1945. The toxicity of Lewisite for the eye. In: National Research Council. Division of Medical Sciences, Committee on Treatment of Gas Casualties, comp. Fasciculus on Chemical Warfare Medicine. Volume I, Eye. Washington, DC: Prepared for the Committee on Medical Research of the Office of Scientific Research and Development. 279-288.

Adler FH, Fry WE, Leopold IH. 1947. Pathologic study of ocular lesions due to Lewisite (beta-chlorovinyldichloroarsine). Archives of Ophthalmology 38:89-108.

Beauvieux. 1920. Ocular lesions caused by vesicant gases. Archives d'Ophtalmologie 37:597-619. [In French]

Friedenwald JS. 1945. The pathological physiology of mustard damage to the cornea. In: National Research Council. Division of Medical Sciences, Committee on Treatment of Gas Casualties, comp. Fasciculus on Chemical Warfare Medicine. Volume I, Eye. Washington, DC: Prepared for the Committee on Medical Research of the Office of Scientific Research and Development. 214-235.

Gates M, Moore S. 1946. Mustard gas and other sulfur mustards. In: Division 9, National Defense Research Committee. Chemical Warfare Agents, and Related Chemical Problems. Summary Technical Report of Division 9, NDRC. Washington, DC: Office of Scientific Research and Development. 30-58.

Gates M, Williams JW, Zapp JA. 1946. Arsenicals. In: Division 9, National Defense Research Committee. Chemical Warfare Agents, and Related Chemical Problems. Summary Technical Report of Division 9, NDRC. Washington, DC: Office of Scientific Research and Development. 83-114.

Genet L. 1925. Ocular burns from yperite gas: persistent sequelae seven years after injury. Case study. Lyon Medical 136:388-391. [In French]

Gilchrist HL, Matz PB. 1933. The Residual Effects of Warfare Gases. Washington, DC: U.S. Government Printing Office.

Goldman M, Dacre J. 1989. Lewisite: its chemistry, toxicology, and biological effects. Reviews of Environmental Contamination and Toxicology 110:75-115.

Heckford F 1937. Delayed corneal ulceration following mustard gas burns. Proceedings of the Royal Society of Medicine 30:949.

Hughes WF Jr. 1942. Mustard gas injuries to the eyes. Archives of Ophthalmology 27:582-601.

Hughes WF Jr. 1945a. The importance of mustard burns of the eye as judged by World War I statistics and recent accidents. In: National Research Council. Division of Medical Sciences, Committee on Treatment of Gas Casualties, comp. Fasciculus on Chemical Warfare Medicine. Volume I, Eye. Washington, DC: Prepared for the Committee on Medical Research of the Office of Scientific Research and Development. 79-90.

Hughes WF Jr. 1945b. The symptomatology and diagnosis of mustard gas injuries of the eyes. In: National Research Council. Division of Medical Sciences, Committee on Treatment of Gas Casualties, comp. Fasciculus on Chemical Warfare Medicine. Volume I, Eye. Washington, DC: Prepared for the Committee on Medical Research of the Office of Scientific Research and Development. 91-109.

Hughes WF Jr. 1946. Treatment of Lewisite burns of the eye with BAL. Journal of Clinical Investigation (July):541-548.

Irwin. 1954. As cited in: Solana RP. 1992. Toxicology of Lewisite. Presentation to the

Institute of Medicine Committee to Survey the Health Effects of Mustard Gas and Lewisite, June 11, 1992. Unpublished.

Karnofsky DA, Nolen JT. 1944. Report on mustard vapor casualties occurring at Bushnell, Fla., 6 June 1944. As cited in: Papirmeister B, Feister AJ, Robinson SI, Ford RD. 1991. Medical Defense Against Mustard Gas: Toxic Mechanisms and Pharmacological Implications. Boca Raton, FL: CRC Press. 42.

Kenyon KR, Tseng SC. 1989. Limbal autograft transplantation for ocular surface disorders. Ophthalmology 96:709.

Little AD Inc. 1986. As cited in: Solana RP. 1992. Toxicology of Lewisite. Presentation to the Institute of Medicine Committee to Survey the Health Effects of Mustard Gas and Lewisite, June 11, 1992. Unpublished.

Mann I. 1944. A study of eighty-four cases of delayed mustard gas keratitis fitted with contact lenses. British Journal of Ophthalmology 28:441-447.

Mann I, Pullinger BD. 1944. A study of mustard gas lesions of the eyes of rabbits and men. American Journal of Ophthalmology 26:1253-1277.

Mann I, Pirie A, Pullinger BD. 1946. Study of Lewisite lesions of the eyes of rabbits. American Journal of Ophthalmology 29:1215-1227.

Maumenee AE, Scholz RO. 1948. The histopathology of the ocular lesions produced by sulfur and nitrogen mustards. Bulletin of Johns Hopkins Hospital 82:121-147.

McNamara BP, Owens EJ, Christensen MK, Vocci FJ, Ford DF, Rozimarek H. 1975. Toxicological Basis for Controlling Levels of Mustard in the Environment. Edgewood Arsenal Special Publication EB-SP-74030. Aberdeen Proving Ground, Maryland: U.S. Army Armament Command. Edgewood Arsenal Biomedical Laboratory.

Moore RF, Heckford F. 1929. Delayed corneal ulceration from mustard gas. British Medical Journal 1:497-498.

Oswald A. 1920. Collapse of the double-sided closure of the central artery as a consequence of blister gas poisoning. Klinische Monatsblatter fur Augenheilkunde 64:381387. [In German]

Ottinger RS, Blumenthal JL, Dal Porto DF, Gruber GI, Santy MJ, Smith CC. 1973. Recommended Methods of Reduction, Neutralization, Recovery or Disposal of Hazardous Waste. Vol. 2, Toxicologic Summary. EPA-670/2-73-053-b. As cited in: Solana RP. 1992. Toxicology of Lewisite. Presentation to the Institute of Medicine Committee to Survey the Health Effects of Mustard Gas and Lewisite, June 11, 1992. Unpublished.

Papirmeister B, Feister AJ, Robinson SI, Ford RD. 1991. Medical Defense Against Mustard Gas: Toxic Mechanisms and Pharmacological Implications. Boca Raton, FL: CRC Press.

Pfister RR. 1986. The biology of persistent epithelial defects of the cornea. In: Brightbill, FS, ed. Corneal Surgery: Theory, Technique, and Tissue. Papers from the First International Cornea and Eye BankingSymposium . St. Louis: Mosby. 582-593.

Pfister RR, Paterson CA. 1980. Ascorbic acid in the treatment of alkali burns of the eye. Ophthalmology 87:1050-1057.

Pfister RR, Nicloaro ML, Paterson CA. 1981. Sodium citrate reduces the incidence of corneal ulcerations and perforations in extreme alkali burned eyes: acetylcysteine and ascorbate have no favorable effect. Investigative Ophthalmology 21:486-490.

Pfister RR, Haddox J, Paterson CA. 1982. The efficacy of sodium citrate in the treatment of severe alkali burns of the eye is influenced by the route of administration. Cornea 1:205-211.

Phillips TJ. 1940. The delayed action of mustard gas and the treatment. Proceedings of the Royal Society of Medicine 33:229-232.

Proceedings of the Royal Society of Medicine. 1940. Discussion on gas injuries to the eye. 33:225-236.

Project Coordination Staff, Chemical Warfare Service. 1946. Technical Aspects of Chemical Warfare in the Field. 2 vols. Washington, DC: Chemical Warfare Service.

Reed CI. 1920. The minimum concentration of dichlorethylsulphide (mustard gas) effective for the eyes of man. Journal of Pharmacology and Experimental Therapeutics 15:77-80.

Rohrschneider W. 1937. Late ocular lesions (keratitis) following yellow cross (dichloroethylsulfide) poisoning. Klinische Monatsblatter fur Augenheilkunde 99:447-455. [In German]

Scholz RO. 1945. Clinical and pathological studies of ocular mustard gas burns. In: National Research Council. Division of Medical Sciences, Committee on Treatment of Gas Casualties, comp. Fasciculus on Chemical Warfare Medicine. Volume I, Eye. Washington, DC: Prepared for the Committee on Medical Research of the Office of Scientific Research and Development. 155-191.

Scholz RO, Woods AC. 1945. Relapsing and chronic mustard gas lesions of the eyes. In: National Research Council. Division of Medical Sciences, Committee on Treatment of Gas Casualties, comp. Fasciculus on Chemical Warfare Medicine. Volume I, Eye. Washington, DC: Prepared for the Committee on Medical Research of the Office of Scientific Research and Development. 260-278.

Scholz RO, Woods AC. 1947. Relapsing and chronic ocular lesions following mustard gas burns. Archives of Ophthalmology 37:137-148.

Solana RP. 1992. Toxicology of Lewisite. Presentation to the Instituteof Medicine Committee to Survey the Health Effects of Mustard Gas and Lewisite, June 11, 1992. Unpublished .

Sourdille GP. 1936. Lesions of cornea, especially marginal vascular lesions by yperite. Bulletin de la Societe d'Ophtalmologie de Paris 799-801. [In French]

Stepanov AA, Popov VN. 1962. [Chemical Weapons and Principles of Antichemical Defense]. Translated by Joint Publications Research Service. JPRS 15107. Washington, DC: JPRS.

Stockholm International Peace Research Institute (SIPRI). 1971. The Problem of Chemical and Biological Warfare: A Study of the Historical, Technical, Military, Legal, and Political Aspects of Chemical and Biological Warfare and Possible Disarmament Measures. Vol. 1, The Rise of Chemical and Biological Weapons. Stockholm: Almqvist & Wiksell.

Uhde GJ. 1946. Mustard gas (dichloroethyl sulfide) burns of human eyes in World War II. American Journal of Ophthalmology 29:929.

Urbanetti JS. 1988. Battlefield chemical injury. In: Loke J, ed. Pathophysiology and Treatment of Inhalation Injuries. New York: Marcel Dekker.

U.S. Army. 1974. Chemical agent data sheets, Vol. 1. Technical Report EO-SR-74001. Edgewood Arsenal, Maryland.

U.S. Army Chemical Research, Development, and Engineering Center (CRDEC). 1990. HD and THD. Material Safety Data Sheet. HCSDS No. 20058A. Aberdeen Proving Ground, MD: U.S. Army Chemical Research, Development and Engineering Center.

U.S. Army and U.S. Air Force. 1975. Military chemistry and chemical compounds. Field Manual No. FM3-9, Regulation No. AFR 355-7.

Wallen LJ, Horton RG, Ferguson RL. 1943. L, HN-1, H and HQ: effects of 0.1 mg drops on eyes of rabbits. Project A10.3. TRLR 18. Edgewood Arsenal, MD: Toxicological Research Laboratory. As cited in: Solana RP. 1992. Toxicology of Lewisite. Presentation to the Institute of Medicine Committee to Survey the Health Effects of Mustard Gas and Lewisite, June 11, 1992. Unpublished.

Watson AP, Griffin GD. 1992. Toxicity of vesicant agents scheduled for destruction by the Chemical Stockpile Disposal Program. Environmental Health Perspectives. Publication in process.

Weill G. 1939. Corneal ulcers after dichloroethylsulfide poisoning: case. Bulletin de la Societe d'Ophtalmologie de Paris 51:281-282. [In French]

9

Dermatological Effects of Mustard Agents and Lewisite

Probably more research has been done on the dermatological effects of sulfur mustard than any of its other effects. This chapter reflects this extensive research in its length and in the level of detail presented in certain sections. Exposure to both sulfur mustard and Lewisite causes acute injury to the skin, including redness, swelling, blisters, ulceration, and necrosis. Sulfur mustard is more effective under conditions of heat and moisture; it appears to damage the skin by disrupting cell proliferation. The arsenic in Lewisite disrupts enzyme activity.

A considerable body of evidence links acute and chronic exposure to sulfur mustard with the long-term development of pigmentary disorders and skin ulcers in humans. Evidence also links sulfur mustard with the development of cutaneous cancers and precancers in both animals and humans. There is insufficient information, however, regarding the long-term effects of Lewisite on the skin.

ANATOMY AND PHYSIOLOGY OF SKIN

The skin is the largest organ in the body, making up approximately 18 percent of the total body mass. Anatomically, the skin is divided into three layers: the epidermis, dermis, and subcutaneous fat (Figure9-1). The severity of burns to the skin is classified according to how many of these layers are damaged (Figure 9-1). The skin serves to protect all of the vital organs of the body from external trauma, invasion by infectious agents, and invasion by noxious substances. It also serves to prevent outward movement of body fluids and other vital substances. Through special anatomical arrangement of the cutaneous circulation and the biochemical and physiological activity of the adnexal structures (eccrine sweat glands, sebaceous glands, and apocrine glands), the skin assists in the regulation of body temperature and the excretion, manufacture, and absorption of electrolytes, vitamins, nitrogenous matter, and other organic substances.

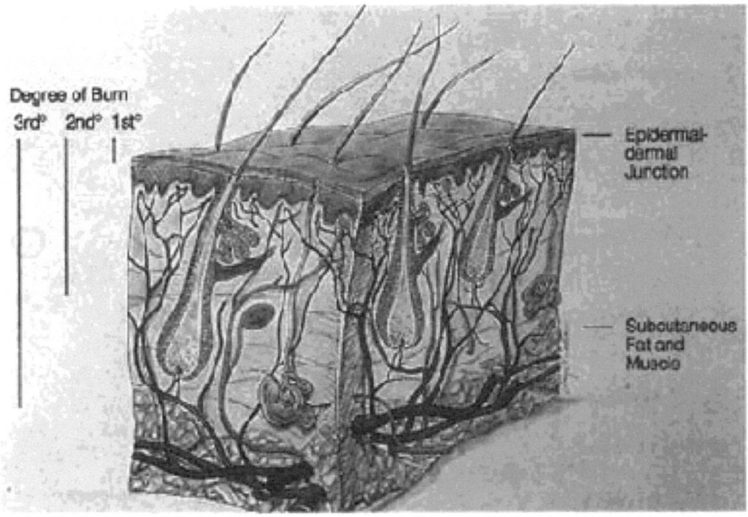

FIGURE 9-1 Anatomy of human skin showing skin layers, hair follicles, sebaceous and sweat glands. The degree of the burn is determined by the depth of damage. First degree burns involve only the epidermis, and cause edema and erythema without vesiculation. Second degree burns involve the epidermis and dermis, and usually result in blisters or dermal necrosis. Third degree burns extend into the subcutaneous fat, muscle, or bone and often cause substantial scarring. SOURCE: Reprinted from Sams and Lynch, 1990, with permission from Churchill Livingstone.

Epidermis

The epidermis occupies the outermost layer of skin and is paramount to maintenance of mammalian homeostasis. It, among the three layers of skin, offers the human body considerable protection from entry of noxious chemicals and microorganisms; it prevents uncontrolled outward movement of fluids, electrolytes, and many organic substances. Large burns due to thermal, chemical, or ultraviolet injury, if they destroy large amounts of epidermis, can lead to an enormous loss of fluid, electrolytes, proteins, and other organic materials through an unprotected dermis. Such burns are quite painful because loss of an intact epidermis exposes cutaneous nerve endings to air, heat, cold, and other direct stimuli.

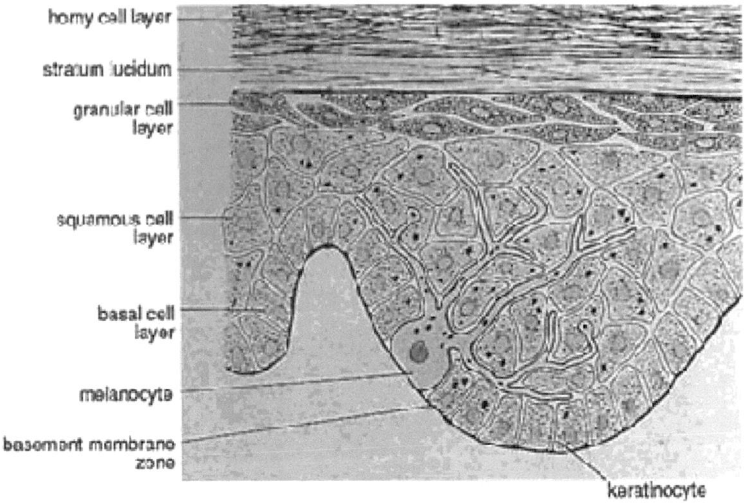

FIGURE 9-2 Layers of the epidermis. As specific cells in the basal layer replicate and differentiate, they move toward the skin surface. The major change in the cells is called "keratinization" in which the cells become filled with a fibrous protein called keratin. Fully keratinized cells are constantly sloughed off at the skin surface. Other types of epidermal cells, such as melanocytes, have other functions. SOURCE: Reprinted from Sams and Lynch, 1990, with permission from Churchill Livingstone.

The epidermis contains several resident cell populations whose responses to certain stimuli can be protective or destructive. The Langerhans cell participates actively in recognition of and presentation of antigens. Lymphocytes respond to signals from Langerhans cells and other macrophages/ monocytes and act in an appropriate fashion to antigens. Melanocytes, pigment-forming cells, protect the skin from harmful ultraviolet (UV) radiation.

The epidermis is composed of four biologically distinct layers of stratified squamous epithelium (Figure 9-2). The deepest layer, usually one cell thick, contains cells that continuously replicate and produce new cells at a rate sufficient to maintain an appropriate number of cells in the upper three layers of the epidermis. Based on location and function, this layer has been called the basal cell layer or germinative layer of the epidermis. Basal cells produce large quantities of the nucleic acids and nucleoproteins required in the process of cell division. Much like the rapidly dividing cells of bone marrow, the intestinal tract, and hair matrix, basal cells are very sensitive to chemicals that affect nucleic acid synthesis. Sulfur mustard is one such agent.

Thickness of the epidermis varies greatly depending principally on the body site and the number of cornified cell layers within the stratum. The stratum corneum is thinnest on the scrotum, on the flexor surfaces of the forearms, within the axillae, and around the eyes. These are body sites through which sulfur mustard penetrates best and exerts its most profound effects after acute exposure. It has been estimated that the entire epidermis renews itself every 45 to 75 days. Sulfur mustard inhibits cell replication within the basal layer of the epidermis and thus disrupts this pattern, resulting in blister formation.

Basal cells of the epidermis are attached to the dermis through the basal lamina, which is often referred to as the basement membrane zone, or epidermal-dermal junction (Figure 9-3). A variety of collagen-like fibrils within basal cells, traverse and attach to the basal lamina within the dermis. Other collagen-like filaments are thought to serve as "anchoring rods" between dermis and epidermis, and dermis and basal lamina. Injury to or destruction of one or more types of anchoring structures causes separation of cells, giving rise to the formation of vesicles and blisters. It has been postulated that proteases released by sulfur mustard, acting on attachment structures between basal cells and basal lamina, give rise to blister formation (Papirmeister et al., 1991). Destruction of the epidermis followed by "shedding" exposes underlying tissue that is devoid of pigmentation and color. When exposed, the underlying tissue, or dermis, imparts a glistening "whitish" appearance to skin even in the darkest of races.

Dermis

The dermis makes up the greatest mass of human skin. It contains cells and fibers that contribute to the skin's elasticity and resiliency (elastic fibers and collagen fibers) and serves as a major force in protecting the internal organs from injury due to external mechanical forces. It is a true supporting structure for cutaneous blood vessels, nerves, and epidermal adnexal structures. Blister fluid is made up principally from fluid released from the dermis. The depth of injury to the dermis and underlying subcutaneous tissues will determine the depth and extent of skin ulceration. Injury to the upper levels of the dermis results in superficial, rapidly healing ulcers. Injury to the entire dermis results in deep, slow-healing skin ulcers.

The predominant cell type found within the dermis, the fibrocyte, is limited in distribution in the normal active dermis; so is the metaboli cally active cell, the fibroblast. Only after injury and during the process of wound healing do fibroblasts proliferate. Large numbers of lymphocytes and monocytes also accumulate in the dermis after injury. Through the production of lymphokines/monokines and other soluble proteins, lymphocytes and monocytes stimulate fibroblast and endothelial cell proliferation and migration, the first step in wound healing and ultimately scar formation. Proliferation of fibroblasts is accompanied by an accelerated production of collagen and mucoproteins and by scar formation. In the normal healing of a cutaneous wound, the accelerated production and degradation of collagen are regulated, through a process of "remodeling," and the degree of scar formation is limited. Unlimited or unrestrained wound healing results in the formation of hypertrophic scars and keloids. Wound infection, which may follow skin injury from sulfur mustard exposure, can cause continued and uncontrolled stimulation of collagen production and ultimately hypertrophic scar and keloid formation.

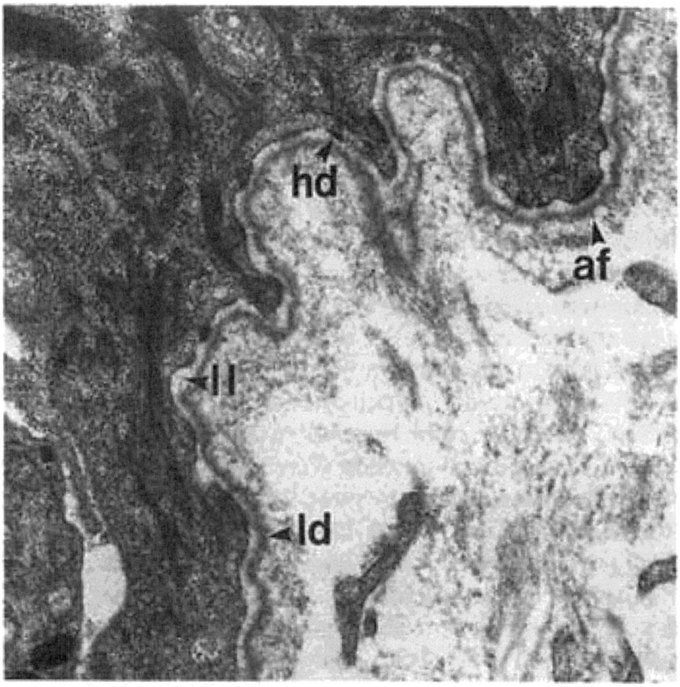

FIGURE 9-3 Epidermal-dermal junction. This electron micrograph of human skin shows the specialized attachments anchoring the epidermis to the underlying dermis. Structures include anchoring fibrils (af), hemidesmosomes (hd), lamina leucida (ll), and lamina densa (Id). Micrograph provided by Henry Bogaars, Brown University.

Eccrine Sweat Glands

Eccrine sweat glands are tube-like invaginations of the epidermis that lie within the normal dermis and are distributed over the entire human body surface. Although exact numbers appear to be related to individual and adaptive factors, there are from 2 million to 4 million glands in the skin of each human being. In an average human, the palms and soles contain the largest number of sweat glands per unit surface area; the back and buttocks contain the least. Excretion of sweat is under emotional and thermoregulatory control, except under resting conditions. Under resting conditions sweating is periodic and involves alternating groups of sweat glands. This form of sweating is invisible or inapparent, and is described as "insensible sweating."

Environmental temperatures above 31°C to 32°C provoke thermoregulatory sweating—a generalized outbreak of sweating and an increase in the number of functioning glands. Areas supplied with few sweat glands may, at this time, be more physiologically active. Thus, at high temperatures, glands of the trunk, thighs, and extremities that respond to thermal stimuli excrete large amounts of sweat. Eccrine sweat glands of the palms, soles, axillae, groin, and forehead respond maximally to emotional stimuli. Under conditions that are stressful these glands are stimulated to produce large volumes of sweat. Sulfur mustard-induced injury to the skin, under wartime conditions, is seen most often in areas that contain thermally and emotionally stimulated sweat glands.

Sweat that is excreted intermittently contains large quantities of chlorides, urea, uric acid, and ammonia. Profuse sweat contains considerably less of these substances, including sodium chloride, and is often pure water. Sulfur mustard is activated by water, yet in the presence of 5 percent sodium chloride it has a markedly reduced effect on human skin (Renshaw, 1946). The decrease in sodium chloride in profuse sweat may account for sulfur mustard's profound cutaneous effects under conditions of high temperature and high humidity. In contrast, the most profound effects of Lewisite, which is deactivated by water, occur under conditions of low temperature, low humidity, and dry skin.

Apocrine Glands

Apocrine sweat glands develop from the follicular epithelium of the pilosebaceous unit, as do the sebaceous glands. The viscous secretions of this gland differ markedly from those of the eccrine sweat gland and are emptied into the canal of hair follicles, rather than directly onto the surface of the skin. In human beings, the apocrine glands are limited in their distribution to the armpit, groin, and pubic regions, around the anus and umbilicus, in a linear band above the umbilicus, and in the external auditory canal. Physiologically, apocrine glands perform little or no useful function. Their secretions serve as rich culture media for gram-negative bacteria. The action of bacteria on apocrine secretions is in part the cause of "offensive" body odors. Subsequent to sulfur mustard skin injury, the large numbers of gram-negative bacteria residing in apocrine areas are often responsible for secondary bacterial infections. Infection, as stated earlier, is frequently associated with hypertrophic scar formation, a common occurrence in the scrotal area of men after sulfur mustard exposure.

Melanocyte System

Skin color in humans is determined by a number of factors, the most important of which is the overall epidermal cell content of melanin. Hemoglobin, the tissue content of carotenoids, keratin, collagen, and the thickness of the keratinizing layers of the epidermis also contribute to the coloring of the skin. Yet, the total color contribution of all other factors combined does not equal that made by melanin. Ultraviolet light, heat, trauma, and a variety of topically applied chemicals can stimulate melanin production and increase skin pigmentation, usually at the site of exposure. Some systemically administered agents can cause increased generalized skin pigmentation.

Melanin is a dense, insoluble polymer derived, in part, from conversion of the amino acid tyrosine by the copper-containing enzyme tyrosinase into an alkyl-insoluble brown chromoprotein. Melanogenesis, the formation of melanin, occurs within specialized cells, called melanocytes (Figure 9-4). Each melanocyte synthesizes a specialized cytoplasmic organelle, called a melanosome, on which the hydroxylation and polymerization of tyrosine to dopa and then to melanin occur. Darkly pigmented races produce large quantities of melanin; Northern European races produce very little melanin and incompletely melanized melanosomes. Darkly pigmented races respond to minimal external and internal stimuli with sharply increased skin pigmentation.

Melanin pigmentation is of substantial benefit in skin. Intense UV light exposure can cause varying degrees of burn damage to unprotected epidermal keratinocytes, adversely affecting cell nuclei, DNA, RNA, structural and enzymatic proteins, and cell membranes. Metabolic alterations caused by UV light injury stimulate epidermal cells either to attempt self-repair or to die, depending on the degree of injury. Severe injury to the epidermal basal cells can lead to faulty cell repair, mutation, and ultimately the development of cancer. Death of epidermal basal cells causes loss of cell-to-basement membrane and cell-to-cell adhesions. Separation of cells from the basement membrane results in subepidermal blister formation. Sulfur mustard-induced blisters are subepidermal in location.

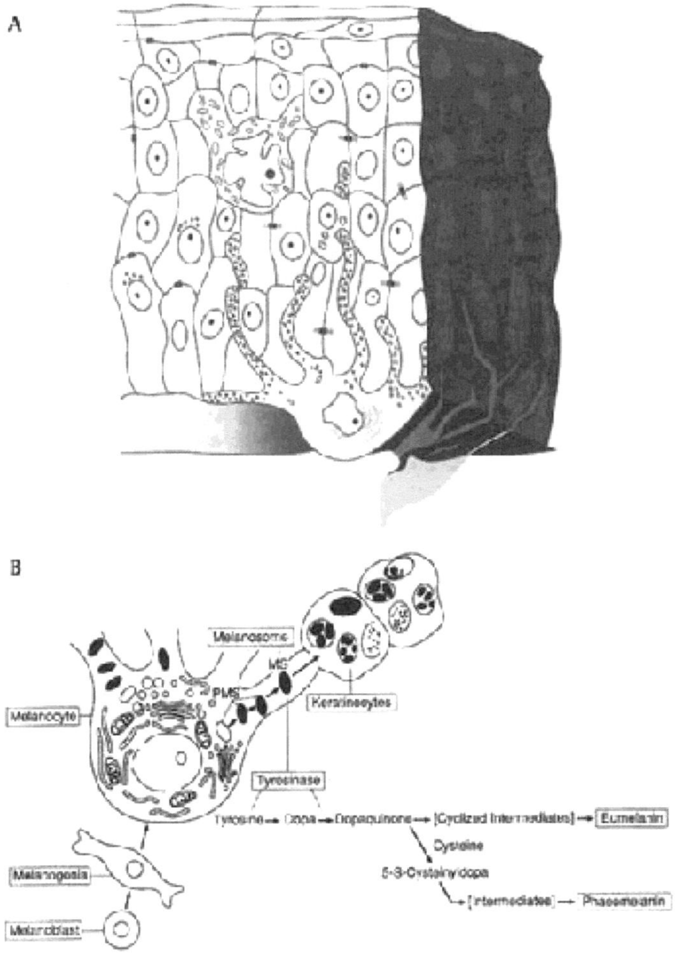

FIGURE 9-4 (A) Epidermal melanin unit. Melanocytes, located in the epidermal layer of the skin, have numerous cellular extensions. These extensions are filled with granules containing the pigment melanin. Melanin granules are "injected" into surrounding epithelial cells from the extensions. (B) Summary of major events in melanocyte development. A cascade of biochemical reactions produces melanin, which protects epidermal cells from damage from the sun. SOURCE: (A) Reprinted from Quevedo, 1969, with permission from American Zoologist. (B) Reprinted from Jimbow et al., 1976, with permission from Williams &Wilkins.

Radiomagnetic emissions from the sun, principally UV light of the A and B spectra, exert a significant influence on skin pigmentation. Polypeptide hormones from the anterior pituitary gland, especially melanocyte-stimulating hormone (MSH), also enhance melanin pigmentation. In women, estrogens stimulate an increase in melanocyte and pigmentary responses in facial, genital, and areolar skin. Chemicals such as theophylline, caffeine, cholera toxin, and prostaglandin E increase the effect of MSH on skin pigmentation. Finally, heat, inflammation, and mechanical injury also stimulate increased pigment formation, especially in the skin of darkly pigmented persons. Based on clinical descriptions of individuals exposed to toxic doses of sulfur mustard, sulfur mustard can also potentiate skin pigmentation. Topical nitrogen mustard, when applied to the skin in the treatment of psoriasis and cutaneous T-cell lymphoma, causes increased skin pigmentation through mechanisms that are as yet unknown.

Injury to the skin of sufficient intensity to cause destruction of melanocytes will result in skin that is devoid of pigmentation. White patches of skin will be noted in even the darkest of pigmented races (leukoderma). Thus, skin that has been subjected to injuries with locally varying intensities, such as after sulfur mustard and Lewisite exposure, will characteristically show areas of depigmentation alternating with areas of hyperpigmentation. In fact, over time the process of healing reveals significant changes in the patterns of pigmentation.

Following skin injury, epithelial cells surrounding the external orifices of the hair follicle and other adnexa proliferate and migrate outward from their source to repopulate skin devoid of epithelium. Epithelial cells from surrounding normal skin also contribute to the process of repair in skin devoid of epithelium. Melanocytes surrounding uninjured hair follicles are stimulated to replicate and increase the production of melanin. Regenerated epithelial cells surrounding the orifices of the hair follicles, then, are the first cells to receive new pigment. The clinical picture of melanocytes repopulating skin is referred to as having a "salt and pepper" appearance. The salt and pepper appearance of skin after sulfur mustard exposure is often written about with some degree of bewilderment. Yet, this is a process that occurs commonly after mechanical and chemical injury to skin.

As melanocytes grow and repopulate normal skin, there is a tendency for overmelanization of any given area. Overmelanized skin at the edges of a healing wound is characteristically darker than skin distal to the healing site. As healing progresses, such skin will eventually return to its normal color and appearance. Normalization of this process often takes 6 to 12 months. The inherent skin color of the affected individual usually determines the amount of time required to return to a normal state.

In many ways, acute and chronic sulfur mustard skin injury mimics injury caused by a variety of toxic chemicals, and mechanical devices. Unsophisticated and untrained observations of sulfur mustard skin injury have often led to distorted accounts of such injuries. Any interpretations of published and unpublished data should be made based on a knowledge of normal and abnormal morphologic, biochemical, and physiological responses of normal and injured human skin.

ACUTE EFFECTS AND BIOLOGICAL MECHANISMS

Sulfur Mustard

Sulfur mustard is an oily substance that is freely soluble in animal oils, fats, and organic solvents (lipophilic). It is only slightly soluble in water, yet water is required for activation. When delivered as a liquid or vapor, the skin plays a very important role as a portal of entry for sulfur mustard. The lipophilic nature of sulfur mustard and the affinity of skin for lipophilic substances make the skin a fairly good transport system for this agent. After cutaneous exposure to sulfur mustard, high levels appear immediately, but transiently, within the skin. A portion of a given dose passes rapidly from the skin into the bloodstream to elicit toxicity at distant sites. However, even under the most ideal circumstances, only a very small portion, probably only 20 percent, of a single dose of sulfur mustard penetrates human skin (Cullumbine, 1947; Renshaw, 1946). Of this amount, about 12 percent reacts with components in the skin, principally within the epidermis. The remainder (about 8 percent) is absorbed systemically. At a temperature of 21°C, sulfur mustard rapidly penetrates human skin. Renshaw (1946) noted that sulfur mustard liquid or saturated vapor penetrates human skin at a rate of 1 to 4 $mg/cm^2/min$ at 21°C. Any increase in ambient temperature causes increased penetration.

There is substantial individual variation in the cutaneous response to sulfur mustard. In general, however, the effects of sulfur mustard on the skin depend on a number of factors including the dose of drug delivered, delivery medium (vapor or liquid), length of exposure of skin cells, degree of hydration of the skin, temperature of the atmosphere, thickness and surface area of the exposed skin, presence or absence of infection, and the intactness of exposed skin. Large dosages of sulfur mustard vapor delivered at 1,000-10,000 $mg{\cdot}min/m^3$ (Ct), or liquid at 40-100 mg/cm^2 over a long exposure time, will yield significant systemic toxicity, including death. Small vapor dosages at 50 Ct, or liquid at 10-20 mg/cm^2 and a short exposure time, yield limited local toxicity. Local toxicity is manifest not only in the skin, but also in the eye and mucous membrane of the respiratory tract.

The time of onset of visible cutaneous effects is related to dose and method of delivery. Microscopically, cutaneous effects begin to appear almost immediately after sulfur mustard contact with skin. Large dosages yield an immediate and profound effect in 1 to 2 hours. Necrosis of the skin following the delivery of large vapor dosages does not appear instantaneously, usually occurring after variable periods of latency. Total necrosis of the skin may occur. Small vapor dosages yield delayed skin effects that may occur 7 to 14 days after exposure. Interestingly, the timing of the onset of sulfur mustard cutaneous reactions is not unlike that observed in cutaneous reactions associated with common *Rhus* (poison ivy oleoresin) dermatitis.

The precise mechanism whereby increased humidity or increased moisture on the skin potentiates sulfur mustard effect is unknown. However, it is possible to assume that wetting the skin alters the permeability of skin cells, thereby increasing the ability of sulfur mustard to penetrate to metabolically active layers; 5 percent, but not 4 percent, sodium chloride-containing water reduces sulfur mustard effects on the skin. It is postulated that 5 percent sodium chloride decreases the solubility of sulfur mustard in water, decreases the overall rate at which sulfur mustard molecules become activated, or alters sulfur mustard penetrability through skin (Renshaw, 1947).

The mechanism whereby an increase in environmental temperature increases the adverse effect of sulfur mustard is also unknown. Increased environmental temperature may simply increase body temperature, stimulating eccrine sweat gland activity and a concomitant increase in hydration of the skin. It is also possible that profuse sweating and a concomitant increase in the amount of pure water on the skin cause activation of greater quantities of sulfur mustard at the site. Blistering of the skin by exposure to UV light is also enhanced by increased temperature and humidity.

The site of skin exposure and the thickness of skin may often determine the type of cutaneous responses experienced upon exposure to sulfur mustard. Thick skin is purported to be less affected by the irritant effects of sulfur mustard. Likewise, young skin, thought to be morphologically thinner than old skin, and female skin, thought to be thinner and more delicate than male skin, are suspected of reacting more severely at all sites upon a given exposure to sulfur mustard (Mathias, 1987; Renshaw, 1946). Racial factors and skin color appear to have even less well defined relationships. It has been generally accepted that black skin reacts in a different manner than white skin to contact allergens and irritants. Yet there are no good experimental data to support the concept that there are substantial differences in the cutaneous response of black or white skin to antigen and injury. Weigand and colleagues (1974) described a difference between black and white skin in the number of cell layers in the stratum corneum, but these differences are no greater than those variations seen in the different body regions of a single individual. According to Nagy and colleagues (1946), there is no difference in the rate of sulfur mustard penetration between the skin of whites and blacks. It is suggested that well-designed contemporary studies are needed to better define the resistance or susceptibility of specific skin thicknesses and types to injury.

Clinical and Microscopic Observations

The most useful clinical descriptions of acute sulfur mustard effects on human skin have been reported by Momeni and colleagues (1992), Smith and Dunn (1991), and a group of clinicians at the Hospital of the University of Ghent Medical School, Belgium (Willems, 1989). These reports were based principally on the examination of patients and patient records of individuals exposed to sulfur mustard during the Iran-Iraq conflict of 1980-1988.

Gross morphologic changes in the skin induced by sulfur mustard are characterized by the appearance of an intense period of itching followed by erythema and edema (signs of inflammation), as well as blister (vesicle) formation, denudation of skin, ulceration, and necrosis. Immediate color changes are often described in affected skin and occur in response to stimulation of melanogenesis, probably an effect akin to the "immediate darkening" effect seen after a specific type of acute UV exposure, and from the darkening effect of "cooked" protein within epidermal cells. Increased darkening of the skin from increased melanogenesis at the periphery of mustard-induced blisters is also characteristically observed. Peripheral darkening is often associated with larger areas of hyperpigmented skin that occur at sites where erythema and edema without previous blister formation had existed. Exfoliation and deep ulceration may occur at these sites. Tissue dosages of sulfur mustard vapor required to induce erythema vary between 0.1 and 1 mg/cm^2. Vesication can be expected in 50 percent of a population at tissue dosages of 1-2 mg/cm^2.

It should be remembered that the site of exposure may be associated with variations in the skin's response to the same amount and extent of sulfur mustard exposure. At the same dosage and time of exposure, loose tissue (less compact dermis) as seen on the face, especially around the eye, and on the genitalia may respond with edema without blistering. However, tissue sites having a very dense dermis, as on the back, may respond with erythema and blister formation without edema.

Healing of the skin is variable and, in the absence of secondary bacterial infection, may proceed without residual defects. The minimally injured hair follicles and other adnexal structures contribute greatly to the healing wound. Tissue reepithelialization often begins and spreads from these sites, as well as from epithelium at the periphery of the injured site. Time for healing varies depending on the degree of tissue injury, residual skin necrosis, and presence or absence of infection.

Scar formation following sulfur mustard injury in specific anatomical areas may be profound and disabling. As has been stated, the genital regions are especially susceptible to sulfur mustard injury. Severe scarring of scrotal and penile tissue can cause deformity and impair sexual performance.

Microscopically, the epidermis and dermis respond adversely to the irritant effects of sulfur mustard. The National Defense Research Committee group at Harvard made an extensive histological study of a large series of experimental sulfur mustard burns of varying severity in human subjects (Renshaw, 1946). Within the epidermis, and within three hours after sulfur mustard dosages that produce erythema, only a few scattered basal cells show nuclear changes consisting of swelling, loss of chromatin, and dispersion of chromatin to the nuclear periphery, clear vesiculation, vacuolization of the cytoplasm around the nucleus, and in some cells vacuolar or hydropic degeneration of the cytoplasm and pyknosis of nuclei. Later, disintegration of the cytoplasmic membrane of basal cells becomes prominent. In some areas, but more often in more severely damaged skin, these degenerative changes may be seen throughout the basal layer of the epidermis.

Ultrastructural studies of human skin have been supplemented with studies of mustard-exposed human skin grafts on athymic nude mice. Ultrastructurally, the type of cellular injury seen in human skin does not appear to differ from that observed subsequent to a wide variety of toxic insults that lead ultimately to epithelial cell degeneration. The sequence of events begins within basal keratinocytes and always within the cell nucleus. Extensive condensation and margination of heterochromatins and loss of euchromatins are followed by blebbing of the nuclear membrane (blebbing of the nuclear membrane is also evident on light microscopy at this stage). Cell lysis begins with the formation of paranuclear vacuoles, swelling of rough and smooth surface endoplasmic reticulum, dissociation of free rosettes of polyribosomes, loss of mitochondrial structure, cytoplasmic vacuolization, and eventual disruption of the plasma membrane.

These changes are probably confined principally to basal cells of the epidermis, because basal cells are the most active metabolic cells, actively and continuously synthesizing nucleic acids and nucleoproteins that are vital for cell growth and division. Antineoplastic agents such as sulfur mustard and nitrogen mustard derivatives exert their most prominent cytotoxic effects on cells that are actively producing large quantities of nucleic acids and nucleoproteins in preparation for cell division. Cross-linking of DNA is one of the most important cytopathic effects caused by sulfur mustard and other alkylating agents (Wheeler, 1967). Cells whose repair systems are overwhelmed by large concentrations of alkylating agents and are unable to sustain effective nucleic acid repair ultimately succumb to cell injury and then death.

Vesiculating and necrotizing dosages of sulfur mustard produce microscopic effects similar to erythemogenic dosages, differing only in the extent of injury to the epithelium. Basal cell degeneration is more widespread, and liquefaction necrosis involves multiple neighboring cells rather than isolated foci. Initially limited dermal-epidermal separation followed by extensive separation causes formation of microscopic then macroscopic vesicles. Gross blistering then becomes obvious.

Although residing within the dermis, adnexal structures such as hair follicles, sebaceous glands, and eccrine glands are morphologic derivatives of epidermis. Therefore, tissue injury within these structures occurs in a manner similar to that seen in normal epidermis. Tissue injury is considerably less within the epidermis of hair follicles; sweat glands show only minor defects, as do sebaceous glands.

In some situations, necrotic degeneration of the entire epidermis has been described following blister formation and after the delivery of large dosages of sulfur mustard. Such an occurrence is to be expected after complete degeneration of the basal cell layer has taken place. An intact basal cell layer is important to survival of the entire epidermis. Cells above the level of the basal cell layer are engaged in the process of differentiation and are not programmed for survival, as are normal basal cells. The end product of the process of differentiation is a nonviable, completely cornified cell. Unlike Lewisite, sulfur mustard has been shown to be relatively inactive in interfering with protein and enzyme synthesis. The action of sulfur mustard on protein and enzymes required in the formation of keratin protein does not appear to be effective enough to cause injury to differentiating epithelial cells. This assumption, however, does not rule out an overwhelming dosage of sulfur mustard acting as individual cell poisons in replicating and differentiating epithelial cells.

Factors responsible for damage to the underlying dermis after injury induced by sulfur mustard are not totally understood. Unlike lower mammalian species, dermal injury in human skin is not often as extensive as epidermal (Renshaw, 1946). During the early erythema and edema stages of skin injury, dilatation of papillary dermal capillaries, thickening of the capillary wall, and endothelial cell swelling are noted. Dermal cellular infiltrates are sparse, accumulating principally in a perivascular location. Lymphocytes predominate early, followed by an invasion of polymorphonuclear leukocytes. Perivascular edema is fairly prominent.

As injury progresses and vesiculation occurs, capillaries beneath vesicles show signs of necrosis. Capillary thrombosis and disruption of capillary wall integrity do not occur until there is evidence of dermal necrosis. Inflammation is increased in the presence of vesiculation and, in fact, may become quite extensive throughout the papillary and reticular dermis. Mononuclear and polymorphonuclear leukocytes appear equally prominent in the presence of epidermal and dermal necrosis. Langerhans cells appear to migrate from the epidermis and collect in clusters at the dermal-epidermal junction.

Fibroblasts, cells that are principally engaged in collagen formation and are basically resting cells in the absence of tissue injury, appear to experience some injury even during the erythematous stage. Pyknosis of the nuclei of fibroblasts becomes prominent and increases as local tissues enter the stage of epidermal and dermal necrosis. Collagen bundles appear not to be affected by injury, retaining a fairly normal morphologic appearance through, and up to, the appearance of blisters.

Mechanisms of Skin Injury

Our ability to explain or understand the mechanisms that lead to sulfur mustard injury in human skin has been encumbered by the lack of an appropriate animal model system. There are no animals in which it has been possible to reproduce, in its entirety, the effects of sulfur mustard on human skin (Mitcheltree et al., 1989; Renshaw, 1946). Numerous investigators using rabbits, pigs, cows, rats, mice, and guinea pigs, both normal and hairless varieties, have made attempts to replicate sulfur mustard human skin injury. Some gross and microscopic similarities to human skin responses have been noted in a few animal species, but for as-yet-undiscovered anatomical and physiological reasons human and animal injury differ to a significant degree. Sulfur mustard penetration of mammalian skin, other than human, occurs rapidly and to fairly deep levels. For a given dosage, higher dermal concentrations are achieved in nonhuman mammalian skin, and hence more profound tissue damage is noted in the dermis rather than epidermis. Injuries to animal skin develop and heal more rapidly than injuries to human skin, despite the same degree and severity of injury. Microblisters rather than macroblisters characteristically appear in the skin of all laboratory species tested. This response seems to occur at all effective dosage levels. The lack of a suitable animal model has also severely limited research directed toward the development of potential prophylactic or treatment agents.

To date, *in vitro* studies using primary cultures of newborn rat epidermal keratinocytes, or human skin grafts placed on athymic nude mice, have provided the largest body of useful information relative to sulfur mustard toxicity in skin (Papirmeister et al., 1991). These studies have uncovered a number of clues to the possible mechanism of sulfur mustard skin toxicity. Based on data reported from these systems two postulates have emerged to explain the pathogenesis of sulfur mustard skin toxicity.

DNA Alkylation and Protein Activation. Using cultures of primary epidermal human keratinocytes, Papirmeister and coworkers (1991) have uncovered evidence that alkylation of DNA, reduced tissue levels of oxidized nicotinamide adenine dinucleotide (NAD^+), and activation of cellular proteases may account for sulfur mustard-induced blister formation in human skin. According to the authors, the key reaction begins with the intramolecular cyclization of sulfur mustard to form an electrophilic ethylene episulfonium intermediate, which, in the presence of water, alkylates a number of tissue macromolecules including DNA, RNA, and protein. Alkylation of DNA ultimately results in single-strand breaks within basal cells of the epidermis, which in turn trigger activation of DNA repair enzymes, particularly poly-(adenosine diphosphateribose) polymerase (PADPRP). Excessive PADPRP activity causes inhibition of glycolytic and respiratory enzymes, impaired glucose uptake, and depleted stores of oxidized NAD^+ . NAD^+ depletion results in inhibition of glycolysis, buildup of glucose-6-phosphate (a substrate in the hexose monophase shunt), and, ultimately, activation of cellular proteases. Proteases released from epidermal cells are thought to cause disruption of dermal-epidermal attachments and resultant blister formation.

In studies using cultured human epidermal cells, Martens (1991) has accumulated some preliminary data to show that after exposure to 0.3 mM sulfur mustard, glucose utilization was 50 percent inhibited, and NAD^+ content within cells decreased to 50-60 percent of control. The onset of sulfur mustard-induced NAD^+ depletion preceded the inhibition of glycolysis. Martens states that his data support the Papirmeister group's postulated mechanism of sulfur mustard damage to human skin. Smith and Dunn (1991) have stated also that studies at their institute are consistent with the Papirmeister proposal. They add, however that "the process would appear to require an active inflammatory response and altered fluid dynamics in the affected tissue to generate . . . blisters."

DNA Cross-Linking. A simpler postulate is advanced by Bernstein and colleagues (1985, 1987). Using rat keratinocytes cultivated on nylon membranes, their experiments—in which low-dose sulfur mustard (0.01-0.5 nmol/cm^2), caused significant inhibition of ^3H

thymidine uptake—show that DNA is the most important target of sulfur mustard. Uptake of ^{3}H uridine (RNA synthesis) and ^{14}C leucine (protein synthesis) was not affected by low-dose sulfur mustard in this system. Concentrations on the order of 10-500 nM/cm² were required to inhibit ^{3}H uridine and ^{14}C leucine incorporation. Bernstein proposes that alkylation of nucleophilic residues of macromolecules creates interstrand and intrastrand cross-linking. Synthesis of DNA in particular but also RNA and protein is inhibited, thereby blocking cells at the interface of G_2/M phases of the cell cycle. Disruption of cell proliferation in the germinative cell population of the skin occurs, eventually death of the affected cells takes place, and separation of the epidermis from the dermis causes blister formation. Yet to be explained in this postulate are the ultimate fate of alkylated DNA and how alkylation is related to cell death and vesication.

Other investigators (Crathorn and Roberts, 1966; Lawley and Brookes, 1965) have found that cytotoxicity of sulfur mustard was not associated with inhibition of RNA or protein synthesis in a certain type of tumor cell (Hela) and in the bacteria *Escherichia coli,* respectively, and that disturbance of growth was specifically associated with inhibition of DNA synthesis. Their data may be advanced in support of Bernstein's theory of blister formation.

That sulfur mustard does indeed induce cross-links in DNA has been shown by Sorscher and Conolly (1989), using primary cultures of newborn rat epidermal cells. DNA cross-linking in sulfur mustard-exposed cells was seen at 5 mm sulfur mustard, but not at a lower dose of 2.5 mm sulfur mustard. Increased cross-linking was not observed in cultures treated with 10 and 20 mm sulfur mustard.

Lewisite

Lewisite is a lipophilic substance; therefore, absorption through the skin is a primary route of entry into the body. Percutaneous absorption may be associated with systemic toxicity, manifested by pulmonary edema, diarrhea, agitation, weakness, hypothermia, and hypotension. Systemic toxicity secondary to Lewisite exposure occurs more rapidly, and toxicity is more severe, than after exposure to sulfur mustard.

Cutaneous reactions after exposure to Lewisite vary depending on the atmospheric temperature, relative humidity, presence or absence of water, physical state of the agent (vapor or liquid), and concentration and duration of agent delivered to the skin. The maximum effect of Lewisite on skin takes place under conditions of low temperature, low humidity, and dry nonalkaline terrain (Gates et al., 1946).

Under standard conditions of temperature and humidity, Lewisite vapor (12 mg/m^3 or less) produces erythema of the skin. Higher concentrations (1,000 $mg{\cdot}min/m^3$) under conditions that yield hot, dry skin may induce small, shallow, turbid vesicles that coalesce to form larger blisters (Goldman and Dacre, 1989). Liquid Lewisite (0.2 mg/cm^2) on the skin produces an immediate (10-12 seconds) stinging and burning sensation at the site of application (Davis, 1944). This is followed within 5-15 minutes by a cooked-skin appearance, characteristically dull dead-white or grayish skin similar to that seen after an acid burn. Shortly thereafter, erythema develops around the contaminated site. The central region of the burn becomes urticaria-like and lemon in color. Later, puckering of the skin occurs around adnexal orifices to give the appearance of "tanned pigskin." Six to eight hours later, pinpoint vesiculation appears, shortly replaced by large bullae that cover the entire erythematous area. Characteristically there is a very sharp line of demarcation between Lewisite-burned skin and normal skin, making it fairly simple to distinguish between the burn of Lewisite and sulfur mustard. Delivery of a large quantity of Lewisite to intact skin over a prolonged period can result in deep penetration of agent through subcutaneous tissue into muscle, with attendant edema and necrosis.

Although Lewisite injuries are often described as healing at a more rapid rate than those due to sulfur mustard, Davis (1944) states that "it has been thought by some observers to heal more rapidly than mustard gas burns, but I cannot assent to this opinion." Davis also notes that residual pigmentation in small Lewisite burns is not as characteristic as in sulfur mustard burns. Rather, Lewisite is more prone to leave residual atrophic scarring.

Lewisite lethality in man when delivered via skin is also related to physical state on delivery. According to data presented to this committee by Colonel Richard Solana of the U.S. Army Medical Research Institute of Chemical Defense (Appendix A), the LD_{50} for humans is estimated to be 40 mg/kg for liquid Lewisite, and 100,000 $mg{\cdot}min/m^3$ for Lewisite vapor. The LD_{50} often quoted for liquid Lewisite is considered low by many investigators (Goldman and Dacre, 1989).

Less is known about Lewisite than sulfur mustard penetration through human skin. Axelrod and Hamilton (1947), in studies using 10 microcuries of radioactive arsenic (^{74}As) per milligram of Lewisite, applied 475 μg of Lewisite to the skin of humans for 10 and 15 minutes. During each experiment 0.43 cm^2 of skin was exposed, and biopsy specimens were taken 24 hours after exposure. Radioautographs of exposed tissue showed Lewisite to be confined principally to the epidermis; very little was found in the dermis, mostly around blood vessels and in some but not all hair follicles. Radioactivity within the epidermis was confined to dead epithelial cells. There appeared to exist massive necrosis of most of the involved epidermis. In similar experi ments using pigskin, labeled Lewisite was deposited primarily in the hair and hair follicle, with a small amount within the epidermis. Ferguson and Silver (1947) described similar experiments using the skin of guinea pigs: Lewisite could be found within the epidermis in 2 minutes and the dermis within 10; it remained concentrated within the dermis for about 30 minutes and then began to disappear; only traces were detectable in the skin after 24 hours.

The histopathologic changes in skin after Lewisite exposure have been described by Davis (1944), Cameron and colleagues (1946), and McGown and colleagues (1985, 1987). Unlike sulfur mustard exposure, Lewisite causes early and complete necrosis of the epidermis in humans. The necrotic process also involves the dermis where it is principally vascular in location. Capillary degeneration and perivascular leukocyte infiltration accompany Lewisite vesiculation. Feister and colleagues (1989) state that there is evidence to show that, like vesication after sulfur mustard exposure, vesication subsequent to Lewisite injury occurs within the lamina lucida. However, it is not clear which anatomical structures are disrupted to cause epidermal-dermal separation. Studies in the human skin-grafted nude mouse system suggest that epidermal-dermal necrosis precedes epidermal-dermal separation.

It is assumed that upon entry of Lewisite into the aqueous medium of the intact skin it is rapidly hydrolyzed to a stable, water-soluble, but highly toxic derivative 2-chlorovinylarsine oxide (Lewisite oxide) and hydrochloric acid. Feister and colleagues (1989) postulate that Lewisite oxide may be the principal metabolite and major cytotoxic form of Lewisite within tissues. It is also believed that the trivalent form of arsenic, which is highly reactive in biological systems, is responsible for the overt toxicity of all arsenicals, including Lewisite, to living systems. Trivalent arsenicals exert their toxic effects through interactions with active tissue sulfhydryl groups (Peters, 1955; Peters et al., 1946; Squibb and Fowler, 1983). Trivalent arsenicals interact directly with protein sulfhydryl or thiol (sulfhydryl attached to a carbon) groups.

Because a vast array of critical enzymes contain thiol groups that interact with arsenicals, the end result of this interaction is enzyme inactivity. This ability of arsenicals to inhibit tissue enzyme activity in a variety of animal systems has made them valuable tools in the study of the biochemistry of specific enzymes, their mechanisms of action, and sites of action. A body of work, beginning with the separate investigations of R.A. Peters and C. Voegtlin in the 1920s, supports the concept that cell death from Lewisite results from the inhibition of the pyruvate-dehydrogenase complex, causing energy depletion within the cell (Peters, 1955; also see Feister et al., 1989). Numerous studies have shown that addition of arsenic to isolated mitochondria produces an inhibition of cellular respiration, the oxidation of tricar boxylic acid cycle substrates, and oxidative phosphorylation (Squibb and Fowler, 1983). An alternative and as yet unproven theory of Lewisite toxicity postulates inhibition of glycolysis secondary to arsenical inhibition of the hexokinase enzyme. The resultant inhibition of each of the subsequent biochemical steps in energy metabolism leads to energy depletion and cell death.

EVIDENCE OF LONG-TERM HEALTH EFFECTS OF MUSTARD AGENTS

Animal Studies and Cellular Bioassays

Animal studies of sulfur mustard effect on skin have been directed principally at defining a role for this agent in the process of carcinogenesis (Fox and Scott, 1980; Heston, 1953; McNamara et al., 1975). Such studies have been performed in a variety of animals including dogs, guinea pigs, rabbits, rats, and mice, as described in Chapter 6. Although several routes of administration have been used, subcutaneous injection of fairly large dosages of sulfur mustard was the most successful route producing cutaneous papillomas and sarcomas. Long-term exposure to sulfur mustard vapor also produced squamous cell and basal cell cancers in rats. These experiments, although crude, suggest that similar acute and chronic exposure in humans may be carcinogenic.

Human Studies

The persistence of lesions following sulfur mustard exposure is directly related to the duration and severity of the exposure. Secondary infection will often influence the process of healing and, in turn, may influence the eventual outcome of skin injury. Injury that results in erythema and edema without vesicle formation is almost always followed by complete healing and no residual cutaneous defects. Sites characterized by early and diffuse hyperpigmentation are exceptions. In such sites, erythema may be followed by exfoliation and skin necrosis. Blistering wounds and necrotic wounds, which characteristically leave large areas of skin devoid of protective epithelium, melanocyte, and intact adnexal structures, are often followed by permanent residual skin defects. Skin damage is intensified in the presence of secondary bacterial infection of unepithelialized skin. The effect of infection is intensified by the action of systemically absorbed sulfur mustard on bone marrow. Intense skin exposure sufficient to cause severe vesiculation and skin necrosis is almost always associated with systemic toxicity. Destroyed or diminished bone marrow activity denotes reduced numbers of or destruction of replicating marrow stem cells (also see Chapter 10).

Reduced granulocyte and other marrow-derived cells in the peripheral blood cause a diminished protective effect from polymorphonuclear leukocytes, macrophages, monocytes, and other cell types that are active in the destruction and scavenging of organisms that invade and impede healing of wounds.

Residual cutaneous lesions most often take the form of scars that result from uncontrolled fibroblastic activity and overgrowth of connective tissue during the process of wound repair. Even well cared for wounds over body sites and parts that are not easily immobilized, such as shoulders, knees, elbows, and male genitalia, often heal with severe residual scar formation. Pigmentation is often altered (either increased or decreased) at these sites, although the degree of alteration does not differ from that observed in injuries caused by burns and other forms of physical and chemical insult. In the absence of melanocyte destruction, hyperpigmentation will predominate. If melanocytes are locally destroyed, and inward migration from destroyed adnexal structures does not occur, depigmentation will predominate. Some previously injured sites have been described as "sensitive" to subsequent mechanical injury. These sites may show recurrent blisters after mild injury.

Skin tumors (basal cell, squamous cell, and Bowen's intraepidermal squamous cell cancer) and rapidly spreading skin ulcers that are resistant to therapy have been reported (Inada et al., 1978; Jackson and Adams, 1973; Klehr, 1984; Wada et al., 1963). To date, the number of cutaneous cancers reported subsequent to acute and chronic sulfur mustard exposure is low, and it is unclear whether some of these cutaneous cancers are related to the carcinogenic effects of sulfur mustard or are related to the presence of chronic skin ulcers (Jackson and Adams, 1973). The occurrence of skin cancers at the site of old scar formation is an acknowledged biological phenomenon (Novick et al., 1977; Treves and Pack, 1930). It appears that cutaneous cancers following acute sulfur mustard exposure usually localize in cutaneous scars, whereas those following chronic exposure can occur on any exposed site (Inada et al., 1978). Many questions remain unanswered in sulfur mustard human carcinogenicity. Some subjects who develop cutaneous cancers after chronic sulfur mustard exposure, particularly Bowen's disease, have had exposure to multiple potential cancer-causing agents, including Lewisite (Inada et al., 1978; Wada et al., 1963). Yet, Kravitz and McDonald (1978) have reported cutaneous cancer induction following chronic topical application of nitrogen mustard in the treatment of cutaneous T-cell lymphoma.

Occupational Exposure

Long-term effects of sulfur mustard exposure have been most frequently described in people previously employed in the manufacture of mustard gas (Büscher, 1932; Easton et al., 1988; Inada et al., 1978; Klehr, 1984). Pigmentary disorders, skin ulcers, and cutaneous cancers and precancers have been the most common entities described. Klehr described a group of World War II (WWII) German mustard gas workers as having multiple skin tumors, even in unexposed skin, and numerous painful ulcerations that tend to spread. Klehr's report is principally descriptive and is without valid comparative data or control populations, as are most other reports of occupational diseases associated with sulfur mustard exposure. Of 53 workers remaining alive and examined, 34 percent experienced multiple skin tumors, and 45 percent experienced arterial vascular ulcers of the lower extremities. Klehr's clinical description of the kinds of ulcers seen gives the impression that they are not unlike leg ulcers commonly found in individuals with chronic arterial and venous disease. The lack of a control population and the very general descriptive and retrospective nature of this report make it difficult to place value on the content.

Wada and colleagues (1962, 1963) and Inada and colleagues (1978) describe findings in a group of former workers exposed for variable periods to a variety of "war gases" manufactured on Okuno-jima island in Japan (see Chapter 6). Of 488 workers, 115 showed pigmentary skin changes consisting of hyperpigmented and depigmented raindrop spots, mostly on covered skin of the trunk and extremities. Another 22 cases with Bowen's disease, basal cell carcinoma, and other hyperkeratotic skin lesions were described. Of 5 cases extensively described, the average time between initial exposure to sulfur mustard and the development of tumors varied between 31 and 46 years. The number of years worked in the facility varied from 3 to just over 15 years. Most workers in this facility wore protective clothing, which was described as "defective and ill-fitting" most of the time: numerous instance of skin burns, blisters, and other cutaneous injuries were reported. Although an adequate control population was not simultaneously studied, 77 workers who were engaged in clerical and guard duty at the same facility, did not develop evidence of long- or short-term defects.

The principal drawback to assigning a cause and effect relationship solely to sulfur mustard exposure at this facility may be found in the background comments: before 1937, workers whose tasks were limited to the production of "war gases" worked in the manufacture of gases of all types, including mustard gas and Lewisite (Wada et al., 1962). Also missing from these studies is a statistical comparison of the number of cases of Bowen's disease one could expect to find in a comparable population of nonexposed Japanese. In a nationwide 5-year study of skin cancer, the overall incidence of skin cancer among 851,685 new patient visits to dermatologic clinics at major Japanese universities was

0.10 percent (Miyaji, 1963). Comparable figures have been reported by other Japanese authors (Kitamura, 1954), although a study in Hiroshima, the site of a Japanese pre-WWII war gas factory, reported an incidence of 0.16 percent (Hosokawa, 1961). When the geographical distribution of skin cancers in Japan was examined, there was a greater incidence in southern and western Japan (including the prefecturate of Hiroshima), areas where the largest amount of annual sunlight is seen (Miyaji, 1963).

In a report on the British occupational experience, Easton and colleagues (1988) looked at mortality data from a World War II mustard gas manufacturing site in Cheshire, England. The observed number of deaths from skin cancer was zero, versus the expected number of two. The implication from this study is that there is a low or nonexistent death rate from skin cancer in this cohort of individuals, and certainly the death rate is lower than generally expected in a group of exposed workers. Indeed, the incidence of all expected diseases and deaths has been less in the British workers than in workers from other nations, a difference attributed to better worker protection measures in British war gas factories.

Battlefield Exposure

To date, there has been only a single report describing delayed toxic effects of sulfur mustard exposure during battlefield operations. Balali (1986), in a prospective study of delayed toxic effects, has followed a cohort of Iranian solders exposed to mustard gas during the Iran-Iraq war. After two years of observation, 41 percent of the exposed victims are experiencing pigmentary disorders. No other abnormalities have as yet surfaced.

Medical Therapeutic Exposure

For a number of years, Russian and Eastern European physicians have studied the effects of a topical preparation containing sulfur mustard 0.005 percent in petrolatum (psoriasin) on a hyperproliferative disease of the skin, psoriasis. The delivery of therapeutic dosages requires about 0.01 μg psoriasin/cm^2 of skin. This amount results in inhibition of DNA synthesis sufficient to reduce basal cell replication, causing a return of the bulk of cells back to a state comparable to normal, yet the cells' ability to repair DNA cross-linking is not impaired. This dosage level is 10-100 times lower than that required to cause erythema in normal skin (Renshaw, 1946). Short-term (15 days) observation of patients treated with psoriasin reveals cutaneous hyperpigmentation like that seen after the application of nitrogen mustard to the skin, stimulation of hair growth, and cutaneous sensitization (Turanow et al., 1977). Long-term effects are yet to be reported.

Experimental Exposure

Renshaw (1946) has reported on the development of contact sensitivity in man following localized exposure to liquid sulfur mustard. Cutaneous sensitivity may be seen within 8 days following the first application, and a more pronounced effect is seen after four weeks. The incidence of hypersensitivity varies between 30 and 65 percent of exposed individuals. Sensitivity may be immediate (urticaria) or delayed (dermatitis) and appears to last for a lifetime. Sensitivity also includes flares of old, healed sulfur mustard injured sites after a fresh application of sulfur mustard to normal unaffected skin.

EVIDENCE OF LONG-TERM HEALTH EFFECTS OF LEWISITE

The long-term health effects of Lewisite on skin are unknown. There is an extensive bibliography on the long-term effects of arsenicals on specific organ systems, but with few exceptions the skin is omitted from these studies. There is considerable controversy over which arsenicals are most toxic to human and animal tissues (e.g., inorganic arsenicals versus organic arsenicals, versus trivalent, versus pentavalent). The more recent literature leans toward the conclusion that most long-term effects attributable to arsenicals are due to exposure to inorganic trivalent arsenic (Goldman and Dacre, 1989; Squibb and Fowler, 1983).

Epidemiologic studies have clearly demonstrated a real association between chronic adverse reactions and occupational exposure to inorganic arsenic in pesticides, herbicides, fungicides, and animal disinfectants, and in smelter workers. In medicine, preparations such as Fowler's solution, asiatic pills, and Donovan's solution that contain trivalent elemental arsenic have been associated with long-term effects including dermatitis, hyperpigmentation, loss of hair, disseminated cutaneous keratoses, palmar hyperkeratosis, and cutaneous cancer, including basal cell, squamous cell, and Bowen's intraepidermal squamous cell cancer. Yet the long-term administration of organic arsenicals in the treatment of syphilis, trypanosomal diseases, parasitic infestations, relapsing fever, and yaws has not been associated with any of the adverse reactions outlined above.

Animal Studies and Cellular Bioassays

Most animal studies of the long-term effects of Lewisite and other arsenicals on skin and other organs have been directed toward the elucidation of carcinogenesis (Fraumeni, 1975; IARC, 1980; Kennaway, 1942; National Research Council, 1977; Pershagen, 1981). Squibb and Fowler (1983) state emphatically, "The question as to whether arsenic is a direct carcinogen ... remains unanswered at this time. Epidemiological data clearly indicate that exposure to arsenic increases the incidence of skin, lung, liver and lymphoid cancer in humans, however, animal studies designed to confirm the carcinogenic potential of arsenic and its compounds have been, for the most part, negative." Goldman and Dacre (1989) state, "There is still reservation about accepting arsenic as a carcinogen because of the failure to demonstrate that arsenic in any form has resulted in an increased incidence in the production of tumors in experimental animals."

Human Studies

Occupational Exposure

The most often quoted evidence of Lewisite-induced cutaneous cancer is a case of Bowen's disease that developed 8 years after Lewisite-produced injury (Krause and Grussendorf, 1978) and the multiple keratoses and skin cancers in the group of Japanese "war-gas" factory workers described above (Inada et al., 1978). The questions regarding the Japanese study in terms of sulfur mustard effects also apply here.

Arsenic has been linked to the production of human cancer by many investigators (Allen, 1967; Graham and Helwig, 1959; Graham et al., 1961; Montgomery and Waisman, 1941). Roth (1956) described arsenic-induced cancers among vineyard workers, as well as a striking multiplicity of arsenic-induced cancers. Sommers and McManus (1953) also called attention to the multiplicity of lesions and the involvement of internal organs as well as skin. Arsenic may produce keratoses (keratinized protuberances of skin, particularly on the palms and soles), squamous cell cancer, basal cell cancer, multicentric intraepithelial basal cell carcinoma, and Bowen's intraepidermal squamous cell cancer.

Convincing proof of the etiologic linkage of arsenic to neoplasms is the demonstration of arsenic in tumors. Two fairly simple tests have been used: (1) the Osborne test, which demonstrates the presence of yellowish-green crystals in specially stained histologic sections of skin; and (2) the direct differential chemical analysis of fragments of tissue for arsenic. Montgomery and Waisman (1941) have shown that normal skin will contain 0.00008 mg of arsenic per gram of tissue; whereas, cancerous skin will contain 0.00024 to 4.3 mg of arsenic per gram of tissue. Arsenic has been recovered from human skin up to 30 or more years after administration of the compound. Graham and colleagues (1961) in an analysis of normal and involved skin found arsenic in increased amounts in a significantly greater proportion of Bowen's disease patients than in lesions of control patients with other dermatoses. These findings were based on a review of material from 15 patients with keratoses and a definite history of ingestion or contact with arsenic, versus patients without such history. In a review of data from another series of studies, Graham and Helwig (1959) conclude that "our observations strongly suggest that arsenic could be one of the causes of Bowen's disease and that the systemic and cutaneous cancers in these patients may well represent the systemic manifestations of this strong chemical carcinogen."

Battlefield Exposure

It has been stated that the value of Lewisite as a military agent depends in large degree on whether the necessary dosages can be "set up in the field." Field experiences indicate that dosages sufficiently large enough to impact on military operations "are probably not attainable with any reasonable expenditure of munitions" (Gates et al., 1946). Neither saturation of fields, nor delivery of thickened and unthickened Lewisite vapor through bombs and airplane spray, has proven of value. The casualty-producing properties of sulfur mustard far outweigh those of Lewisite, and for this reason there has been no known battlefield use of Lewisite.

Medical Therapeutic Exposure

As stated above, inorganic as well as organic arsenicals have been used for medicinal purposes. Inorganic arsenicals have been used since the time of Hippocrates (460-377 B.C.). Due to their low comparative toxicities, organic arsenicals supplanted the general use of inorganic arsenicals in medicine during the early 1900s. However, the use of inorganic arsenicals was not totally eliminated, and many products were sold as "over-the-counter" home remedies and tonics through the latter half of this century. In individuals exposed to inorganic arsenicals through this route, all of the adverse reactions described earlier have been seen. Cutaneous cancers, basal cell, squamous cell, and Bowen's disease have been well described in these populations. In numerous instances, systemic metastatic cancers of the internal organs have been associated with a large number and variety of cutaneous cancers (Sommers and McManus, 1953).

Experimental Exposure

Other than sensitization subsequent to Lewisite application to skin, described above, there is a paucity of information regarding long-term effects of acute or chronic exposure to Lewisite from experimental situations.

SUMMARY

Gaps in the Literature

There is a wealth of information on the acute and short-term effects of sulfur mustard on human and animal skin. However, there is a paucity of literature describing delayed or long-term effects. Based on a small body of information derived from fairly crude data, observation periods as long as 35-45 years may be required to produce meaningful human data. To our knowledge, the only prospective study of long-term cutaneous effects of acute sulfur mustard exposure on human skin is that of Balali (1986). This study is now in its fourth or fifth year and should provide very valuable information in 15-20 years.

Human data derived from patients previously treated in Russian and Eastern European studies of the agent psoriasin may also be useful in determining the delayed effect of short-term administration of suberythema dosages of sulfur mustard. We are now approaching 20 to 25 years from the beginning of these studies. Follow-up of those participants, if properly done, now would be of invaluable help in determining delayed effects of acute sulfur mustard exposure. It is possible that some studies were designed using chronic dosing and adequate control populations. If so, patients in this category may aid in determining if chronic sulfur mustard administration in subinjury dosages, like nitrogen mustard, may lead to the development of cutaneous cancer.

Ideally, if one were able to determine successfully who participated, and when, in the variety of experiments carried out by the U.S. Armed Forces and its Allies in World War II, an examination of this group, potentially numbering in the thousands, would serve as an excellent source of data on the long-term effects of sulfur mustard on the skin.

There are also numerous gaps in the literature relative to the acute and long-term effects of Lewisite skin exposure. Lewisite has been subjected to much less intense investigation than sulfur mustard. Very little is known regarding its specific effect on skin; data on such basic areas as absorption, disposition, and excretion after skin exposure are minimal. Although much is known of Lewisite's biochemical interactions, little is known of the morphologic sites of these interactions. Microscopic examination of affected skin has yet to be pursued in depth. Most studies have been impaired, as has been work on sulfur mustard exposure, by the lack of good animal model systems.

Studies on the carcinogenicity or noncarcinogenicity of Lewisite need to be broadened and pursued with greater intensity. The information obtained from these studies, unlike studies of sulfur mustard exposure, will have broad application in industry, farming, and medicine.

Conclusions

Despite the many years that the problem of acute sulfur mustard toxicity to human skin has been known and observed, its long-term effects after acute and chronic exposure remain obscure. Unfortunately, large volumes of pertinent literature on experimental studies of human exposure remain obscure or destroyed. Despite the flaws in the literature explored to date, it is possible to conclude that (1) **the evidence indicates a causal relation between acute, severe exposure to mustard agents and increased pigmentation and depigmentation in human skin; (2) acute and severe exposure can lead to chronic skin ulceration, scar formation, and the development of cutaneous cancer; and (3) chronic exposure to minimally toxic and even subtoxic doses can lead to skin pigmentation abnormalities and cutaneous cancer.** The evidence would nevertheless be strengthened by (a) intensive data review; (b) physical examination of identifiable victims of experimentation during and preceding former wars, and the comparison of these individuals with matched cohorts of nonexposed persons; and (c) continued prospective evaluation of individuals with recent battlefield and experimental exposure. It should also be emphasized that scarring of scrotal and penile tissue, quite likely in mustard agent exposure, can impair sexual performance.

There is insufficient information, however, to establish a causal relationship between Lewisite exposure and long-term adverse effects on skin.

REFERENCES

Allen AC. 1967. The Skin: A Clinicopathological Treatise. 2nd ed. New York: Grune and Stratton.

Axelrod DJ, Hamilton JG. 1947. Radio-autographic studies of the distribution of Lewisite and mustard gas in skin and eye tissues. American Journal of Pathology 23:389-411.

Balali M. 1986. First report of delayed toxic effects of yperite poisoning in Iranian fighters. In: Heyndricks B, ed. Terrorism: Analysis and Detection of Explosives. Proceedings of the Second World Congress on New Compounds in Biological and Chemical Warfare. Gent, Belgium: Rijksuniversiteit. 489-495.

Bernstein IA, Brabec MJ, Conolly RC, Gray RH, Kulkarni A, Mitra R, Vaughan FL. 1985. Chemical Blistering: Cellular and Macromolecular Components. AD-A190 313. Ann Arbor, MI: University of Michigan.

Bernstein IA, Bernstam L, Brown R, Fan L, Feng HW, Ku W, Locey B, Ribeiro P, Scavarelli R, Vaughan FL, Zaman S. 1987. Macromolecular and cellular effects of sulfur mustard on keratinocyte cultures. In: Proceedings of the 6th Chemical Defense Bioscience Review. Frederick, MD. 243-250.

Büscher H. 1932. Green and Yellow Cross. Hamburg, Germany: Himmelheber. Translated from the German in 1944 by N Conway, Kettering Laboratory of Applied Physiology, Cincinnati, OH.

Cameron GR, Carleton HM, Short RHD. 1946. Pathological changes induced by Lewisite and allied compounds. Journal of Pathology and Bacteriology 58:411-422.

Crathorn AR, Roberts JJ. 1966. Mechanism of the cytotoxic action of alkylating agents in mammalian cells and evidence for the removal of alkylated groups from deoxyribonucleic acid. Nature 211:150-153.

Cullumbine H. 1947. Medical aspects of mustard gas poisoning. Nature 159:151-153.

Davis J. 1944. Dermatologic aspects of vesicant war gases (dichloroethyl sulfide and dicholorovinylarsine). Journal of the American Medical Association 126:209.

Easton D, Peto J, Doll R. 1988. Cancers of the respiratory tract in mustard gas workers. British Journal of Industrial Medicine 45:652-659.

Feister A, Papirmeister B, Robinson S, Kiebzak G, McNally R, Ford R, Baggett J, Gottlieb J, Bareis D. 1989. Sulfur Mustard and Lewisite: Current Perspectives and Future Directions. Prepared for the U.S. Army Medical Research Institute of Chemical Defense. Unpublished.

Ferguson RL, Silver SD. 1947. A method for the visual demonstration of Lewisite in skin. American Journal of Clinical Pathology 17:37-38.

Fitzpatrick TB, Eisen AZ, Wolff K, Freedberg IM, Austeru KF. 1979. Dermatology in General Medicine: Textbook and Atlas. 2nd ed. New York: McGraw-Hill.

Fox M, Scott D. 1980. The genetic toxicology of nitrogen and sulphur mustard. Mutation Research 75:131-168.

Fraumeni JF Jr. 1975. Respiratory carcinogenesis: an epidemiologic appraisal. Journal of the National Cancer Institute 55:1039-1046.

Gates M, Williams JW, Zapp JA. 1946. Arsenicals. In: Division 9, National Defense Research Committee, comp. Chemical Warfare Agents, and Related Chemical Problems. Summary Technical Report of Division 9, NDRC. Washington, DC: Office of Scientific Research and Development.

Goldman M, Dacre J. 1989. Lewisite: its chemistry, toxicology, and biological effects. Reviews of Environmental Contamination and Toxicology 110:75-115.

Graham JH, Helwig EB. 1959. Bowen's disease and its relationship to systemic cancer. AMA Archives of Dermatology 80:133-159.

Graham JH, Mazzanti GR, Helwig EB. 1961. Chemistry of Bowen's disease: relationship to arsenic. Journal of Investigative Dermatology 37:317-332.

Heston WE. 1953. Occurrence of tumors in mice injected subcutaneously with sulfur mustard and nitrogen mustard. Journal of the National Cancer Institute 14:131-140.

Hosokawa T. 1961. Studies on skin tumors: statistical observation on skin tumors. Hiroshima Medical Journal.

Inada S, Hiragun K, Seo K, Yamura T. 1978. Multiple Bowen's disease observed in former workers of a poison gas factory in Japan with special reference to mustard gas exposure. Journal of Dermatology 5:49-60.

International Agency for Research on Cancer (IARC). 1980. IARC Monographs on the Evaluation of Carcinogenic Risk of Chemicals to Humans. Vol. 23, Some Metals and Metallic Compounds. Lyon: IARC.

Jackson R, Adams RH. 1973. Horrifying basal cell carcinoma: a study of 33 cases and a comparison with 435 non-horror cases and a report on four metastatic cases. Journal of Surgical Oncology 5:431-463.

Jimbow K, et al. 1976. Some aspects of melanin biology: 1950-1975. Journal of Investigative Dermatology 67:72-89.

Kennaway E. 1942. A contribution of the mythology of cancer research. Lancet 2:769-772.
Kitamura K. 1954. Tumors of skin. J Dermatol and Venereal Dis 64:303-312.
Klehr NW. 1984. Late manifestations in former mustard gas workers, with special consideration of the cutaneous findings. Zeitschrift fur Hautkrankheiten 59:1161-1170. [In German]
Krause H, Grussendorf El. 1978. Syntopy of Bowen's disease and "Lost"-induced scar. Hautarzt 29:490-493. [In German]
Kravitz P, McDonald CJ. 1978. Topical nitrogen mustard-induced carcinogenesis. Acta Dermato-Venereologica 58:421-425.
Lawley P, Brookes P. 1965. Molecular mechanism of the cytotoxic action of difunctional alkylating agents and of resistance to this action. Nature 206:480-483.
Martens ME. 1991. Glucose metabolism and NAD^+ content in cultured human epidermal keratinocytes exposed to sulfur mustard (abstract). FASEB Journal 5:A823.
Mathias CGT. 1987. Clinical and experimental aspects of cutaneous irritation. In: Marzulli FN, Maibach HI, eds. Dermatotoxicology. 3rd ed. Washington, DC: Hemisphere Publishing.
McGown EL, Van Ravenswaay T, Damlao CR. 1985. Histologic changes caused by application of Lewisite analogs to mouse skin and human skin xenografts. San Francisco: Letterman Army Institute of Research. AD-A159-554.
McGown EL, Van Ravenswaay T, Dumlao CR. 1987. Histologic changes in nude mouse skin and human skin xenografts following exposure to sulfhydryl reagents: arsenicals. Toxicologic Pathology 15:149-156.
McNamara BP, Owens EJ, Christensen MK, Vocci FJ, Ford DF, Rozimarek H. 1975. Toxicological Basis for Controlling Levels of Mustard in the Environment. Edgewood Arsenal Special Publication EB-SP-74030. Aberdeen Proving Ground, Maryland: U.S. Army Armament Command. Edgewood Arsenal Biomedical Laboratory.
Mitcheltree LW, Mershon MM, Wall HG, Pulliam JD, Manthei JH. 1989. Microblister formation in vesicant-exposed pig skin. Journal of Toxicology—Cutaneous and Ocular Toxicology 8:309-319.
Miyaji T. 1963. Skin cancer in Japan: a nationwide 5 year survey. In: Urbach F, ed. Conference on Biology of Cutaneous Cancer. National Cancer Institute Monograph No. 10. 55-70.
Momeni AZ, Enshaeih S, Meghadi M, Amindjavaheri M. 1992. Skin manifestations of mustard gas. A clinical study of 535 patients exposed to mustard gas. Archives of Dermatology 128:775-580.
Montgomery H, Waisman M. 1941. Epithelioma attributable to arsenic. Journal of Investigative Dermatology 4:365-383.
Nagy SM, Golumbic C, Stein WH, Fruton JS, Bergmann M. 1946. The penetration of vesicant vapors into human skin. Journal of General Physiology 29:441-469.
National Research Council. Division of Medical Sciences. 1977. Arsenic. Medical and Biologic Effects of Environmental Pollutants. Washington, DC: National Academy of Sciences.
Novick M, Gard DH, Hardy SB, Spira M. 1977. Burn scar carcinoma: a review and analysis of 46 cases. Journal of Trauma 17:809-817.
Papirmeister B, Feister AJ, Robinson SI, Ford RD. 1991. Medical Defense Against Mustard Gas: Toxic Mechanisms and Pharmacological Implications. Boca Raton, FL: CRC Press.
Pershagen G. 1981. The carcinogenicity of arsenic. Environmental Health Perspectives 40:93-100.
Peters RA. 1955. Biochemistry of toxic agents: I. Present status of knowledge of biochemical lesions induced by trivalent arsenical poisoning. Bulletin of Johns Hopkins Hospital 97:1-20.
Peters RA, Sinclair HM, Thompson RHS. 1946. An analysis of the inhibition of pyruvate

oxidation by arsenicals in relation to the enzyme theory of vesication. Biochemical Journal 40:516-524.

Quevedo WC, Jr. 1969. The control of color in mammals. American Zoologist 9:531-540.

Renshaw B. 1946. Mechanisms in production of cutaneous injuries by sulfur and nitrogen mustards. In: Division 9, National Defense Research Committee, comp. Chemical Warfare Agents, and Related Chemical Problems. Summary Technical Report of Division 9, NDRC. Washington, DC: Office of Scientific Research and Development.

Renshaw B. 1947. Observations on the role of water in the susceptibility of human skin to vesicant injury. Journal of Investigative Dermatology 9:75-85.

Roth F. 1956. Chronic arsenicism and cancers among vineyard workers in the Moselle Valley. Zeitschrift fur Krebsforschung und Klinische Onkologie 61:287-319.

Sams WM, Lynch PJ. 1990. Principles and Practice of Dermatology. New York: Churchill Livingstone.

Smith WJ, Dunn MA. 1991. Medical defense against blistering chemical warfare agents. Archives of Dermatology 127:1207-1213.

Sommers S, McManus RG. 1953. Multiple arsenical cancers of skin and internal organs. Cancer 6:347-359.

Sorscher D, Conolly R. 1989. Pretreatment of primary rat cutaneous epidermal keratinocyte culture with a low concentration of MNNG: effect on DNA cross-linking measured *in situ* after challenge with bis-2-chloroethyl sulfide. Journal of Toxicology and Environmental Health 27:367-379.

Squibb KS, Fowler BA. 1983. The toxicity of arsenic and its compounds. In: Fowler BA, ed. Biological and Environmental Effects of Arsenic. Vol. 6. Amsterdam: Elsevier Science. 233-269.

Stedman's Medical Dictionary. 1976. 23rd ed. Baltimore: Williams & Wilkins.

Treves N, Pack GT. 1930. Development of cancer in burn scars: analysis and report of 34 cases. Surgery, Gynecology, and Obstetrics 51:749-782.

Turanow NM, Trofimonwa LJ, Bolschakowa GM, Chapilowa WI. 1977. Experience report from the USSR on the management of psoriasis patients using "Psoriasin." Zeitschrift fur Hautkrankheiten 52:1045-1049. [In German]

Wada S, Nishimoto Y, Miyanishi M, Katsuta S, Nishiki M. 1962. Review of Okuno-Jima poison gas factory regarding occupational environment. Hiroshima Journal of Medical Sciences 11:75-80.

Wada S, Yamada A, Nishimoto Y, Tokuoka S, Miyanishi M, Katsuta S, Umisa H. 1963. Neoplasms of the respiratory tract among poison gas workers. Journal of the Hiroshima Medical Association 16:728-745.

Weigand DA, Haygood C, Gaylor G. 1974. Cell layers and density ofNegro and Caucasian stratum corneum . Journal of Investigative Dermatology 62:563-568.

Wheeler GP. 1967. Some biochemical effects of alkylating agents. Proceedings of the Society for Experimental Biology and Medicine 26:885-892.

Willems JL. 1989. Clinical management of mustard gas casualties. Annales Medicinae Militaris Belgicae 3:S1-61.

10

Other Physiological Effects of Mustard Agents and Lewisite

Exposure to mustard agents and Lewisite may have long-term effects on a number of other physiological systems. This chapter reviews the animal experimental data and human epidemiological studies that have focused on the following systems:

- immune system, including blood lymphocytes, lymphatic tissues, and bone marrow;
- gastrointestinal system;
- blood system;
- nervous system, including autonomic and higher-order functions; and
- reproductive risks.

EFFECTS ON THE IMMUNE SYSTEM

Anatomy and Physiology

The human body has the ability to resist many types of organisms or toxic agents to which it may be exposed. This capacity is called immunity. The immune response includes specific actions of blood lymphocytes (one type of white blood cells) and is facilitated by other white blood cells including neutrophils, monocytes, macrophages, eosinophils, and basophils.

The leukocytes are the mobile units of the body's protective system. They circulate throughout the body, moving in and out of tissues via the circulatory system and lymphatic system. The leukocytes are formed partially in the bone marrow and partially in the lymph tissue, but after formation they are transported in the blood to different parts of the body where they are to be used.

The development of the human immune system begins late in the fetal period, is functioning at birth, and reaches maximum capacity near the time of puberty. In human adults the majority of circulating lymphocytes are T cells and the remainder are B cells and NK (natural killer) cells (a lymphocyte that can nonspecifically destroy certain virally infected or tumor cells). The production of B cells and T cells continues, albeit at a reduced rate, throughout life (Twomey, 1982).

The normal function of the immune system involves a complex sequence of cellular and biochemical events. After exposure to an antigen (a molecule that stimulates a specific immune response), there is phagocytosis (ingestion) of the antigen by macrophages during which the antigen undergoes intracellular breakdown by enzymatic hydrolysis. After hydrolysis, the fragments of the antigen move to the surface of the macrophage for reaction with specific T lymphocytes called helper-inducer T cells. Activation of these T lymphocytes occurs only if the interacting lymphocytes have specific receptors that bind to a complex of antigen fragments and a special protein derived from the major histocompatibility complex (Twomey, 1982).

The generation of antibody-producing plasma cells (B cells) and cytotoxic cells (T cells) requires the presence of biochemical factors (lymphokines and cytokines) secreted by T cells and macrophages. Clonal expansion increases the number of these specifically reactive T and B cells, so that subsequent exposure to the same antigen leads to a rapid specific immune response. As an immune response occurs, a decrease of the T cells is likely, and negative feedback into earlier phases prevents excessive reaction. Thus, the specific antibody reacts with the offending antigen to cause neutralization or inactivation while effecter T cells inactivate or destroy cellular targets. When these mechanisms are functioning properly, the immune system recognizes and eliminates foreign agents quickly and efficiently. Opportunities for dysfunction can occur at any point along the pathway of cellular and biochemical processes, resulting in a variety of immunological effects from hypersensitivity to immunodeficiency, as illustrated in Figure 10-1.

For example, exposure to immunotoxicants can cause immunosuppression, resulting in altered host resistance. The outcome of immune suppression is influenced by the dose and mechanism of action of the immunotoxicants, along with concomitant exposure to other agents such as bacteria, viruses, parasites, or chemicals at levels that might normally be innocuous. In its suppressed condition, the immune system does not respond adequately to hazardous agents. Adverse consequences are those of severe disseminated infectious diseases caused by a variety of agents that are usually not pathogenic. Age, poor nutrition, and stress (physiologic and psychologic) can exacerbate the development of such immunologic disease (Golub, 1987).

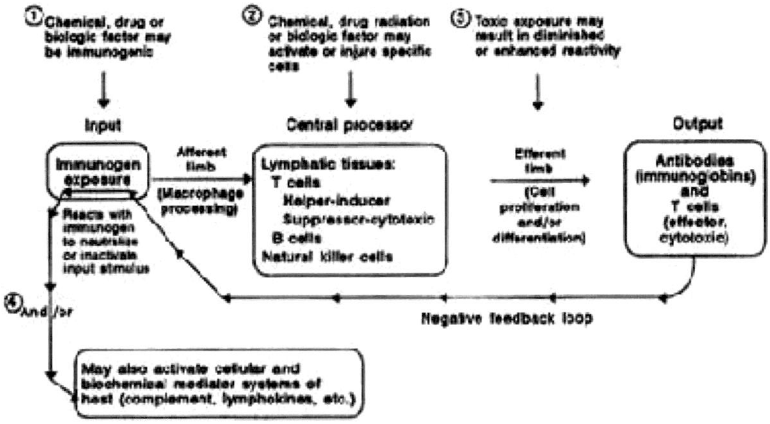

FIGURE 10-1 Model of the competent immune system depicting sites of potential effects of the major components by toxic factors. SOURCE: National Research Council, 1992.

Xenobiotics can also act as sensitizers to stimulate the immune system as antigens by provoking a substantial immune response that leads to hypersensitivity. Immunologic tissue damage can result from activation of the cellular and biochemical systems of the host. The interactions of an antigen with a specific antibody or with effecter lymphocytes trigger the sequence of humoral and cellular events to produce the pathophysiologic effects that lead to tissue injury or disease (Vos, 1977).

Chemicals that suppress bone marrow function can affect reserves of stem cells that are needed for cell replacement. Blood line cells are derived from stem cells, which can develop into many cell types (pluripotent) and which, in adult humans, are primarily in bone marrow. Within the marrow microenvironment, these self-renewing cells mature into committed progenitors, which are in peripheral cells. Stem cells often appear to be sensitive targets for therapeutic and environmental toxicants, most likely because of their rapid proliferation. Xenobiotics or various drugs that are toxic to the myelocytes of the bone marrow can cause profound immunosuppression due to loss of stem cells.

Animal Studies

In a review of the literature, a distinct impression arises that sulfur mustard has been consistently observed to cause pathogenic states of decreased immunoresponsiveness in animals. One of the earliest clues of this effect emerged from a series of animal experiments by Hektoen and Corper (1920). They observed that dogs and rabbits experienced depressed antibody formation after intravenous and intraperitoneal (IP) exposure to sulfur mustard. The sulfur mustard had a restraining effect on precipitation and on lysis—two of the principal ways antibodies can inactivate invading foreign agents—and profoundly modified the leukocyte count of the blood in experimental animals.

This series of studies placed sulfur mustard in a class with other leukocytic toxins such as benzene, which has frequently been associated with myelotoxicity expressed as leukopenia, pancytopenia, anemia, and aplastic or hypoplastic bone marrow (Dean and Murray, 1991).

Later studies with rabbits demonstrated that the number of leukocytes increased immediately after inhalation exposure but later diminished morphologically. The polymorphonuclear basophil (a type of white blood cell) showed abnormal developments of the nucleus and dissolution of the granules. The lymphocytes, which produce acquired immunity, also showed degenerative changes (Hektoen and Corper, 1920).

Similar results were achieved in experiments in which the route of absorption was intravenous: polymorphonuclear basophils increased following injection and then diminished rapidly, apparently disappearing from the peripheral blood. Zimmerman (1942) reported that lymphocyte disintegration began within 5 hours of intravenous injection of sulfur mustard; within 24 hours most of the lymphocytes had disappeared.

Quantitative histologic investigation of the effects of intravenously injected sulfur and nitrogen mustard on albino rats suggested a decreased immunoresponsiveness, expressed as leukopenia, lymphopenia, and neutropenia (the disappearance of the respective blood cells), as well as hypoplasia and hyperemia of bone marrow. In the lymphoid organs, tissue decreased in volume because of the destruction of lymphocytes (Kindred, 1947). The author further noted that the bone marrow reacted more slowly to the mustards than did the lymphoid organs, but it became hyperplastic. There was some destruction of cells, particularly of the mature granulocytes that protect the body against invading agents by ingesting them (Kindred, 1949).

These data on the albino rat have been supplemented by studies showing that dogs exposed to sulfur and nitrogen mustard experience toxic effects on lymphoid organs and bone marrow. The results of this action were observed in the quantitative decrease of cells in the peripheral blood of poisoned animals. The extent of cellular intoxication was directly related to the amount of the mustard injected (Kindred, 1949). Spurr (1947) further elucidated the influence of mustard com pounds on immune function by simultaneous intramuscular injection of typhoid vaccine and nitrogen mustard into rabbits. The data suggest that the toxicity of nitrogen interferes with or suppresses antibody-forming mechanisms of the lymphocyte.

More recent studies have generally confirmed earlier evidence that laboratory animals exposed to mustards experience changes in cells of the immune system that result in undesirable effects (i.e., immunosuppression, alteration of host defense mechanisms against pathogens and neoplasia). Coutelier and colleagues (1991) noted a marked decrease in the number of spleen cells in mice one week after receiving a relatively high dose of sulfur mustard. B lymphocytes were relatively more affected than T lymphocytes by sulfur mustard; similar results were seen in humans, where toxicity for B lymphocytes led to a decrease in B-cell number following exposure to nitrogen mustard compounds.

Blank and colleagues (1991) compared the immunotoxicity of sulfur mustard and nitrogen mustard on humoral and cell-mediated immunity of mice. The effects on thymic and splenic weight, spleen cell number, and the formation of antibody were similar to earlier laboratory results. Both compounds induced splenic and thymic weight loss. When splenic cellularity was depressed, the total number of cells producing antibody response was decreased. Only when sulfur mustard reached lethal levels were the total spleen cells producing antibody response at a level equivalent to that observed following nitrogen mustard administration. Hence, the immunotoxic effect of nitrogen mustard could be distinguished from general toxicity and was tolerated at a higher dose than was sulfur mustard. Nitrogen mustard had an additional immunotoxic effect of decreasing host resistance to tumor cells that was not observed with sulfur mustard. The reason for these differences is unclear.

Human Studies

Evidence that sulfur mustard causes immunosuppression in humans has emerged from several lines of investigation. The earliest evidence came from clinical observations of humans directly exposed to sulfur mustard during World War I (WWI), who showed significant quantitative and qualitative changes in the circulating elements of the immune system. Stewart (1918) studied 10 fatal cases of mustard poisoning and observed striking depression of bone marrow production of white blood cells. For example, in one case the patient showed a total leukocyte count of 7,630/mm^3 on the second day after exposure (gassing), 6,650 on the third day, and 270 on the sixth day, 24 hours before death. In another case a total leukocyte count of 35,000 was measured on the second day after gassing, which dropped to 16,000 on the third day, and to 172 on the seventh day, six hours before death.

Krumbhaar (1919a,b) observed that among the first changes in the circulating blood of patients exposed to sulfur mustard was an exhaustion of leukocyte-forming centers. This downward trend in leukocyte count ultimately leads to severe leukopenia. Thus the leukocyte counts in four fatal cases were observed to fall steadily from (1) 10,200 to 2,900; (2) 17,800 to 3,200; (3) 20,400 to 7,600; and (4) 36,000 to 14,000/mm^3. Postmortem examination of the femoral bone marrow revealed only a slight mottling, shown histologically to be due to primordial cells and megablasts with a greater or lesser disappearance of normoblast, myelocytes, and adult forms. This was interpreted as an inadequate attempt at blood regeneration, and the resulting lack of leukocytes in the bloodstream suggested a weakening of the immune system. Even in bronchopneumonia that frequently supervenes in heavily gassed patients, little or no reactive rise in leukocyte count was observed.

In another study, the femoral bone marrow in 55 of 75 autopsies of mustard gas-exposed patients was examined (Krumbhaar and Krumbhaar, 1919). The results: 14 marrows were classified as showing almost no regenerative potential; 8 showed only slight reaction; and only 13 showed moderate reaction. In no case was the marrow as hyperplastic as in ordinary lobar pneumonia or acute infections accompanied by leukopenia. The authors concluded that

> the blood and bone marrow changes are due to direct action of the poison and not the secondary infections (1) because they have been found well marked in cases where infection was slight or absent; (2) because influenza, typhoid, malaria, and such leucopenia infections played no part in these cases; and (3) because the kind of infection found to be present (pyogenic) does not lead to leucopenia or impaired bone marrow function.

In 1946, Anslow and Houck reviewed classified literature concerning the pharmacological action of sulfur and nitrogen mustard. They reported that evidence from soldiers gassed or burned in World War I was essentially the same as seen in experimental animals. Marked leukopenia and loss of reactivity of the bone marrow were observed in severe cases of mustard intoxication. In mask volunteers, sublethally exposed to sulfur mustard under temperate or tropical climate conditions, there was a moderate to marked leukocytosis appearing as early as four hours after exposure. This was followed by a moderate reduction in the number of leukocytes in the blood.

Another set of clinical observations comes from reports on over 600 sulfur mustard casualties following the release of sulfur mustard in Bari harbor, Italy, in December 1943 (Alexander, 1947). The effects upon the leukocytes in the circulating blood were most severe: white blood cell counts of 100 cells/mm^3 or less were recorded; lymphocytes were the first to disappear; and granulocytes were also severely affected but lagged behind the lymphocytes in their rate of decrease. Not all cases demonstrated a sharp decline in white blood cell count, but all casualties with an extremely low leukocyte count died. Infection was a dominant feature, as it was among sulfur mustard casualties during the Iran-Iraq conflict in 1984 and 1986. These patients experienced leukopenia accompanied by total bone marrow aplasia, which included extensive losses of myeloid stem cells (Balali, 1986; Eisenmenger et al., 1991). These findings are further evidence of an association between suppression of immunologic functions and an increased incidence of infectious disease.

In one of the few studies of long-term effects, Zandieh and colleagues (1990) measured the cell-mediated immunity in three groups of Iranians exposed to sulfur mustard: (1) three months to two years after exposure; (2) one to two years after exposure; and (3) two years after exposure. In comparison to normal controls, T lymphocytes showed a significant decrease of 50 percent in all three groups; helper-inducer T cells were significantly decreased in 52 percent of the first and second groups; and T suppressor cells were increased in 53 percent of the first group and 22 percent of the second and third groups. These measurements indicate that depression of the cell-mediated immunity was observed one, two, and three years after exposure to sulfur mustard. Several recent case reports from the Bahar Medical Laboratory in Tehran, Iran, describe similar long-term effects: about 100 patients exposed to sulfur mustard were observed for a year, and the investigator found alterations in B and T lymphocytes. In addition, the phagocytic activity of these patients was reduced to 20 percent of that of a normally functioning immune system (Balali, 1986).

The search for chemical agents with antitumor activities has provided another clue that sulfur and nitrogen mustards are immunotoxic. As alkylating agents, they form covalent linkages with biologically important molecules, resulting in disruption of cell function, especially cell division. As a result, these agents are particularly toxic to rapidly proliferating cells including neoplastic, lymphoid, and bone marrow cells. Because of these immunosuppressive effects, sulfur and nitrogen mustards have particular clinical importance. Nitrogen mustards were the first nonhormonal agents to show significant antitumor activities in humans, producing dramatic tumor regression in lymphoma patients (Colvin and Chabner, 1990). The significance of the immunosuppression produced by alkylating agents in the setting of cancer therapy is uncertain. The major concerns are (1) the danger of increased susceptibility to infection in the immunosuppressed host and (2) the potential interference with a host immune response to the tumor. Spitz (1948) conducted a histological analysis of postmortem tissue of 57 cases consisting of a variety of lymphomas, leukemias, and other malignant tumors treated with nitrogen mustards. Attributable to mustard therapy in these cases was a consistent, apparently accumulative hypoplasia of bone marrow. This hypoplasia was correlated with the degree of leukopenia and thrombocytopenia that had existed prior to death.

The immunosuppressive effects of alkylating agents have also been demonstrated in treating autoimmune disease and as adjunctive therapy in organ transplantation procedures, to prevent rejection by the recipient. Immunosuppressive treatments, like exposure to sulfur mustard, result in increased incidence of infectious disease. There is a well-established association between the therapeutic use of chemical immunosuppressants, such as those used in organ transplant therapy, and an increased incidence of infectious disease in humans (Ehrke and Mihich, 1985).

Conclusions

Evidence from animal experiments consistently confirms that mustard agents affect immune system functions. Animal models are most valuable in studying the physiological and molecular mechanisms involved in mustard agent effects. In general, this review accorded greater weight to data derived from studies in more than one animal species or test system, on results that have been reproduced in different laboratories and on data that indicated a dose-response relationship. However, results from animal studies cannot be used alone either to affirm or to negate relationships between exposure to mustard agents and chronic disorders, nor can they be used to estimate accurately the size of the effects in humans.

Clinical observations in humans provide the most direct evidence of the immunologic effects of mustard agents. In this review, greater significance was accorded to observations in humans that provide clear evidence that mustard agent exposure is associated with bone marrow toxicity expressed as leukopenia, pancytopenia, or a plastic or hypoplastic bone marrow. Underrepresented in these works is information on chronic or delayed effects. This may be attributable to the fact that patients, if they survive the acute effects, experience a number of secondary infections and may in fact die from them. Finally, the data presented here indicate that clinical studies as a whole support a close parallelism between animal experiments and observations in humans regarding the immunosuppressive properties of mustard agents.

EFFECTS ON SYSTEMS OTHER THAN THE IMMUNE SYSTEM

Gastrointestinal Effects

Although not the major focus of most studies of mustard agents, gastrointestinal effects have been documented in some animal studies. The most common findings have been intestinal histopathological changes, including destruction of the mucosa and shedding of epithelial elements (Graef et al., 1948; Papirmeister et al., 1991). Gavage studies have also shown epithelial hyperplasia of the forestomach (Sasser et al., 1989b).

In humans, gastrointestinal effects of mustard agents are commonly seen in people with acute high exposures. Common symptoms include nausea and vomiting, both immediately after exposure and as a delayed effect (Papirmeister et al., 1991; Schonwald, 1992). Several mechanisms may account for these effects including (1) a direct and immediate cholinergic effect, (2) an inflammatory reaction of the upper gastrointestinal mucosa, (3) a delayed radiomimetic effect on the small intestine, and (4) physical stress secondary to skin and other effects from mustard agent exposure (Papirmeister et al., 1991). In some individuals there may be chronic gastrointestinal symptoms, but these appear to occur only secondarily to other chronic health problems due to acute high exposures to mustard agents.

Exposure of dogs to high doses of Lewisite produces some injury to the intestinal mucosa and focal necrosis of the liver with peribiliary hemorrhage (Cameron et al., 1946, 1947; Gates et al., 1946). Exposure of rats by lavage produced lesions of the forestomach including necrosis of the stratified epithelium (Sasser et al., 1989a). No information could be found on the effects of Lewisite on the human gastrointestinal systems.

Hematological Effects

In animals exposed to very high doses (i.e., LD_{50}) of mustard agents, aplastic changes occur in the bone marrow. Lymphatic damage also occurs, including cellular depletion of the splenic sinuses and resultant disappearance of lymphocytes from the blood (Graef et al., 1948; Papirmeister et al., 1991). Similar hematological changes occur in humans after very high exposures to mustard agents. Marrow suppression including anemia, thrombocytopenia, and leukopenia has been seen (Papirmeister et al., 1991; Willems, 1989). Severe leukopenia appears to occur only after very high exposures, and secondary infections can be a significant contributor to the mortality in this group.

Laboratory studies have also demonstrated that exposure to Lewisite may cause a hemolytic anemia similar to that seen with arsine exposure (Goldman and Dacre, 1989). As before, there are no data regarding hematological effects on humans.

Neurological Effects

Extremely high exposures to mustard agents can cause central nervous system excitation leading to convulsions in animals (Anslow and Houck, 1946). Cardiac rhythm irregularities including atrioventricular block may also occur at high levels of exposure (Anslow and Houck, 1946). Renal changes have also been reported, but usually as a late complication in fatal exposures (Papirmeister et al., 1991).

In humans, acute neurological symptoms are common with high exposures to mustard agents, including severe depression and changes in mentation. These symptoms are produced both directly by the chemical and secondarily to other physiological changes (see also Chapter 11). Follow-up of workers in German chemical warfare plants showed a high prevalence of various neurological disorders including impaired concentration, diminished libido, and sensory hypersensitivity (Lohs, 1975). To what extent these can be attributed to mustard agents rather than other chemical warfare agents is unclear, but other exposures at this facility included known nerve agents. Renal and cardiac effects do not appear to occur after human exposures to mustard gas other than as secondary effects in severely affected individuals.

Neurological effects have not been documented in animals after acute exposures to Lewisite, but this apparently has not been well researched. Acute exposure to high levels of Lewisite leads to a shock syndrome due to increased capillary permeability (Goldman and Dacre, 1989). Some direct cardiotoxic effects can be seen with intravenous administration. No direct evidence exists that Lewisite may cause neurological problems in humans. However, arsenic is a well-known neurotoxin, and peripheral neuropathy has been described in humans after a single exposure to arsenic (LeQuesne and McLeod, 1977).

Conclusions

Gastrointestinal, hematological, and neurological effects are common after acute high exposures to mustard agents.

These effects can be attributed primarily to the known toxicological effects of these agents and secondarily to effects on other organ systems (i.e., shock, burns). Although effects on these other organ systems have not been a focus of follow-up studies of people exposed to mustard agent, the available evidence provides no indication of persistent or delayed effects other than those secondary to other conditions related to exposure to mustard agents.

There is insufficient information to link Lewisite with long-term health effects on the hematological, gastrointestinal, and neurological systems.

REPRODUCTIVE RISKS

Sulfur Mustards

Sulfur mustard causes cross-linking of DNA and is known to alkylate DNA at the 0-6 position of guanine. This observation is consistent with the known mutagenic potential of sulfur mustard (see Chapter 6). It is a bacterial and mammalian mutagen. Sulfur mustard causes chromosome breakage and induces sister chromatid exchanges in a wide variety of cells. Epidemiologic studies have also led the International Agency for Research on Cancer to classify sulfur mustard as a human carcinogen. These observations underscore the potential of this compound to induce genetic damage. They also suggest that sulfur mustards could be a reproductive toxin.

Animal Studies

Sulfur mustards induce dominant lethal mutations in *Drosophila.* Luening (1952) observed gross deletions in chromosomes in *Drosophila* offspring. Sonbati and Auerbach (1960) also showed that sulfur mustard can cause heritable mutations in germ cells in the offspring of treated *Drosophila.* Sasser and colleagues (1989c) administered sulfur mustard orally to rats in sesame oil for 5 days per week for 10 weeks at doses of 0, 0.08, 0.2, and 0.5 mg/kg. Treated and untreated males were mated with treated females and untreated females and their fetuses evaluated after 14 days. The observed effects included early fetal resorptions and preimplantation losses, as well as decreases in live embryo implants. A significant increase in abnormal sperm was detected in males exposed to the highest dose. The authors concluded that the timing of these effects was consistent with an effect during the post-meiotic stages of spermatogenesis, possibly involving the generally sensitive spermatids. In an earlier experiment, Rozmiarek and colleagues (1973) exposed pregnant female rats to 0.1 mg/m^3 of sulfur mustard by inhalation; exposure during the interval of gestation failed to produce any fetal malformations.

In an extensive study, Hackett and colleagues (1987) studied maternal toxicity, intrauterine mortality, and developmental toxicology in rats and rabbits. Distilled sulfur mustard, diluted in sesame oil, was delivered by intragastric intubation in a volume of 1 ml/300 g rat and 1 ml/4 kg rabbit. Rats were dosed daily from 6 to 15 days with 0, 0.5, 1.0, and 2.0 mg/kg; rabbits received 0, 0.4, 0.6, and 0.8 mg/kg on 6-19 days. In rats, indicators of maternal toxicity were observed at all dose levels, but significant fetal effects (decreased weights, reduced ossification, and skeletal anomalies) occurred only at the highest dose (2.0 mg/kg). In rabbits, maternal toxicity occurred at the highest dose level, body weights of fetuses were reduced, but no other fetal effects were observed.

Finally, Sasser and colleagues (1989c) studied rats to evaluate the effects of sulfur mustards on reproduction. Groups of rats (27 females and 20 males per group per generation) were force-fed with 0, 0.03, 0.1, or 0.4 mg/kg sulfur mustard for 13 weeks prior to mating and throughout gestation, parturition, and lactation in a 42-week two-generation study. There were no significant effects on reproductive function or pregnancy outcome in either generation. Growth of adult F1 rats of both sexes was reduced by the highest exposure, as was the growth of their F1 and F2 offspring during lactation. A dose-related lesion of the squamous epithelial mucosa of the forestomach was observed in both sexes, and benign neoplasms of the forestomach were found in about 10 percent of the intermediate (8/94) and high-dose (10/94) groups. The NOEL in this study was <0.03 mg/kg for toxicity and > 0.4 mg/kg for reproductive effects.

Human Studies

In humans, two studies have attempted to evaluate the potential of sulfur mustard to induce adverse reproductive outcomes. Yamakido and colleagues (1985) used electrophoresis to study blood protein variants in 456 children of 325 workers exposed occupationally to sulfur mustard and Lewisite at the Okuno-jima poison gas factory; children were divided into three exposure groups based upon parental job category within the plant. The blood protein analysis revealed 6 types of protein variants in 11 children, and 11 variant erythrocyte proteins in 25 children. Examination of 32 of the 36 families of interest showed all of the detected variants to be familial and not exposure induced. One protein variant was found in one child that did not differ electrophoretically, but did differ in its enzymatic activity. This variant was not inherited. The female child in question was mentally retarded and had a normal g-banded karyotype but was born with a cleft palate and hypomyotonia (significant decrement in normal muscle tone). The significance of these findings is unclear, as this variant might have been the result of either (1) a germ cell mutation in an intron sequence (a DNA sequence outside of the region coding for the RNA and subsequent protein) that controls expression of the gene of interest, or (2) aberrant translation or another mechanism unrelated to exposure-induced reproductive anomalies.

In consultation with Dr. Sanford S. Leffingwell (Special Programs Group, Center for Environmental Health, Centers for Disease Control, U.S. PHS) the committee attempted to estimate the power of the Yamakido and colleagues (1985) study to detect induced mutations in the offspring of exposed workers. Of the 456 subjects, 87 were children of workers with the highest likelihood for exposure, 75 were children of workers with intermediate likelihood for exposure, and 294 were children of workers with relatively low likelihood of exposure. The results suggest that an effect could be detected with reasonable probability only if the mutation risk were increased at least 100-fold in the least-exposed group (i.e., from 10^{-6} to 10^{-4}) or 1,000-fold in the smaller, more-expos*ed*groups (i.e., from 10^{-6} to 10^{-3}). Thus, the Yamakido study did not evaluate a sufficiently large sample to detect induced mutations in the offspring of exposed workers of the Okuno-jima war gas factory.

One additional report suggests that sulfur mustard may be responsible for the induction of cleft lip and palate in the offspring of exposed parents. Taher (1992) studied 21,138 live births at Najmeia Hospital in Tehran between 1983 and 1988, following the use of sulfur mustards in the Iraq-Iran conflict. He asserts that parental sulfur mustard exposures were associated with 30 of the 79 cases of cleft lip and palate that were recorded among newborns in this hospital during this period. No actual exposure data exist, although parents were questioned about their exposure to mustard gas, as well as any family history of clefts, rubella during pregnancy, dietary deficiency, and drug use during pregnancy. No information is presented regarding whether mustard exposure was maternal or paternal or both. It is also unclear if the population captured by this hospital represents the geographic area of heaviest mustard exposure during the Iran-Iraq conflict. Further, it is unclear if this incidence of clefts is truly elevated relative to other regions in this country or other parts of the world. Therefore, this study is hardly definitive. It appears to contain, however, the only other human data that attempt to address the reproductive toxicity of sulfur mustards.

It should also be noted that other animal studies have shown that the structurally similar nitrogen mustards are potent teratogens (Danforth and Center, 1954; Haskin, 1948; Murphy et al., 1958; Sanyal et al., 1981).

Summary and Conclusions

Data exist on the reproductive toxicity of sulfur mustards in more than one animal species, but it would be useful to have additional studies to examine the extent of the variability between species. More inhalation and cutaneous exposure studies would also be very helpful, as these exposure results would more accurately mimic human exposure. Certainly studying larger numbers of animals would provide a more sensitive measure of the possible magnitude of any reproductive risk associated with exposure to sulfur mustards. Short-term, high-dose exposures would also be helpful in attempting to examine any dose-rate effects. It should also be noted that the quality of the human data on the reproductive toxicity of sulfur mustards is quite poor. There has been insufficient follow-up of the occupational or battlefield cohorts to determine the nature of any reproductive toxicity or teratogenic effects attributable to these exposures.

Evidence suggests a causal relationship between sulfur mustard exposure and reproductive toxicity in laboratory animals, but the database is far too small and uncertain to allow a clear understanding of human reproductive risk from exposure to sulfur mustards. Sulfur mustards can cause genetic alterations in the sperm of male rats after inhalation or gastric exposure, but rodent and rabbit studies showed that sulfur mustards are not detectable teratogens in animals. The human data are difficult to interpret: it is unclear if it is significant that the Japanese child with one nonheritable variant protein had the same congenital malformation (cleft palate) that was reported in association with parental mustard exposure in Iran.

Lewisite

Introduction

The literature addressing the reproductive toxicity of Lewisite is small; in fact, few data exist that advance the precise biologic fate of Lewisite in humans. Arsenic itself is known to be embryotoxic and teratogenic, and while Lewisite is an organic arsenical, in highly alkaline environments inorganic arsenic could be formed. These facts led the committee to examine the reproductive effects of inorganic and organic arsenicals, as well as Lewisite itself.

Inorganic Arsenicals. Ancel (1946) reported that sodium arsenate induced a significant reduction in size of offspring when chickens were given this compound orally. James and colleagues (1966) also reported a similar decreased size in offspring of ewes fed sodium arsenite during gestation. Ferm and Carpenter (1968) studied intravenous administration of sodium arsenate (20 mg/kg) in pregnant golden hamsters; the treatment resulted in exencephaly when injected on the eighth day of gestation. Genitourinary abnormalities, cleft lip, cleft palate, microanophthalmia, and ear deformities also were observed when exposure occurred at varying times during gestation (Ferm, 1974; Ferm et al., 1971; Holmberg and Ferm, 1969). Ferm and Saxon (1971) also reported renal agenesis in hamsters treated with arsenic. Hood and Bishop (1972), studying IP injection of pregnant mice using 45 mg/kg of sodium arsenate, found a variety of malformations in the surviving fetuses.

Finally, Luego and colleagues (1969) report the case of a 17-year-old girl who ingested "a dose" of inorganic arsenic during the third trimester of pregnancy. She gave birth to a baby weighing 1.1 kg who died 11 hours later. The child had very high levels of arsenic in the liver, brain, and kidney. This clearly shows that anionic arsenic can pass through the mammalian placental barrier. This is consistent with work by Morris and colleagues (1938).

Organic Arsenicals. Organic arsenicals appear to be stored in the placenta at high concentrations, but they appear not to cross the placental membrane with any great ease in humans, cats, or rabbits (Eastman, 1931; Underhill and Amatruda, 1923). Several studies of the teratogenicity of a variety of organic arsenicals have been completed. Sodium dimethylarsonate was given by gavage to CD rats (7.5-60 mg/kg per day) and CD-1 mice (200-600 mg/kg per day) on day 7-16 of gestation. Sacrifice of the animals on day 18 (mice) and day 21 (rats) showed that there was both maternal and fetal toxicity in both species (Rogers et al., 1981). Cleft palate was seen at the middle and highest doses in the mouse. In the rat, genesis of irregular rugae (stomach folds) was observed to be dose related. Both species had maternal toxicity at doses below the maximum.

Frost and colleagues (1964) found no teratogenic effects in feeding studies (0.01, 0.02, and 0.05 percent arsanilic acid) following seven generations of rats. Harrison and colleagues (1980) studied intravenous administration of sodium dimethylarsonate in pregnant CD-1 mice. On day 9 and 10 they observed increased fetal resorption and mortality rates, with increased incidence of skeletal malformations and exencephaly. Hamsters given 900-1,000 mg/kg of sodium dimethylarsonate on day 9, 11, and 12 of pregnancy had a significant dose-related incidence of fetal resorptions (Hood et al., 1982a,b). The same work showed that IP injection of methanearsonic acid into hamsters (500 mg/kg) also induced both fetotoxicity and teratogenicity.

The sodium salts of the methanearsonates administered by IP injection in rodents (500-1,500 mg/g) also caused maternal mortality (Harrison et al., 1980; Hood, 1985). Inorganic arsenic injected at doses of 5-15 mg/kg induced developmental defects in mice (Hood, 1972; Hood and Bishop, 1972). Finally, Willhite (1981) injected 20-100 mg/kg of methylarsonic acid or 20-100 mg of dimethylarsinic acid intravenously into golden hamsters on the eighth day of gestation. In this work there was no teratogenic response.

Animal Studies

Two animal studies of the teratology of Lewisite appear to exist in the literature. Hackett and colleagues (1987) studied the reproductive effects of Lewisite administered daily by intragastric intubation at 0.5, 1.0, and 1.5 mg/kg in rats and 0.07, 0.2, and 0.6 mg/kg in rabbits. These doses were chosen after careful study of the toxic effects of larger dose ranges in these animals. The highest dose used in the study of rats (1.5 mg/kg) did not cause any toxicity or teratogenicity in the maternal animals or their fetuses. In the rabbits, maternal mortality occurred in all but one of the Lewisite treatment groups. This mortality ranged from 14 percent (0.07 mg/kg) to 100 percent (1.5 mg/kg). At the 0.6-mg/kg dose a decrease in maternal body weight gains and an increase in fetal stunting occurred, with decreased fetal weights.

The second study (Sasser et al., 1989a) reported similar negative results. Rats were administered 0, 0.10, 0.25, or 0.60 mg/kg of Lewisite per day in sesame oil by gavage prior to mating, during mating, and after mating until the birth of offspring. The dams received Lewisite during lactation. At weaning, a selected group of both genders continued to receive Lewisite during adolescence, mating, and throughout gestation. Lewisite did not affect reproductive organ weights, performance, or fertility in the two generations. No adverse effects on any offspring were observed. Histologic study showed no changes in male or female reproductive organs.

The committee also received information from Colonel Richard Solana of the U.S. Army Research Institute of Chemical Defense concerning unpublished data on the reproductive toxicity of Lewisite (Appendix A). The litters of approximately 140 rats were tested. Preconception maternal exposures were to either 0.045 or 0.002 mg/cm^3 by inhalation for 4 hours per day, 5 days per week for 4 months. Subsequently, the exposed females were mated to unexposed males. After 21 days the pregnant rats were sacrificed and reproductive outcomes assessed. The numbers of corpus lutea and implantations, intrauterine mortality, number and physical dimensions of fetuses, and the degree of ossification of the long bones were measured. The interpretation of the data concluded that no significant exposure-related differences were observed.

Human Studies

The work of Yamakido and colleagues (1985) on the offspring of workers at the Okuno-jima poison gas factory includes individuals exposed to Lewisite. As is noted above, this study is difficult to interpret as clearly negative, since one child was found to have an activity-variant enzymatic protein that was not familial. If this represents a significant finding, it cannot be attributed with any certainty to mustard exposure alone, Lewisite exposure alone, or the combination of the two.

Summary and Conclusions

Serious gaps clearly exist in our knowledge concerning the potential of Lewisite to cause reproductive problems. There appears to be no measurable reproductive toxicity of Lewisite in rodents when exposure was by gavage or inhalation. Our ability to extrapolate from these data to humans is limited, however. The conversion of Lewisite to other toxic forms of arsenic *in vivo* in animals and man is poorly understood. The comparability of the biochemical conversion of the Lewisite to potentially teratogenic arsenicals between animals and man is unknown.

More studies of multiple species are needed to define more accurately any differences between species. Also clearly needed are studies of the biologic fate of Lewisite in man. The precise nature of the products of bioconversion of Lewisite should be more thoroughly studied. The precise form of arsenic that is a potent teratogen in animals is not entirely clear. This should be combined with studies of the mechanism of action of teratogenic arsenicals. It then would be more clear that the differences in reproductive toxicity between some arsenicals and Lewisite are not related to the bioconversion of Lewisite or to select species-specific action of these compounds.

The studies of the reproductive toxicity of Lewisite in laboratory animals are negative and are therefore insufficient to indicate a causal relationship between exposure and adverse reproductive outcomes.

REFERENCES

Alexander SF. 1947. Medical report of the Bari harbor mustard casualties. Military Surgeon 101:1-17.

Ancel P. 1946. Experimental research on spina bifida. Archives d'Anatomie Microscopique et de Morphologie Experimentale 36:45-68. [In French]

Anslow WP, Houck CR. 1946. Systemic pharmacology and pathology of sulfur and nitrogen mustards. In: Division 9, National Defense Research Committee. Chemical Warfare Agents, and Related Chemical Problems. Summary Technical Report of Division 9, NDRC. Washington, DC: Office of Scientific Research and Development.

Balali M. 1986. First report of delayed toxic effects of yperite poisoning in Iranian fighters. In: Heyndricks B, ed. Terrorism: Analysis and Detection of Explosives. Proceedings of the Second World Congress on New Compounds in Biological and Chemical Warfare. Gent: Rijksuniversiteit. 489-495.

Blank JA, Joiner RL, Houchens DP, Dill GS, Hobson DW. 1991. Comparative immunotoxicity of 2,2'-dichlorodiethyl sulfide and cyclophosphamide: evaluation of L1210 tumor cell resistance, cell-mediated immunity, and humoral immunity. International Journal of Immunopharmacology 13:251-257.

Cameron GR, Carleton HM, Short RHD. 1946. Pathological changes induced by Lewisite and allied compounds. Journal of Pathology and Bacteriology 58:411-422.

Cameron GR, Courtice FC, Short RHD. 1947. Disturbances of function induced by Lewisite (2-chlorvinyldichlorarsine). Quarterly Journal of Experimental Physiology 34:1-28.

Colvin M, Chabner BA. 1990. Alkylating agents. In: Chabner BA, Collins JM, eds. Cancer

Chemotherapy: Principles and Practice. Philadelphia: J.B. Lippincott.

Coutelier JP, Lison D, Simon O, Willems J. 1991. Effect of sulfur mustard on murine lymphocytes. Toxicology Letters 58:143-148.

Danforth CH, Center E. 1954. Nitrogen mustard as a teratogenic agent in the mouse. Proceedings of the Society for Experimental Biology and Medicine 86:705-707.

Dean JH, Murray MJ. 1991. Toxic responses of the immune system. In: Amdur MO, Doull J, Klaassen CD, eds. Casarett and Doull's Toxicology. New York: Pergamon.

Eastman NJ. 1931. The arsenic content of the human placenta following arsphenamine therapy. American Journal of Obstetrics and Gynecology 21:60-64.

Ehrke MJ, Mihich E. 1985. Effects of anticancer agents on immune response. Trends in Pharmacological Sciences 6:412-417.

Eisenmenger W, Drasch G, von Clarmann M, Kretschmer E, Rolder G. 1991. Clinical and morphological findings on mustard gas [bis(2-chloroethyl)sulfide] poisoning. Journal of Forensic Science 36:1688-1698.

Ferm VH. 1974. Effects of metal pollutants upon embryonic development. Reviews on Environmental Health 1:237-259.

Ferm VH, Carpenter S. 1968. Malformations induced by sodium arsenate. Journal of Reproduction and Fertility 17:199-201.

Ferm VH, Saxon A. 1971. Amniotic fluid volume in experimentally induced renal agenesis and anencephaly. Experientia 27:1066-1068.

Ferm VH, Saxon A, Smith BW. 1971. The teratogenic profile of sodium arsenate in the golden hamster. Archives of Environmental Health 22:557-560.

Frost DV, Main BT, Cole J, Sanders PG, Perdue HS. 1964. Reproduction studies in rats with arsanilic acid. Federation Proceedings 3:291.

Gates M, Williams JW, Zapp JA. 1946. Arsenicals. In: Division 9, National Defense Research Committee, comp. Chemical Warfare Agents, and Related Chemical Problems. Summary Technical Report of Division 9, NDRC. Washington, DC: Office of Scientific Research and Development.

Goldman M, Dacre J. 1989. Lewisite: its chemistry, toxicology, and biological effects. Reviews of Environmental Contamination and Toxicology 110:75-115.

Golub ES. 1987. Immunology, A Synthesis. Sunderland, MA: Sinauer Associates. 55 p.

Graef I, Karnofsky DA, Jager VB, Krichesky B, Smith HW. 1948. The clinical and pathologic effects of the nitrogen and sulfur mustards in laboratory animals. American Journal of Pathology 24:1-47.

Hackett PL, Sasser LB, Rommereim RL, Cushing JA, Buschbom RL, Kalkwarf DR. 1987. Teratology studies of Lewisite and sulfur mustard agents: effects of Lewisite in rats and rabbits. PNL-6408, Pt. 1. Richland, WA: Pacific Northwest Laboratory. AD-A198-423.

Harrison WP, Frazier WP, Maxxanti EM, Hood RD. 1980. Teratogenicity of disodium methanarsonate and sodium dimethylarsonate (sodium cacodylate) in mice. Teratology 21:43A.

Haskin D. 1948. Some effects of nitrogen mustard on the development of external body form in the fetal rat. Anatomical Record 102:493-511.

Hektoen L, Corper HC. 1920. The effect of mustard gas (dichloroethylsulphid[e]) on antibody formation. Journal of Infectious Diseases 28:279-285.

Holmberg RE, Ferm VH. 1969. Interrelationship of selenium, cadmium, and arsenic in mammalian teratogenesis. Archives of Environmental Health 18:873-877.

Hood RD. 1972. Effects of sodium arsenite on fetal development. Bulletin of Environmental Contamination and Toxicology 7:216-220.

Hood RD. 1985. Cacodylic Acid: Agricultural Uses, Biological Effects, and Environmental Fate. VA Monograph Series. Washington, DC: U.S. Veterans Administration.

Hood RD, Bishop SL. 1972. Teratogenic effects of sodium arsenatein mice . Archives of Environmental Health 24:62-65.

Hood RD, Harrison WP, Vedel GC. 1982a. Evaluation of arsenic metabolites for prenatal effects in the hamster. Bulletin of Environmental Contamination and Toxicology 29:679-687.

Hood RD, Vedel GC, Zaworotko MJ, Tatum FM. 1982b. Distribution of arsenite and methylated metabolics in pregnant mice (abstract). Teratology 25:50A.

James LF, Lazor VA, Binns W. 1966. Effects of sublethal doses of certain minerals on pregnant ewes and fetal development. American Journal of Veterinary Research 27:132-135.

Kindred JE. 1947. Histologic changes occurring in the hemopoietic organs of albino rats after single injections of 2-chloroethyl vesicants: a quantitative study. Archives of Pathology 43:253-295.

Kindred JE. 1949. The blood cells and the hemopoietic and other organs of dogs given intravenous injections of 2-chloroethyl vesicants. Archives of Pathology 47:378-398.

Krumbhaar EB. 1919a. Role of the blood and the bone marrow in certain forms of gas poisoning. Journal of the American Medical Association 72:39-41.

Krumbhaar EB. 1919b. Bone marrow changes in mustard gas poisoning. Journal of the American Medical Association 73:715.

Krumbhaar EB, Krumbhaar HD. 1919. The blood and bone marrow in yellow cross gas (mustard gas) poisoning. Journal of Medical Research 40:497-506.

LeQuesne PM, McLeod JG. 1977. Peripheral neuropathy following a single exposure to arsenic. Journal of Neurological Sciences 32:437-451.

Lohs K. 1975. Delayed Toxic Effects of Chemical Warfare Agents. Stockholm International Peace Research Institute Monograph. Stockholm: Almqvist & Wilksell International.

Luego G, Cassady G, Palmisano P. 1969. Acute maternal arsenic intoxication with neonatal death. American Journal of Diseases of Children 117:328-330.

Luening KG. 1952. Studies on the origin of apparent gene mutations in *Drosophila melanogaster.* Acta Zoologica 33:193.

Morris HP, Laug EP, Morris HJ, Grant RL. 1938. The growth and reproduction of rats fed diets containing lead acetate and arsenic trioxide. Journal of Pharmacology and Experimental Therapeutics 64:420-445.

Murphy ML, Del Moro A, Lacon C. 1958. The comparative effects of five polyfunctional alkylating agents on the rat fetus, with additional notes on the chick embryo. Annals of the New York Academy of Sciences 68:762-782.

National Research Council. Committee on Biologic Markers. Subcommittee on Immunotoxicology. 1992. Biologic Markers in Immunotoxicology. Washington, DC: National Academy Press.

Papirmeister B, Feister AJ, Robinson SI, Ford RD. 1991. Medical Defense Against Mustard Gas: Toxic Mechanisms and Pharmacological Implications. Boca Raton, FL: CRC Press.

Rogers EH, Chernoff N, Kavlock RJ. 1981. The teratogenic potential of cacodylic acid in the rat and mouse. Drug and Chemical Toxicology 4:49-61.

Rozmiarek H, Capizzi RL, Papirmeister B, Fuhrman WH, Smith WJ. 1973. Mutagenic activity in somatic and germ cells following chronic inhalation of sulfur mustard (abstract). Mutation Research 21:13-14.

Sanyal MK, Kitchin KT, Dixon RL. 1981. Rat conceptus development *in vitro:* comparative effects of alkylating agents. Toxicology and Applied Pharmacology 57:14-19.

Sasser LB, Cushing JA, Kalkwarf DR, Mellick PW, Buschbom RL. 1989a. Toxicology studies on Lewisite and sulfur mustard agents: subchronic toxicity study of Lewisite in rats. AD-A217 886. Richland, WA: Pacific Northwest Laboratory.

Sasser LB, Miller RA, Kalkwarf DR, Buschbom RL, Cushing JA. 1989b. Toxicology studies on Lewisite and sulfur mustard agents: subchronic toxicity of sulfur mustard (HD) in rats. AD-A214-555. Richland, WA: Pacific Northwest Laboratory.

Sasser LB, Miller RA, Kalkwarf DR, Buschbom RL, Cushing JA. 1989c. Toxicology studies

on Lewisite and sulfur mustard agents: two-generation reproduction study of sulfur mustard (HD) in rats. Richland, WA: Pacific Northwest Laboratory. AD-A216-423.
Schonwald S. 1992. Mustard gas. PSR Quarterly 2:92-97.
Sonbati EM, Auerbach C. 1960. The brood pattern for intragenic and intergenic changes after mustard gas treatment of *Drosophila* males. Zeitschrift fur Vererbungslehre 91:253-258.
Spitz S. 1948. The histological effects of nitrogen mustards on human tumors and tissues. Cancer 1:383-398.
Spurr CL. 1947. Influence of nitrogen mustards on the antibody response. Proceedings of the Society for Experimental Biology and Medicine 64:259.
Stewart MJ. 1918. Report on cases of poisoning by mustard gas (dichlorethylsulphide) with special reference to the histological changes and to alterations in the leukocyte count. Chemical Warfare Medical Committee (London). Report 17. As cited in: Smith H. 1943. Review of the literature on the systemic action of mustard gas to August 1, 1943. OSRD Report No. 1717. New York University. Prepared for the Office of Scientific Research and Development.
Taher AA. 1992. Cleft lip and palate in Tehran. Cleft Palate Craniofacial Journal 29:15-16.
Twomey JJ, ed. 1982. The Pathophysiology of Human Immunologic Disorders. Baltimore: Urban & Schwarzenberg.
Underhill FP, Amatruda FG. 1923. The transmission of arsenic from mother to fetus. Journal of the American Medical Association 81:2009-2015.
Vos JG. 1977. Immune suppression as related to toxicology. CRC Critical Reviews in Toxicology 5:67-101.
Willems JL. 1989. Clinical management of mustard gas casualties. Annales Medicinae Militaris Belgicae 3:S1-61.
Willhite CC. 1981. Arsenic-induced axial skeletal (dysraphic) disorders. Experimental and Molecular Pathology 34:145-158.
Yamakido M, Nishimoto Y, Shigenobu T, Onari K, Satoh C, Goriki K, Fujita M. 1985. Study of genetic effects of sulphur mustard gas on former workers of Okuno-jima poison gas factory and their offspring. Hiroshima Journal of Medical Sciences 34:311322.
Zandieh T, Marzban S, Tarabadi F, Ansari H. 1990. Defects of cell-mediated immunity in mustard gas injury after years (abstract). Scandinavian Journal of Immunology 32:423.
Zimmerman T. 1942. As cited in: Smith HW. 1943. Review of the literature on the systemic action of mustard gas to August 1, 1943. OSRD Report No. 1717. New York University. Prepared for the Office of Scientific Research and Development.

11

Relationship of Mustard Agent and Lewisite Exposure to Psychological Dysfunction

During the course of this study the committee decided to consider the psychological, as well as the physiological, health effects of exposure to mustard agents and Lewisite. This decision was based on three areas of inquiry. The first was examination of the experimental protocols used in the World War II (WWII) chemical warfare testing programs. It became clear through this examination that numerous aspects of the experiments could be expected to cause moderate to extreme stress to the human subjects involved. The second area of inquiry was the public hearing process (also see Chapter 4). Nearly two months before the April 15, 1992, hearing, more than 50 veterans had already contacted the committee. Among their reported health problems, long-standing psychological problems were common, especially problems with depression and anxiety, and some individuals had been diagnosed with post-traumatic stress disorder (PTSD; see Chapter 4 and Appendix G). Finally, the committee investigated whether or not any scientific literature existed that might apply specifically to a possible causal relation between exposure to these warfare agents, under circumstances that existed in WWII, and the development of psychological dysfunction. On April 15, 1992, Robert Ursano, who heads the Department of Psychiatry at the Uniformed Services University of the Health Sciences and has studied the psychological effects of chemical and biological warfare environments, presented information about what characteristics of such environments are important in the etiology of stress reactions (see Appendix A).

Following this presentation and consideration of evidence from the other areas of inquiry, the committee appointed an additional committee member to review the information already gathered, along with appro priate scientific literature relating to the development of psychological dysfunction as a result of exposure to environmental toxins or experiences in chemical and biological warfare environments. Although the amount of scientific literature in this narrow area of focus was found to be quite small, it can be assessed against the background provided by intensive research into PTSD and other stress-related syndromes.

This chapter begins with a description of the historical development of the concept of PTSD. It also relates the findings from the literature and places those findings in context with what is known about the chemical warfare testing programs in WWII. Finally, this chapter outlines the committee's conclusions on the likelihood of adverse psychological health effects from exposure to mustard agents and Lewisite, particularly exposures such as those experienced in WWII testing programs.

The emphasis is on PTSD because the majority of work on the psychological sequelae of "war" experiences, such as those experienced by veterans contacting this committee, emphasizes PTSD. In this context, depression and anxiety are most often considered to be part of a constellation of psychological and psychiatric symptoms that comprise PTSD. In addition, mood disorders (e.g., major depressive disorder, bipolar disorder, and dysthymia) and anxiety disorders (e.g., generalized anxiety disorder, panic disorder, obsessive-compulsive disorder, and phobia) can also occur independently or as diagnostic entities coincident with PTSD (defined as comorbidity).

The precise definition of PTSD comes from the Diagnostic and Statistical Manual of the American Psychiatric Association (APA), 3rd Revised Edition (DSM-III-R, 1987; Box 11-1). The DSM is constantly undergoing revision by panels of experts who reexamine each diagnostic category and refine definitions and criteria based on the latest research data. It is also important to note that the diagnostic categories must be broad enough to cover PTSD caused by combat stress, sexual abuse and violence, environmental disasters, and many other types of stress. In addition, the exact combination of symptoms may also be dependent on age, gender, and other variables. At present there are four diagnostic criteria for PTSD:

- the existence of a recognizable stressor that would evoke significant symptoms of distress in almost anyone and is outside the range of usual human experience;
- reexperiencing of the trauma and intensification of symptoms with exposure to events that symbolize or resemble the traumatic event;
- numbing of responsiveness to, or reduced involvement with, the world, beginning after the trauma and including avoidance of activities that arouse recollection of the traumatic event; and

BOX 11-1 DIAGNOSTIC CRITERIA FOR 309.89 POST-TRAUMATIC STRESS DISORDERS

A. The person has experienced an event that is outside the range of usual human experience and that would be markedly distressing to almost anyone, e.g., serious threat to one's life or physical integrity; serious threat or harm to one's children, spouse, or other close relatives and friends; sudden destruction of one's home or community; or seeing another person who has recently been, or is being, seriously injured or killed as the result of an accident or physical violence.

B. The traumatic event is persistently reexperienced in at least one of the following ways:

(1) recurrent and intrusive distressing recollections of the event (in young children, repetitive play in which themes or aspects of the trauma are expressed)

(2) recurrent distressing dreams of the event

(3) sudden acting or feeling as if the traumatic event were recurring (includes a sense of reliving the experience, illusions, hallucinations and dissociative [flashback] episodes, even those that occur upon awakening or when intoxicated)

(4) intense psychological distress at exposure to events that symbolize or resemble an aspect of the traumatic event, including anniversaries of the trauma

C. Persistent avoidance of stimuli associated with the trauma or numbing of general responsiveness (not present before the trauma), as indicated by at least three of the following:

(1) efforts to avoid thoughts or feelings associated with the trauma

(2) efforts to avoid activities or situations that arouse recollections of the trauma

(3) inability to recall an important aspect of the trauma (psychogenic amnesia)

(4) markedly diminished interest in significant activities (in young children, loss of recently acquired developmental skills such as toilet training or language skills)

(5) feeling of detachment or estrangement from others

(6) restricted range of affect, e.g., unable to have loving feelings

(7) sense of a foreshortened future, e.g., does not expect to have a career, marriage, or children, or a long life

D. Persistent symptoms of increased arousal (not present before the trauma), as indicated by at least two of the following:

(1) difficulty falling or staying asleep

(2) irritability or outbursts of anger

(3) difficulty concentrating

(4) hypervigilance

(5) exaggerated startle response

(6) physiologic reactivity upon exposure to events that symbolize or resemble an aspect of the traumatic event (e.g., a woman who was raped in an elevator breaks out in a sweat when entering any elevator)

E. Duration of the disturbance (symptoms in B, C, and D) of at least one month.

Specify delayed onset if the onset of symptoms was at least six months after the trauma.

SOURCE: American Psychiatric Association, 1987.

- a miscellaneous group of symptoms including symptoms of increased arousal, hypervigilance, irritability, and exaggerated startle responses.

A total of 17 symptoms are encompassed by the final three of these four categories, and specific numbers of each symptom type are required for a diagnosis of PTSD. These criteria are a matter of ongoing revision and controversy, however, and failure to meet the full criteria for PTSD does not preclude the presence of other psychiatric disorders or other forms of traumatic stress disorder. Further modification of these characteristics is likely in the future publication of DSM-IV (American Psychiatric Association, 1991; Davidson and Foa, 1991). Possible changes include the incorporation of subjective experiences of a stressor, greater distinctions between the anxiety symptoms in PTSD and those associated with panic disorder, and more precise definition of subtypes of PTSD.

HISTORICAL DEVELOPMENT OF THE CONCEPT OF PTSD

Since time immemorial, human beings have faced major to maximal stressors—floods, fires, earthquakes, plagues, and wars. While the vastness of human misery is overwhelming, however, the consequences of stressors are uniquely experienced by individuals. At least since the 1600s, the diaries kept by literary men of the day have recorded these individual responses to trauma. Trimble (1985) reported several. For example, Samuel Pepys described the Great Fire of London in his Diary, relating his experiences on September 2, 1666, as the fire approached his home and he saw the terror in others. Subsequently, Pepys developed dreams of the fire and falling houses. Six months later, he wrote that he was still unable to sleep "without great terrors of fire" and in his diary referred to the sequelae of the disaster for others, including attempted suicide. In 1865, Charles Dickens was involved in a railroad accident in which he was "shaken up" from viewing the dead and dying around him. He wrote, "I am curiously weak—weak as if I were recovering from a long illness." Later, Dickens developed a phobia of railway travel. Since the late 1800s and early 1900s, industrial accidents, major natural and man-made disasters, concentration camps, terrorist activities, and especially wars have been the source of observations about human responses to major traumatic events. (Andreasen, 1980, 1985; Boehnlein et al., 1985; Giller, 1990; Horowitz, 1976, 1986; Hulbert, 1920; Idelson, 1923; Jewett, 1942; Kardiner, 1941; Kinzie, 1989; Kolb, 1990; Krystal, 1968; Macleod, 1991; Nemiah, 1980; Rennie and Small, 1943; Ross, 1966; Solomon, 1989; Tabatabai et al., 1988; Trimble, 1985; Ursano and Holloway, 1985; Ursano et al., 1981).

Anxiety syndromes in war settings were reported by military sur geons in the Franco-Prussian War of 1870, in the Civil War (DaCosta, 1871), and in the Boer War of 1899-1902 (Nemiah, 1980). Various labels have been given what is now called PTSD. In the Civil War, PTSD was called "soldier's heart" (Horowitz, 1976); later in that century Oppenheim introduced the term "traumatic neurosis"; and Mott used the World War I (WWI) term "shell shock" (Trimble, 1985). Idelson (1923), writing about his observations on one of the larger samples of men with "traumatic neuroses" from WWI, attempted to distinguish between "toxic sequelae" and "psychic or psychogenic sequelae." Without achieving the separation, yet in the best tradition of clinical observation, he described clearly and convincingly in men exposed to toxic gases, those behaviors and responses that are now termed PTSD.

There was little understanding of combat stress reactions in WWI, however, and many of the disagreements about labels were really disagreements about whether or not certain individuals were "predisposed" to develop adverse psychological effects following traumatic events (Horowitz, 1976). Further, there was uncertainty regarding the contributions of physical trauma and the precise type of mental disturbances that could be caused by stressful experiences. A paper by Hulbert in 1920 deemed "gas neurosis" to be the reaction of "discontented soldiers with a morbid, ignorant fear of being gassed." Later analysis by Jewett (1942) distinguished between panic and anxiety states in groups of WWI combat casualties. He further defined a subgroup that exhibited "psychoneuroses" typified by anxiety with psychosomatic components, conversion reactions, and dissociation (amnesia). In one field hospital, Jewett reported that 500 troops suffered psychological symptoms, which were accepted as real effects of combat (as opposed to malingering) because these were all seasoned troops. In his conclusion, Jewett called for "intelligent handling" of psychological casualties.

In WWII (Andreasen, 1980; Kinzie, 1989), many labels appeared: "traumatic war neurosis, combat neurosis, combat fatigue, combat exhaustion, battle stress, operational fatigue, and gross stress reaction." Early in WWII, Kardiner (1941) wrote of the traumatic neuroses of war, warning that while some men were responding rapidly to treatment of acute stress symptoms (now recognizable as PTSD), studies were needed of all phases of such responses, including chronic phases. The importance of the chronic phase of PTSD in WWII veterans was recently affirmed by Kolb (1990), who wrote, "We now know that such symptoms may persist up to 40 years." Yet as van Kammen and Ver Ellen (1990) noted, "After every war, disaster, or publicized atrocity of the last 125 years, the renewed public interest in the stress response syndromes quickly wanes, whereas the victims continue to suffer."

Labels continued to change through the 1950s and 1960s. In 1952, the first edition of APA's Diagnostic and Statistical Manual (DSM-I) used the term "gross stress reaction." Interestingly, this term was omitted in the revised DSM-II released in 1968, a year synonymous with the peak of the Vietnam War. That war was destined to produce great numbers of veterans suffering from PTSD, so many that there are now Centers for Post Traumatic Stress Disorder in a number of locations, established and funded by the Department of Veterans Affairs. It was not until 1980, with the release of DSM-III, that stress response syndromes, called PTSD for the first time, became a diagnostic category once again. This category and its accompanying criteria represented a landmark in the development of the concept of PTSD in that the first criterion, "existence of a recognizable stressor that would evoke significant symptoms of distress in almost anyone and is outside the range of usual human experience," clearly indicates that all individuals are at risk, not just those with pre-existing emotional problems.

CURRENT RESEARCH IN PTSD

There are intensive programs of research into the etiology, course, and treatment of PTSD. In considering all the possible causes of PTSD and the variables that affect the course and treatment of PTSD in specific individuals, it is not surprising that the body of PTSD research varies tremendously in focus. In addition, many theoretical approaches are brought to bear on this issue. For example, much of the research is based on classical stress research and animal studies. Other research is focused on various treatment types, from cognitive and behavioral approaches to drug therapies. One of the most active areas of PTSD research is the biological assessment of PTSD and the design of therapies that deal with the psychological, as well as physiological, manifestations of the disorder. These studies employ multivariate techniques that integrate neurochemistry, hormonal status, and other physiological conditions with sophisticated psychological assessment strategies (see Giller, 1990; Kulka et al., 1990). Perusal of the proceedings of the Seventh Annual Convention of the International Society for Traumatic Stress Studies (1991) also underscores the diversity of this research: over 260 presentations were made at this meeting, in 12 different categories.

EVIDENCE FROM STUDIES OF MILITARY PERSONNEL AND VETERANS

Chemical and Biological Warfare Environments

In the mid-1980s, a project was undertaken at the Uniformed Services University of the Health Sciences (USUHS) to examine the psychological and behavioral responses to chemical and biological warfare (CBW) environments (Ursano, 1987a,b; 1988a,b,c). The project, inspired by the continuing threat of CBW in world conflicts and specifically aimed toward development of effective strategies for use by the U.S. Air Force, had three components. The first component was the development of a database containing available knowledge about the psychological and behavioral responses to stressful environments, CBW environments, and the effects of such responses on performance and health. A second project component was direct observation of CBW training exercises and facilities. Finally, two major conferences were held, supplemented by a variety of small group discussions. Although largely applicable to all CBW agents, the project's major focus was on nerve agents. The final recommendations of this project were published in 1988, covering command, medical care, performance, and training issues (Ursano, 1988d). Similar findings, based on the above project and a review of additional literature, were published in 1990 (Fullerton and Ursano, 1990).

A number of recommendations from the USUHS project are interesting when viewed in reference to the handling of subjects in earlier CBW testing programs. Under the category of command issues, recommendations were made about the value of effective communications between command levels and troops to reduce the likelihood of adverse psychological and behavioral responses to stress. This included ensuring that troops had adequate knowledge of what was happening. The communication issues were so important to the troops' perception of risk and overall fear response that they were reiterated in the medical care, performance, and training categories. A further recommendation from this project was that officers, medical personnel, and others in command roles be sensitive to the problems and concerns of their troops, especially to those problems that were "unique" to CBW environments. These include heat stress,[1] concern about exposure levels, stress of confinement in protective structures holding many people, and many others. Another recommendation warned commanders that individuals may appear healthy initially in a psychological sense, but can develop delayed symptoms that warrant attention. It was also determined that an adverse response in one individual, or small group, can be "contagious" to the rest of a group under CBW conditions, and that "over-dedication" to the mission can cause commanders and individual troops to surpass their physiological capabilities and lead to exhaustion, as well as faulty reasoning and decision making.

[1] Heat stress is a particular concern due to the CBW protective gear alone. This group estimated that 5 to 20 percent of the casualties from a CBW exercise or incident will be from the characteristics of the environment and protective gear.

Numerous issues emerged as relevant to appropriate medical care in CBW environments. For example, it was found that, at least for nerve agents, stress and fear increased with prolonged and low-dose exposures. The need for health workers trained to deal with stress reactions was emphasized, as well as the need for commanders and supervisors to be sensitive to the manifestations of stress among their troops. To reduce the incidence of severe adverse reactions, certain strategies were recommended, including making certain troops had privacy and time during rest periods to interact with others about their concerns and experiences. Attention was also specifically directed at those who may witness grotesque injuries and handle the dead. Such activities were judged to be among the most stressful experiences possible in these environments. In cases of actual CBW exposures, it was also recommended that spouses of the affected troops be briefed about what had occurred so that the spouse could be an active participant in the post-trauma support efforts.

The recommendations outline the possibility of major decreases in performance and concentration arising from the psychological reactions to CBW environments, decrements that could in some cases cause life-threatening mistakes. Such decrements could also result from the physiological effects of nerve agents, or from heat stress alone, so the importance of distinguishing the causes and treating them appropriately was emphasized.

Under issues of training, the recommendations underscored the importance of prior training to reduce stress and to strengthen group cohesiveness, which further reduces stress. These training issues were quite varied, from training to deal with injuries and death to training in handling the protective gear effectively. An interesting recommendation addressed the costs versus benefits of the use of live agents in training and concluded that further investigation was needed to resolve this question. Finally, CBW trainers were warned to expect a certain number of individuals to develop claustrophobia during training exercises and were advised to attempt "talk down" procedures or remove the individual from the area.

In summary, then, numerous aspects of CBW environments can cause stress reactions of varying magnitudes in those involved. Although no direct statements are made regarding the long-term effects of such stress reactions, some of the characteristics of CBW environments fall into the description of experiences "outside the usual human experience ... [and] ... stressful to almost anyone" (DSM-III-R). Thus, it may be assumed that experiences in these environments may result in adverse psychological effects, including depression, anxiety, and PTSD-like symptoms in certain individuals.

Studies of PTSD in Veteran Populations

It is difficult to estimate how many veterans in the United States suffer from PTSD, but estimates range from 500,000 to 1.5 million (Kulka et al., 1990). The large numbers of veterans affected and a greater appreciation for the disabling effects of PTSD have inspired an increased research effort on the causes, psychological and physiological mechanisms, diagnostic criteria, course, and treatment of PTSD. Although it is impossible to review all of this work here, examples of this research bear directly on some of the key aspects of the experiences of veterans who contacted this committee.

A recent study of Vietnam veterans examined the kind of stressors most likely to result in PTSD in combination with other psychological disorders, including depression, panic disorder, phobic disorder, and alcoholism (Green et al., 1989). Over 52 percent of the sample of 196 individuals met the criteria for some type of psychiatric disorder: 29 percent met the criteria for PTSD, 21 percent were found to have phobic disorder, 15 percent suffered from major depression, 11 percent were alcoholic, and 7 percent exhibited panic disorder. Interestingly, a little more than 10 percent met criteria for both PTSD and major depression, and over 14 percent had both PTSD and phobic disorder. When these disorders were correlated with specific types of military experiences, some intriguing associations emerged. For example, the younger the individuals were entering the military, the higher was the percentage of PTSD found. In addition, the most potent predictor of PTSD and anxiety disorder comorbidity (PTSD with phobic or panic disorder) was involvement in dangerous and high-pressure "special assignments." Exposure to grotesque scenes of death and mutilation was associated with higher incidence of multiple disorders (e.g., PTSD and some other disorder, but no one disorder in particular). None of the results could be reliably attributed to psychiatric conditions prior to the war experiences.

Reasonable arguments may be made that Vietnam veterans, as a group, are quite different from WWII veterans due to a difference in the public perception of the respective conflicts, generational value differences, and other variables. Davidson and colleagues (1990) have compared Vietnam veterans with WWII and Korean War veterans on measures of depression, anxiety, severity and types of symptoms, and intensity and nature of combat experiences. The two groups of veterans were generally the same in terms of the sequential emergence of diagnoses and the age of onset of diagnoses: PTSD was followed by a general anxiety, which was then followed by panic disorder, major depression, and intermittent depressions. Alcoholism tended to emerge later in World War II and Korean War veterans than in Vietnam veterans. In addition, Vietnam veterans had, among other differences, more severe symptoms, higher depression scores, greater survivor guilt feelings, and more work disruption than World War II and Korean War veterans. In addition, despite similar rates of comorbidity overall, the Vietnam veterans were more likely to have panic disorder and PTSD. The differences in the veterans' reports of intensity of their experiences were interesting: WWII and Korean War veterans were most upset by general fear, fear of physical injury, and fear of incapacitation; Vietnam veterans, on the other hand, were traumatized most by witnessing brutality, sight of mutilated bodies, and loss of a friend in combat.

That stress reactions can set the stage for lifetime psychological difficulties is important in view of the nearly 50 years that have passed since the WWII testing programs with mustard agents and Lewisite. Indeed, a constellation of psychological problems may result from traumatic experiences, including depression, anxiety, panic disorders, and PTSD. Almost no data exist regarding the natural course of these adverse psychological effects over decades in the absence of treatment. A collection of case studies published by Macleod (1991) is pertinent to this issue.

Eighteen WWII veterans from New Zealand, identified as suffering from chronic PTSD, were interviewed extensively following routine psychiatric review for the War Pensions Claims Panel. Most of these men had not been treated for PTSD, but some had been treated for other psychiatric illnesses. The majority of the men recalled traumatic experiences during the war with great vividness and detail. Key among their emotional responses to these events was fear, physiological arousal, and helplessness. Two thirds of the men reported lifelong troubles with, or distance from, their spouses and families. Most interesting was the commonality of the reported long-term course of their emotional problems. This course generally began with a controlled response at the time of the trauma, followed by superficial attempts at coping with the incident. After the war ended, the men experienced significantly greater emotional difficulty for a period of time, but the schedule and routine of their working years was associated with a moderation of symptoms. A second escalation of symptoms arose in these men after retirement. The author concluded from this sequence that work became a distraction for the men and the cessation of working caused a reemergence of underlying problems. Although the conclusions drawn from this case study must be tentative, the paper supports the concept that PTSD symptoms, untreated and unrecognized over long periods of time, may nevertheless be traceable and diagnosable. Further, the suggestion is unavoidable that there may be many WWII veterans who have struggled unknowingly with PTSD or PTSD-like symptoms for decades.

EVIDENCE FROM STUDIES OF THE PSYCHOLOGICAL EFFECTS OF ENVIRONMENTAL CONTAMINATIONS

Scientists interested in stress reactions have also begun to look at groups of people who have been exposed to environmental contaminations of various kinds, or who have lived in close proximity to contaminated areas. It is useful to consider the findings of such studies, because some of the characteristic stressors of these experiences are relevant to the WWII testing programs.

Longitudinal studies of people living near the ill-fated Three Mile Island (TMI) nuclear power plant have been reported by Baum and colleagues (1981, 1987). A number of measures have been employed in these studies, including performance tests, depression and anxiety measurements, interviews, and physiological assessments. In addition, the TMI group was compared to other communities located nearby, or distant from, toxic waste dumps. One of the main problems for members of the TMI group has been the high level of uncertainty they have lived with since the accident. There has been uncertainty about the level of exposure they experienced, whether radioactivity is still present, and what the short-and long-term health effects will be. The uncertainty has been exacerbated by conflicting reports from government officials, the press, and members of the community themselves. The authors further argue that the very presence of the "smokestacks," so identified with the accident in the minds of all, serves as a constant reminder to those who continue to live in the area.

Among the TMI group, there is a widespread concern about cancer and somatic complaints are common, especially headaches, backaches, and gastric distress. Measures of anxiety have increased with time, as have specific physiological measures of stress. Blood pressure measurements in the TMI group, assessed by examination of personal physician records from before the accident, show increases by one year after the accident and have remained high; 10 to 15 percent have developed hypertension since the accident, a percentage not explained by smoking or other controlled variables. Compared to other groups studied, the TMI population also exhibits higher concentrations of urinary catecholamines, which are indicative of physiological stress responses.

Some of the individuals in the TMI group scored high on all three types of measures: performance tasks, depression scales, and physiological measures. Such individuals were categorized as a "high-stress" group and studied further in comparison to "low-stress" and control groups. It was found that, although the entire population showed decreased lymphocyte counts (monocytes, B cells, T cells, T helper, and T suppressor cells), the high-stress group exhibited clear indications of immune system depression by these measures. Interestingly, it was found that individuals who had strong social supports reported fewer symptoms of any kind, but these individuals nevertheless exhibited differences in biochemical and physiological measurements. Finally, there were no differences reported in any of the groups between those individuals who were actually exposed and those who believed they were exposed.

Indeed, the entire issue of exposure to toxic chemicals or radiation is complex in terms of psychological, physiological, and social responses. Uncertainty is probably the most important characteristic in increased perception of risk and level of stress, yet uncertainty on the part of the involved health professionals can also decrease the effectiveness of health care for those exposed. The interplay of such variables has been the subject of Henry Vyner's 1988 analysis of the psychosocial correlates of exposure to toxic chemicals and radiation, so-called invisible environmental contaminants. In his book, Vyner analyzed numerous studies done with various "exposed" populations, including people living at Love Canal, veterans exposed to radiation during the A-bomb tests, people in Michigan who were affected when a toxic chemical known as PBB was mistakenly put into cattle feed, and the TMI community.

According to Vyner, the psychological effects of such exposures proceed in sequential fashion, from uncertainty arising from the individual's attempt to adapt to the possible health threats of the exposure, to hypervigilance about one's health and the development of nonempirical belief systems, to "traumatic neuroses."

Vyner argues that this sequence can become a vicious circle in which these individuals get caught. The more hypervigilant they are, the more they believe their health is threatened, the less seriously their complaints are taken by the medical establishment or other groups charged with the medical care or compensation of these individuals. Institutional responses to persons exposed to toxic chemicals or radiation often show a tendency to blame the victims and view them as hypochondriacs or malingerers. When this happens, the individual becomes more threatened and more vigilant. Vyner writes that the challenge to physicians and health care providers is to recognize such an individual's uncertainty and vigilance, and to provide the patient with as much information and control as possible, in order to increase the effectiveness of health care for that person and, likely, the eventual outcome.

CONCLUSIONS

The experiences of the human subjects represent a combination of variables seen in combat stress and environmental contaminations. Such a combination fits quite well into several of the broad categories of reactions to stressful life events outlined above and would be expected to increase the likelihood of adverse psychological effects for a certain percentage of individuals. **It is the judgment of this committee that the best available evidence indicates a causal relation between the experiences of the subjects in chamber and field tests of mustard agents and Lewisite and the development of adverse psychological effects. These effects may be highly individual, but diagnosable, and may include long-term mood and anxiety disorders, PTSD, or other traumatic stress disorder responses.** In addition, some of the experiences of those who worked with chemical warfare agents or who were exposed to sulfur mustard at Bari harbor, such as explosions, injuries, and witnessing of injury and death, may also have resulted in development of such adverse psychological effects.

There are many aspects of the chamber and field test situations that individually may have been sufficient to produce adverse psychological effects in certain human subjects. These include (1) the number and duration of chamber trials in the presence of live agent and under hot, humid conditions; (2) the inability to escape the chamber without fear of severe reprisals; (3) the sight and/or experience of severe blistering, especially to genital areas; (4) the young age and lack of adequate preparation of the subjects; (5) the commands of secrecy and its resultant isolation of subjects; and (6) certainty that an exposure occurred (in the chamber tests, evidence of exposure was the end point of the experiments). Such conditions certainly qualify as "outside the range of normal experience" and they would be upsetting to almost anyone. For those involved with handling warfare gases, who witnessed or were injured by explosions and other types of accidents, these would also qualify as stressful experiences outside the normal range of human experiences.

The aspects above are, however, only part of the total experience of many of the human subjects. The passage of time, the imposed silence, the lack of medical follow-up, and the institutional denials have almost certainly complicated the original trauma and worsened its effects. Such effects would include hypervigilance and the establishment of long-held beliefs regarding health problems and their causes.

Despite intensive research, it is not possible to know the degree to which PTSD and other psychological disorders are accompanied or caused by physiological perturbations such as changes in hormone levels, tone of the autonomic nervous system, levels of circulating lymphocytes, and other measures of physiological function. It is not possible, therefore, to draw any conclusions about specific physiological conditions and their possible psychological concomitants or causes.

REFERENCES

American Psychiatric Association. 1952. Diagnostic and Statistical Manual of Mental Disorders. Washington, DC: American Psychiatric Association.

American Psychiatric Association. 1968. Diagnostic and Statistical Manual of Mental Disorders (DSM-II). Washington, DC: American Psychiatric Association.
American Psychiatric Association. 1980. Diagnostic and Statistical Manual of Mental Disorders (DSM-III). 3rd ed. Washington, DC: American Psychiatric Association.
American Psychiatric Association. 1987. Diagnostic and Statistical Manual of Mental Disorders (DSM-IIIR). 3rd ed., rev. Washington, DC: American Psychiatric Association.
American Psychiatric Association, Task Force on DSM-IV. 1991. DSM-IV Options Book: Work in Progress 9/1/91. Draft publication.
Andreasen NC, 1980. Posttraumatic stress disorder. In: Kaplan HI, Freedman, AM, Saddock, BJ, eds. Comprehensive Textbook of Psychiatry, 3rd ed. Baltimore: willems and Wilkins. 1517-1525.
Andreasen NC. 1985. Posttraumatic stress disorder. In: Kaplan HI, Freedman AM, Saddock BJ, eds. Comprehensive Textbook of Psychiatry. 4th ed. Baltimore: Williams and Wilkins. 918-924.
Baum A. 1981. Chronic and acute stress associated with the Three Mile Island accident and decontamination: preliminary findings of a longitudinal study. As cited in: Vyner HM. 1988. The psychological dimensions of health care for patients exposed to radiation and the other invisible environmental contaminants. Social Science in Medicine 27:10971103.
Baum A. 1987. Biopsychosocial effects of disasters: Three Mile Island. In: Ursano RJ, ed. Individual and Group Behavior in Toxic and ContainedEnvironments . A Conference to Explore the Psychological Effects of Chemical and Biological Warfare. Bethesda, Maryland: Uniformed Services University of the Health Sciences.
Boehnlein JK, Kinzie JD, Ben R, Fleck J. 1985. One-year follow-up study of posttraumatic stress disorder among survivors of Cambodian concentration camps. American Journal of Psychiatry 142(8):956-959.
Da Costa JM. 1871. On irritable heart: a clinical study of a form of functional cardiac disorder and its consequence. American Journal of Medical Science 61:17.
Davidson JR, Foa EB. 1991. Refining criteria for posttraumatic stress disorder. Hospital and Community Psychiatry 42:259-261.
Davidson JR, Kudler HS, Saunders WB, Smith RD. 1990. Symptom and comorbidity patterns in World War II and Vietnam veterans with posttraumatic stress disorder. Comprehensive Psychiatry 31:162-170.
Fullerton CS, Ursano RJ. 1990. Behavioral and psychological responses to chemical and biological warfare. Military Medicine 155:54-59.
Giller EL Jr, ed. 1990. Biological Assessment and Treatment of Posttraumatic Stress Disorder. Washington, DC: American Psychiatric Press.
Green BL, Lindy JD, Grace MC, Gleser GC. 1989. Multiple diagnosis in posttraumatic stress disorder. The role of war stressors. Journal of Nervous and Mental Disease 177:329-335.
Horowitz MJ. 1976. Stress Response Syndromes. New York: Jacob Aronson.
Horowitz MJ. 1986. Stress Response Syndromes. 2nd ed. Northvale, NJ: Jacob Aronson.
Hulbert HS. 1920. Gas neurosis syndrome. American Journal of Insanity 75:213-216.
Idelson H. 1923. Nervous disorders caused by toxic gases and their relationship to traumatic neuroses. Revue Neurologique 30:140-151. [In French]
International Society for Traumatic Stress Studies. 1991. Proceedings. The Reality of Trauma in Everyday Life: Implications for Intervention and Policy. Washington, DC, October 24-27, 1991.
Jewett S. 1942. Neuropsychiatric chemical warfare casualties. Bulletin of the New York Medical College (June-October):107-113.
Kardiner A. 1941. The Traumatic Neuroses of War. New York: Hoeber.

Kinzie JD. 1989. Post-traumatic stress disorder. In: Kaplan HI, Saddock BJ, eds. Comprehensive Textbook of Psychiatry. 5th ed. Baltimore: Williams and Wilkins. 1000-1008.

Kolb LW. 1990. Foreword. In: Giller EL. Biological Assessment and Treatment of Posttraumatic Stress Disorder. Washington, DC: American Psychiatric Press.

Krystal H. 1968. Massive Psychic Trauma. New York: International Universities Press.

Kulka RA, Schlenger WE, Fairbank JA, Hough RL, Jordan BK, Marmar CR, Weiss DS. 1990. Trauma and the Vietnam War Generation. New York: Brunner Mazel.

Macleod AD. 1991. Posttraumatic stress disorder in World War Two veterans. New Zealand Medical Journal 104:285-288.

Nemiah JC. 1980. Anxiety state. In: Kaplan HI, Freedman AM, Saddock BJ, eds. Comprehensive Textbook of Psychiatry. Baltimore: Williams and Wilkins. 1483-1493.

Rennie TAC, Small SM. 1943. Psychological Aspects of Chemical Warfare. Review Series, Vol. 1, No. 1. New York: Josiah Macy, Jr. Foundation.

Ross WD. 1966. Neuroses following trauma and their relation to compensation. In: Arieti S, ed. American Handbook of Psychiatry. New York: Basic Books. 131-147.

Solomon Z. 1989. Characteristic psychiatric symptomatology of post-traumatic stress disorder in veterans: a three year follow-up. Psychological Medicine 19:927-936.

Tabatabai SM, Rizabi N, Kamkardel MH, Balali M. 1988. Study of psychiatric complications of poisonings with chemical warfare agents. In: Abstracts of the First International Medical Congress on Chemical Warfare Agents in Iran. June 13-16, 1988, Mashhad University of Medical Sciences. Mashhad, Iran. No. 66.

Trimble MR. 1985. Post-traumatic stress disorder: history of a concept. In: Figley CR, ed. Trauma and its Wake. New York: Brunner Mazel. 5-14.

Ursano RJ, ed. 1987a. Groups and Organizations in War, Disasters, and Trauma. Bethesda, MD: Uniformed Services University of the Health Sciences.

Ursano RJ, ed. 1987b. Individual and Group Behavior in Toxic and Contained Environments. A Conference to Explore the Psychological Effects of Chemical and Biological Warfare. December 12-14, 1986, Airlie, Virginia. Bethesda, MD: Uniformed Services University of the Health Sciences.

Ursano RJ. 1988a. Combat stress in the chemical and biological warfare environment. Aviation, Space, and Environmental Medicine 59:1123-1124.

Ursano RJ, ed. 1988b. Performance and Operations in Toxic Environments. 2nd ed. Bethesda, MD: Uniformed Services University of the Health Sciences.

Ursano RJ, ed. 1988c. Training for the Psychological and Behavioral Effects of the CBW Environment: A Conference to Explore Training for Operational and Medical Personnel for Coping, Adaptation and Performance in the High Stress Environment of Chemical and Biological Warfare. November 6-8, 1987, Airlie, Virginia. Bethesda, MD: Uniformed Services University of the Health Sciences.

Ursano RJ, ed. 1988d. Psychological and Behavioral Responses to a Chemical and Biological Warfare Environment: Final Recommendations. Bethesda, MD: Uniformed Services University of the Health Sciences. Ursano RJ, Holloway HC. 1985. Military psychiatry. In: Kaplan HI and Saddock BJ, eds. Comprehensive Textbook of Psychiatry. 4th ed. Baltimore: Williams and Wilkins.

Ursano RJ, Boydstun JA, Wheatley RD. 1981. Psychiatric illness in U.S. Air Force Vietnam prisoners of war: a 5 year follow-up. American Journal of Psychiatry 138:310-314.

Van Kammen DP, Ver Ellen P. 1990. Afterword. In: Giller EL. Biological Assessment and Treatment of Posttraumatic Stress Disorder.

Vyner HM. 1988. Invisible Trauma: The Psychosocial Effects of Invisible Environmental Contaminants. Lexington, MA: Lexington Books.

12

Summary of Findings and Recommendations

GENERAL CONCLUSIONS

In the course of this study the committee reviewed almost 2,000 papers, monographs, abstracts, and technical summaries in search of information regarding the long-term health effects of exposure to mustard agents and Lewisite. The committee found a "stunted" body of literature, clearly focused on the acute effects of these agents and the prevention or treatment of these effects. Certainly, protection of lives in combat situations is an important and necessary effort. Yet the narrow focus of the literature presented a major barrier to this committee, concerned as it was with surveying the scientific and medical literature to assess the health risks incurred by anyone exposed to these agents, but especially the human subjects in the World War II (WWII) testing programs. Thus, **the lack of follow-up health assessments of the human subjects in WWII gas chamber and field tests severely diminished the amount and quality of information that could be applied in the assessment of long-term health consequences of exposure to mustard agents and Lewisite.**

The lack of follow-up of these subjects particularly dismayed the committee for a number of reasons. For example, the end point of the chamber and field tests was tissue injury, but it was already known by 1933 that certain long-term health problems resulted from sulfur mustard exposure. Further, it was documented that numerous subjects suffered severe injuries that required up to a month of treatment. Finally, the exposure levels were sufficiently high that even the most efficient gas mask could have leaked enough mustard agent or Lewisite to cause inhalation and eye injuries.

There was, in fact, no long-term follow-up of any of the thousands of individuals exposed to these agents during WWII as evidenced by the accompanying **lack of epidemiological studies of chemical warfare production workers, war gas handlers and trainers, and combat casualties of the Bari harbor bombing.** The committee was particularly dismayed at this lack of epidemiological and follow-up data from the United States, despite the availability of a large cohort of civilian workers and military personnel who were involved in chemical warfare production and training, as well as the individuals who served as human subjects in chemical warfare testing programs. The committee was forced to rely on studies done in Japan and Great Britain to assess what was known about the long-term health risks from occupational exposure to mustard agents and Lewisite. As demonstrated in Chapters 3 and 7, such occupational data are directly relevant to the assessment of the potential effects of mustard agent and Lewisite exposure in the experimental testing programs, because **the levels of exposure to mustard agents or Lewisite experienced by the human subjects may have been much higher than inferred in the summaries of the** gas **chamber and field tests.**

These exposures were likely as high as those estimated for battlefield and occupational exposures, due to cumulative skin exposure compounded by inhalation exposure. Numerous lines of evidence demonstrate that inhalation exposures did indeed occur (see Chapter 3 and 7). First, modern gas masks have efficiency ratings (or PF) between 50 and 100 (a PF of 100 means that 1 percent of the contaminant in the atmosphere will penetrate a mask's filter canister); however, the efficiency achieved in actual use has been demonstrated to be much lower. Even if a much higher PF of 1,000 is assumed for the gas masks used in the WWII testing programs, penetration of sufficient amounts of the agents to cause respiratory and ocular signs and symptoms would have been expected at many of the concentrations used in the experiments. Second, there is documentation in the actual records of these experiments, as well as official histories of production settings, that respiratory and ocular symptoms and injuries did occur, and that problems were encountered with gas masks leaking after repeated use. Third, the specific diaphragm type of gas mask used in the gas chamber tests was eventually shown to be leaky due to penetration of the diaphragm element, independent of the filter canister employed.

The reasons for the lack of follow-up of human subjects and combat casualties, as well as gas production, handling, and training personnel, can only be surmised, but the climate of secrecy within which the WWII chemical warfare production and testing programs were conducted is probably a key factor.

CONCLUSIONS REGARDING THE CAUSAL RELATIONSHIPS OF EXPOSURE TO THE DEVELOPMENT OF SPECIFIC DISEASES

The major conclusions reached by the committee regarding the association of exposure to mustard agents or Lewisite to specific diseases in different organ systems are summarized in Table 12-1. In some cases, the data examined were found to indicate a causal relationship between exposure and a particular disease or health problem. For other health problems, the data were suggestive, but not completely clear. Finally, there were certain health problems for which very little or no data existed regarding the possible contributions of exposure to mustard agents or Lewisite. By the same token, however, there was no condition evaluated that could be removed from consideration as a health consequence of exposure to these agents. Thus, for many diseases and health problems, there remains significant doubt about whether or not exposure to these agents is a key etiological factor.

> The evidence indicates a causal relation between sulfur mustard exposure and the occurrence of excess respiratory and skin cancer, and possibly leukemia. This conclusion is based upon estimates of exposure to sulfur mustard during the chamber tests, which may have approximated the battlefield exposure of surviving World War I (WWI) soldiers and WWII production workers in Japan and Great Britain. Inadequate exposure information, however, limits precise estimation of the cancer excesses that may be expected. The evidence is insufficient to indicate a causal relationship for Lewisite carcinogenesis.

Mustard agents are DNA-alkylating agents and are extremely cytotoxic at low doses. DNA alkylation is probably responsible for the mutagenicity of mustard agents. These agents also alkylate RNA and proteins and can, at moderate to high doses, produce nonrepairable DNA lesions (genotoxicity). The sulfur mustards induce a wide variety of genetic lesions in many types of mammalian cells *in vitro* in a dose-related fashion. They also induce genetic damage *in vivo* in peripheral blood lymphocytes from exposed individuals at low doses. The toxicology of Lewisite has been poorly studied.

Chamber exposure to sulfur mustard has produced skin malignancies in rats, and intravenous injection has produced a significant increase in pulmonary tumors in highly susceptible strain A mice. Subcutaneous injection of sulfur mustard has been shown to cause sarcomas and other tumors at the injection site in C3H, C3Hf, and strain A mice, but did not produce an increase of tumors at other sites.

Nitrogen mustard, particularly HN2, has been more widely tested than sulfur mustard and has been found to be a carcinogen, producing pulmonary tumors from both intravenous and intraperitoneal injections in strain A mice. Subcutaneous exposures produced injection site tumors and pulmonary tumors in selected strains of mice. The carcino

TABLE 12-1 Summary of Findings Regarding Specific Health Problems

Evidence Indicates Causal Relationship to Mustard Agent Exposure [a]
Respiratory cancers
nasopharyngeal
laryngeal
lung
Skin cancer
Pigmentation abnormalities of the skin
Chronic skin ulceration and scar formation
Leukemia (typically acute nonlymphocytic type, nitrogen mustard only)
Chronic respiratory diseases (also Lewisite)
asthma
chronic bronchitis
emphysema
chronic obstructive pulmonary disease
laryngitis
Recurrent corneal ulcerative disease[b]
Delayed recurrent keratitis of the eye
Chronic conjunctivitis
Bone marrow depression and immunosuppression[c]
Psychological disorders[d]
mood disorders
anxiety disorders (including post-traumatic stress disorder)
other traumatic stress disorder responses
Sexual dysfunction[e]
Evidence Suggestive of Causal Relationship to Mustard Agent Exposure
Leukemia (sulfur mustard)
Reproductive dysfunction (genotoxicity, mutagenicity, etc.)
Insufficient Evidence of Causal Relationship to Mustard Agent Exposure
Gastrointestinal diseases
Hematologic diseases
Neurological diseases
Reproductive dysfunction (Lewisite)
Cardiovascular disease[f]

[a] Includes Lewisite only when indicated.

[b] Includes corneal opacities; acute severe injuries to eye from Lewisite will persist.

[c] An acute effect that may result in greater susceptibility to serious infections with secondary permanent damage to vital organ systems.

[d] These may result from traumatic or stressful features of the exposure experience, not a toxic effect of the agents themselves.

[e] Scrotal and penile scarring may prevent or inhibit normal sexual performance or activity. Decreased sexual function may adversely affect reproductive success.

[f] Except when caused by serious infection (e.g., rheumatic fever) closely following an exposure that produced bone marrow depression and immunosuppression.

genic potency of nitrogen mustard appears to be similar to sulfur mustard. In addition, nitrogen mustard has been shown to be one of the most potent carcinogens amongst the alkylating agents tested in the strain A bioassay program of the National Cancer Institute.

Studies of these agents in humans have involved occupational, battlefield, and therapeutic exposures. Occupational exposure to sulfur mustard has been associated with respiratory tract cancer. The data from battlefield exposures, however, have been somewhat more equivocal: an excess of lung cancer was observed, but the excess was not statistically significant. Follow-up of cancer patients treated with nitrogen mustard derivatives has clearly indicated a causal association with skin cancer and leukemia, particularly the acute nonlymphocytic type. Although an excess of skin cancer or leukemia was not evident in the occupational or battlefield studies, the discrepancy may result from differences in amount of exposure; the leukemias or skin cancer may have occurred prior to the start of observation of the occupational and battlefield cohorts; or nonfatal cases of skin cancer may not have been detected in mortality studies. It is also possible that skin cancers did not occur in the studied populations, or that there was a difference in effects between sulfur and nitrogen mustards. Although nitrogen mustard-associated leukemia and skin cancer occur usually within a decade of therapeutic exposure, the occurrence of an excess of such cases among the WWII human subjects, Bari casualties, or workers would not be surprising.

> The evidence indicates a causal relation between exposure to sufficient concentrations of sulfur mustard (and presumably nitrogen mustard and Lewisite) and chronic nonreversible respiratory effects in humans.

Follow-up of WWI battlefield casualties has demonstrated the association between exposure to sulfur mustard and development of chronic bronchitis, emphysema, and asthma. Chronic respiratory effects have also been shown in workers from WWII chemical weapons factories and casualties of the Iran-Iraq war. These results are well supported by studies in laboratory animals. Given the concentrations of mustard agents and Lewisite used in the WWII experiments, prior research predicts the development of chronic nonreversible lung diseases. Further, indirect evidence, based on a review of the relationships between acute and chronic effects caused by other substances, suggests that these long-term respiratory effects may occur in the absence of an acute respiratory response.

> The evidence indicates a causal relation between exposure to sulfur mustard and recurrent corneal ulcerative disease (including corneal opacities), delayed recurrent keratitis, and chronic

> intractable conjunctivitis. Evidence in laboratory animals indicates no causal relation between exposure to Lewisite and any long-term ocular disease process. However, any corneal scarring or vascularization that occurs soon after injury from Lewisite will persist.

There is an extensive base of knowledge from studies in laboratory animals and humans regarding the long-term effects of mustard agents on the eye. Thus, acute, severe injury of the eye with sulfur mustard, resulting in corneal scarring among other effects, can result in recurrent corneal ulcerative disease. The maximum incidence of this disease occurs 15 to 20 years after the injury. Acute severe injury from sulfur mustard has also been shown to result in the development of delayed recurrent keratitis and corneal opacities. The conjunctiva of the eye has been shown to be more vulnerable than the cornea to sulfur mustard exposure, explaining the development of intractable, prolonged conjunctivitis even in the absence of severe injury to the cornea.

> The evidence indicates a causal relation between acute, severe exposure to mustard agents and increased skin pigmentation and depigmentation, chronic skin ulceration, scar formation, and the development of cancer in human skin. A causal relationship also exists between chronic exposure to minimally toxic, and even subtoxic, doses and skin pigmentation abnormalities and cutaneous cancer. There is insufficient evidence, however, to establish a causal relationship between Lewisite exposure and long-term adverse effects on skin.

There has been much research on the toxic mechanisms of acute skin injury from mustard agents, especially in laboratory animals and tissue cultures. Injuries from mustard agents have been shown in these models to result in a complex cascade of biochemical reactions that cause cell death and genotoxicity. Studies of carcinogenesis were positive in laboratory animals, but these studies employed outdated methods and are relatively crude by today's standards. Studies in humans after battlefield or occupational exposure also vary tremendously in quality.

Nevertheless, skin cancers have been observed in many of these studies. That fact, coupled with documented and plausible biological mechanisms, indicates cancer as a likely consequence of acute, severe (or chronic, mild to moderate) mustard agent injury to the skin. Scar formation and chronic ulceration of the skin following mustard agent exposure have been well documented in the literature. Genital regions are especially sensitive to exposure, and scarring of the scrotum and penis can seriously impair sexual performance and capability. In those studies in which pigmentation abnormalities were reported, including recent observations in casualties of the Iran-Iraq war, the abnormalities are completely consistent with known effects of skin damage. Despite data highly suggestive of a link between skin diseases and arsenic exposure, very little data exist that can be directly extrapolated to exposure to the organic arsenical Lewisite and its consequences in the skin.

> The evidence indicates a causal relationship between exposure to mustard agents and bone marrow depression and immune system dysfunction. These acute effects would render individuals highly susceptible to infections, including pneumonia, rheumatic fever, and tuberculosis, which in severe cases may cause permanent damage to vital organs. There is insufficient evidence with which to draw conclusions regarding the effects of Lewisite on immune system function.

Animal studies clearly demonstrate a causal relationship between exposure to mustard agents and immunotoxicity. Evidence from observations in humans indicates a causal relationship between mustard agent exposure and acute bone marrow toxicity expressed as leukopenia, pancytopenia, and aplastic or hypoplastic bone marrow. However, underrepresented in human studies is information on chronic or delayed effects. The data examined, however, indicate that clinical studies as a whole support a close parallelism between animal experiments and observations in humans regarding the immunosuppressive properties of mustard agents.

> There is insufficient evidence to demonstrate a causal relationship between exposure to mustard agents and the development of long-term gastrointestinal, hematologic, or neurological diseases or dysfunctions, other than those secondary to other conditions related to exposure to mustard agents. In addition, there is insufficient information to link Lewisite with long-term health effects on the hematological, gastrointestinal, and neurological systems.

Gastrointestinal, hematological, and neurological effects are common after acute high exposures to mustard agents and can be attributed primarily to the known toxicological effects of these agents and secondarily to effects on other organ systems (e.g., from shock or burns). However, effects on these organ systems have not been a focus of any follow-up studies of humans exposed to mustard agents or Lewisite.

> There is insufficient evidence to demonstrate a causal relationship between exposure to mustard agents or Lewisite and toxicity to the reproductive system.

The database is too small and uncertain to allow a clear understanding of human reproductive risk from exposure to sulfur mustards.

However, there is evidence to suggest a causal relationship between sulfur mustard exposure and reproductive toxicity in laboratory animals. Reproductive success, however, can be adversely affected by impaired sexual function caused by scarring of penile tissue.

The studies of the reproductive toxicity of Lewisite in laboratory animals are negative. Such data, however, are not complete and thus are insufficient to support or deny a causal relationship between exposure and adverse reproductive outcomes.

> The evidence indicates a causal relationship between characteristic aspects of the chamber and field experiences and the development of adverse psychological effects. These effects may be highly individual, but diagnosable, and may include long-term mood and anxiety disorders, PTSD, or other traumatic stress disorder responses. Data are insufficient, however, to associate the presence of adverse psychological disorders with any physiological disease or dysfunction.

Many elements of the gas chamber and field experiments were highly stressful. These include lack of prior knowledge about what to expect, the duration and conditions of the chamber trials, the experience of skin injury from the exposures, the threats of punishment if the experiments were revealed, and other elements. Any stress reaction from the experiences may have well been magnified by subsequent secrecy, fears about the health risks, and institutional denials. Data from studies of chemical and biological warfare environments and environmental exposures to toxic chemicals or radiation support the assertion that, in certain individuals, these experiences would have been sufficient to cause adverse psychological effects, resulting in long-term dysfunction. It is likely that such effects also occurred in some production workers, gas handlers and trainers, and Bari harbor survivors as a result of traumatic episodes including explosions, accidents, personal exposure injury, or the witnessing of severe injury or death of others. Current investigations of the physiological concomitants of psychological disorders are compelling and of great interest for future research. However, it is not possible to predict from these studies what adverse physiological effects may be attributable to long-standing psychological problems.

GAPS IN THE LITERATURE REGARDING MUSTARD AGENTS AND LEWISITE

Human Studies

Clearly the most important gap in studies assessing the effect of agent exposure on humans is the lack of epidemiological studies of occupa tional exposure. Only limited cohorts of workers in Japan and Great Britain have been studied, and the value of these studies has been diminished by a lack of precise exposure information. The few exposure measurements made were usually from specific plant regions or during particularly troublesome parts of the manufacturing process. More useful would have been exposure information according to specific job categories. Finally, no attempts were made in such studies to determine the likely dose-response relationships. Such prediction would require quantitative risk assessments for which adequate data are not available.

The focus on carcinogenicity in epidemiological studies also left large gaps in the literature pertinent to the development of nonmalignant diseases. Although sufficient studies exist to associate the development of nonmalignant respiratory diseases, eye damage, and certain skin diseases with exposure, little to nothing is known regarding the effect of exposure on the development of gastrointestinal, immunological, and neurological diseases in humans. Further, no human data exist concerning the possible adverse reproductive effects of exposure to mustard agents or Lewisite.

The most extensively studied organ systems in terms of human pathology are the eye and the skin. Much of this research is quite old and, for the skin, research into the long-term effects in humans of exposure is still lacking. The recent casualties of sulfur mustard exposure, such as those injured in the Iran-Iraq war, are now being followed to assess long-term cutaneous effects of acute sulfur mustard exposure. Yet, observation periods as long as 35-45 years may be required to produce meaningful human data. Because these studies are only in their fourth or fifth year, conclusive results will probably not be available for another 15 to 20 years. In the short term for this cohort, however, investigative application of modern methods of ophthalmological treatment, such as the use of soft contact lenses to reduce the effects of chronic relapsing keratitis of the eyes, may yield some benefits. There are also human data derived from patients previously treated in Russian and Eastern European studies of the sulfur mustard-containing agent psoriasin that may be useful in determining the delayed effect of short term administration of sub-erythema dosages of sulfur mustard. Because these studies began 20 to 25 years ago, follow-up of the psoriasin-treated patients now, if properly done, would be of invaluable help in determining delayed effects of acute sulfur mustard exposure.

Animal Studies

The most critical gap in animal studies is the lack of more extensive carcinogenicity and toxicity studies of mustard agents and Lewisite. Particularly lacking are studies, employing modern methods, of the long-term effects of varying levels of exposure to these agents. For example, investigations of the systemic toxicity of short-term exposures to these agents to the gastrointestinal, immunological, and neurological systems are nonexistent. Further studies of the mechanisms of long-term damage to the respiratory system would also be useful. Despite intensive research on the mechanisms of sulfur mustard injury to the skin in animal models, little data exist regarding the long-term consequences of such injury.

In addition, the mechanisms of eye injury from sulfur mustard remain to be elucidated. For example, research on the effect of sulfur mustard on the ciliary body would disclose any changes in its vascular, nutritive, and transport functions. Research to determine the effect of sulfur mustard and Lewisite on limbal stem cells might be useful in determining whether stem cell replacement could reverse some of the adverse effects of injury. Finally, research directed at protection of the stromal component of the cornea might reduce the incidence of ulceration and perforation so common after chemical injuries to the eyes.

Although data exist on the reproductive toxicity of sulfur mustards in more than one animal species, it would be useful to have additional studies to examine the extent of the variability between species. More inhalation and cutaneous exposure studies would also be very helpful, as these exposure results would more accurately mimic human exposure. Certainly studying larger numbers of animals would provide a more sensitive measure of the possible magnitude of any reproductive risk associated with exposure to sulfur mustards. Short-term, high-dose exposures might also be helpful in attempting to examine any dose-rate effects.

The gaps in our knowledge of the toxicity and carcinogenicity of Lewisite based on animal studies are especially prominent, even in the most basic types of research. There is little information available in the literature concerning the reactions of Lewisite with biologically important molecules. Studies on the carcinogenicity or noncarcinogenicity of Lewisite need to be broadened and pursued with greater intensity. Much of the information obtained from these studies, unlike studies of sulfur mustard exposure, will have broad application in industry, farming, and medicine, because arsenic-containing chemicals are in wide use today.

There are also numerous gaps in the literature relative to the acute and long-term effects of Lewisite skin exposure. Very little is known regarding its specific effect on skin; data on such basic areas as absorption, disposition, and excretion after skin exposure are minimal. In addition, the morphological sites vulnerable to Lewisite are not known. Microscopic examination of affected skin has yet to be pursued in depth, although most studies have been impaired, as has been work on sulfur mustard exposure, by the lack of good animal model systems.

Serious gaps also exist in our knowledge concerning the potential of Lewisite to cause reproductive problems. The reproductive toxicity of Lewisite in males is unclear. Our ability to extrapolate from the animal studies to humans is limited. The kinetics of absorption through the skin are unclear, as is the potential of this exposure to induce long-term storage of potentially teratogenic arsenic in doses high enough to induce reproductive problems later in life. Even the form of arsenic that is a potent teratogen in animals, and the ability of Lewisite to yield this form as a metabolite in man, are not entirely clear. More studies of multiple species would be helpful in understanding the potential reproductive toxicity of this compound.

RECOMMENDATIONS

With the immense gaps in the knowledge base about the long-term health risks associated with exposure to mustard agents and Lewisite, and after serious consideration of the historical analyses of the WWII testing programs and the likely exposure levels to the human subjects involved, this committee believes certain recommendations are necessary and justified. First, **the committee recommends that the Department of Veterans Affairs (VA) institute a program to identify each human subject in the WWII testing programs (chamber and field tests, and to the degree possible, patch tests), so that these individuals can be notified of their exposures and the likely health risks associated with those exposures. Further, all subjects so identified, if still living, should be medically evaluated and followed by the VA as to their health status in the future. These individuals should also, if they request it, be treated by the VA for any exposure-related health problems discovered. Morbidity and mortality studies should be accomplished by the VA, comparing chamber, field, and patch test cohorts to appropriate control groups, in order to resolve some of the remaining questions about the health risks associated with exposure to these agents.**

The only way to answer some of the key remaining questions is to establish a base of knowledge based on human exposures. There is precedent for this recommendation in the later identification and follow-up of veterans exposed to chemicals, including hallucinogenic drugs, in the military testing programs between 1950 and 1975. The committee is also well aware that a half century has now passed and that many of those who might have benefited from a broader understanding of the toxicity and carcinogenicity of mustard agents and Lewisite are already dead. Nevertheless, these individuals' surviving family mem bers deserve to know about the testing programs, the exposures, and the potential results of those exposures. For those veterans still living, diseases such as skin and lung cancer may still appear. Treating these cancers with full knowledge of their likely cause should be the responsibility of the VA and may be life-saving; for example, the likelihood of survival from skin cancer is greatly increased by early diagnosis and treatment.

In the case of the human subjects of the WWII testing programs, it is reasonable to assume that the secrecy surrounding the experiments may have kept individuals and entire families from successful resolution and treatment of any adverse psychological effects that may have been caused. Given this possibility and the special problems of ambiguity, health fears, and institutional denials encountered by many of those exposed to these agents, **the committee recommends that careful attention be paid by health care providers to the special problems and concerns of the affected veterans and their families. This attention may include the convening of a special task force of experts in stress disorders and risk perception to aid the VA, further than this committee is able, in the establishment of comprehensive guidelines for handling of these cases.**

The above recommendations are not meant to ignore the fact that thousands, probably tens of thousands, of other military and civilian personnel were exposed to mustard agents and Lewisite in occupational and training settings, and in combat in the Bari harbor disaster. Some of these exposures will have resulted in one or more of the exposure-related health problems identified in this report. The committee is also aware that some military personnel who served in the Chemical Warfare Service have qualified for service-connected disability as a result of such exposures. However, many more military personnel were exposed to significant levels of mustard agents or Lewisite than is obvious from service records, because of job classifications, inadequate documentation of "live agent" training and accidents, and other factors. Therefore, **the committee additionally recommends that the Department of Defense (DoD) should use all means at its disposal, including public channels, to identify cohorts of chemical warfare production workers (military or civilian) and individuals exposed to mustard agents or Lewisite from gas handling, training, the Bari harbor disaster, or other circumstances. Records of former military personnel could be turned over to the VA for notification, inclusion in morbidity and mortality studies, and health status evaluation. Records of the civilian personnel should be used by the DoD to notify the former workers. These workers should also be advised as to their health risks and options for seeking appropriate compensation for any illnesses that resulted from their exposures.**

This committee discovered that an atmosphere of secrecy still exists to some extent regarding the WWII testing programs. Although many documents pertaining to the WWII testing programs were declassified shortly after the war ended, others were not. Of those declassified, many remained "restricted" to the present day and are not released to the public. As a result, the committee often had great difficulty obtaining information. For example, only one of the three major chamber test locations, the Naval Research Laboratory, freely shared technical reports and detailed summaries with the committee from the beginning of the study. For other locations, such information only arrived as the study was in its final stages, despite months of requests and inquiries to a variety of offices. The committee is certain that other relevant information exists that was never obtained. It is also clear that there may be many exposed veterans and workers who took an oath of secrecy during WWII and remain true to that oath even today. Veterans, who had just heard about the study and thought it might now be permissible to reveal their experiences, were still contacting the committee for information up until the very end of the study. Such continuing secrecy, in the committee's view, has impeded well-informed health care for thousands of people. Therefore, **the committee recommends that the VA and DoD publicly announce and widely advertise that personnel exposed to mustard agents or Lewisite during their service are released from** any **oath of secrecy taken at the time. In addition, professional educational materials should be prepared by the DoD or the VA, or both, and made available for physicians who may be treating affected individuals. These materials should incorporate the latest information regarding the long-term health effects of exposure to mustard agents and Lewisite.**

There is no doubt that the long-term health consequences of exposure to mustard agents or Lewisite can be serious and, in some cases, devastating. This report has demonstrated that complete knowledge of these long-term consequences has been and still is sorely lacking, resulting in great costs to some of those exposed in WWII. The lack of knowledge, however, has ongoing ramifications as nations will probably continue to use these chemical weapons in battle or begin to grapple with their disposal. Thus, accidental and deliberate human exposures to mustard agents and Lewisite can only be expected to continue in the foreseeable future.

Bibliography

PUBLISHED LITERATURE

Aasted A, Wulf HC, Darre E, Niebuhr E. 1985. Fishermen exposed to mustard gas. Clinical experience and evaluation of the cancer risk. Ugeskrift for Laeger 147:2213-2216. [In Danish]

Aasted A, Darre E, Wulf HC. 1987. Mustard gas: clinical, toxicological, and mutagenic aspects based on modern experience. Annals of Plastic Surgery 19:330-333.

Abbott WN. 1935. Late clinical and pathological sequels of gas exposure. New Zealand Medical Journal 34:166-170.

Abbruzzetti A. 1936. Phosgene and dichloroethyl sulfide, two most powerful war gases. Giornale Veneto di Scienze Mediche 10:383-391. [In Italian]

Abell CW, Falk HL., Shimkin MB. 1965. Uracil mustard: a potent inducer of lung tumors in mice. Science 147:1443-1445.

Abramov TV, Lebedev AV, Rait AS. 1988. Investigation of diastereomersof non-ionic oligonucleotide analogs v. alkylation of nucleic acids in living cells by ethyl phosphotriester derivatives of oligonucleotides containing residue of nitrous yperite on the effect of the phosphotriester fragment configuration . Molekuliarnaia Biologia 22:1285-1292. [In Russian]

Achard C. 1918-1919. Study of respiratory exchanges in war gas intoxications. Archives de Medecine Experimentale et d'Anatomie Pathologique 28:468-525. [In French]

Achard C. 1919. The sequelae of intoxication by war gases. Bulletin de l'Academie Nationale de Medecine 81:135-150. [In French]

Acheson ED. 1983. Nasal cancer, occupational. In: Parmeggiani L, ed. Encylcopaedia of Occupational Health and Safety. 3rd rev. ed. Vol. 2. Geneva: International Labour Office. 1426-1427.

Acheson ED. 1986. Epidemiology of nasal cancer. In: Barrow CS, ed. Toxicology of the Nasal Passages. Chemical Industry Institute of Toxicology Series. New York: Hemisphere. 135-142.

Adachi S, Takemoto K. 1987. Occupational lung cancer: a comparison between humans and experimental animals. Sangyo Igaku 29:345-357. [In Japanese]

Adair FE, Bagg HJ. 1931. Experimental and clinical studies on the treatment of cancer by mustard gas. Annals of Surgery 190-199.

Adelheim R. 1921. Classification of gases according to the pathology produced in the lung. Zentralblatt fur Pathologie 25:261. [In German]

Adler FH, Fry WE, Leopold IH. 1947. Pathologic study of ocular lesions due to Lewisite (b-chlorovinyldichloroarsine). Archives of Ophthalmology 38:89-108.

Adley FE, Uhle RJ. 1969. Protection factors and self-contained compressed-air breathing apparatus. American International Hygiene Association Journal 30:355-359.

Agency for Toxic Substances and Disease Registry. 1991. Draft. Toxicological Profile of Mustard "Gas." Washington, DC: U.S. Department of Health and Human Services.

Aitken RS. 1943. Effects of accidental exposure to mustard gas vapor. Lancet 2:602-603.

Akamizu H, Nambu S, Egawa H, Tokuoka S. 1981. Cytological and histological studies on atypical bronchial epithelium in mustard gas ex-workers. Haigan 21:318. [In Japanese]

Albert A. 1987. Xenobiosis: Foods, Drugs, and Poisons in the Human Body. New York: Chapman and Hall.

Albrecht GJ, Kiese M, Szinicz L, Sies H, Weger N. 1975. Problems of mustard gas poisoning: on the molecular understanding of the alkylating reaction with nitrogen mustard compounds. Wehrmedizinische Monatsschrift 19:12-14. [In German]

Alexander SF. 1947. Medical report of the Bari Harbor mustard casualties. Military Surgeon 101:1-17.

Alho J, Kauppinen T, Sundquist E. 1988. Use of exposure registration in the prevention of occupational cancer in Finland. American Journal of Industrial Medicine 13:581-592.

Allen AC. 1967. The Skin: A Clinicopathological Treatise. 2nd ed. New York: Grune and Stratton.

Alpert LK. 1958. Preliminary studies with sulfur mustard in human neoplastic diseases. Annals of the New York Academy of Sciences 68:1223-1224.

Althouse R, Huff J, Tomatis L, Wilbourn J. 1979. Chemicals and Industrial Processes Associated with Cancer in Humans. Supplement 1. Lyon: International Agency for Research on Cancer.

AMA Archives of Industrial Health. 1959. Beryllium disease and its control. Conference held at Massachusetts Institute of Technology, Sept. 30-Oct. 1, 1958. AMA Archives of Industrial Health 19.

Amagai T, Tsutsumi S, Matsumoto Y. 1981. Effects of arsenic trioxide and its antidotes on the uptake of tritium labeled leucine of mouse lymphocytes (abstract). Japanese Journal of Pharmacology 31:209P.

Amalric P, Bessou P, Farenc M. 1965. Delayed relapsing mustard gas keratitis. Bulletin des Societes d'Ophtalmologie 65:101-106. [In French]

Amantea G. 1918. On a new method of curing lesions from dichloroethyl sulfide. Giornale de Medicina Militare 66:893-896. [In Italian]

American Psychiatric Association. 1952. Diagnostic and Statistical Manual of Mental Disorders. Washington, DC: American Psychiatric Association.

American Psychiatric Association. 1968. Diagnostic and Statistical Manual of Mental Disorders (DSM-II). 2nd ed. Washington, DC: American Psychiatric Association.

American Psychiatric Association. 1980. Diagnostic and Statistical Manual of Mental Disorders (DSM-III). 3rd ed. Washington, DC: American Psychiatric Association.

American Psychiatric Association. 1987. Diagnostic and Statistical Manual of Mental Disorders (DSM-IIIR). 3rd ed., rev. Washington, DC: American Psychiatric Association.

American Psychiatric Association, Task Force on DSM-IV. Draft publication. DSM-IV Options Book: Work in Progress 9/1/91.

Ancel P. 1946. Experimental research on spina bifida. Archives d'Anatomie Microscopique et de Morphologie Experimentale 36:45-68. [In French]

Anders MW. 1988. Glutathione-dependent toxicity: biosynthesis and bioactivation of cytotoxic S-conjugates. ISI Atlas of Science: Pharmacology 2:99-104.

Anders MW. 1991. Glutathione-dependent bioactivation of xenobiotics. In: Witmer CM, ed. Molecular and Cellular Effects and Their Impact on Human Health. 4th International Symposium on Biological Reactive Intermediates. New York: Plenum.

Anders MW, Lash L, Dekant W, Elfarra A, Dohn D. 1988. Biosynthesisand biotransformation of glutathione S-conjugates to toxic metabolites . Critical Reviews in Toxicology 18:311-341.

Andersson M, Philip P, Pedersen-Bjergaard J. 1990. High risk of therapy-related leukemia and preleukemia after therapy with prednimustine, methotrexate, 5-fluorouracil, mitoxantrone, and tamoxifen for advanced breast cancer. Cancer 65:2460-2464.

Andreasen NC, 1980. Posttraumatic stress disorder. In: Kaplan HI, Freedman, AM, Saddock, BJ, eds. Comprehensive Textbook of Psychiatry. 3rd ed. Baltimore: Williams and Wilkins. 1517-1525.

Andreasen NC. 1985. Posttraumatic stress disorder. In: Kaplan HI, Freedman AM, Saddock BJ, eds. Comprehensive Textbook of Psychiatry. 4th ed. Baltimore: Williams and Wilkins. 918-924.

Andreassi L. 1991. Chemical warfare and the skin. International Journal of Dermatology 30:252-253.

Andreoli I. 1918. A case of intoxication by yperite. Gazzetta Medica Lombarda 77:108-110. [In Italian]

Andreoni R. 1939. Toxicological research on liquid yperite. Clinica Veterinaria 62:421-428. [In Italian]

Andrus EC, Bronk DW, Carden GA Jr, Keefer CS, Lockwood JS, Wearn JT, Winternitz MC, eds. 1948. Advances in Military Medicine. Science in World War II: Office of Scientific Research and Development. Boston: Little, Brown.

Andrzejewski S, Scianowski J. 1975. Recovery pattern of lactate dehydrogenase (LDH) total activity changes induced with sulfur mustard in internal organ tissue and serum of rats. Bulletin de l'Academie Polonaise des Sciences, Serie des Sciences Biologiques 23:773-776.

Andrzejewski S, Scianowski J. 1978. Investigation on the toxic influence of sulfur yperite on selected internal rat organs and a histoautoradiographic assay of its distribution in internal organs damaged earlier with sulfur yperite. Pathologia Polska 29:51-60. [In Polish]

Ankel EG, Ring BJ, Lai CS, Holcenberg JS. 1986. The lack of effects of alkylating agents on mammalian cell membranes. International Journal of Tissue Reactions 8:347-354.

Annals of the New York Academy of Sciences. 1969. Biological effects of alkylating agents. 163:589-1029.

Ansarin K, Salehi R. 1988. Flow-volume loop changes in persons exposed to mustard gas up to two years after exposure. In: Abstracts of the First International Medical Congress on Chemical Warfare Agents in Iran. June 13-16, 1988, Mashhad University of Medical Sciences. Mashhad, Iran. No. 67.

Anslow WP, Karnofsky DA, Jager BV, Smith HW. 1948. The intravenous, subcutaneous and cutaneous toxicity of bis(b-chloroethyl) sulfide (mustard gas) and of various derivatives. Journal of Pharmacology and Experimental Therapeutics 93:1-9.

Aposhian HV, Aposhian MM. 1989. Newer developments in arsenic toxicity. Journal of the American College of Toxicology 8:1297-1305.

Arbuckle LD. 1932. The physiological action and therapeutic potentialities of mustard gas. Chemical Warfare Bulletin 18:1078-1086.

Armengaud. 1919. Laryngitis from gases and the de Cauterets cure. Pr Med 32:313. [In French]

Armengaud. 1920. Persistent bronchitis in yperite victims. Revista de Medicina y Cirugia de la Habana 25:379-382. [In Spanish]

Arseneau JC, Sponzo RW, Levin DL, Schnipper LE, Bonner H, Young RC, Canellos GP,

Johnson RE, DeVita VT. 1972. Non-lymphomatous malignant tumours complicating Hodgkin's disease. New England Journal of Medicine 287:1119-1122.

Asboe S, Fledelius H. 1978. Mustard gas: a medical-ecological problem. Eye and skin injuries in three Ostersjo fisherman. Ugeskrift for Laeger 140:2048-2050 . [In Danish]

Ashby J, Tinwell H, Callander RD, Clare N. 1991. Genetic activity of the human carcinogen sulfur mustard toward *Salmonella* and the mouse bone marrow. Mutation Research 257:307-312.

Ashkenazi I, Blumenthal M, Avni I, Belkin M. 1991. Mustard gas injuries of the eyes. Harefuah 120:279-283. [In Hebrew]

Atkinson WS. 1948. Delayed keratitis due to mustard gas (dichlorodiethyl sulfide) burns. Archives of Ophthalmology 40:291-301.

Auerbach C. 1943. Chemically induced mutations as rearrangements. Drosophila Information Service 17:48-50.

Auerbach C. 1946. The induction by mustard gas of chromosomal instabilities in *Drosophila melanogaster.* Proceedings of the Royal Society of Edinburgh. Section B: Biology 62:307-320.

Auerbach C. 1949. Chemical induction of mutations. Proceedings of the Eighth International Congress of Genetics. 128-147.

Auerbach C. 1950. Differences between effects of chemical and physical mutagens. Pubblicazioni della Stazione Zoologica di Napoli 22:1-21.

Auerbach C. 1951. The effect of oxygen concentration on the mutagenic action of mustard gas. Kurze Mitteilungen 15:341-342.

Auerbach C. 1953. Sensitivity of *Drosophila* germ cells to mutagens. Hereditas 6:247-257.

Auerbach C. 1958. Mutagenic effects of alkylating agents. Annals of the New York Academy of Sciences 68:731-749.

Auerbach C. 1973. History of research on chemical mutagenesis. Chemical Mutagens: Principles and Methods for Their Detection 3:1-19.

Auerbach C. 1976. Mutation Research: Problems, Results and Perspectives. London: Chapman and Hall.

Auerbach C, Moser H. 1950. Production of mutations by monochloro-mustards. Nature 166:1019-1020.

Auerbach C, Moser H. 1950. The effect of oxygen concentration on the mutagenic action of mustard gas. Experientia 7:341-342.

Auerbach C, Robson JM. 1946. Action of mustard gas (dichloroethyl sulfide) on bone marrow (letter). Nature 158:878-879.

Auerbach C, Robson JM. 1946. Chemical production of mutations. Nature 157:302.

Auerbach C, Robson JM. 1947. Production of mutations by chemical substances. Proceedings of the Royal Society of Edinburgh. Section B: Biology 62:271-283.

Auerbach C, Robson JM. 1947. Tests of chemical substances for mutagenicaction . Proceedings of the Royal Society of Edinburgh. Section B: Biology 62:284-291.

Auerbach C, Sonbati EM. 1960. Sensitivity of the *Drosophila* testis to the mutagenic action of mustard gas. Zeitschrift fur Vererbungslehre 91:237-252.

Auerbach C, Robson JM, Carr JG. 1947. Chemical production of mutations. Science 105:243-247.

Axelrod DJ, Hamilton JG. 1947. Radio-autographic studies of the distribution of Lewisite and mustard gas in skin and eye tissues. American Journal of Pathology 23:389-411.

Axford AT, McKerrow CB, Jones AP, Le Quesne PM. 1976. Accidental exposure to isocyanate fumes in a group of firemen. British Journal of Industrial Medicine 33:65-71.

Azizi F, Nafarabadi M, Azartash P. 1988. Serum testosterone concentrations in young men exposed to chemical weapons containing mustards. In: Abstracts of the First International Medical Congress on Chemical Warfare Agents in Iran. June 13-16, 1988, Mashhad University of Medical Sciences. Mashhad, Iran. No. 52.

Azizi F, Keshavarz A, Elyasi H, Roshanzamir F, Rahimi F. 1988. Testicular function in men exposed to chemical weapons containing mustards. In: Abstracts of the First International Medical Congress on Chemical Warfare Agents in Iran. June 13-16, 1988, Mashhad University of Medical Sciences. Mashhad, Iran. No. 27.

Azizi F, Elyasi H, Sohrabpour H, Jalali N, Nafarabadi M. 1989. Serum concentrations of various hormones following exposure to chemical weapons containing sulfur mustard. Medical Journal of the Islamic Republic of Iran 3:105-107.

Azizi F, Jalali N, Nafarabadi M. 1989. The effect of chemical weapons on serum concentrations of various hormones. Iranian Journal of Medical Sciences 14:46-50.

Baburina ND. 1972. Effects of mustard gas on *Allium fistulosum* cells. Genetika 8:174-177.

Bacq ZM. 1947. [Recent Work on War Gases] Trauvaux Recents sur les Toxiques de Guerre. Paris: Masson. [In French]

Bacq ZM. 1971. The concept of the biochemical lesion. In: Fundamentals of Biochemical Pharmacology. Oxford: Pergamon Press. 139-142.

Bacq ZM, Angenot P. 1940. Inhibition of lactic fermentation of muscles by vesicant war gases. Comptes Rendus des Seances de la Societe de Biologie et de Filiales 134:105-107. [In French]

Bacq ZM, Angenot P. 1941. On the relationship between inhibition of glycolysis and vesicant action. Enzymologia 10:48-60. [In French]

Bacq ZM, Goffart M. 1940. Lusndgaard effect produced on frog muscles by vesicant gases. Comptes Rendus des Seances de la Societe de Biologie et de Filiales 133:696-697. [In French]

Bacq ZM, Charlier R, Philippot E. 1948. Action of dichloroethyl sulfide and its derivatives on heart and respiratory exchange of anesthetized dog. Archives Internationales de Pharmacodynamie et de Therapie 77:353-368. [In French]

Bacq ZM, Charlier R, Klutz A, Mazzella H. 1952. Action of dichloroethyl sulfide and methyl-bis-(b-chloroethyl) amine on electrocardiogram of dog. Archives Internationales de Pharmacodynamie et de Therapie 89:357-363.

Badger GM. 1956. Miscellaneous chemical carcinogens: chemical constitution and carcinogenic activity. British Journal of Cancer 10:330-356.

Badolle, Bonnemour. 1928. Bronchial dilation from gas. Bulletin de l'Academie Nationale de Medecine 99:235-237. [In French]

Bagirova N. 1939. Sequels of poisoning with war gases. Klinicheskaia Meditsina 17:116-117. [In Russian]

Bahadori M, Shakoor A. 1988. Autopsy findings on Iranian victims of chemical warfare. In: Abstracts of the First International Medical Congress on Chemical Warfare Agents in Iran. June 13-16, 1988, Mashhad University of Medical Sciences. Mashhad, Iran. No. 31.

Bahadori M, Pishva-Bahadori N, Masyedi MR. 1988. Histology of the lung on the Iranian victims of chemical warfare. In: Abstracts of the First International Medical Congress on Chemical Warfare Agents in Iran. June 13-16, 1988. Mashhad University of Medical Sciences. Mashhad, Iran. No. 24.

Bahar K, Dayhimi I, Eliasy H. 1988. Tumor markers in sulfur mustard gas (SMG) injured patients. In: Abstracts of the First International Medical Congress on Chemical Warfare Agents in Iran. June 13-16, 1988. Mashhad University of Medical Sciences. Mashhad, Iran. No. 13.

Baker DB, Landrigan PJ. 1990. Occupationally related disorders. Medical Clinics of North America 74:441-460.

Balali M. 1984. Clinical and laboratory findings in Iranian fighters with chemical gas poisoning. Archives Belges (Supplement):254-259.

Balali M. 1986. First report of delayed toxic effects of yperite poisoning in Iranian fighters. In: Heyndricks B, ed. Terrorism: Analysis and Detection of Explosives. Proceedings of the Second World Congress on New Compounds in Biological and Chemical Warfare. Gent: Rijksuniversiteit. 489-495.

Balali M, Moodi JR. 1988. Report of third study on late toxic effects of sulfur mustard poisoning. In: Abstracts of the First International Medical Congress on Chemical Warfare Agents in Iran. June 13-16, 1988. Mashhad University of Medical Sciences. Mashhad, Iran. No. 65.

Balali M, Farhoudi M, Navaeian A, Panjvani FA, Tabarestani M. 1988. Evaluation of clinicopathological findings of sulfur mustard poisoning in chemically injured combatants. In: Abstracts of the First International Medical Congress on Chemical Warfare Agents in Iran. June 13-16, 1988. Mashhad University of Medical Sciences. Mashhad, Iran. No. 29.

Balali M, Seddigh M, Akhavain F. 1988. Report of second study on late toxic effects of sulfur mustard poisoning. In: Abstracts of the First International Medical Congress on Chemical Warfare Agents in Iran. June 13-16, 1988. Mashhad University of Medical Sciences. Mashhad, Iran. No. 64.

Balali-Mood M. 1990. Sulfur mustard poisoning in the Iran-Iraq war (abstract). Clinical Pharmacology and Therapeutics 47:184.

Balali-Mood M, Navaeian A. 1986. Clinical and paraclinical findings in 233 patients with sulfur mustard poisoning. In: Heyndricks B, ed. Terrorism: Analysis and Detection of Explosives. Proceedings of the Second World Congress on New Compounds in Biological and Chemical Warfare. Gent: Rijksuniversiteit. 464-473.

Balanescu I, Balaban I. 1936. Pulmonary tuberculosis in war gas victims. Revista Sanitara Militara 35:632-638. [In Romanian]

Ball CR. 1971. The role of DNA repair mechanisms in resistance toalkylating agents . In: Semonsky M, Hejzlar M, Masak S, eds. Advances in Antimicrobial and Antineoplastic Chemotherapy: Progress in Research and Clinical Application. Vol. II , Antineoplastic Chemotherapy Radioprotectives. Baltimore: University Park Press. 11-12.

Ball CR, Roberts JJ. 1970. DNA repair after mustard gas alkylation by sensitive and resistant Yoshida sarcoma cells *in-vitro.* Chemico-Biological Interactions 2:321-329.

Ball CR, Roberts JJ. 1972. Estimation of interstrand DNA cross-linking resulting from mustard gas alkylation of HeLa cells. Chemico-Biological Interactions 4:297-303.

Ballantyne B. 1986. Intramuscular prednisolone pretreatment does not influence the dermal lesions induced by topical sulfur mustard. Veterinary and Human Toxicology 28:204-206.

Banks TE. 1946. The preparation of radiosulphur (^{35}S) and its determination in biological tracer experiments. British Journal of Radiology 19:333-338.

Banks TE, Boursnell JC, Francis GE, Hopwood FL, Wormall A. 1946. Studies on mustard gas (b,b dichlordiethyl sulphide) and some related compounds. 1. General introduction and acknowledgements. Biochemical Journal 40:734-736.

Banks TE, Boursnell JC, Francis GE, Hopwood FL, Wormall A. 1946. Studies on mustard gas (b,b dichlordiethyl sulphide) and some related compounds. 4. Their action on proteins (studied with the aid of radioactive sulphur). Biochemical Journal 40:745-756.

Baptista A. 1981. Contribution to the study of Lewisite and tripuhyite. Anais da Academia Brasileira de Ciencias 53:283-287. [In Portuguese]

Baradaran H, Farid R, Amina H, Tavakoli J, Balali M. 1988. Immunological studies in one-hundred cases of combatants poisoned with chemical warfare agents. In: Abstracts of the First International Medical Congress on Chemical Warfare Agents in Iran. June 13-16, 1988. Mashhad University of Medical Sciences. Mashhad, Iran. No. 57.

Barnaby F. 1988. Iran-Iraq war: the use of chemical weapons against the Kurds. Ambio 17:407-408.

Barranco VP. 1991. Mustard gas and the dermatologist. International Journal of Dermatology 30:684-686.

Barre C. 1936. The military significance of bis(2-chloroethyl) sulfide. Tidskrift I Militar Halsovard 61:97-113. [In Swedish]

Barre C. 1938. [Treatment of disorders due to war gases]. Hygeia 100:433-457.

Barrett JC, Lamb PW, Wang TC, Lee TC. 1989. Mechanisms of arsenic-induced cell transformation. Biological Trace Element Research 21:421-429.

Barrieu C. 1940. Insidious action of yperite on buccal mucous membrane. Avenir Medical 37:76-77. [In French]

Barron ESG, Miller ZB, Barlett GR, Meyer J, Singer TP. 1947. Reactivation of dithiols of enzymes inhibited by Lewisite. Biochemical Journal 41:69-74.

Barron ESG, Harlett GR, Miller ZB, Meyer J, Seegmiller JE. 1948. The effect of nitrogen mustards on enzymes and tissue metabolism. II. The effect on tissue metabolism. Journal of Experimental Medicine 87:503-519.

Barron ESG, Meyer J, Miller ZB. 1948. The metabolism of skin. Effect of vesicant agents. Journal of Investigative Dermatology 11:97-118.

Bartsch H, Tomatis L, Malaveille C. 1982. Qualitative and quantitative comparisons between mutagenic and carcinogenic activities of chemicals. In: Heddle JA, ed. Mutagenicity: New Horizons in Genetic Toxicology. New York: Academic Press. 35-72.

Bashiri MH. 1988. Effects of chemical warfare agents on pregnancy. Report of a case. In: Abstracts of the First International Medical Congress on Chemical Warfare Agents in Iran. June 13-16, 1988. Mashhad University of Medical Sciences. Mashhad, Iran. No. 70.

Baskerville C. 1919. Certain war gases and health. Science 49:50.

Batt SC, Peterson PJ. 1988. Risk assessment techniques for carcinogenic chemicals. In: Richardson ML, ed. Risk Assessment of Chemicals in the Environment. London: Royal Society of Chemistry. 153-176.

Bauer KH. 1963. [The Cancer Problem] Das Krebs problem. 2nd ed. Berlin: Springer-Verlag. [In German]

Baxter JP. 1946. Scientists Against Time. Boston: Little, Brown.

Bazaz SF, Shahreiari Z, Homauni M. 1988. A new approach for the detection of the sulfur mustard mutagenicity by Ames test. In: Abstracts of the First International Medical Congress on Chemical Warfare Agents in Iran. June 13-16, 1988. Mashhad University of Medical Sciences. Mashhad, Iran. No. 46.

Beauvieux. 1920. Ocular lesions caused by vesicant gases. Archives d'Ophtalmologie 37:597-619. [In French]

Becklake MR, Bourbeau J, Menzies R, Ernst P. 1988. The relationship between acute and chronic airway responses to occupational exposure. In: Current Pulmonology. Chicago: Year Book Medical Publishers.

Beckman L, Nordenson I. 1986. Interaction between some common genotoxic agents. Human Heredity 36:397-401.

Beebe G. 1960. Lung cancer in World War I veterans: possible relation to mustard gas injury and 1918 influenza epidemic. Journal of the National Cancer Institute 25:1231-1252.

Belitskii GA, Khudolei VV. 1986. Short-term tests in the system of detecting chemical compounds carcinogenic for humans. Voprosy Onkologii 32:3-11. [In Russian]

Benschop HP, Moes GH, Fidder A, Scheffer AG, Van der Schans GP. 1989. Immunochemical detection of mustard gas adducts with DNA: identification of adducts. In: Proceedings of the Medical Defense Bioscience Review, Aberdeen Proving Ground, MD. 1-8.

Berenblum I. 1929. The modifying influence of dichloroethyl sulphide on the induction of tumors in mice by tar. Journal of Pathology and Bacteriology 32:425-434.

Berenblum I. 1931. The anti-carcinogenic action of dichlorodiethylsulphide (mustard gas). Journal of Pathology and Bacteriology 34:731-746.

Berenblum I. 1935. Experimental inhibition of tumour induction by mustard gas and other compounds. Journal of Pathology and Bacteriology 40:549-558.

Berenblum I. 1947. Action of mustard gas (b,b-dichlordiethyl sulfide) on nucleoproteins. Nature 159:727-729.

Berenblum I, Wormall A. 1939. The immunological properties of proteinstreated with b,b-dichlorodiethyl sulphide (mustard gas) and b,b-dichlorodiethyl sulfone . Biochemical Journal 33:75-80.

Berenblum I, Kendal LP, Orr JW. 1936. Tumour metabolism in the presence of anticarcinogenic substances. Biochemical Journal 30:709-715.

Berens C, Hartmann E. 1943. The effects of war gases and other chemicals on the eyes of the civilian population. Bulletin of the New York Academy of Sciences 19:356-367.

Berger NA. 1985. Poly(ADP-ribose) in the cellular response to DNA damage. Radiation Research 101:4-15.

Berghoff RS. 1919. The more common gases: their effect on the respiratory tract. Observation on two thousand cases. Archives of Internal Medicine 24:678-684.

Bergmeister R. 1941. Chemical warfare from the viewpoint of an ophthalmologist. Wiener Medizinische Wochenschrift 91:765-771. [In German]

Bergsagel DE, Bailey AJ, Langley GR, MacDonald RN, White DF, Miller AB. 1979. The chemotherapy of plasma-cell myeloma and the incidenceof acute leukemia . New England Journal of Medicine 301:743-748.

Bergsagel DE, Alison RE, Bean HA, Brown TC, Bush RS, Clark RM, Chua T, Dalley D, DeBoer G, Gospodarowicz M, Hasselback R, Perrault D, Rideout DF. 1982. Results of treating Hodgkin's disease without a policy of laparotomy staging. Cancer Treatment Reports 66:717-731.

Berk PD, Goldberg JD, Silverstein MN, Weinfeld A, Donovan PB, Ellis JT, Landaw SA, Laszlo J, Najean Y, Pisciotta AV, Wasserman LR. 1981. Increased incidence of acute leukemia in polycythemia vera associated with chlorambucil therapy. New England Journal of Medicine 304:441-447.

Bernstein IA, Mitra RS, Connolly RB, Vaughn FL, Gray RH, Brabec MJ, Bernstam L, Brown R, Sorscher DA, Ribeiro P, Zidell RH. 1985. Effects of sulfur mustard on cutaneous keratinocytes *in vitro.* In: Proceedings of the 5th Chemical Defense Bioscience Review, Frederick, MD.

Bernstein IA, Bernstam L, Brown R, Fan L, Feng HW, Ku W, Locey B, Ribeiro P, Scavarelli R, Vaughan FL, Zaman S. 1987. Macromolecular and cellular effects of sulfur mustard on keratinocyte cultures. In: Proceedings of the 6th Chemical Defense Bioscience Review, Frederick, MD. 243-250.

Bernstein IA, Bernstam L, Fan L, Ku W, Ribeiro P, Scavarelli R, Vaughan FL, Zaman S. 1989. Early molecular and cellular effects of mustard studied in vitro. In: Proceedings of the 1989 Medical Defense Bioscience Review, Aberdeen Proving Ground, MD.

Bertein, Baudet. 1925. Sequelae of conjunctivitis from yperite. Soc De Med Mil Franc Bull Mems (Paris) 19:171-175. [In French]

Bertolero F, Marafante E, Rade JE, Pietra R, Sabbioni E. 1981. Biotransformation and intracellular binding of arsenic in tissues of rabbits after intraperitoneal administration of ^{74}As labelled arsenite. Toxicology 20:35-44.

Berton P. 1919. Two observations on cutaneous wounds caused by yperite. Bulletin Societe Centrale de Medecine Veterinaire 72:186-190. [In French]

Bertrand M, Deen DF. 1980. Factors influencing the recovery from potentially lethal damage (PLD) in mammalian cells *in vitro* and *in vivo.* Cancer Treatment Reviews 7:1-15.

Beruard M, Maret J, Armand J. 1973. Acute arsenical poisoning with renal impairment treated with hemodialysis and forced diuresis. Lyon Medical 230:185-190. [In French]

Bhuyan BK, Fraser TJ. 1974. Cytotoxicity of antitumor agents in a synchronous mammalian cell system. Cancer Chemotherapy Reports 58:149-155.

Bickerton RE. 1934. New cases of war blindness due to mustard gas. British Medical Journal 2:769-770.

Bird MJ, Fahmy OG. 1953. Cytogenetic analysis of the action of carcinogens and tumor inhibitors in *Drosophila melanogaster*.Part 1. 1:2,3:4-diepoxybutane. Proceedings of the Royal Society of London. Series B: Biological Sciences 140B:556-578.

Black JE, Glenny ET, McNee JW. 1915. Observations on 685 cases of poisoning by noxious gases used by the enemy. British Medical Journal 165-167.

Black RM, Brewster K, Clarke RJ, Hambrook JL, Harrison JM, Howells DJ. 1992. Biological fate of sulfur mustard, 1,1'-thiobis(2-chloroethane). Isolation and identification of urinary metabolites following intraperitoneal administration to rat. Xenobiotica 22:405-418.

Black RM, Hambrook JL, Howells DJ, Read RW. 1992. Biological fate of sulfur mustard, 1,1'-thiobis(2-chloroethane). Urinary excretion profiles of hydrolysis products and beta-lyase metabolites of sulfur mustard after cutaneous application in rats. Journal of Analytical Toxicology 16:79-84.

Black S, Thomson JF. 1947. Metabolic effects produced by large doses of di-b-chloroethyl sulfide in rats. Journal of Biological Chemistry 167:283-289.

Blackmore HS. 1925-1926. The pathology of mustard gas burns and its relation to problems of prevention and treatment. Proceedings of the Royal Society of Medicine 19:25-29.

Blank JA, Hobson DW, Dill GS, Joiner RL. 1990. Comparative effects of sulfur mustard on NAD levels and viability of peripheral blood lymphocytes, human keratinocytes, and human lymphocytic cell lines (abstract). Toxicologist 331.

Blank JA, Joiner RL, Houchens DP, Dill GS, Hobson DW. 1991. Comparative immunotoxicity of 2,2'-dichlorodiethyl sulfide and cyclophosphamide: evaluation of L1210 tumor cell resistance, cell-mediated immunity, and humoral immunity. International Journal of Immunopharmacology 13:251-257.

Blayney DW, Longo DL, Young RC, Greene MH, Hubbard SM, Postal MC, Duffey PL, DeVita VT Jr. 1987. Decreasing risk of leukemia with prolonged followup after chemotherapy and radiotherapy for Hodgkin's disease. New England Journal of Medicine 316:710-714.

Blechmann G. 1919. Eclampsia: loss of consciousness and arterial hypertension in a patient intoxicated by vesicant gas. Bulletins et Memoires de la Societe Medicale des Hopitaux de Paris 3:507-510. [In French]

Blewett WK. 1986. Tactical weapons: is mustard still king? Nuclear, Biological, and Chemical Defense and Technology International 1:64-66.

Blodi F. 1971. Mustard gas keratopathy. International Ophthalmology Clinics 11:1-13.

Blumenfeld F 1919-1920. The behavior of the respiratory passages in poisoning by war gases. Zeitschrift fur Laryngologie, Rhinologie, Otologie und Ihre Grenzgebiete 9:21. [In German]

Bockmeyer M. 1985. Documentation of mucous membrane alterations after contact with poison gas (Lost). Laryngologie, Rhinologie, Otologie 64:532-534. [In German]

Bodansky O. 1945. Contributions of medical research in chemical warfare to medicine. Science 102:517-521.

Bodell WJ. 1990. Molecular dosimetry for sister chromatid exchange induction and cytotoxicity by monofunctional and bifunctional alkylating agents. Mutation Research 233:203-210.

Bodnar T. 1937. Fatal dichloroethyl sulfide poisoning. Orvosi Hetilap 81:1112-1113. [In Hungarian]

Boehnlein JK, Kinzie JD, Ben R, Fleck J. 1985. One-year follow-up study of posttraumatic stress disorder among survivors of Cambodian concentration camps. American Journal of Psychiatry 142(8):956-959.

Boice JD, Greene MH, Keehn RJ, Higgins GA, Fraumeni JF Jr. 1980. Late effects of low-dose adjuvant chemotherapy in colorectal cancer. Journal of the National Cancer Institute 64:501-511.

Boice JD, Greene MH, Killen JY, Ellenberg SS, Keehn RJ, McFadden E, Chen TT, Fraumeni JF Jr. 1983. Leukemia and preleukemia after adjuvant treatment of gastrointestinal cancer with semustine (methyl-CCNU). New England Journal of Medicine 309:1079-1084.

Boice JD, Day NE, Andersen A, Brinton LA, Choi NW, Clarke EA, Coleman MP, Curtis RE, Flannery JT, Hakama M, Hakulinen T,Howe GR, Jensen OM, Kleinerman RA, Magnin D, Magnus K, Makela K, Malker B, Miller AB, Nelson N, Patterson CC, Pettersson F, Pompe-Kirn V, Primic-Zakelj M, Prior P, Ravnihar B,Skeet RG, Skjerven JE, Smith PG, Spengler RF, Storm HH, Stovall M, Tomkins GWO, Wall C. 1985. Second cancers following radiation treatment for cervical cancer. An international collaboration among cancer registries. Journal of the National Cancer Institute 74:955-975.

Boiteau H, Paulet G. 1955. Effect of dichloroethyl sulfide (mustard gas) on division and growth of cells of bakers' yeast. Journal of Physiology 47:605-619.

Boivin JF, Hutchinson GB, Lyden M, Godbold J, Chorosh J, Schottenfeld D. 1984. Second primary cancers following treatment of Hodgkin's disease. Journal of the National Cancer Institute 72:233-241.

Bolla-Wilson K, Bleecker M. 1987. Neuropsychological impairment following inorganic arsenic exposure. Journal of Occupational Medicine 29:500-503.

Bol'Shakova EN, Borzov VV, Bondarev GN. 1979. Use of acridine yperite in selection of *Actinomyces nodosus* producing amphotericin B. Antibiotiki (Moscow) 24:3-6. [In Russian]

Bonhoff. 1917. On poisonous gases. Munchener Medizinische Wochenschrift 64:762. [In German]

Bonnefon G. 1919. Action of hypertonic solution on the ocular mucous membrane impregnated by dichloroethyl sulfide (yperite). Comptes Rendus des Seances de la Societe de Biologie et de Filiales 82:1089-1091. [In French]

Bonnefon G. 1919. The eye exposed to yperite: clinical and therapeutic observations. Annales d'Oculistique 156:577-597. [In French]

Bonnel F. 1930. Chemical warfare: effects of war gases. Journal de Medecine de Bordeaux et du Sud-Ouest 107:415-422. [In French]

Bonnet P. 1939. Corneal dystrophy with torpid ulcer as late manifestation of dichloroethylsulfide burn: case. Bulletin de la Societe d'Ophtalmologie de Paris 51:432-433. [In French]

Bonnet P. 1939. Vascular ectasis of conjunctiva following burn by dichloroethylsulfide. Bulletin de la Societe d'Ophtalmologie de Paris 51:407-409. [In French]

Borak J, Sidell FR. 1992. Agents of chemical warfare: sulfur mustard. Annals of Emergency Medicine 21:303-308.

Boswell W. 1919. Notes on the pathological changes occurring as a result of poisoning by mustard gas. Transactions of the Royal Academy of Medicine in Ireland 37:335-338.

Boudansky Y. 1987. Soviet military involvement in Afghanistan. In: Klass R, ed. Afghanistan, the Great Game Revisited. Lanham, MD: Freedom House. 229-285.

Boursnell JC. 1948. Some reactions of mustard gas (b-b-dichlorodiethyl sulphide) with proteins. In: Williams RT, ed. The Biochemical Reactions of Chemical Warfare Agents. Biochemical Society Symposia No. 2. Cambridge: Cambridge University Press. 8-15.

Boursnell JC, Francis GE, Wormall A. 1946. Studies on mustard gas (b,b-dichlorodiethylsulphide) and some related compounds. 2. The action of mustard gas b,b-dichlorodiethyl sulphone and divinyl sulphone on amino acids. Biochemical Journal 40:737-742.

Boursnell JC, Francis GE, Wormall A. 1946. Studies on mustard gas (b,b-dichlorodiethylsulphide) and some related compounds. 3. The preparation and use of mustard gas containing radioactive sulphur and deuterium. Biochemical Journal 40:743-745.

Boursnell JC, Cohen JA, Dixon M, Francis GE, Greville GD, Needham DM, Wormall A. 1946. Studies on mustard gas (b,b dichlorodiethylsulphide) and some related compounds. 5. The fate of injected mustard gas (containing radioactive sulphur) in the animal body. Biochemical Journal 40:756-764.

Boursnell JC, Francis GE, Wormall A. 1946. Studies on mustard gas (b,b-dichlorodiethylsulphide) and some related compounds. 6. The fate of injected b,b-dichlordiethyl sulphone and b,b-dichlorodiethyl sulphoxide (containing radioactive sulphur) in the animal body. Biochemical Journal 40:765-768.

Boursnell JC, Francis GE, Wormall A. 1946. Studies on mustard gas (b,b-dichlorodiethylsulphide) and some related compounds. 7. The immunological properties of proteins treated with mustard gas and some related compounds. Biochemical Journal 40:768-774.

Boursnell JC, Francis GE, Wormall A. 1946. Studies on mustard gas (b,b-dichlorodiethylsulphide) and some related compounds. 8. The action of mustard gas, divinyl sulphone and b,b-dichlorodiethyl sulphone on complement. Biochemical Journal 40:774-778.

Boxwell W. 1919. Poisoning by mustard gas. Dublin Journal of Medical Sciences 147:7-14.

Boyland E. 1952. The effect of radiation and radiomimetic substances. Endeavour 11:87-91. [In German]

Boyland E. 1969. The correlation of experimental carcinogenesis and cancer in man. Progress in Experimental Tumor Research 11:222-234.

Boyland E. 1975. Mutagenesis and carcinogenesis. Archives fur Geschwulstforschung 45:625-627.

Boyland E. 1977. Biochemistry of occupational cancer. Journal of the Society of Occupational Medicine 27:97-101.

Boyland E, Horing ES. 1949. The induction of tumours with nitrogen mustards. British Journal of Cancer 3:118-123.

Boyland E, Sargent S. 1951. The local greying of hair in mice treatedwith x-rays and radiometic drugs . British Journal of Cancer 5:433440.

Breastup A. 1940. [Renal threshold for sugar during poisoning with gas]. Nordisk Medicin 5:51-52.

Brill NQ, Beebe GW. 1955. A Follow-up Study of War Neuroses. VA Medical Monograph. Washington, DC: U.S. Government Printing Office.

Brockhausen G, Amelung B. 1991. Bronchial carcinoma and exposure to mustard gas: case reports. Pneumologie 45:181-182. [In German]

Brodsky CM. 1983. Psychological factors contributing to somatoform diseases attributed to the workplace. Journal of Occupational Medicine 25:459-464.

Brookes P. 1975. Covalent interaction of carcinogens with DNA. Life Sciences 16:331-344.

Brookes P. 1978. The importance of a detailed knowledge of the nature of DNA damage by chemical mutagens and carcinogens. Heredity 40:333-334.

Brookes P. 1990. The early history of the biological alkylating agents. Mutation Research 233:3-14.

Brookes P, Lawley PD. 1960. The reaction of mustard gas with nucleic acids *in vitro* and *in vivo.* Biochemical Journal 77:478-484.

Brookes P, Lawley PD. 1961. The reaction of mono- and di-functional alkylating agents with nucleic acids. Biochemical Journal 80:496-503.

Brookes P, Lawley PD. 1963. Effects of alkylating agents on T2 and T4 bacteriophages. Biochemical Journal 89:138-144.

Brookes P, Lawley PD. 1963. Evidence for the action of alkylating agents on deoxyribonucleic acid. Experimental Cell Research Supplement 9:521-524.

Brookes P, Lawley PD. 1964. Alkylating agents. British Medical Bulletin 20:91-95.

Brooks SM, Weiss MA, Bernstein IL. 1985. Reactive airways dysfunction syndrome (RADS). Chest 88:376-384.

Brophy LP, Fisher G. 1959. The Chemical Warfare Service: Organizing for War. United States Army in World War II: The Technical Services. Washington, DC: Office of the Chief of Military History, Department of the Army.

Brophy LP, Miles WD, Cochrane RC. 1959. The Chemical Warfare Service: From Laboratory to Field. United States Army in World War II: The Technical Services. Washington, DC: Office of the Chief of Military History, Department of the Army.

Brown SS. Nitrogen mustards and related alkylating agents. Advances in Pharmacology 2:243-295.

Bruckert H, Schoene K, Wodtke G. 1987. The effect of mustard gas on skin: quantitative correlation of exposition, uptake and erythema. In: Proceedings of the 6th Chemical Defense Bioscience Review, Frederick, MD, U.S. Army Medical Research and Development Command.

Bryden J. 1989. Deadly Allies: Canada's Secret War 1937-1947. Toronto: McClelland & Stewart.

Budavari S, ed. 1989. The Merck Index. 11th ed. Rahway, NJ: Merck & Co.

Budelmann G. 1938. Clinical aspects of intoxication due to inhalation of gases (diphenylchlorarsine, dichloroethylsulfide, carbon monoxide and benzene. Medizinische Klinik 34:1029.

Bujadoux L. 1926. Conjunctival lesions following yperite burns. Lyon Medical 1:652. [In French]

Bulletin of the New York Medical College. 1942. Symposium of medical aspects of chemical warfare. (June-October).

Bulychev NB, Vorog'ev N, Koshkin AA, Shishkin GV. 1991. Modification of nucleic acids by reactive oligonucleotide derivatives having a 5'-terminal nitrogen yperite residue, covalently bound with linkers of varying length. Conformation dynamics of group reactivity in complementary complexes. Bioorganicheskaya Khimiya 17:795-805. [In Russian]

Bursuk GG. 1936. On the issue concerning infection of the eye afflicted with dichlorodiethylsulfide. Zentralblatt fur die Gesamte Ophthalmologie und Ihre Grenzgebiete 36:513. [In German]

Buruiana LM, Pavlu V, Niculescu V. 1967. On the radiomimetic effect of nitrogen yperite. Revue Roumaine de Biochimie 4:97-102. [In French]

Buscher H. 1932. [Green and Yellow Cross]. Hamburg, Germany: Himmelheber. Translated from the German in 1944 by Conway N, Kettering Laboratory of Applied Physiology, Cincinnati, OH.

Bussy. 1921. Note on treatment of ophthalmia caused by yperite. Archives d'Ophtalmologie 38:169-171. [In French]

Buu-Hoi, Ratsimamanga R, Pacault A. 1945. Mechanism of action of substances of dichloroethyl sulfide group. Bulletin de la Societe de Chimie Biologique 27:259-264.

Caballero LR, Moro BH. 1988. Cutaneous lesions by mustard gas. In: Abstracts of the First International Medical Congress on Chemical Warfare Agents in Iran. June 13-16, 1988. Mashhad University of Medical Sciences. Mashhad, Iran. No. 91.

Cahan WG, Woodard HQ, Higinbotham NL, Stewart FW, Coley BL. 1948. Sarcoma arising in irradiated bone. Cancer 1:3-29.

Callaway S, Pearce KA. 1958. Protection against systemic poisoning by mustard gas, di(2-chloroethyl) sulphide, by sodium thiosulphate and thiocit in the albino rat. British Journal of Pharmacology 13:395-398.

Cameron GR, Courtice FC. 1948. Effects of skin contamination with liquid mustard on water balance in animals. Quarterly Journal of Experimental Physiology 34:165.

Cameron GR, Carleton HM, Short RHD. 1946. Pathological changes induced by Lewisite and allied compounds. Journal of Pathology and Bacteriology 58:411-422.

Cameron GR, Gaddum JH, Short RHD. 1946. The absorption of war gases by the nose. Journal of Pathology and Bacteriology 58:449-455.

Cameron GR, Courtice FC, Short RHD. 1947. Disturbances of function induced by Lewisite (2-chlorvinyldichlorarsine). Quarterly Journal of Experimental Physiology 34:1-28.

Camras CB, Bito LZ. 1980. The pathophysiological effects of nitrogen mustard on the rabbit eye. I. The biphasic intraocular pressure response and the role of prostaglandins. Experimental Eye Research 30:41-52.

Camras CB, Bito LZ. 1980. The pathophysiological effects of nitrogen mustard on the rabbit eye. II. The inhibition of the initial hypertensive phase by capsaicin and the apparent role of substance P. Investigative Ophthalmology &Visual Science 19:423-428.

Canelli AF. 1918. Contribution to the knowledge about acute yperite intoxication particularly anatomical-pathological reports. Riv Osped 8:2-7. [In Italian]

Capizzi RL, Smith WJ, Field R, Papirmeister B. 1973. A host mediated assay for chemical mutagens using the L5178Y/Asn murine leukemia (abstract). Mutation Research 21:6.

Capizzi RL, Papirmeister B, Mullins JM, Cheng E. 1974. The detection of chemical mutagens using the L5178Y/Asn murine leukemia *in vitro* and in a host mediated assay. Cancer Research 34:3073-3082.

Carbery AD. 1935. Residual effects of warfare gases. New Zealand Medical Journal 34:160-166.

Carnes SA. 1989. Disposing of chemical weapons: a desired end in search of an acceptable means. Environmental Professional 11:279-290.

Cames SA, Watson AP. 1989. Disposing of the U.S. chemical weapons stockpile: an approaching reality. Journal of the American Medical Association 262:653-659.

Carpenter FH, Wood JL, Stevens CM, du Vigneaud V. 1948. Chemical studies on vesicant-treated proteins. Journal of the American Chemical Society 70:2551-2553.

Carr D, Denman WE, Skinner WF. 1947. Noxious gases (dichloroethyl sulfide) and bronchiectasis. Diseases of the Chest 13:596-601.

Carr FJ, Fox BW. 1982. The effects of bifunctional alkylating agents on DNA synthesis in sensitive and resistant Yoshida cells. Mutation Research 95:441-456.

Carr WP. 1918. Mustard gas burns. Transactions of the Southern Surgical Association 31:243-252.

Carr WP. 1919. Mustard gas bums. American journal of Surgery 33:251-253.

Carter RL, Roe FJC. 1975. Chemical carcinogens in industry. Journal of the Society of Occupational Medicine 25:86-94.

Case RM, Lea AJ. 1955. Mustard gas poisoning, chronic bronchitis, and lung cancer: an investigation into the possibility that poisoning by mustard gas in the 1914-1918 war might be a factor in the production of neoplasia. British Journal of Preventive and Social Medicine 9:62-72.

Castellan RM, Olenchock SA, Kinsley KB, Hankinson JL. 1987. Inhaled endotoxin and decreased spirometric values. New England Journal of Medicine 317:605-610.

Casto BC, Meyers J, DiPaolo JA. 1979. Enhancement of viral transformation for evaluation of the carcinogenic and mutagenic potential of inorganic metal salts. Cancer Research 39:193-198.

Castro JA. 1968. Effects of alkylating agents on human plasma cholinesterase. Biochemical Pharmacology 17:195-303.

Castro M. 1943. War and vesicant gases. Impresna Med (Rio de Janiero) 19:95-104.

Castro V, Dawson K, Field E. 1966. Response of cells to combined action of x-radiation and mustard gas. International Journal of Radiation Biology and Related Studies in Physics, Chemistry and Medicine 11:513-515.

Catalini N, Gibertini G. 1972. Comparative study of the action of ionizing radiation and radiomimetic agents on the involutive phase and recovery of the thymus in young rats *Rattus-norvegicus.* Part 2: Effect of yperite 2,2 dichlorodiethyl sulfide and azotoyperite 2,2,2 trichlorotrithylamine. Atti della Accademia Nazionale dei Lincei, Classe di Scienze Fisiche, Matematiche e Naturali, Rendiconti 53:630-636.

Catalani N, Gibertini G, Margotta V, Filoni S. 1971. Comparative study of the biological effects of x-rays and of mustard gas b-b-dichloroethylsulfur on the thymus of young rats *Rattus-norvegicus.* Atti della AccademiaNazionale dei Lincei, Classe di Scienze Fisiche, Matematiche e Naturali, Rendiconti 51:86-92.

Cattelain E. 1935. The physician Cesar Despretz and the discovery of yperite (mustard gas). Journal de Pharmacie et de Chimie 22:512-514. [In French]

Cattelain E. 1938. Yperite or mustard gas: its history, the secret of its power, its future. Revue Generale des Sciences Pures et Appliquees et Bulletin de l'Association Francais pour l'Avancement des Sciences 49:205-211. [In French]

Cattelain E. 1940. L'yperite ou Gaz Moutarde: Historique, Production, Proprietes, Action, Protection, Therapeutique. Paris: G. Doin. [In French]

Catton J. 1919. Gas warfare: its aftermath. Military Surgeon 45:65-74.

CBW News, Quarterly Bulletin of Chemical and Biological Weapons Issues. 1992. 9:(July)8.

Cella DF, Perry SW, Kulchycky S, Goodwin C. 1988. Stress and coping in relatives of burn patients: a longitudinal study. Hospital and Community Psychiatry 39:159-166.

Centers for Disease Control. 1988. Final recommendations for protecting the health and safety against potential adverse effects of long-term exposure to low doses of agents: GA, GB, VX, Mustard Agent (H, HD,T) and Lewisite (L). Federal Register 53:8504-8507.

Cersosimo RJ, Licciardello JTW, Matthews SJ, Bromer R, Hong WK. 1984. Acute pneumonitis associated with MOPP chemotherapy of Hodgkin's disease. Drug Intelligence and Clinical Pharmacy 18:609-611.

Chaieb J, Spolidoro J, Caleffi M. 1982. Bronchial carcinoma: etiologic factors in occupational exposure. Jornal de Pneumologia 8:15-19. [In Portuguese]

Chailleux E. 1982. Bronchial pulmonary cancers and professional factors. Archives des Maladies Professionnelles de Medecine du Travail et de Securite Sociale 43:687-689. [In French]

Chak LY, Sikic BI, Tucker MA, Horns RC Jr, Cox RS. 1984. Increased incidence of acute nonlymphocytic leukemia following therapy in patients with small cell carcinoma of the lung. Journal of Clinical Oncology 2:385-390.

Chambers SK, Chopyk RL, Chambers JT, Schwartz PE, Duffy TP. 1989. Development of leukemia after doxorubicin and cisplatin treatment for ovarian cancer. Cancer 64:2459-2461.

Chan P-K, Hayes AW. 1985. Assessment of chemically induced ocular toxicity: a survey of methods. In: Hayes AW, ed. Toxicology of the Eye, Ear, and Other Special Senses. New York: Raven Press.

Charan NB, Myers CG, Lakshminarayan S, Spencer TM. 1979. Pulmonary injuries associated with acute sulfur dioxide inhalation. American Review of Respiratory Disease 119:555-560.

Chen JJ, Tseng SC. 1991. Corneal epithelial wound healing in the absence of limbal epithelium. Investigative Ophthalmology & Visual Sciences 32:2219-2233.

Cherey BW, McBride OW, Chen C, Alkhatib H, Bhatia K, Hensley P, Smulson ME. 1987. cDNA sequence, protein structure, and chromosomal location of the human gene for poly(ADP-ribose) polymerase. Proceedings of the National Academy of Sciences (USA) 84:8370-8374.

Chetverikova V. 1930. Eye disease due to war gases. Vrachebnaia Gazeta 34:1220-1223. [In Russian]

Chetverikova V. 1937. Rare case of eye injury due to dichloroethylsulfide poisoning. Vestnik Oftalmologii 11:895-897. [In Russian]

Chhuttani PN, Chawla LS, Sharma TD. 1967. Arsenical neuropathology. Neurology 17:269-274.

Chiesman WE. 1944. Diagnosis and treatment of lesions due to vesicants. British Medical Journal 2:109-112.

Chilow KL, Zassossow RA. 1931. On the clinical and prophylatic effect of chloramine on the nasal mucosa injured with yperite. Milit Med Z Leningr u Moskau 2:508-515. [In German]

Chirurco GA. 1934. Effects of war gases. Terapia 24:204. [In Italian]

Chovil A. 1979. Occupational lung cancer and smoking: a review in the light of current theories of carcinogenesis. Canadian Medical Association Journal 121:548-555.

Chovil A. 1981. Laryngeal cancer: an explanation for the apparent occupational association. Medical Hypotheses 7:951-956.

Christophe L. 1933. Experimental research on belated death from burns. Journal de Chirurgie et Annales de la Societe Belge de Chirurgie 30-32:356-375. [In French]

Chusid JG, Marquardt GH. 1946. Onset of Guillain-Barre syndrome following exposure to mustard gas (dichloroethyl sulfide). Archives of Neurology and Psychiatry 55:57-58.

Clayson DB. 1977. Principles underlying testing for carcinogenicity. Cancer Bulletin 29:161-166.

Cleaver JE, Goth R. 1976. Value of measurements of DNA repair levels in predicting carcinogenic potential of chemicals. In: Montesano R, Bartsch H, Tomatis L, eds. Screening Tests in Chemical Carcinogenesis. IARC Scientific Publication No. 12. Lyon: International Agency for Research on Cancer. 639-655.

Clemedson CJ, Kristoffersson H, Sorbo B, Ullberg S. 1963. Distribution of 35S labelled mustard gas in the body studied by means of whole body autoradiography in mice. Nordisk Hygienisk Tidskrift 44:3-7. [In Swedish]

Clemedson CJ, Kristoffersson H, Sorbo B, Ullberg S. 1963. Whole bodyautoradiographic studies of the distribution of sulphur 35-labelled mustard gas in mice . Acta Radiologica: Therapy, Physics, Biology 1:314-320.

Clerc A, Roudinesco A. 1919. Pulmonary sequelae of yperite. Comptes Rendus des Seances de la Societe de Biologie et de Filiales 82:787. [In French]

Clerc A, Rousselot. 1919. Cardiovascular symptoms with yperite. Bulletins et Memoires de la Societe Medicale des Hopitaux de Paris 3:593-597. [In French]

Clerc A, Ramond L, Guilhaume H. 1919. A clinical study of the pulmonary sequelae in yperite poisoning. Presse Medical 27:477-478. [In French]

Clerc A, Guilhaume H, Rousselot. 1919. Genital sequelae from yperite. Progres Medicale 34:351-352. [In French]

Clerc A, Ramond L, Guilhaume H. 1919. Pulmonary abscesses caused by yperite. Progres Medicale 34:222. [In French]

Cogan DG. 1943. Lewisite burns of the eye. Journal of the American Medical Society 122:435-436.

Cohen BM. 1953. Methodology of record follow-up studies on veterans. American Journal of Public Health 43:1292-1298.

Cohen BM, Cooper MZ. 1954. A Follow-up Study of World War II Prisoners of War. Washington, DC: U.S. Government Printing Office.

Colardyn F, de Bersaques J. 1984. Clinical observation and therapy of injuries with vesicants. Archives Belges (Supplement):298-301.

Colardyn F, de Keyser H, Ringoir S, de Bersaques J. 1986. Clinical observation and therapy of injuries with vesicants. Journal de Toxicologie Clinique et Experimentale 6:237-246.

Colardyn F, Verbeke J, Vogelaers D, Vandenbogaerde J. 1988. The clinics and therapy of victims of mustard gas intoxication. In: Abstracts of the First International Medical

Congress on Chemical Warfare Agents in Iran. June 13-16, 1988. Mashhad University of Medical Sciences. Mashhad, Iran. No. 88.

Cole HH. 1923. A clinical study of the gassed ex-soldier with special reference to pulmonary tuberculosis. American Review of Tuberculosis 7:230-256.

Cole HN, Driver JR, Bowen SS, Cooper G. 1939. War gases and industrial hazards in their manufacture. Archives of Dermatology and Syphilology 39:45.

Cole P, Merletti F. 1980. Chemical agents and occupational cancer. Journal of Environmental Pathology and Toxicology 3:399-417.

Coleman CN, Kaplan HS, Cox R, Varghese A, Butterfield P, Rosenberg SA. 1982. Leukemias, non-Hodgkin's lymphomas and solid tumours in patients treated for Hodgkin's disease. Cancer Surveys 1:734-744.

Coleman CN, Williams CJ, Flint A, Glatstein EJ, Rosenberg SA, Kaplan HS. 1977. Hematologic neoplasia in patients treated for Hodgkin's disease. New England Journal of Medicine 297:1249-1252.

Collins DJ. 1928. Several medical aspects of chemical warfare. RevueInternationale de la Croix-Rouge 10:850-858. [In French]

Collins JM, Zaharko DS, Dedrick RL, Chabner BA. 1986. Potential roles for preclinical pharmacology in phase I clinical trials. Cancer Treatment Reports 70:73-80.

Collins JM, Grieshaber CK, Chabner BA. 1990. Pharmacologically guided phase I clinical trials based upon preclinical drug development. Journal of the National Cancer Institute 82:1321-1326.

Colman M, Easton DF, Horwich A, Peckham MJ. 1988. Second malignancies and Hodgkin's disease: the Royal Marsden Hospital experience. Radiotherapy and Oncology 11:229-238.

Coltman CA, Dixon DO. 1982. Second malignancies complicating Hodgkin's disease: a Southwest Oncology Group 10-year followup. Cancer Treatment Reports 66:1023-1033.

Colvin M, Chabner BA. 1990. Alkylating agents. In: Chabner BA, Collins JM, eds. Cancer Chemotherapy: Principles and Practice. Philadelphia: J.B. Lippincott.

Cone J. 1987. Occupational lung cancer. Occupational Medicine 2:273-295.

Conen PE, Lansky GS. 1961. Chromosome damage during nitrogen mustard therapy. British Medical Journal 529:1055-1057.

Conklin JW, Upton AC, Christenberry KW. 1965. Further observations on late somatic effects of radiomimetic chemicals and X-rays in mice. Cancer Research 25:20-28.

Connors T. 1974. Mechanism of action of 2-chloroethylamine derivatives, sulfur mustards, epoxides, and aziridines. Handbuch der Experimentellen Pharmakologie 38:18-34.

Coombs MM. 1980. Chemical carcinogenesis: a view at the end of the first half-century. Journal of Pathology 130:117-146.

Coppens M, Roels H, Van den Heede M, Heyndrickx A. 1986. Clinical history and autopsy observations associated with the toxicological findings in an Iranian soldier exposed to yperite (mustard gas). In: Heyndricks B, ed. Terrorism: Analysis and Detection of Explosives. Proceedings of the Second World Congress on New Compounds in Biological and Chemical Warfare. Gent: Rijksuniversiteit. 542-552.

Cordone M. 1938. Do you know the action of mustard gas? Protar 4:132-136.

Corper HJ, Rensch OB. 1921. The effect of mustard gas (dichlorethylsulphide) on experimental tuberculosis. Journal of Infectious Diseases 28:286.

Corper HJ, Black LT, Moore M. 1922. The effect of the roentgen ray and mustard gas (dichloroethylsulphide) on active anaphylaxis in the guinea-pig. Journal of Infectious Diseases 30:50-57.

Cortini A. 1918. On the morbid effects of yperite. Giorale de Medicina Militare 66:235-243. [In Italian]

Coutelier JP, Lison D, Willems J. 1990. Lymphocytic markers in the evaluation of the immune system: observations on a patient exposed to yperite. Annales Medicinae Militaris Belgicae 4:83-85. [In French]

Coutelier JP, Lison D, Simon O, Willems J. 1991. Effect of sulfur mustard on murine lymphocytes. Toxicology Letters 58:143-148.

Covey GW, Barron M. 1919. Pathology of mustard gas inhalation. American Journal of the Medical Sciences 157:808-826.

Cowan FM, Broomfield CA, Smith WJ. 1991. Effect of sulfur exposure on protease activity in human peripheral blood lymphocytes. Cell Biology and Toxicology 7:239-248.

Cowan FM, Broomfield CA, Smith WJ. 1991. Inhibition of sulfur mustard-increased protease activity by niacinamide (abstract). FASEB Journal 5:A828.

Cowen SO. 1919. The after-effects of gas-poisoning, with special reference to lung lesions. Medical Journal of Australia 2:369-372.

Cowles S. 1983. Cancer of the larynx: occupational and environmental associations. Southern Medical Journal 76:894-898.

Cox GH. 1918. Gas poisoning in warfare: a study of the effect of so-called mustard gas upon the upper respiratory tract. Transactions of the Annual Meeting of the American Laryngological Association 40:344-367.

Cox GH. 1919. A study of the effects of mustard gas poisoning upon the upper air tract. Transactions of the American Laryngological, Rhinological and Otological Society 25:201-209.

Craighead JE, Vallyathan NV. 1980. Cryptic pulmonary lesions in workers occupationally exposed to dust containing silica. Journal of the American Medical Association 244:1939-1941.

Crathorn AR, Roberts JJ. 1966. Mechanism of the cytotoxic action of alkylating agents in mammalian cells and evidence for the removal of alkylated groups from deoxyribonucleic acid. Nature 211:150-153.

Creech HJ, Preston RK, Peck RM, O'Connell AP, Ames BN. 1972. Antitumor and mutagenic properties of a variety of heterocyclic nitrogen and sulfur mustards. Journal of Medicinal Chemistry 15:739-746.

Cressman EH. 1942. Vesicants in chemical warfare. Journal of the American Osteopathic Association 42:145-149.

Cristiani M. 1939. Preliminary research on the possibility of neutralizing yperite in the respiratory tract by endotracheal injections. Biochimica e Terapia Sperimentale 26:465-477. [In Italian]

Cross AG. 1948. Effects of mustard gas on eyes. Medical Press 219:281-283.

Crowe RS. 1961. Effect of Mustard Gas on Some Genes in *Drosophila melanogaster* (dissertation). Ames: Iowa State University.

Csapody-Mocsy M. 1940. Experiments on therapy of dichloroethylsulfide poisoning of the eye. Gyogyaszat 80:555-571. [In Hungarian]

Cullumbine H. 1946. Chemical warfare experiment using human subjects. British Medical Journal 2:576-578.

Cullumbine H. 1946. The mode of penetration of the skin by mustard gas. British Journal of Dermatology and Syphilis 58:291-294.

Cullumbine H. 1947. Medical aspects of mustard gas poisoning. Nature 159:151-153.

Cullumbine H. 1948. Factors influencing the penetration of the skin by chemical agents. Quarterly Journal of Experimental Physiology 34:83-89.

Cullumbine H. 1954. The physiological and biochemical disturbances produced by war gases. British Medical Bulletin 10:18-21.

Curtis RE, Boice JD Jr. 1988. Second cancers after radiotherapy for Hodgkin's disease (letter). New England Journal of Medicine 319:244-245.

Curtis RE, Hankey BF, Myers MH, Young JL Jr. 1984. Risk of leukemia associated with the

first course of cancer treatment: an analysis of the surveillance, epidemiology, and end results program experience. Journal of the National Cancer Institute 72:531-544.
Curtis RE, Hoover RN, Kleinerman RA, Harvey EB. 1985. Second cancer following cancer of the female genital system in Connecticut, 1935-82. National Cancer Institute Monographs 68:113-138.
Curtis RE, Boice JD Jr, Stovall M, Flannery JT, Moloney WC. 1989. Leukemia risk following radiotherapy for breast cancer. Journal of Clinical Oncology 7:21-29.
Curtis RE, Boice JD Jr, Moloney WC, Ries LG, Flannery JT. 1990. Leukemia following chemotherapy for breast cancer. Cancer Research 50:2741-2746.
Cushing JA, Sasser LB, Mellick PW, Dacre JC. 1989. Subchronic oral toxicity study of Lewisite in rats (abstract). Toxicologist 9:198.
Czerny K, Ciszewska-PopioLek B, Mitura K. 1990. Histochemical studies of the kidney of white rats after experimental external application of sulfur mustard gas. Gegenbaurs Morphologisches Jahrbuch 136:89-94. [In German]
D'Arman S. 1935. Respiratory lesions from the gas yperite: study of fifty cases. Rassegna della Previdenza Sociale 22:56-74. [In Italian]
D'Halluin F, Roels H. 1984. Autopsy observations in an Iranian soldier exposed to war gases. Archives Belges (Supplement):284-290.
Da Costa JM. 1871. On irritable heart: a clinical study of a form of functional cardiac disorder and its consequence. American Journal of Medical Science 61:17.
Dadlez J, Koskowski W. 1934. War gases and their effect on the human organism. Polska Gazeta Lekarska 13:50-74. [In Polish]
Dahl AR. 1990. Contemporary issues in toxicology. Toxicology and Applied Pharmacology 103:185-197.
Dahl AR, Schlesinger RB, Heck HD, Medinsky MA, Lucier GW. 1991. Comparative dosimetry of inhaled materials: differences among animal species and extrapolation to man. Fundamental and Applied Toxicology 16:1-13.
Dahl H, Gluud B, Vangsted P, Norn M. 1985. Eye lesions induced by mustard gas. Acta Ophthalmologica Supplement 173:30-31.
Danforth CH, Center E. 1954. Nitrogen mustard as a teratogenic agent in the mouse. Proceedings of the Society for Experimental Biology and Medicine 86:705-707.
Danieli Y. 1985. The treatment and prevention of long-term effects and intergenerational transmission of victimization: a lesson from Holocaust survivors and their children. In: Figley CR, ed. Trauma and Its Wake. New York: Brunner Mazel. 295-313.
Dannenberg AM Jr, Pula PJ, Liu LH, Harada S, Tanaka F, Vogt RF Jr, Kajiki A, Higuchi K. 1985. Inflammatory mediators and modulators released in organ culture from rabbit skin lesions produced *in vivo* by sulfur mustard. I. Quantitative histopathology; PMN, basophil,and mononuclear cell survival, and unbound (serum) protein content . American Journal of Pathology 121:15-27.
Dannenberg AM Jr, Masahiro MD, Nakamura MD, Moore KG, Pula PJ. 1987. Basic research studies on the pathogenesis of dermal sulfur mustard lesions. In: Proceedings of the 6th Chemical Defense Bioscience Review, Frederick, MD. 751-754.
Dannenberg AM Jr, Moore KG, Shofield BH, Higuchi K, Kajiki A, Au KW, Pula PJ, Bassett DP. 1987. Two new *in vitro* methods for evaluating toxicity to skin (employing short-term organ culture). I. Paranuclear vacuolization, seen in glycol methacrylate tissue sections. II. Inhibition of ^{14}C-leucine incorporation. Alternative Methods in Toxicology 5:115-127.
Darbois P. 1919. The respiratory sequelae of war gas intoxications. Bulletins et Memoires de la Societe de Radiologie Medicale de France 43:730. [In French]
David G, Gyarmati L. 1962. Comparative study of the serum tyrosine level in the early stages of experimental radiation injury and mustard gas poisoning. Radiobiologia, Radiotherapia 3:191-194. [In German]

Davidson JR, Foa EB. 1991. Refining criteria for posttraumatic stress disorder. Hospital and Community Psychiatry 42:259-261.

Davidson JR, Kudler HS, Saunders WB, Smith RD. 1990. Symptom and comorbidity patterns in World War II and Vietnam veterans with posttraumatic stress disorder. Comprehensive Psychiatry 31:162-170.

Davis J. 1944. Dermatologic aspects of vesicant war gases (dichloroethyl sulfide and dicholorovinylarsine). Journal of the American Medical Association 126:209.

Davison C, Rozman RS, Bliss L, Smith PK. 1957. Studies on the metabolic fate of bis(2-chloroethyl) sulfide (mustard gas) in the mouse and human. Proceedings of the American Association for Cancer Research 2:195.

Davison C, Rozman RS, Smith PK. 1961. Metabolism of bis-b-chloroethyl sulfide (sulfur mustard gas). Biochemical Pharmacology 7:65-74.

Dayhimi I, Bahar K, Eliasy H. 1988. The effect of sulfur mustard gas (SMG) on the immune system. In: Abstracts of the First International Medical Congress on Chemical Warfare Agents in Iran. June 13-16, 1988. Mashhad University of Medical Sciences. Mashhad, Iran. No. 12.

de Keyser H, Geerts M, Colardyn F, de Bersaques J. 1986. Skin damage caused by the effect of mustard gas. Hautarzt 37:467-471. [In German]

de Laguna J. 1979. Possible environmental factors in the etiology of cancer. Gaceta Medica de Mexico 115:258-263. [In Spanish]

de Luna C. 1920. A case of gastritis from yperite with histological findings. Marseille Medical 67:43. [In French]

de Luna C. 1921. A case of gastritis caused by yperite with histological findings. Archives des Maladies de l'Appareil Digestif et des Maladies de la Nutrition 11:208-212. [In French]

de Quidt J, Furon D, Haguenoer J-M. 1971. Concerning an unusual case of arsenic poisoning. Bulletin de la Societe de Pharmacie de Lille 2:95-96. [In French]

de Saracibar JM. 1945. Conjunctival lesions caused by gases such as dichloroethyl sulfide. Archivos de la Sociedad Oftalmologica Hispano-Americana 5:50-53.

de Young LM, Mufson RA, Boutwell RK. 1977. An apparent inactivation of initiated cells by the potent inhibitor of two-stage mouse skin tumorigenesis, bis(2 chloroethyl) sulfide. Cancer Research 37:4590-4594.

Dean JH, Murray MJ. 1991. Toxic responses of the immune system. In: Amdur MO, Doull J, Klaassen CD, eds. Casarett and Doull's Toxicology. New York: Pergamon.

Deaton MA, Jones GP, Bowman PD. 1990. (^{14}C)Mechlorethamine binding to proteins of the human keratinocyte. Military Medicine 155:477-480.

Deaton MA, Bowman PD, Jones GP, Powanda MP. 1990. Stress protein synthesis in human keratinocytes treated with sodium arsenite, phenyldichloroarsine, and nitrogen mustard. Fundamental and Applied Toxicology 14:471-476.

Dedrick RL, Morrison PF. 1992. Carcinogenic potency of alkylating agents in rodents and humans. Cancer Research 52:2464-2467.

Delarue J, Mignot J, Simard C. 1960. Yperite burns of the jugal pouch of the golden hamster. Study of the vascular changes. Comptes Rendus des Seances de la Societe de Biologie et de Filiales 154:526-528. [In French]

Deloof J. 1931. Destruction of carcinogenic power of tar by yperite. Le Cancer 8:89-98. [In French]

Den Engelse L. 1973. Toxic, mutagenic and carcinogenic effects of alkylating agents. Chemisch Weekblad 69:9-12. [In Dutch]

Den Engelse L. 1974. Mechanisms of tumourigenesis by chemicals. Jaarboek van Kankeronderz in Nederland 23:103-110. [In Dutch]

Deneauve-Lockhart C, Sauvaget P, Touron C, Chariot P. 1992. Acute occupational exposure to mustard gas. Archives des Maladies Professionnelles de Medicine du Travail et de Securite Sociale 53:121-124. [In French]

Dennis CE. 1919. Pulmonary fibrosis after gassing, as shown by x-rays. Medical Journal of Australia 2:372-373.

Derby GS. 1918. Effects of mustard gas on the eyes. American Journal of Medical Sciences 156:733-736.

Derby GS. 1920. Ocular manifestations following exposure to various types of poisonous gases. Archives of Ophthalmology 49:119-130.

Despretz M. 1822. Chlorine compounds. Ann Chim Phys 21:428. [In French]

Dewhurst F. 1981. Laboratory usage of some suspect carcinogens. British Journal of Cancer 44:304.

Dickel H, Fincke A, Gohr H. 1952. Poisoning of workers due to dichloroethyl sulfide. Medizinische. 686-690. [In German]

Dickey F, Cleland G, Lotz C. 1949. Role of organic peroxides in the induction of mutations. Proceedings of the National Academy of Sciences (USA) 35:581-586.

Dickman S. 1988. Nerve gas cloud hangs over West German firms. Nature 332:573.

Dill K, Adams ER, Davis SD, O'Connor RJ, McGown EL, Hallowell SF. 1989. Reaction of trans-2-chlorovinylarsine oxide with polydeoxynucleotides. Drug and Chemical Toxicology 12:337-343.

Dill K, Huang LH, McGown EL, Youn SH, O'Connor RJ. 1991. Substituent effects on the binding constants of arsenical-dithiol adducts. Research Communications in Chemical Pathology and Pharmacology 72:367-370.

Dixon M, Needham DM. 1946. Biochemical research on chemical warfare agents. Nature 158:432-438.

Doll R. 1970. Practical steps towards the prevention of bronchial carcinoma. Scottish Medical Journal 15:433-447.

Doll R. 1975. Part III: 7th Walter Hubert Lecture. Pott and the prospects for prevention. British Journal of Cancer 32:263-274.

Doll R. 1977. The prevention of cancer. Journal of the Royal College of Physicians of London 11:125-140.

Doll R. 1978. Atmospheric pollution and lung cancer. Environmental Health Perspectives 22:23-31.

Doll R. 1981. Avoidable cancer: attribution of risk. South African Cancer Bulletin 25:125-146.

Donofrio P, Wilbourn A, Albers J, Rogers L, Salanga V, Greenberg H. 1987. Acute arsenic intoxication presenting as Guillain-Barre like syndrome. Muscle and Nerve 10:114-120.

Dor L, Fouassier. 1922. Late effects of conjunctival burs from yperite. Clinique Ophtalmologique 26:183-191. [In French]

Dor L, Fouassier. 1922. Ocular sequelae of yperite bums. Lyon Medical 131:625-627. [In French]

Dorffel I, Popping. 1935. Animal experiments to investigate changes in the skin after corrosion with dichlordiethylsulfide (mustard gas) and mineral acids. Virchows Archiv fur Pathologische Anatomie und Physiologie und fur Klinische Medizin 295:1-20. [In German]

Doskocil J, Sormova Z. 1965. The reaction of DNA with mustards. II. The reaction kinetics. Collection of Czechoslovak Chemical Communications 30:492-506.

Douglas DB, Owen R. 1976. Occupational cancer. In: Brennan MJ, Krakoff IH, eds. The Physiopathology of Cancer. 3rd ed. Vol. 2, Diagnosis, Treatment, Prevention. Basel: S. Krager. 323-344.

Draize JH, Woodard G, Calvery HO. 1944. Methods for the study of irritation and toxicity of substances applied topically to the skin and mucous membranes. Journal of Pharmacology and Experimental Therapeutics 82:377-390.

Drasch G, Kretschmer E, Pharm M, Kauert G, Von Meyer L. 1987. Concentrations of mustard gas [bis(2-chloroethyl)sulfide] in the tissues of a victim of a vesicant exposure. Journal of Forensic Sciences 32:1788-1793.

Drasch G, Kretschmer E, Kauert G, Von Meyer L. 1988. Distribution of bis(2-chlorethyl) sulfide [mustard gas, yperite, S-lost] in the tissues of a victim of vesicant exposure. In: Abstracts of the First International Medical Congress on Chemical Warfare Agents in Iran. June 13-16, 1988. Mashhad University of Medical Sciences. Mashhad, Iran. No. 82.

Drasch G, Kauert G, Von Meyer L. 1988. The detection of the chemical warfare agent S-Lost bis-2 chloroethylsulfide in human organs and body fluids. Archiv der Pharmazie 321:611. [In German]

Drews M. 1939. Studies on the clinically provable resorptive effect for skin damages caused by dichlorodiethylsulfide in cats. Zeitschrift fur die Gesamte Experimentelle Medizin 105:29-45. [In German]

Druckrey H. 1973. Chemical structure and action in transplacental carcinogenesis and teratogenesis. In: Transplacental Carcinogenesis. IARC Scientific Publication No. 4. Lyon: International Agency for Research on Cancer. 45-58.

Drugov YV, Sentyurin BS. 1937. Arsenic poisons, their effects on the organism and the therapy of the poisoning. Voenno-Sanitarno Delo 9:55-64. [In Bulgarian]

Dudley HC, Jones BF. 1938. Toxicology of phenyldichloroarsine. I. Experiments with animals. Public Health Reports 53:338-347.

Duke-Elder WS. 1954. The offensive vesicants. In: Textbook of Ophthalmology. Vol. VI, Injuries. St. Louis: C.V. Mosby. 6696-6713.

Duke-Elder WS, MacFaul PA. 1972. Chemical injuries. In: System of Ophthalmology. Vol. 14, Injuries, Part 2: Non-mechanical injuries. St. Louis: CV Mosby.

Dunn P. 1986. The chemical war: Iran revisited: 1986. Nuclear, Biological and Chemical Defense and Technology International 1(3):32-39.

Dunn P. 1986. The chemical war: journey to Iran. Nuclear, Biological and Chemical Defense and Technology International 1(2):28-37.

Dustin P. 1947. Some new aspects of mitotic poisoning. Nature 159:794-797.

Eagle H, Magnuson HJ, Fleischman R. 1946. Clinical uses of 2,3-dimercaptopropanol (BAL). I. The systemic treatment of experimental arsenic poisoning (marpharsen, Lewisite, phenyl arsenoxide) with BAL. Journal of Clinical Investigation 25:451-466.

Eastman NJ. 1931. The arsenic content of the human placenta following arsphenamine therapy. American Journal of Obstetrics and Gynecology 21:60-64.

Easton D, Peto J, Doll R. 1988. Cancers of the respiratory tract in mustard gas workers. British Journal of Industrial Medicine 45:652-659.

Eckert-Mobius A. 1940. Therapy of diseases of upper respiratory tract due to war gases. Medizinische Welt 14:313-315. [In German]

Edwards PA, Shooter KV. 1971. Sedimentation characteristics of DNA multiply crosslinked by a difunctional alkylating agent, mustard gas. Biopolymers 10:2079-2082.

Egawa H. 1982. Histological studies on precarcinomatous and early carcinomatous lesions of the tracheobronchial epithelium in mustard gas ex-workers. Gencho Hiroshima Igaku 30:741-786.

Egert G, Lendle L. 1969. Reversibility of the damage induced in female gonads by cytostatics. Arzneimittel-Forschung 19:1-5. [In German]

Egert G, Schuster R, Lendle L. 1969. Mutual tolerance induction in preliminary treatment with sublethal doses of nitrogen mustard gas or irradiation. Strahlentherapie 137:675-679. [In German]

Ehrke MJ, Mihich E. 1985. Effects of anticancer agents on immune response. Trends in Pharmacological Sciences 6:412-417.

Einhor J. 1985. Nitrogen mustard: the origin of chemotherapy for cancer. International Journal of Radiation Oncology, Biology, Physics 11:1375-1378.

Einhorn N. 1978. Acute leukemia after chemotherapy. Cancer 41:444.

Eisenmenger W, Drasch G, von Clarmann M, Kretschmer E. 1988. Clinical, pathological and morphological findings of a mortal yperite [bis(2-chloroethyl)sulfide] intoxication. In: Abstracts of the First International Medical Congress on Chemical Warfare Agents in Iran. June 13-16, 1988, Mashhad University of Medical Sciences. Mashhad, Iran. No. 83.

Eisenmenger W, Drasch G, von Clarmann M, Kretschmer E, Rolder G. 1991. Clinical and morphological findings on mustard gas [bis(2-chloroethyl)sulfide] poisoning. Journal of Forensic Science 36:1688-1698.

Elder GH, Clipp EC. 1989. Combat experience and emotional health: impairment and resilience in later life. Journal of Personality 57:311-341.

Elmore DT, Guiland JM, Jordan DO, Taylor HFW. 1948. Reaction of nucleic acids with mustard gas (dichloroethyl sulfide). Biochemical Journal 42:308-316.

Elsayed NM, Omaye ST, Klain GJ, Inase JL, Wheeler CW, Korte DW Jr. 1988. Mouse lung response to subcutaneous administration of the vesicant, butyl 2-chloroethyl sulfide (abstract). American Review of Respiratory Disease 137:521.

Elsayed NM, Omaye ST, Klain GJ, Inase JL, Dahlberg ET, Korte DW Jr. 1988. Response of mouse brain to subcutaneous administration of butyl 2-chloroethyl sulfide (abstract). Toxicologist 8:54.

Elsayed NM, Ta PN, Korte DW Jr. 1989. Biochemical alterations in mouse liver induced by nitrogen mustard (abstract). Toxicologist 9:26.

Elsayed NM, Omaye ST, Klain GJ, Inase JL, Dahlberg ET, Wheeler CR, Korte DW Jr. 1989. Response of mouse brain to a single subcutaneous injection of the monofunctional sulfur mustard, butyl 2-chloroethyl sulfide (BCS). Toxicology 58:11-20.

Elsayed NM, Nakashima JM, Wheeler CR, Omaye ST, Korte DW Jr. 1991. Biochemical changes in mouse lung and brain after single subcutaneous injection of a sulfur mustard (abstract). Toxicologist 215.

Elsayed NM, Omaye ST, Klain GJ, Korte DW Jr. 1992. Free radical-mediated lung response to the monofunctional sulfur mustard butyl 2-chloroethyl sulfide after subcutaneous injection. Toxicology 72:153-165.

Engel R. 1919. Laryngeal sequelae from gas. Gaz D Hop Paris 92:697-702. [In French]

English F, Bennett Y. 1990. The challenge of mustard gas keratopathy (letter). Medical Journal of Australia 152:55-56.

Enshayeh S, Meghdadi M, Momeni A. 1988. Skin manifestations of the mustard gases. In: Abstracts of the First International Medical Congress on Chemical Warfare Agents in Iran. June 13-16, 1988. Mashhad Universityof Medical Sciences . Mashhad, Iran. No. 37.

Epstein J. 1984. Nitrogen mustard (mechlorethamine) and UVB photocarcinogenesis: a dose-response effect. Journal of Investigative Dermatology 83:320-322.

Evans CG, Bodell WJ, Tokuda K, Doane-Setzer P, Smith MT. 1987. Glutathione and related enzymes in rat brain tumor cell resistance to 1,4-bis(2-chloroethyl)-l-nitrosourea and nitrogen mustard. Cancer Research 47:2525-2530.

Evers A. 1931. Relative capacity of experimental animals and man for absorbing poisonous gases. Archiv fur Hygiene und Bakteriologie 106:255-270. [In German]

Eyster JAE, Maver ME. 1920-1921. An apparatus for the exposure of skin or mouse membrane to the vapor of toxic substances, with observations on dichlorethylsulphide. Journal of Pharmacology and Experimental Therapeutics 15:95.

Fahmy MJ, Fahmy OG. 1969. The genetic effects of the biological alkylating agents with reference to pesticides. Annals of the New York Academy of Sciences 160:228-243.

Fahmy OG, Fahmy MJ. 1971. Mutability at specific euchromatic and heterochromatic loci with alkylating and nitroso compounds in *Drosophila melanogaster*. Mutation Research 13:19-34.

Fahmy OG, Fahmy MJ. 1972. Mutagenic selectivity for the RNA-forming genes in relation to the carcinogenicity of alkylating agents and polycyclic aromatics. Cancer Research 32:550-557.

Fairclough WA. 1937. Gas keratitis. Australian and New Zealand Journal of Surgery 7:163-166.

Fan L. 1990. Effects of Sulfur Mustard on Base Mismatch Repair of DNA in Monkey Kidney Cells (dissertation). University of Michigan.

Fan L , Bernstein IA. 1991. Effect of bis(b-chloroethyl)sulfide (BCES) on base mismatch repair of DNA in monkey kidney cells. Toxicology and Applied Pharmacology 111:233-241.

Farber E. 1973. Chemical carcinogenesis. In: Anfinsen CB, Potter M, Schechter AN, eds. Current Research in Oncology-1972. New York: Academic Press. 95-123.

Farber E. 1981. Chemical carcinogenesis. New England Journal of Medicine 305:1379-1389.

Farhoodi M, Panjvani FA, Balali M. 1988. A case of bronchogenic carcinoma a year after exposure to sulfur mustard. In: Abstracts of the First International Medical Congress on Chemical Warfare Agents in Iran. June 13-16, 1988. Mashhad University of Medical Sciences. Mashhad, Iran. No. 72.

Farhoodi M, Nehmatpour E, Panjvani F. 1988. Bronchoscopic and biopsy evaluation of four patients exposed to chemical war gases and suffering from chronic respiratory symptoms. In: Abstracts of the First International Medical Congress on Chemical Warfare Agents in Iran. June 13-16, 1988. Mashhad University of Medical Sciences. Mashhad, Iran. No. 25.

Farhoodi M, Panjvani F, Tabarastani M, Balali M, Bahrami A, Hoseini RF. 1988. Clinical and pathological findings in nine patients who died as a result of sulphur mustard poisoning. In: Abstracts of the First International Medical Congress on Chemical Warfare Agents in Iran. June 13-16, 1988. Mashhad University of Medical Sciences. Mashhad, Iran. No. 20.

Farhoodi M, Keshmiri M, Balali M, Naghibi M, Taberstani M, Saghabie S. 1988. Successful treatment of patient with 90% skin burns and leukopenia due to chemical warfare agents. In: Abstracts of the First International Medical Congress on Chemical Warfare Agents in Iran. June 13-16, 1988. Mashhad University of Medical Sciences. Mashhad,Iran . No. 32.

Farid R, Balali M, Farhoodi M, Navaian A. 1988. Respiratory complications of sulfur mustard poisoning in Iranian combatants. In: Abstracts of the First International Medical Congress on Chemical Warfare Agents in Iran. June 13-16, 1988. Mashhad University of Medical Sciences. Mashhad, Iran. No. 23.

Farney RJ, Morris AH, Armstrong JD Jr, Hammer S. 1977. Diffuse pulmonary disease after therapy with nitrogen mustard, vincristine, procarbazine, and prednisone. American Review of Respiratory Disease 115:135-145.

Fasilati M, Taghikhani M, Riazi GH. 1988. Effect of mustard gas on enzymatic level in liver and spleen. In: Abstracts of the First International Medical Congress on Chemical Warfare Agents in Iran. June 13-16, 1988, Mashhad University of Medical Sciences. Mashhad, Iran. No. 43.

Faure-Fremiet, Guieysse, Magne, Mayer A. 1920. Skin lesions caused by certain vesicant compounds. Comptes Rendus Hebdomadaires des Seances de l'Academie des Sciences 170:1476. [In French]

Fell HB, Allsopp CB. 1946. The effects of Lewisite and of Lewisiteoxide on living cells *in vitro* . British Journal of Experimental Pathology 27:305-309.

Fell HB, Allsopp CB. 1948. Action of mustard gas on living cells *in vitro:* immediate cytologic effects of mustard gas and of its hydrolysis products, effect on cell growth of

adding small concentrations of mustard gas to culture medium. Cancer Research 8:145-161.
Fell HB, Allsopp CB. 1948. The effect of repeated applications of minute quantities of mustard gas (b,b'-dichlorodiethylsulfide) on the skin of mice. Cancer Research 8:177-181.
Feng H-W, Ribeiro P, Scavarelli RM, Bernstein IA. 1987. Differential effects of bis(2-chloroethyl) sulfide on the differentiation of keratinocytes in culture. Journal of Toxicology—Cutaneous and Ocular Toxicology 6:273-282.
Feng X, Zhao X, Li F, Cheng J. 1987. The effect of sulfur mustard on the self-renewal and differentiation capacity of hemopoietic stem cells in mice. Chinese Journal of Pharmacology and Toxicology 1:116-122.
Ferguson RL, Silver SD. 1947. A method for the visual demonstration of Lewisite in skin. American Journal of Clinical Pathology 17:37-38.
Ferm VH. 1974. Effects of metal pollutants upon embryonic development. Reviews on Environmental Health 1:237-259.
Ferm VH, Carpenter S. 1968. Malformations induced by sodium arsenate. Journal of Reproduction and Fertility 17:199-201.
Ferm VH, Saxon A. 1971. Amniotic fluid volume in experimentally induced renal agenesis and anencephaly. Experientia 27:1066-1068.
Ferm VH, Saxon A, Smith BW. 1971. The teratogenic profile of sodium arsenate in the golden hamster. Archives of Environmental Health 22:557-560.
Ferraloro G. 1936. Caustic and toxic effects of dichloroethylsulfide. Pensiero Medico 25:157-159. [In Italian]
Ferraloro G. 1937. Gulltadiaphot method in diagnosis of dichloroethyl sulfide poisoning: experimental study. Giornale de Medicina Militare 85:385-394. [In Italian]
Ferri G. 1937. Investigations on the individual sensitivity of the human skin to mustard gas, and some factors capable of modifying it. Giorale de Medicina Militare (September):919-933. [In Italian]
Fex T. 1940. Gas injuries from the dermatologic standpoint. Tidskrift I Militar Halsovard 65:19-25. [In Swedish]
Fields RM. 1980. Victims of terrorism: the effects of prolonged stress. Evaluation and Change Special Issue:7684.
Figley CR. 1985. Trauma and Its Wake. New York: Brunner Mazel.
Fischbeck R. 1969. Pathogenesis, clinic and therapy of local injuries caused by skin irritants with special consideration of yperite poisoning. Zeitschrift fur Militarmedizin 10:83-90. [In German]
Fischer B, Goldschmid E. 1920. Changes in the respiratory tract following vesicant gas poisoning and burns. Zentralblatt fur Pathologie 23:11-33. [In German]
Fischer H. 1969. Accidents caused by combat and smoke chemicals and their sequelae. Wehrmedizinische Monatsschrift 13:355-359. [In German]
Fisher B, Rockette II, Fisher ER, Wickerham DL, Redmond C, Brown A. 1985. Leukemia in breast cancer patients following adjuvant chemotherapy or postoperative radiation: the NSABP experience. Journal of Clinical Oncology 3:1640-1658.
Fisher P. 1982. New toxicology: carcinogenesis. II. Cytotoxics and cocarcinogens. British Homeopathic Journal 71:183-196.
Fitzpatrick TB, Eisen AZ, Wolff K, Freedberg IM, Austeru KF. 1979. Dermatology in General Medicine: Textbook and Atlas. 2nd ed. New York: McGraw-Hill.
Fitzsimmons CE. 1926. X-ray findings and combination chest conditions of chronic gassed cases versus tuberculosis. Medical Bulletin (US Veterans Administration) 2:764-767.
Flamm WG, Bernheim NJ, Spalding J. 1969. Selective inhibition of the semiconservative replication of mouse satellite DNA. Biochimica et Biophysica Acta 195:273-275.

Flamm WG, Berheim NJ, Fishbein L. 1970. On the existence of intrastrand crosslinks in DNA alkylated with sulfur mustard. Biochimica et Biophysica Acta 224:657-659.

Fletcher C, Peto R, Tinker C, Speizer FE. 1976. The Natural History of Chronic Bronchitis and Emphysema. Oxford: Oxford University Press.

Flury F. 1921. War-gas poisoning. IX. Local action of arsenic compounds. Zeitschrift fur die Gesamte Experimentelle Medizin 13:523-578. [In German]

Flury F. 1928. Moder occupational intoxications. Archiv fur Experimentelle Pathologie und Pharmakologie 138:65-82. [In German]

Flury F. 1937. Chemical warfare. Gasschutz und Luftschutz 7:57-73. [In German]

Flury F, Wieland. 1921. War-gas poisoning. VII. The pharmacological action of dichloroethylsulfide. Zeitschrift fur die Gesamte Experimentelle Medizin 13:367-483. [In German]

Flury F, Zerick F. 1931. [Harmful Gases, Vapors, Fogs, Kinds of Smoke and Dust] Schadliche Gase, Dampfe, Nebel, rauch-und Staubarten. Berlin: Springer. [In German]

Foa V, Bertolero F. 1983. Arsines. In: Parmeggiani L, ed. Encyclopaedia of Occupational Health and Safety. 3rd rev. ed. Vol. 1. Geneva: International Labour Office. 183-184.

Fong J. 1953. Sulfur mustard (dichloroethyl sulfide)-inactivated influenza virus as interfering agent. Journal of Immunology 71:241-245.

Fong J, Bernal E. 1953. *In vitro* action of mustards on infective and toxic components of influenza A (PR8) virus. Journal of Immunology 70:89-96.

Fong J, Bernal E. 1953. *In vitro* action of sulfur mustard and chloroethylamine derivatives upon antigenicity of influenza A(PR8) virus. Journal of Immunology 70:97-102.

Fonseca M, Lunt G, Aguilar J. 1991. Inhibition of muscarinic cholinergic receptors by disulfide reducing agents and arsenicals: differential effect on locust and rat. Biochemical Pharmacology 41:735-742.

Foster J. 1939. Ophthalmic injuries from mustard gas. British Medical Journal 83:1181-1183.

Foulhoux P. 1963. Contemporary chemical agents: their toxic and therapeutic values. Revue des Corps de Sante des Armees 4:693-722.

Foussereau J, Benezra C, Maibach H. 1982. Occupational Contact Dermatitis: Clinical and Chemical Agents. Philadelphia: Munksgaard.

Fowler BA, Weissberg JF. 1974. Arsine poisoning. New England Journal of Medicine 291:1171-1174.

Fox BW. 1975. Mechanisms of DNA repair in mammalian cells after alkylating agents. Excerpta Medica International Cancer Congress Series. No. 349, Cell Biology and Tumor Immunology 1:153-156.

Fox BW, Fox M. 1973. The nature of the resistance to methylene dimethane sulphonate in the Yoshida sarcoma (abstract). British Journal of Cancer 28:81.

Fox M, Scott D. 1980. The genetic toxicology of nitrogen and sulphur mustard. Mutation Research 75:131-168.

Fraenkel-Conrat H. 1961. Chemical modification of viral ribonucleic acid. 1. Alkylating agents. Biochimica et Biophysica Acta 49:169-180.

Francini AP, Gilchrist HL. 1924. Mustard gas and tuberculosis. Military Surgeon 54:470-485.

Francois J, Van Oye R. 1977. Immunodepressors in ophthalmology. Annales d'Oculistique 210:89-110. [In French]

Frank AL. 1978. Occupational lung cancer. In: Harris CC, ed. Pathogenesis and Therapy of Lung Cancer. Vol. 10, Lung Biology in Health and Disease. New York: Marcel Dekker. 25-51.

Frank AL. 1982. The epidemiology and etiology of lung cancer. Clinics in Chest Medicine 3:219-228.

Fraumeni JF Jr. 1975. Chemicals in the induction of respiratory tract tumors. Excerpta Medica International Cancer Congress Series No. 351. 3:327-335.

Fraumeni JF Jr. 1975. Respiratory carcinogenesis: an epidemiologic appraisal. Journal of the National Cancer Institute 55:1039-1046.

Fraumeni JF Jr. 1979. Epidemiological studies of cancer. In: Griffin AC, Shaw CR, eds. Carcinogens: Identification and Mechanisms of Action. New York: Raven Press.

Fredriksson T, Skog E. 1959. Skin-sensitizing capacity of mustard gas. Tidskrift I Militar Halsovard 84:151-156. [In Swedish]

Freeman K. 1991. The unfought chemical war. Bulletin of the Atomic Scientists 47:30-39.

Freilinger G, Pauser G. 1988. New experience with intoxicated victims by chemical warfare. In: Abstracts of the First International Medical Congress on Chemical Warfare Agents in Iran. June 13-16, 1988. Mashhad University of Medical Sciences. Mashhad, Iran. No. 87.

Freitag A, Firusian N, Stamatis G, Greschuchna D. 1991. The role of bronchoscopy in pulmonary complications due to mustard gas inhalation. Chest 100:1436-1441.

French H. 1918. Permanent emphysema in the young after gas-poisoning. Clinical Journal (London) 47:157-158.

Frescoln LD. 1918. Mustard (yellow cross) burs. Journal of the AmericanMedical Association 71:1911-1912.

Friberg L, Cederlof R. 1978. Late effects of air pollution with special reference to lung cancer. Environmental Health Perspectives 22:45-66.

Friedberg K, Mengel K, Schlick E. 1983. The action of azimexone on the cells of the hemopoietic system in mice, especially after damage with x-rays. Radiation and Environmental Biophysics 22:117-131.

Friedenwald JS. 1948. Note on karyolysis of the corneal stroma cells. Bulletin of the Johns Hopkins Hospital 82:178-181.

Friedenwald JS. 1948. Studies on the physiology, biochemistry, and cytopathology of the cornea in relation to injury by mustard gas and allied toxic agents. XVII. Summary and some possible interpretations. Bulletin of the Johns Hopkins Hospital 82:326-337.

Friedenwald JS, Busche W. 1948. Nuclear fragmentation produced by mustard and nitrogen mustards in the corneal epithelium. Bulletin of the Johns Hopkins Hospital 82:161-177.

Friedenwald JS, Busche W, Moses SG. 1948. Comparison of the effects of mustard, ultraviolet and x-radiation, and colchicine on the cornea. Bulletin of the Johns Hopkins Hospital 82:312-325.

Friedenwald JS, Busche W, Scholz RO. 1948. Effects of mustard and nitrogen on mitotic and wound healing activities of the corneal epithelium. Bulletin of the Johns Hopkins Hospital 82:148-160.

Friedenwald JS, Scholz RO, Snell AJ, Moses SG. 1948. Primary reaction of mustard with the corneal epithelium. Bulletin of the Johns Hopkins Hospital 82:102-120.

Fries AA, West CJ. 1921. Chemical Warfare. New York: McGraw-Hill.

Frost DV, Main BT, Cole J, Sanders PG, Perdue HS. 1964. Reproduction studies in rats with arsanilic acid. Federation Proceedings 3:291.

Fuhner H. 1918. Toxicology of arsine. I. Protozoa. Archiv fur Experimentale Pathologie und Pharmakologie 82:44-50. [In German]

Fuhner H. 1922. Toxicology of arsine. I. Toxicity for warm-blooded animals. Archiv fur Experimentale Pathologie und Pharmakologie 92:288-301. [In German]

Fujita M. 1987. Study on the genetic effects of sulfur mustard gas exposure using variant proteins as indicators. Hiroshima Daigaku Igaku Zasshi 35:305-340. [In Japanese]

Fujita M, Goriki K, Satoh C, Hamilton HB, Yamakido M, Inamizu T, Onari K, Nishimoto Y, Shigenobu T. 1983. Study of genetic effects of occupational exposure to mustard gas (abstract). Jinrui Idengaku Zasshi 28:109. [In Japanese]

Fulgosi A. 1956. [Two cases of yperite poisoning]. Vojno Sanitetski Pregled 13:500-502.

Fullerton CS, Ursano RJ. 1990. Behavioral and psychological responses to chemical and biological warfare. Military Medicine 155:54-59.

Fullerton CS, Ursano RJ. Submitted for publication. The chemical and biological warfare environment: psychological responses of a health care delivery team in a high stress environment.
Fullerton CS, Ursano RJ, Kao T-C, Bhartiya VR. In press. The chemical and biological warfare environment: psychological responses and social supports in a high stress environment. Journal of Applied Social Psychology.
Gaffuri E. 1957. Chronic occupational pulmonary lesions due to yperite. Medicina del Lavoro 48:539-544. [In Italian]
Gaffuri E. 1970. Current knowledge of occupational tumors of the respiratory tract. Lotta Contro la Tubercolosi 40:281-302. [In Italian]
Gage EL. 1946. Mustard gas (dichloroethyl sulfide) burs: in clinical experiences. West Virginia Medical Journal 42:180-185.
Galdston M, Leutscher JA, Longcope WT, Ballich NL. 1947. A study of the residual effects of phosgene poisoning in human subjects. I. After acute exposure. Journal of Clinical Investigation 26:145-168.
Gales YA, Gross CL, Krebs RC, Smith WJ. 1989. Flow cytometric analysis of toxicity by alkylating agents in human epidermal keratinocytes. Alternative Methods in Toxicology 7:169-174.
Gallotto WA. 1992. Corps tries to ease Raritan Arsenal cleanup fear. Sunday Star-Ledger, Newark, New Jersey. February 2, 1992.
Galwey WR. 1922. Gas warfare. Adoption, method of use, protection of troops. Journal of the Royal Army Medical Corps 38:62-72.
Galwey WR. 1922. Gas warfare. Effect of poison gases, early and late. Journal of the Royal Army Medical Corps 38:144-154.
Galwey WR. 1922. Gas warfare. Treatment of gas casualties. Journal of the Royal Army Medical Corps 38:223-234.
Galwey WR. 1932. Medical problems of mustard gas poisoning. Journal of the Royal Army Medical Corps 59:125-132.
Ganas P. 1969. New developments in chemical and biological warfare. Forces Aeriennes Francaises 24:449-475.
Garelli F. 1918. On the new German vesicant gases. Riforma Medica 34:313-314. [In Italian]
Garner RC. 1979. Carcinogen prediction in the laboratory: a personal view. Proceedings of the Royal Society of London. Series B: Biological Sciences 205:121-134.
Garrett NE, Stack HF, Gross MR, Waters MD. 1984. An analysis of the spectra of genetic activity produced by known or suspected human carcinogens. Mutation Research 134:89-111.
Gasschutz und Luftschutz. 1932. Properties of war gases of the World War and post-war periods. Gasschutz und Luftschutz 2:264. [In German]
Gastinel P, Sohier R. 1938. Research on the role of the autonomic nervous system in lesions observed over time in an animal intoxicated with dichloroethyl sulfide. Comptes Rendus des Seances de la Societe de Biologie et de Filiales 127:46-49. [In French]
Gawler DR. 1940. Mustard gas, with special reference to eye lesions. Medical Journal of Australia 27:106-108.
Gebert F. 1937. The reaction between arsine and hemoglobin. Biochemische Zeitschrift 293:157-186. [In German]
Geeraets W, Abedi S, Blanke R. 1977. Acute corneal injury by mustard gas. Southern Medical Journal 70:348-350.
Generoso WM, Bishop JB, Gosslee DG, Newell GW, Sheu C, von Halle E. 1980. Heritable translocation test in mice. Mutation Research 76:191-215.
Genet L. 1922. Yperite: delayed ocular sequelae. Lyon Medical 131:786-788. [In French]

Genet L. 1925. Ocular burns from yperite gas: persistent sequelae seven years after injury. Case study. Lyon Medical 136:388-391. [In French]

Genet L. 1937. Conjunctival ischemia from yperite, delayed corneal ulceration. Lyon Medical 160:229-231. [In French]

Genet L, DeLord E. 1936. Delayed symmetrical keratitis caused by yperite. Lyon Medical 158:216-221. [In French]

Gentilhomme E, Neveux Y, Hua A, Thiriot C, Faure M, Thivolet J. 1992. Action of bis (betachloroethyl)sulphide (BCES) on human epidermisreconstituted in culture: morphological alterations and biochemical depletion of glutathione . Toxicology in Vitro 6:139-147.

Gerchik M. 1939. Medical experiences of Americans with chemical warfare agents during the World War. Protar 5:173-179, 199-202.

Gerhardsson L, Dahlgren E, Eriksson A, Lagerkvist B, Lundstrom J, Nordberg G. 1988. Fatal arsenic poisoning: a case report. Scandinavian Journal of Work, Environment and Health 14:130-133.

Gershenson SM, Zil'Berman RA, Levochkina OL, Pashkovskii AM, Sit'Ko PO, Tamavskii ND. 1947. Induction of mutations in *Drosophila* by dichlorodiethylsulfide. Doklady Akademii Nauk SSSR 58:1495-1496. [In Russian]

Gescher A. 1976. Alkylating antineoplastic agents. Pharmazie in Unserer Zeit 5:41-52. [In German]

Gilbert RM, Rowland S, Davison CL, Papirmeister B. 1975. Involvementof separate pathways in the repair of mutational and lethal lesions induced by a monofunctional sulfur mustard . Mutation Research 28:257-275.

Gilchrist HL. 1928. A Comparative Study of World War Casualties from Gas and Other Weapons. Washington, DC: U.S. Government Printing Office.

Gilchrist HL. 1928. Chemical warfare and its medical significance. Military Surgeon 63:477-492.

Gilchrist HL. 1933. The residual effects of warfare gases: the use of arsenical compounds, with reports of cases. Medical Bulletin (US Veterans Administration) 10:79-98.

Gilchrist HL, Matz PB. 1933. The residual effects of warfare gases. Medical Bulletin (US Veterans Administration) 9:339-390.

Gilchrist HL, Matz PB. 1933. The Residual Effects of Warfare Gases. Washington, DC: U.S. Government Printing Office.

Gill GH. 1938. Mustard gas. Pharm J 140:291-292.

Giller EL Jr., ed. 1990. Biological Assessment and Treatment of Posttraumatic Stress Disorder. Washington, DC: American Psychiatric Association Press.

Gillert E. 1944. [Blister Gas Injuries] Die Kampfstoffverletzungen. 11th ed. Berlin: Verlag Union und Schwarzenberg.

Gilman A. 1946. Symposium on advances in pharmacology resulting from war research: therapeutic applications of chemical warfare agents. Federation Proceedings 5:285-292.

Gilman A. 1963. The initial clinical trial of nitrogen mustard. American Journal of Surgery 165:574-578.

Gilman A, Phillips FS. 1946. The biological actions and therapeutic applications of the b-chloroethyl amines and sulfides. Science 103:409-415.

Giraud. 1917. The first symptoms of intoxication from mustard gas. Journal de Medecine et de Chirurgie Pratiques 88:890-894. [In French]

Gjessing EC, Chanutin A. 1946. Electrophoretic analyses of sera after treating dogs with p-chloroethyl vesicants. Journal of Biological Chemistry 165:413-420.

Glass LR, Easterly CE, Jones TD, Walsh PJ. 1991. Ranking of carcinogenic potency using a relative potency approach. Archives of Environmental Contamination and Toxicology 21:169-176.

Gluud B, Dahl H, Vangsted P, Nom M. 1985. Mustard gas: a returning threat. Nordisk Medicin 100:222-223, 243. [In Danish]

Goebel HH, Schmidt PF, Bohl J, Tettenborn B, Kramer G, Gutmann L. 1990. Polyneuropathy due to acute arsenic intoxication: biopsy studies. Journal of Neuropathology and Experimental Neurology 49:137-149.

Goffart M. 1947. Reaction of vesicants, oxidizing agents and metallic ions with sulfhydryl radicals of skin. Archives Internationales de Pharmacodynamie et de Therapie 74:9-30.

Gohlke H, Ullerich K. 1951. Skin and eye injuries caused by dichlorodiethyl sulfide (mustard gas). Hautarzt 2:404-407. [In German]

Gold LS, Sawyer CB, Magaw R, Backman GM, de Veciana M, Levinson R, Hooper NK, Havender WR, Bernstein L, Peto R. 1984. A carcinogenic potency database of the standardized results of animal bioassays. Environmental Health Perspectives 58:9-319.

Gol'dfarb DM, Cherin LS, Gol'dberg GI, Akatova NS, Gukova LA. 1970. Some properties of nitrogenous yperite inhibiting factor produced by male strains of *Escherichia coli* K-12. Soviet Genetics (English Translation of Genetika) 6: 1360-1366.

Goldman L. 1943. Some medical problems of vesicant chemical warfare agents as affecting civilian populations. Bulletin of the New York. Academy of Medicine 19:57-64.

Goldman L, Cullen GE. 1940. Some medical aspects of chemical warfare agents. Journal of the American Medical Association 114:2200-2204.

Goldman L, Cullen GE. 1940. The vesicant chemical warfare agents. Archives of Dermatology and Syphilology 42:123-136.

Goldman M, Dacre J. 1989. Lewisite: its chemistry, toxicology, and biological effects. Reviews of Environmental Contamination and Toxicology 110:75-115.

Goldschlag S. 1942. Experiments on the improvement of the treatment of mustard gas lesions of the skin. Medical Journal of Australia 1:620-622.

Golub ES. 1987. Immunology, A Synthesis. Sunderland, MA: Sinauer Associates.

Goodman L, Gilman A. 1941. The Pharmacological Basis of Therapeutics. New York: Macmillan. 714-716.

Gordonoff T. 1941. Toxicology of dichloroethylsulfide. Schweizerische Medizinische Wochenschrift 71:446-448. [In German]

Gortler L. 1979. Mustard gas and mechanisms. The impact of World War II on the emergence of physical organic chemistry (abstract). Abstracts of the American Chemical Society's 178th Annual Meeting, Washington, DC.

Goss BC. 1919. Mustard gas: weapon and shield. Nat Scrv 6:85-89.

Gougerot, Clara. 1919. Genital effects of yperite simulating syphilis. Annales des Maladies Veneriennes 14:257-262. [In French]

Graef I, Karnofsky DA, Jager VB, Krichesky B, Smith HW. 1948. The clinical and pathologic effects of the nitrogen and sulfur mustards in laboratory animals. American Journal of Pathology 24:1-47.

Graham AF, Levvy GA, Chance AC. 1947. Fate of arsenical vesicants in skin and effect of BAL. Biochemical Journal 41:352-357.

Graham JH, Helwig EB. 1959. Bowen's disease and its relationship to systemic cancer. AMA Archives of Dermatology 80:133-159.

Graham JH, Mazzanti GR, Helwig EB. 1961. Chemistry of Bowen's disease: relationship to arsenic. Journal of Investigative Dermatology 37:317-332.

Granone P, Montalto G, Laudati A, Nallo S, Panebianco V. 1980. Medico-social problems of bronchopulmonary carcinoma in Italy. Note IIA. Etiopathogenetic aspects and the possibility of preventive intervention. Igiene e Sanita Pubblica 36:208-219. [In Italian]

Grant J, Ritchie TF. 1942. Mustard gas burns. British Medical Journal 2:217-218.

Grant WM. 1974. Toxicology of the Eye. Springfield, IL: Charles C. Thomas. 375-378.

Green AG. 1919. History of mustard gas. Journal of the Society of Chemical Industry 38:363-364, 469.

Green BL, Lindy JD, Grace MC, Gleser GC. 1989. Multiple diagnosis in posttraumatic stress disorder. The role of war stressors. Journal of Nervous and Mental Disease 177:329-335.

Green SJ, Price TS. 1921. The chlorovinylchloroarsines. J Chem Soc Transactions 119:448-453.

Greene MH, Wilson J. 1985. Second cancer following lymphatic and hematopoietic cancers in Connecticut, 1935-82. National Cancer Institute Monographs 68:191-217.

Greene MH, Young TI. 1981. Malignant melanoma in renal transplant recipients. Lancet 1:1196-1198.

Greene MH, Boice JD Jr, Greer BE, Blessing JA, Dembo AJ. 1982. Acute nonlymphocytic leukemia after therapy with alkylating agents for ovarian cancer. New England Journal of Medicine 307:1416-1421.

Greene MH, Boice JD Jr, Strike TA. 1982. Carmustine as a cause of acute nonlymphocytic leukemia (letter). New England Journal of Medicine 313:579.

Greene MH, Young RC, Merrill JM, DeVita VT. 1983. Evidence of a treatment dose-response in acute nonlymphocytic leukemias which occur after therapy of non-Hodgkin's lymphoma. Cancer Research 43:1891-1898.

Greene SA, Wolff RK, Hahn FF, Henderson RF, Mauderly JL, Lundgren DL. 1984. Sulfur dioxide induced chronic bronchitis in beagle dogs. Journal of Toxicology and Environmental Health 13:945-958.

Greene MH, Harris EL, Gershenson DM, Malkasian GD Jr, Melton LJ III, Dembo AJ, Bennett JM, Moloney WC, Boice JD Jr. 1986. Melphalan may be a more potent leukemogen than cyclophosphamide. Annals of Internal Medicine 105:360-367.

Griffin AC, Brandt EL, Tatum EL. 1950. Nitrogen mustards as cancer-inducing agents. Journal of the American Medical Association 144:571.

Grivat, Got. 1919. Clinical considerations on the effects of the new German toxic gases on the mucous membranes of the upper respiratory tract. Revue de Laryngologie, Otologie, Rhinologie 40:121-129. [In French]

Gross CL, Watson CV, Petrali J, Papirmeister B. 1981. Effect of sulfur mustards on lysosomes from rat liver *in vitro*. Toxicology and Applied Pharmacology 61:147-151.

Gross CL, Meier HL, Papirmeister B, Brinkley FB. 1984. Sulfur mustard decreases NAD levels in human tissue (abstract). Federation Proceedings 43:2452.

Gross CL, Meier HL, Papirmeister B, Brinkley FB, Johnson JB. 1985. Sulfur mustard lowers NAD concentrations in human skin grafted to athymic nude mice. Toxicology and Applied Pharmacology 81:85-90.

Gross CL, Innace JK, Krebs RC, Smith WJ, Meier HL. 1989. Sulfur mustard (HD) cytotoxicity in lymphocytes can be affected by intracellular glutathione levels (abstract). FASEB Journal 3:A326.

Gross D. 1963. Chemical synovectomy with mustard gas in primary chronic polyarthritis (preliminary report). Zeitschrift fur Rheumaforschung 22:456-459. [In German]

Grunicke H, Putzer H, Scheidl F, Wolff-Schreiner E, Grunewald K. 1982. Inhibition of tumor growth by alkylation of the plasma membrane. Biosci Rep 2:601-604.

Grzhebin ZN. 1931. Sensitivity to dichloroethylsulfide. Voenno-Sanitarno Delo 9:25-33. [In Bulgarian]

Gueit C. 1919-1920. Treatment of broncho-pulmonary lesions due to yperite with intravenous injections of bicarbonate of soda. Montpellier Medical 41:125-127. [In French]

Guglianetti L. 1935. Ocular lesions from toxic chemicals. Giornale de Medicina Militare 83:517-519. [In Italian]

Guittin P, Schorch F, Fontaine JJ, Crespeau F, Parodi AL. 1989. Experimental pathology induced in rat by a single skin application of mustard gas (abstract). Pathology, Research and Practice 185:68-69.

Guthrie F. 1859. On some derivatives of the olefines. Quarterly Journal of the Chemical Society 12:109-126.

Guthrie F. 1860. On some derivatives of the olefines. Quarterly Journal of the Chemical Society 13:129-135.

Gutmann A. 1919. Effects of war gases on the eye. Deutsche Medizinische Wochenschrift 45:1082. [In German]

Gutmann A. 1919. Mustard gas. Moniteur Scientifique 9:149. [In French]

Guy RH, Maibach HI. 1984. Correction factors for determining body exposure from forearm percutaneous absorption data. Journal of Applied Toxicology 4:26-28.

Haber L. 1986. The Poisonous Cloud. Oxford: Clarendon Press.

Habraken Y, Ludlum D. 1989. Release of chloroethyl ethyl sulfide-modified DNA bases by bacterial 3-methyladenine-DNA glycosylases I and II. Carcinogenesis 10:489-492.

Haddow A. 1959. The chemical and genetic mechanisms of carcinogenesis. II. Biologic alkylating agents. In: Physiopathology of Cancer. 2nd ed. New York: Hoeber-Harper. 602-643.

Haddow A. 1973. On the biological alkylating agents. Perspectives in Biology and Medicine 16:503-524.

Hagelsten J. 1991. Chemical warfare agents. Ugeskrift for Laeger 153:795-796. [In Danish]

Hall TC, Wood CH, Stoeckle JD, Tepper LB. 1959. Case data from the beryllium registry. AMA Archives of Industrial Health 19:100-103.

Hambrook JL, Harrison JM, Howells DJ, Schock C. 1992. Biological fate of sulphur mustard (1,1'-thio-bis(2-chlorethane)): urinary and faecal excretion of 35S by rat after injection or cutaneous application of 35S-labelled sulphur mustard. Xenobiotica 22:65-75.

Hankins JL, Klotz WC. 1922. Permanent pulmonary effects of gas in warfare. American Review of Tuberculosis 6:571-574.

Hans H. 1934. Dichloroethylsulfide and its designation by yellow cross. Medizinische Welt 8:599-602. [In German]

Hanski C, Stehlik G. 1980. Methylation of nucleic acids and its relation to carcinogenesis. Wiener Klinische Wochenschrift 92:134-140. [In German]

Hanslian R. 1937. [Chemical Warfare] Der Chemische Krieg. Berlin: E.S. Mittler. [In German]

Hanssen OE. 1946. Relation between direct and indirect action in local lesions caused by dichloroethyl sulfide. Acta Pathologica, Microbiologica, et Immunologica Scandinavica Supplement 91:48-50.

Hanzlik PJ, Tarr J. 1919. The comparative skin irritant properties of dichlorethylsulfide (mustard gas) and other agents. Journal of Pharmacology and Experimental Therapeutics 14:221-228.

Harada S, Dannenberg AM Jr, Kajiki A, Higuchi K, Tanaka F, Pula PJ. 1985. Inflammatory mediators and modulators released in organ culture from rabbit skin lesions produced *in vivo* by sulfur mustard. II. Evans blue dye experiments that determined the rates of entry and turnover of serum protein in developing and healing lesions. American Journal of Pathology 121:28-8.

Harada S, Dannenberg AM Jr., Vogt RF Jr., Myrick JE, Tanaka F, Redding LC, Merkhofer RM, Pula PJ, Scott AL. 1987. Inflammatory mediators and modulators released in organ culture from rabbit skin lesions produced *in vivo* by sulfur mustard. III. Electrophoretic protein fractions, trypsin-inhibitory capacity, a_1-proteinase inhibitor, and a_1- and a_2-macroglobulin proteinase inhibitors of culture fluids and serum. American Journal of Pathology 126:148-163.

Harada Y. 1977. Scanning electron microscopy observation of the tracheal mucous membrane injury. Jibi To Rinsho Otologia 23:660-665. [In Japanese]

Harker WG, Kushlan P, Rosenberg SA. 1984. Combination chemotherapy for advanced Hodgkin's disease after failure of MOPP: ABVD and B-CAVe. Annals of Internal Medicine 101:440446.

Harkins HN. 1938. Acute ulcer of the duodenum (Curling's ulcer) as a complication of burs: relation to sepsis. Surgery 3:608-641.

Harold FM, Ziporin ZZ. 1958. Effect of nitrogen and sulfur mustard on nucleic acid synthesis in *Escherichia coli.* Biochimica et Biophysica Acta 28:482-491.

Harold FM, Ziporin ZZ. 1958. The relationship between the synthesis of DNA and protein in *Escherichia coli* treated with sulfur mustard. Biochimica et Biophysica Acta 28:492-503.

Harris R, Paxman J. 1982. A Higher Form of Killing: The Secret Story of Chemical and Biological Warfare. New York: Hill and Wang.

Harrison HE, Ordway NK, Durlacher SH, Albrink WS, Bunting H. 1946. Poisoning from inhalation of the vapors of Lewisite and phenyldichlorarsine: its pathology in the dog and treatment with 2,3-dimercaptopropanol (BAL). Journal of Pharmacology and Experimental Therapeutics 87:76-80.

Harrison WP, Frazier WP, Maxxanti EM, Hood RD. 1980. Teratogenicity of disodium methanarsonate and sodium dimethylarsonate (sodium cacodylate) in mice. Teratology 21:43A.

Hartung M, Valentin H. 1988. Malignant diseases of the respiratory tract and their occupational causes. Atemswegs und Lungenkrankheiten 14:156-159. [In German]

Hartwell JL. 1946. Reaction of mustard gas and some of its derivatives with proteins and amino acids. Journal of the National Cancer Institute 6:319-324.

Harvey SC. 1975. Arsenic. In: Goodman LS, Gilman A. The Pharmacological Basis of Therapeutics. 5th ed. New York: MacMillan. 924-928.

Haskin D. 1948. Some effects of nitrogen mustard on the development of external body form in the fetal rat. Anatomical Record 102:493-511.

Haskonen H, Nordman H, Korhonen O. 1979. Long-term effects of exposure to sulfur dioxide: lung function four years after a pyrite dust explosion. American Review of Respiratory Diseases 119:555-560.

Hauck L. 1937. Cutaneous lesions caused by mustard gas: diagnosis, first aid, and therapy. Munchener Medizinische Wochenschrift 84:1292-1294. [In German]

Hauptmann S, Poege AW. 1973. Acylation of bis-2 chloroethylamine nitrogen mustard gas. Pharmazie 28:520-522. [In German]

Hawkins MM, Kinnier Wilson LM, Stovall MA, Marsden HB, Kingston JE, Chessells JM. 1992. Epipodophyllotoxins, alkylating agents, and radiation and risk of secondary leukaemia after childhood cancer. British Medical Journal 304:951-958.

Hay A, Roberts G. 1990. The use of poison gas against the Iraqi Kurds: analysis of bomb fragments, soil, and wool samples (letter). Journal of the American Medical Association 263:1065-1066.

Hayashi Y. 1963. Case of laryngeal carcinoma caused by mustard gas poisoning at the former Japanese Army Poison Gas Manufacturing Plant, Okuno-jima Island. Gencho Hiroshima Igaku 11:383-387. [In Japanese]

Haydarnajad H. 1988. Clinical and spirometric evaluation of 400 chemical weapon victims and effect of anticholinergic bronchodilator in improving airway obstruction. In: Abstracts of the First International Medical Congress on Chemical Warfare Agents in Iran. June 13-16, 1988. Mashhad University of Medical Sciences. Mashhad, Iran. No. 7-1.

Heckford F. 1937. Delayed corneal ulceration following mustard gas burns. Proceedings of the Royal Society of Medicine 30:949.

Heidelberger C. 1975. Chemical carcinogenesis. Annual Review of Biochemistry 44:79-121.

Heikkila P, Savela A, Vuorela R, Partanen T. 1988. Employees exposed to carcinogens in Finland in 1986. Helsinki: Institute of Occupational Health.

Heinc A. 1959. Remote sequelae following yperite bums of the eye. Ceskoslovenska Ofthalmologie 15:31-36. [In Czech]

Heinsius E. 1940. Pathology of ocular lesions caused by mustard gas. Klinische Monatsblatter fur Augenheilkunde 105:15-25. [In German]

Heitzmann O. 1921. War-gas poisoning. VIII. The pathological changes following poisoning with bis(b-chlorethyl)sulfide in animal experiments. Zeitschrift fur die Gesamte Experimentelle Medizin 13:484-522. [In German]

Hektoen L, Corper HC. 1920. The effect of mustard gas (dichloroethylsulphide) on antibody formation. Journal of Infectious Diseases 28:279-285.

Hellman U. 1970. Catamnestic Studies of Persons Injured with Blister Gas with Special Consideration of Psychiatric Late and Permanent Injuries (dissertation). Marburg, Germany: University of Marburg. [In German]

Helm U, Weger N. 1980. Fundamentals of military toxicology. In: Military Medicine: A Concise Handbook with Contributions to Disaster Medicine. Munich: Urban & Schwarzenberg.

Helm UK. 1988. Cholinergic and cholinotoxic effects of mustard gases. In: Abstracts of the First International Medical Congress on Chemical Warfare Agents in Iran. June 13-16, 1988. Mashhad University of Medical Sciences. Mashhad, Iran. No. 89.

Helm UK, Balali M. 1988. Cutaneous lesions produced by mustard gas. In: Abstracts of the First International Medical Congress on Chemical Warfare Agents in Iran. June 13-16, 1988. Mashhad University of Medical Sciences. Mashhad, Iran. No. 90.

Henderson Y, Haggard HW. 1943. Noxious Gases and the Principles of Respiration Influencing Their Action. New York: Reinhold.

Hergt W. 1932. Toxic action of war gases: review of literature. Vereinsblatt der Pfalzischen Arzte 44:232, 245, 257, 273. [In German]

Herberg S. 1977. Incidence of cancer in population with environmental exposure to metals. In: Hiatt HH, Watson JD, Winston JA, eds. Origin of Human Cancer. Cold Spring Harbor, NY: Cold Spring Harbor Laboratory. 147-157.

Herriott RM. 1948. Inactivation of viruses and cells by mustard gas. Journal of General Physiology 32:221-239.

Herriott RM. 1951. Nucleic acid synthesis in mustard gas-treated *E. coli* B. Journal of General Physiology 34:761-764.

Herriott RM. 1971. Effects on DNA: transforming principle. Chemical Mutagens: Principles and Methods for Their Detection 1:175-217.

Herriott RM, Price WH. 1948. The formation of bacterial viruses in bacteria rendered non-viable by mustard gas. Journal of General Physiology 32:63-68.

Herriott RM, Anson ML, Northrup JH. 1946. Reaction of enzymes and proteins with mustard gas, bis(b-chloroethyl) sulfide. Journal of General Physiology 30:185-210.

Herrmann GR. 1918-1919. The clinical pathology of mustard gas (dichlorethyl sulphide) poisoning. Journal of Laboratory and Clinical Medicine 4:1-30.

Herrmann H, Hickman FH. 1948. Exploratory studies on corneal metabolism. Bulletin of Johns Hopkins Hospital 82:225-250.

Herrmann H, Hickman FH. 1948. Further experiments on corneal metabolism in respect to glucose and lactic acid. Bulletin of Johns Hopkins Hospital 82:260-272.

Herrmann H, Hickman FH. 1948. Loosening of the corneal epithelium after exposure to mustard. Bulletin of Johns Hopkins Hospital 82:213-224.

Herrmann H, Hickman FH. 1948. The adhesion of epithelium to stroma in the cornea. Bulletin of Johns Hopkins Hospital 82:182-207.

Herrmann H. 1948. The effect of histamine and related substances on the cohesion of the corneal epithelium. Bulletin of Johns Hopkins Hospital 82:208-212.

Herrmann H, Hickman FH. 1948. The consumption of pyruvate, acetoin, acetate, and butyrate by the cornea. Bulletin of Johns Hopkins Hospital 82:273-286.

Herrmann H, Hickman FH. 1948. The effect of mustard on some metabolic processes in the cornea. Bulletin of Johns Hopkins Hospital 82:251-259.

Herrmann H, Hickman FH. 1948. The utilization of ribose and other pentoses by the cornea. Bulletin of Johns Hopkins Hospital 82:287-294.

Herrmann H, Moses SG. 1948. Studies on non-protein nitrogen in the cornea. Bulletin of Johns Hopkins Hospital 82:295-309.

Hertzberg R. 1959. Delayed mustard gas keratitis. A report of 5 cases treated by lamellar keratoplasty. Medical Journal of Australia 1:529-530.

Herve A. 1948. Future possibilities of dichloroethyl sulfide and radioactive isotopes in therapy. Revue Medicale de Liege 3:555-562.

Heston WE. 1949. Induction of pulmonary tumors in strain A mice with methyl-bis(b-chloroethlyl) amine hydrochloride. Journal of the National Cancer Institute 10:125-130.

Heston WE. 1950. Carcinogenic action of the mustards. Journal of the National Cancer Institute 11:415-423.

Heston WE. 1953. Occurrence of tumors in mice injected subcutaneously with sulfur mustard and nitrogen mustard. Journal of the National Cancer Institute 14:131-140.

Heston WE. 1953. Pulmonary tumors in Strain A mice exposed to mustard gas. Proceedings of the Society for Experimental Biology and Medicine 82:457-460.

Heston WE, Lorenz E, Deringer MK. 1953. Occurrence of pulmonary tumors in strain A mice following total body x-radiation and injection of nitrogen mustard. Cancer Research 13:573-577.

Heubner W. 1920. War gas diseases. Naturwissenchaften 8:247. [In German]

Heully F, Gruninger RM, Duroch F. 1956. Collective intoxication caused by the explosion of a mustard gas shell. Annales de Medicine Legale, Criminologie, Police Scientifique et Toxicologie 36:195-204. [In French]

Heusghem C, Charlier R. 1949. Action of dichloroethyl sulfide and chlorethylamine on formed elements of blood and on renal excretion of adrenal 17-ketosteroids in dogs. Revue Belge de Pathologie et de Medecine Experimentale 19:339-345. [In French]

Heyndrickx A, Heyndrickx B. 1984. Treatment of Iranian soldiers attacked by chemical and microbiological war gases. Archives Belges (Supplement):157-159.

Heyndrickx A, Heyndrickx B. 1990. Management of war gas injuries (letter). Lancet 336:1248-1249.

Heyndrickx A, Heyndrickx B. 1990. Treatment of Iranian soldiers attacked by chemical and microbiological war gases. Rivista di Tossicologia Sperimentale 19:3-6.

Heyndrickx A, Van den Heede M. 1986. The toxicological analysis of chemical warfare agents in samples originating from Iranian soldiers. In: Heyndricks B, ed. Terrorism: Analysis and Detection of Explosives. Proceedings of the Second World Congress on New Compounds in Biological and Chemical Warfare. Gent: Rijksuniversiteit. 598-619.

Heyndrickx A, Cordonnier J, De Bock A. 1984. Chromatographic procedures for the toxicological determination of bis (2-chloroethyl) sulfide (mustard gas, yperite) in environmental and human biological samples. Archives Belges (Supplement):102-109.

Heyndrickx A, De Puydt EH, Cordonnier J. 1984. Comparative study of two different field tests for the detection of ypefite in the atmosphere, applied on biological samples of gassed soldiers. Archives Bulges (Supplement):61-68.

Higginson J. 1975. Importance of environmental factors in cancer. In: Rosenfeld C, Davis W, eds. Environmental Pollution and Carcinogenic Risks. IARC Scientific Publication No. 13. Lyon: International Agency for Research on Cancer. 15-23.

Higuchi K, Kajiki A, Nakamura M, Harada S, Pula PJ, Scott AL, Dannenberg AM Jr. 1988.

Proteases released in organ culture by acute dermal inflammatory lesions produced *in vivo* in rabbit skin by sulfur mustard: hydrolysis of synthetic peptide substrates for trypsin-like and chymotrypsin-like enzymes. Inflammation 12:311-334.

Hill AB. 1971. Principles of Medical Statistics. 9th ed. New York: Oxford University Press.

Hill L. 1920. Toxicology of diphenylarsinchlorid. Journal of the Royal Army Medical Corps 35:334.

Hindmarsh JT, McCurdy RF. 1986. Clinical and environmental effects of arsenic toxicity. CRC Critical Reviews in Clinical Laboratory Sciences 23:315-347.

Hirade J, Ninomiya A. 1950. Studies on the mechanism of the toxic action of organic halogen compounds. Journal of Biochemistry 37:19-34.

Hirano T. 1980. Chemical carcinogenesis. Part I: biological, biochemical, and general observations. Yokohama Medical Bulletin 31:67-79.

Hirono M, Shun-ichi S, Toshihiro H, Minoru N, Takao H. 1984. Early gastric cancer with widespread lymph node metastases: a case report. Japanese Journal of Surgery 14:143-145.

Hirose F. 1953. Hematological observations on patients with chronic occupational war-gas poisoning, especially due to "mustard gas" and particularly on its relationship to aplastic anemia. Nippon Haematologica Japonica 16:209.

Hjorth N. 1953. Food poisoning from cod-roe contaminated by mustard gas: report with 5 case histories. Acta Medica Scandinavica 147:237-245.

Hobbs EB. 1944. Fatal case of mustard gas poisoning. British Medical Journal 2:306-307.

Hochmeister M, Vycudilik W. 1989. Morpho-toxicologic findings following war gas effect (S-Lost). Beitrage zur Gerichtlichen Medizin 47:533-538. [In German]

Hoegberg B. 1974. Steroid mustard gas derivative: viewpoints against a background of the working mechanism of hormones. Svensk Farmaceutisk Tidskrift 78:99-105. [In Swedish]

Holm M. L, Holmberg B. 1987. Exposures to carcinogens and consequences of listing of carcinogens in the Swedish working environment. Regulatory Toxicology and Pharmacology 7:185-199.

Holmberg B. 1975. Biological aspects of chemical and biological weapons. Ambio 4:211.

Holmberg RE, Ferm VH. 1969. Interrelationship of selenium, cadmium, and arsenic in mammalian teratogenesis. Archives of Environmental Health 18:873-877.

Hood RD. 1972. Effects of sodium arsenite on fetal development. Bulletin of Environmental Contamination and Toxicology 7:216-220.

Hood RD. 1985. Cacodylic Acid: Agricultural Uses, Biological Effects, and Environmental Fate. VA Monograph Series. Washington, DC: Veterans Administration.

Hood RD, Bishop SL. 1972. Teratogenic effects of sodium arsenate in mice. Archives of Environmental Health 24:62-65.

Hood RD, Vedel GC, Zaworotko MJ, Tatum FM. 1982. Distribution of arsenite and methylated metabolics in pregnant mice (abstract). Teratology 25:50A.

Hood RD, Harrison WP, Vedel GC. 1982. Evaluation of arsenic metabolites for prenatal effects in the hamster. Bulletin of Environmental Contamination and Toxicology 29:679-687.

Hoover R, Fraumeni JF Jr. 1981. Drug-induced cancer. Cancer 47(Supplement):1071-1080.

Horowitz MJ. 1986. Stress Response Syndromes. 2nd ed. Northvale, NJ: Jacob Aronson.

Horowitz NH, Houlahan MB, Hungate MY, Wright B. 1946. Mustard gas mutations in *Neurospora.* Science 104:233-234.

Hosokawa T. 1961. Studies on skin tumors: statistical observation on skin tumors. Hiroshima Medical Journal.

Hosseini ESA, Motamedi F, Semnanian S, MishMast NG. 1988. Study of correlation between age, height, weight, cigarette smoking, duration of injury, type of chemical warfare and lung volumes and capacities. In: Abstracts of the First International

Medical Congress on Chemical Warfare Agents in Iran. June 13-16, 1988. Mashhad University of Medical Sciences. Mashhad, Iran. No. 54.

Hosseini K, Moradi A, Mansouri A, Vessal K. 1989. Pulmonary manifestations of mustard gas injury: a review of 61 cases. Iranian Journal of Medical Sciences 14:20-26.

Howard JK. 1981. Occupationally related cancer. Practitioner 225:810-817.

Hozawa S. 1987. Studies on antibody producing mechanism and cytokines in former poison gas workers. Hiroshima Daigaku Igaku Zasshi 35:737-769. [In Japanese]

Hrusovsky J. 1975. Effect of radiomimetics on the composition of the proteins of dog blood serum. Veterinami Medicina (Prague) 20:701-704. [In Czech]

Hrusovsky J. 1979. Electrophoretic determination of lipoproteins in the blood of dogs after the administration of b,b-dichlorodiethylsulfide. Veterinarni Medicina (Prague) 24:745-749. [In Czech]

Hrusovsky J. 1979. Electrophoretic examination of changes in the protein spectrum of dogs after exposure to b,b-dichlorodiethylsulfide. Veterinami Medicina (Prague) 24:507-512. [In Czech]

Hrusovsky J. 1979. Glycoproteins in the blood of dogs after administration of b,b-dichlorodiethylsulphide. Veterinami Medicina (Prague) 24:699-703. [In Czech]

Huber J. 1930. Pneumonia from gas. Revue Critique de Pathologie et de Therapeutique 1:110-114. [In French]

Hueper WC. 1962. Symposium on chemical carcinogenesis. I. Environmental and occupational cancer hazards. Clinical Pharmacology and Therapeutics 3:776813.

Hueper WC. 1966. Mustard gas-yperite-lost-b,b dichlorodiethyl sulfide-bis(b-chlorethyl)sulfide. In: Recent Results in Cancer Research. Vol. 3. New York: Springer-Verlag. 103-105.

Hughes WF Jr. 1942. Mustard gas injuries to the eyes. Archives of Ophthalmology 27:582-601.

Hughes WF Jr. 1946. Treatment of Lewisite bums of the eye with BAL. Journal of Clinical Investigation Uuly):541-548.

Hughes WF Jr. 1948. The physiology, biochemistry, and cytopathology of the cornea in relation to injury by mustard gas and allied toxic agents. Appendix 1. The tolerance of rabbit cornea for various chemical substances. Bulletin of Johns Hopkins Hospital 82:338-349.

Hulbert HS. 1920. Gas neurosis syndrome. American Journal of Insanity 75:213-216.

Hunt. 1919. The late symptoms of gas poisoning and their treatment. Guys Hospital Gazette 33:200-205.

Huntress WT, Goodridge TH, Bratzel RP. 1963. Survey of sulfur mustards. Cancer Chemotherapy Reports 26:323-338.

Huntstein W, Rehn K. 1975. Tumor induction through cytostatic agents in man. Deutsche Medizinische Wochenschrift 100:155-158. [In German]

Hurd CD, Moe OA, Starke AC Jr. 1972. Observations regarding chloroethyl and chlorovinyl sulfides. International Journal of Sulfur Chemistry, Part A 2:113-119.

Husain I, van Houten B, Thomas DC, Sancar A. 1986. Sequences of E. *coli* uvrA gene and protein reveal two potential ATP binding sites. Journal of Biological Chemistry 261:4895-4901.

Ichinotsubo D, Mower HF, Setliff J, Mandel M. 1977. The use of recbacteria for testing of carcinogenic substances. Mutation Research 46:53-62.

Idelson H. 1923. Nervous disorders caused by toxic gases and their relationship to traumatic neuroses. Revue Neurologique 30:140-151. [In French]

Illig L. 1977. Management of psoriasis vulgaris using an external sulfur mustard compound with special reference to its possible carcinogenic hazard. Carcinogenesis of sulfur mustard compound in animal experiments and in man. Zeitschrift fur Hautkrankheiten 52:1035-1044. [In German]

Illig L. 1977. Treatment of psoriasis vulgaris with external mustard gas with special regard to a possible carcinogenesis risk. Zeitschrift fur Hautkrankheiten 52:973-987. [In German]

Illig L, Paul E, Eyer P, Weger N, Born W. 1979. Treatment of psoriasis vulgaris with external sulfur mustard gas with particular reference to its potential carcinogenic risk. III. Clinical and experimental studies on the extent of percutaneous and inhalational uptake of sulfur mustard gas. Zeitschrift fur Hautkrankheiten 54:941-951. [In German]

Inada S, Hiragun K, Yamura T. 1975. Skin lesions observed in former workers in a poison gas factory (abstract). Nishi Nippon Hifuka 37:868. [In Japanese]

Inada S, Hiragun K, Kanemitsu A, Kimura I, Ueno R, Yamura T. 1977. Clinico-pathological studies on skin tumors observed in former workers of a poison gas factory, with special reference to basal cell carcinoma (abstract). Nishi Nippon Hifuka 39:430-431. [In Japanese]

Inada S, Hiragun K, Ueno R, Kanemitsu A, Yamura T. 1977. Clinico-pathological studies on skin tumors observed in former workers of a poison gas factory, with special reference to Bowen's disease (abstract). Japanese Journal of Dermatology 87:160-161. [In Japanese]

Inada S, Hiragun K, Seo K, Yamura T. 1978. Multiple Bowen's disease observed in former workers of a poison gas factory in Japan with special reference to mustard gas exposure. Journal of Dermatology 5:49-60.

Inada S, Yamura T, Takiyama W, Kamitsuna A, Nishimoto Y. 1979. Basal cell carcinoma of the scrotum in a former employee of a Japanese mustard gas factory. Gan No Rinsho 25:67-70. [In Japanese]

Inai K, Kou E, Nambu S, Tokuoka S. 1987. An altered lectin binding to mucus glycoprotein in goblet cells of human tracheobronchial epithelium among former mustard gas workers. Acta Pathologica Japonica 37:537-548.

Infield GB. 1976. Disaster at Bari. London: New English Library.

Ingram A, Grasso P. 1985. Nuclear enlargement: an early change produced in mouse epidermis by carcinogenic chemicals applied topically in the presence of a promoter. Journal of Applied Toxicology 5:53-60.

Inns RH, Rice P. 1989. Comparison of rabbit tissues after dichloro(2-chlorvinyl)arsine poisoning followed by therapy with chelating agents (abstract). Human Toxicology 8:64-65.

Inns RH, Bright JE, Marrs TC. 1987. Comparison of arsenic levels in rabbit tissues following equitoxic doses of sodium arsenite and dichloro(2-chlorovinyl)arsine (abstract). Human Toxicology 6:434-435.

Inns RH, Bright JE, Marrs TC. 1988. Comparative acute systemic toxicity of sodium arsenite and dichloro(2-chlorovinyl) arsine in rabbits. Toxicology 51:213-222.

Inns RH, Rice P, Bright JE, Marrs TC. 1990. Evaluation of the efficacy of dimercapto chelating agents for the treatment of systematic organic arsenic poisoning in rabbits. Human and Experimental Toxicology 9:215-220.

Insight. 1988. Angola may be using chemical weapons. 4:7.

International Agency for Research on Cancer. 1973. IARC Monographs on the Evaluation of Carcinogenic Risk of Chemicals to Man. Vol. 2, Some Inorganic and Organometallic Compounds. Lyon: IARC.

International Agency for Research on Cancer. 1975. IARC Monographs on the Evaluation of Carcinogenic Risk of Chemicals to Man. Vol. 9, Some Aziridines, N-, S- & OMustards and Selenium. Lyon: IARC.

International Agency for Research on Cancer. 1980. IARC Monographs on the Evaluation of Carcinogenic Risk of Chemicals to Humans. Vol. 23, Some Metals and Metallic Compounds. Lyon: IARC.

International Society for Traumatic Stress Studies. 1991. Proceedings. The Reality of

Trauma in Everyday Life: Implications for Intervention and Policy. Washington, DC, October 24-27, 1991.

Ireland MM. 1926. Medical Aspects of Gas Warfare. Vol. XIV. The Medical Department of the United States Army in the First World War. Washington, DC: U.S. Government Printing Office.

Isobe T. 1957. The influence of poison gas on women. I. General examinations and clinical observations on women who worked at Okuno-jima (so-called Poison Gas Island). Journal of the Hiroshima Medical Association, Special Series 10:582-600. [In Japanese]

Iwaszkiewicz J. 1966. Burns of the upper respiratory tract caused by mustard gas. Polish Medical Journal 5:706-709. [English translation of the Otolaryngologia Polska article]

Iwaszkiewicz J. 1966. Burns of the upper respiratory tract due to mustard gas. Otolaryngologia Polska 20:237-241. [In Polish]

Jabbar MA, Pourismaili F, Riazi GH, Nouri DM. 1989. Detection of the mutagens in urine of civilians exposed to mustard gas (abstract). Environmental and Molecular Mutagenesis 14:236.

Jackson KE. 1936. The history of mustard gas. Journal of the Tennessee Academy of Sciences 11:98-106.

Jackson R, Adams RH. 1973. Horrifying basal cell carcinoma: a study of 33 cases and a comparison with 435 non-horror cases and a report on four metastatic cases. Journal of Surgical Oncology 5:431-463.

Jacobson-Kram D, Montalba D. 1985. The Reproductive Effects Assessment Group's report on the mutagenicity of inorganic arsenic. Environmental Mutagenesis 7:787-804.

Jacqeau MM, Bujadoux L. 1925. Acute and spontaneous infection in the eyes after a two year interval in a victim of previous yperite exposure. Bull Soc Franc Ophtal 38:370-380. [In French]

Jacquelin A, Konechowsky F. 1930. Evolution of respiratory sequelae from war gas. Paris Medical 75:159-166. [In French]

Jacques J. 1991. First victims of yperite. New Journal of Chemistry 15:3-4. [In French]

Jacquot G. 1965. Possibilities of chemical warfare agents. L'Armee 49:1-15. FSTC translation no. FSTC-381-T65-391.

Jain RK, Wei J. 1977. Dynamics of drug transport in solid tumors:distributed parameter model . Journal of Bioengineering 1:313-330.

James LF, Lazor VA, Binns W. 1966. Effects of sublethal doses of certain minerals on pregnant ewes and fetal development. American Journal of Veterinary Research 27:132-135.

Jampol LM, Axelrod A, Tessler H. 1976. Pathways of the eye's response to topical nitrogen mustard. Investigative Ophthalmology 15:486-489.

Janouchek B, Horakova M, Bartak P. 1987. Treatment of psoriasis by dichlordiethylsulfide. Phlebologie 40:171-175. [In French]

Jarman GN. 1959. Chemical Corps experience in the manufacture of Lewisite in metal-organic compounds. In: Advances in Chemistry. Vol. 23. Washington, DC: American Chemical Society.

Jenkins RB. 1966. Inorganic arsenic and the nervous system. Brain 89:479-498.

Jensen BV, Carlsen NLT, Nissen NI. 1990. Influence of age and durationof follow-up on lung function after combined chemotherapy for Hodgkin's disease . European Respiratory Journal 3:1140-1145.

Jensen KA, Kirk I, Westergaard M. 1950. Mutagenic activity of some mustard gas compounds. Nature 166:1020-1021.

Jensen KA, Kirk I, Koelmark G, Westergaard M. 1952. Chemically induced mutations in *Neurospora.* Cold Spring Harbor Symposia on Quantitative Biology 16:245-261.

Jewett S. 1942. Neuropsychiatric chemical warfare casualties. Bulletin of the New York Medical College (June-October):107-113.

Johnson JB, Lusco CT, Gross CL, Graham LM, Meier HL. 1987. Inhibitors of poly (ADP-ribose) polymerase prevent cellular toxicity by the alkylating agent 2,2'-dichlorodiethyl sulfide (sulfur mustard, HD). FASEB Journal.

Johnstone RM. 1963. Sulfhydryl agents: arsenicals. In: Hochster RM, Quastel JH, eds. Metabolic Inhibitors. Vol. II. New York: Academic Press. 99-118.

Jones JS, Weber S, Prakash L. 1988. The *saccharomyces cerevisiae* RAD18 gene involves a protein that contains potential zinc finger domains for nucleic acid binding and a putative nucleotide binding sequence. Nucleic Acids Research 16:7119-7131.

Jones RN, Weill H, Ziskind M. 1975. Pulmonary function in sandblasters' silicosis. Bulletin de Physio-Pathologie Respiratoire 11:589-595.

Jones RN, Hughes JM, Glindmeyer H, Weill H. 1986. Lung function after acute chlorine exposure. American Review of Respiratory Disease 134:1190-1195.

Jones TD, Easterly CE. 1991. On the rodent bioassay currently being conducted on 44 chemicals: a RASH analysis to predict test results from the National Toxicology Program. Mutagenesis 6:507-514.

Jones TD, Walsh PJ, Zeighami EA. 1985. Permissible concentrations of chemicals in air and water derived RTECS entries: a "RASH" chemical scoring system. Toxicology and Industrial Health 1:213-234.

Jones TD, Walsh PJ, Watson AP, Owen BA, Barthouse LW, Sanders DW. 1988. Chemical scoring by a Rapid Screening Hazard (RASH) method. Risk Analysis 8:99-118.

Jorgensen B, Olesen B, Berntsen O. 1985. Accidents with mustard gas near Bornholm. Ugeskrift for Laeger 147:2251-2254. [In Danish]

Joshi S, Hughes JB. 1981. Inhibition of coupling factor B activity by cadmium ion, arsenite-2, 3-dimercaptopropanol, and phenylarsine oxide, and preferential reaction by dithiols. Journal of Biological Chemistry 256:11112-11116.

Jostes RF, Rausch RJ, Miller BM, Sasser LB, Dacre JC. 1989. Geneotoxicity of Lewisite in Chinese hamster ovary cells (abstract). Toxicologist 232.

Jung F. 1939. Arsine and blood. Naturwissenschaften 27:317. [In German]

Jung F. 1939. Solubility of arsine in serum and its reactivity. Biochemische Zeitschrift 302:294-309. [In German]

Kabelik J. 1932. Chemotherapeutic experiments on Ehrlich's mousecancer . Comptes Rendus des Seances de la Societe de Biologie et de Filiales 110:394-397. [In French]

Kadivar H, Adams SC. 1991. Treatment of chemical and biological warfare injuries: insights derived from the 1984 Iraqi attack on Majnoon Island. Military Medicine 156:171-177.

Kajiki A, Higuchi K, Nakamura M, Liu LH, Pula PJ, Dannenberg AM. 1988. Sources of extracellular lysosomal enzymes released in organ-culture by developing and healing inflammatory lesions. Journal of Leukocyte Biology 43:104-116.

Kaldor JM, Day NE, Band P, Choi NW, Clark EA, Coleman MP, Hakama M, Koch M, Langmark F, Neal FE, Pettersson F, Pompe-Kir V, Prior P, Storm HH. 1987. Second malignancies following testicular cancer, ovarian cancer, and Hodgkin's disease: an international collaborative study among cancer registries. International Journal of Cancer 39:571-585.

Kaldor JM, Day NE, Hemminki K. 1988. Quantifying the carcinogenicity of antineoplastic drugs. European Journal of Cancer and Clinical Oncology 24:703-711.

Kaldor JM, Day NE, Pettersson F, Clarke EA, Pedersen D, Mehnert W, Bell J, Host H, Prior P, Karjalainen S, Neal F, Koch M, Band P, Choi W, Kirn VP, Arslan A, Zaren B, Belch AR, Storm H, Kittelman B, Fraser P, Stovall M. 1990. Leukemia following chemotherapy for ovarian cancer. New England Journal of Medicine 322:1-6.

Kaldor JM, Day NE, Clarke EA, Van Leeuwen FE, Henry-Amar MH, Fiorentino MV, Bell J, Pedersen D, Band P, Assouline D, Koch M, Choi W, Prior P, Blair V, Langmark F, Kir VP, Neal F, Peters D, Pfeiffer R, Karjalainen S, Cuzick J, Sutcliffe SB, Somers R,

Pellae-Cosset B, Pappagallo GL, Fraser P, Storm H, Stovall M. 1990. Leukemia following Hodgkin's disease. New England Journal of Medicine 322:7-13.

Kambe S. 1963. Case of carcinoma of the urinary bladder developing in a male occupationally exposed to mustard gas and Lewisite for many years. Gencho Hiroshima Igaku 11:317-321. [In Japanese]

Kantor HI, Levin PM. 1948. Arsenical encephalopathy in pregnancy with recovery. American Journal of Obstetrics and Gynecology 56:370-374.

Kaplan W. 1953. Factors influencing the frequency of mustard gas induced dominant lethals in *D. melanogaster.* Proceedings of the Ninth International Congress of Genetics 9 (Part 2):693-694.

Kardiner A. 1941. The Traumatic Neuroses of War. New York: Hoeber.

Karnofsky DA, Graef J. 1948. Studies on the mechanism of action of nitrogen and sulfur mustards *in vivo.* American Journal of Pathology 24:275.

Kaspar L, Donner A, Balogh A, Kosak D, Slany J. 1985. Intensive care of chemical warfare intoxication. Intensivmedizin 22:243-247. [In German]

Katz J. 1992. The consent principle of the Nuremberg Code: its significance then and now. In: Annas GJ, Grodin MA, eds. The Nazi Doctors and the Nuremberg Code: Human Rights in Human Experimentation. New York: Oxford University Press. 227-239.

Keane WM, Atkins JP Jr, Wetmore R, Vidas M. 1981. Epidemiology of head and neck cancer. Laryngoscope 91:2037-2045.

Keeser E, Oelkers HA, Vincke E. 1936. The prophylaxis and therapy of mustard gas injury of the skin. Archiv fur Experimentale Pathologie und Pharmakologie 180:557-567. [In German]

Kehoe RA, Kitzmiller KV. 1942. Pulmonary irritants. Cincinnati Journal of Medicine 23:423-443.

Keller W. 1942. The sensitivity of the human skin with respect to vesicants of the yellow cross group. Dermatologica 85:1-26. [In German]

Kelly WE, ed. 1985. Post-Traumatic Stress Disorder and the War Veteran Patient. New York: Brunner/Mazel.

Kennaway E. 1942. A contribution of the mythology of cancer research. Lancet 2:769-772.

Kennedy SM. 1992. Acquired airway hyperresponsiveness from nonimmunogenic irritant exposure. Occupational Medicine: State of the Art Reviews 7:287-300.

Kenyon KR, Tseng SC. 1989. Limbal autograft transplantation for ocular surface disorders. Ophthalmology 96:709.

Kershner WE. 1918. Secondary optic atrophy due to gassing. American Journal of Ophthalmology 1:168.

Keshavarz-Dehnow A, Nowroozi M, Noor-Mohamadi I. 1988. Ascorbic acid changes in war injured patients. In: Abstracts of the First InternationalMedical Congress on Chemical Warfare Agents in Iran . June 13-16, 1988. Mashhad University of Medical Sciences. Mashhad, Iran. No. 50.

Keshmiri M. 1989. Pulmonary cause of death from chemical warfare agents: the Halabche experience. Iranian Journal of Medical Sciences 14:10-19.

Khakshoor D. 1988. Ocular injuries of sulfur mustard gas. In: Abstracts of the First International Medical Congress on Chemical Warfare Agents in Iran. June 13-16, 1988. Mashhad University of Medical Sciences. Mashhad, Iran. No. 36.

Khaykin MG, Beletskiy MP. 1933. Dichloroethylsulfide poisoning in civil life. Vrachebnoe Delo 16:865-868. [In Russian]

Kimina SN, Lifyland LM. 1985. Evaluation of the carcinogenic action of chemical substances in humans. Gigiena i Sanitariya 2:57-60. [In Russian]

Kindred JE. 1947. Histologic changes occurring in the hemopoietic organs of albino rats after single injections of 2-chloroethyl vesicants: a quantitative study. Archives of Pathology 43:253-295.

Kindred JE. 1949. The blood cells and the hemopoietic and other organs of dogs given intravenous injections of 2-chloroethyl vesicants. Archives of Pathology 47:378-398.

King JR, Monteiro-Riviere NA. 1990. Cutaneous toxicity of 2-chloroethyl methyl sulfide in isolated perfused porcine skin. Toxicology and Applied Pharmacology 104:167-179.

King JR, Monteiro-Riviere NA. 1991. Dose-response study on the toxicity of the vesicant arsenical, Lewisite, in the isolated perfused porcine skin flap (abstract). Toxicologist. 283.

Kinsey VE, Grant WM. 1946. Determination of the rate of disappearance of mustard gas and mustard intermediates in the corneal tissue. Journal of Clinical Investigation 25:776-779.

Kinsey VE, Grant WM. 1946. Reaction of mustard gas with proteins: biological assay of amino acids affected. Archives of Biochemistry 10:311-320.

Kinsey VE, Grant WM. 1946. Reaction of mustard gas with proteins: nutritional value of casein reacted with mustard gas. Archives of Biochemistry 10:303-309.

Kinsey VE, Grant WM. 1947. Action of mustard gas and other poisons on yeast cells. Correlation between quantity of glutathione bound by mustard and divinyl sulfone and their effect on growth rate. Journal of Cellular and Comparative Physiology 29:289-299.

Kinsey VE, Grant WM. 1947. Action of mustard gas and other poisons on yeast cells. I. Effect of mustard gas on the rate of cell division. Journal of Cellular and Comparative Physiology 29:51-64.

Kinsey VE, Grant WM. 1947. Action of mustard gas and other poisons on yeast cells. II. Effect of mustard gas on the mortality, morphology, carbohydrate metabolism and permeability. Journal of Cellular and Comparative Physiology 29:65-74.

Kinsey VE, Grant WM. 1947. Action of mustard gas and other poisons on yeast cells. III. Distribution of fixed mustard gas in yeast. Journal of Cellular and Comparative Physiology 29:75-84.

Kinsey VE, Grant WM. 1947. Action of mustard gas and other poisons on yeast cells. IV. Study of the relationship between inhibition of carbohydrate metabolism and inhibition of growth by various poisons, and effects of other toxic agents on yeast. Journal of Cellular and Comparative Physiology 29:31-42.

Kinzie JD. 1989. Post-traumatic stress disorder. In: Kaplan HI, Saddock BJ, eds. Comprehensive Textbook of Psychiatry. 5th ed. Baltimore: Williams and Wilkins. 1000-1008.

Kircher M, Brendel M. 1983. DNA alkylation by mustard gas in yeast strains of different repair capacity. Chemico-Biological Interactions 44:27-40.

Kircher M, Fleer R, Ruhland A, Brendel M. 1979. Biological and chemical effects of mustard gas in yeast. Mutation Research 63:273-289.

Kitamura K. 1954. Tumors of skin. J Dermatol and Venereal Dis 64:303-312.

Klain GJ, Schuschereba ST, McKinney LM, Omaye ST. 1988. Ocular toxicity of systemic exposure to butyl 2-chloroethyl sulfide (abstract). Toxicologist 8:130.

Klarenbeek A. 1936. Experiments relating to injury of the skin by mustard gas. Nederlands Tijdschrift voor Geneeskunde 16:2169-2171. [In Dutch]

Klarenbeek A. 1937. Skin protection against mustard gas with cellulose preparations. Nederlands Tijdschrift voor Geneeskunde 81:5905-5906. [In Dutch]

Kleber BE, Birdsell D. 1959. The Chemical Warfare Service: Chemicals in Combat. United States Army in World War II: The Technical Services. Washington, DC: Office of the Chief of Military History, Department of the Army.

Klehr NW. 1984. Late manifestations in former mustard gas workers, with special consideration of the cutaneous findings. Zeitschrift fur Hautkrankheiten 59:1161-1170. [In German]

Klein EM, Bito LZ. 1983. Species variations in the pathophysiologic responses of vertebrate eyes to a chemical irritant, nitrogen mustard. Investigative Ophthalmology & Visual Science 24:184-191.

Klimmek R, Szinicz L, Weger N. 1983. [Chemical Poisons and Blister Gases: Effect and Therapy] Chemische Gifte und Kampfstoffe: Wirkung und Therapie. Stuttgart: Hippokrates. [In German]

Kling A, Lecordier G. 1936. Effect of two war vesicants and their hydrolysis products on the interfacial tensions between lipoids and physiological saline as well as on their hydrophilic qualities. Comptes Rendus Hebdomadaires des Seances de l'Academie des Sciences 203:1544-1546. [In French]

Kling A, de Fonbrune P, Raynal F. 1939. The mechanism by which mustard gas acts on living cells. Comptes Rendus Hebdomadaires des Seances de l'Academie des Sciences 208:1679-1681. [In French]

Knowles FC, Benson AA. 1983. The biochemistry of arsenic. Trends in Biomedical Sciences 8:178.

Kobayashi K, Katsuta S, Miyanishi M. 1953. Investigation on the cause of death of the workers at Okuno-jima Island (so called poison gas island). Journal of the Hiroshima Medical Association, Special Series 6:222-226. [In Japanese]

Kobayashi K, Miyanishi M, Kunihara K. 1962. Investigations of sequels observed among the workers in the so called poison gas factory at Okuno Island. Report I. Clinical observations. Journal of the Hiroshima Medical Association, Special Series 6:218-221. [In Japanese]

Koch W. 1921. War diseases produced directly by the use of chemical agents: gas poisoning. Handbuch der Arztlichen Erfahrungen im Weltkriege. Vol. 8. Leipzig: Barth. 526-536.

Kohn KW, Steibigel NH, Spears CL. 1965. Cross-linking and repair of DNA in sensitive and resistant strains of *E. coli* treated with nitrogen mustard. Proceedings of the National Academy of Sciences (USA) 53:1154-1160.

Koletsky AJ, Bertino JR, Farber LR, Prosnitz LR, Kapp DS, Fischer D, Portlock CS. 1986. Second neoplasms in patients with Hodgkin's disease following combined modality therapy: the Yale experience. Journal of Clinical Oncology 4:311-317.

Koller P. 1958. Comparative effects of alkylating agents on cellular morphology. Annals of the New York Academy of Sciences 68:783801.

Kondrashov V. 1978. Relative hazards of poisoning with fumes and gases of toxic substances with their dermal and inhalation routes of action. Gigiena Truda I Professionalnye Zabolevaniia 2:34-38. [In Russian]

Konetzke GW. 1988. The evaluation of carcinogenic, mutagenic, embryotoxic, and teratogenic substances. In: Schmidt P, ed. [Biological Monitoring in Occupational Medicine] Biologische Kontrollmethoden in der Arbeitsmedizin. Berlin: Veb Verlag Volk und Gesundheit. 91-110. [In German]

Kooij R. 1940. Extensive contamination with mustard gas, case. Nederlands Tijdschrift voor Geneeskunde 84:947-952. [In Dutch]

Koontz AR. 1924. Pathology of phosgene and mustard poisoning. Military Surgeon 54:663-672.

Koontz AR. 1925. When do lungs return to normal following exposureto war gases? Archives of Internal Medicine 36:204-219.

Koontz AR. 1927. War gases and tuberculosis. An experimental study. Archives of Internal medicine 39:833-864.

Koontz AR. 1929. Mustard gas and tuberculosis. Archives of Internal Medicine 43:90-95.

Koontz AR. 1933. Post-war developments in medical aspects of chemical warfare. Military Surgeon 72:277-287.

Korzan VA, Khomchenovskii El, Snyakina IP, Zaretskii II, eds. 1974. Antileukotic activity of amino acid derivatives of bis(2-chloroethyl) sulfide. In: [Morphology, Biochemistry and Clinical Features of Leukemia]. Riga: Zinatne. 34-40.

Kou E. 1986. Pathohistological studies on goblet cell metaplasia in the human bronchial epithelium. Hiroshima Daigaku Igaku Zasshi 33:1105-1128. [In Japanese]

Koupilova M. 1975. Effects of sublethal doses of mustard gas on RES function. Vojenske Zdravotnicke Listy 44:77-81. [In Czech]

Kovacic P, Crawford P, Ryan MD, Nelson VC. 1986. Charge transfer mechanism for carcinogenesis by alkylating and other agents. Bioelectrochemistry and Bioenergetics 15:305-316.

Kramer SP. 1922. Behavior of b,b-dichloroethylsulfide. Kolloid Zeitschrift 31:150-151. [In German]

Kratochvil V, Martinek J. 1969. New method for detecting yperite. Chemicke Zvesti 23:382-390. [In German]

Krause H. 1992. Comment on the short report by H. Krause and E.I. Grussendorf: Syntopy of Bowen's disease and mustard gas (letter). Hautarzt 43:54. [In German]

Krause H, Grussendorf El. 1978. Syntopy of Bowen's disease and "Lost"-induced scar. Hautarzt 29:490-493. [In German]

Krause H, Grussendorf El. 1979. Therapy of morbus Bowen with 5-fluoro-uracil ointment. Dermatosen in Beruf und Umwelt 27:176-178. [In German]

Kravitz P, McDonald CJ. 1978. Topical nitrogen mustard-induced carcinogenesis. Acta Dermato-Venereologica 58:421-425.

Kreyberg L, Hanssen GE. 1949. Vascular responses of mouse skin to mustard gas. Acta Pathologica, Microbiologica, et Immunologica Scandinavica 26:809-820.

Kriek E. 1978. Mechanisms of metabolic activation of carcinogens. Excerpta Medica International Cancer Congress Series No. 420. 51-55.

Krikorian JG, Burke JS, Rosenberg SA, Kaplan HS. 1979. Occurrence of non-Hodgkin's lymphoma after therapy for Hodgkin's disease. New England Journal of Medicine 300:452-458.

Krumbhaar EB. 1919. Bone marrow changes in mustard gas poisoning. Journal of the American Medical Association 73:715.

Krumbhaar EB. 1919. Role of the blood and the bone marrow in certain forms of gas poisoning. Journal of the American Medical Association 72:39-41.

Krumbhaar EB, Krumbhaar HD. 1919. The blood and bone marrow in yellow cross gas (mustard gas) poisoning. Journal of Medical Research 40:497-506.

Krumbhaar EB, Krumbhaar HD. 1987. The blood and bone marrow in yellow cross gas (mustard gas) poisoning. Changes produced in the bone marrow of fatal cases. Cancer Journal 1:331-334. [Reprinted from Journal of Medical Research, 1919]

Krystal H. 1968. Massive Psychic Trauma. New York: International Universities Press.

Ku WW. 1987. Bis(b-chloroethyl) Sulfide (BCES)-Induced Changes in Proliferation and Differentiation Using Primary Monolayer Cultures of Cutaneous Keratinocytes from the Newborn Rat (dissertation). University of Michigan.

Ku WW, Bernstein IA. 1988. Bis-(b-chloroethyl)sulfide (BCES)-induced changes in epidermal cell homeostasis *in vitro*. Toxicology and Applied Pharmacology 95:397-411.

Kubatko E. 1971. Trials of treatment of the optic nerve atrophy by nitrogranulogen. Klinika Oczna 41:215-219. [In Polish]

Kucharik J, Telbisz A. 1939. The role of the oxidation products in the mechanism of the effect of mustard gas. Orvosi Hetilap 83:734-738. [In Hungarian]

Kucharik J, Telbisz A. 1945. Role of adrenals in pathology and therapy of dichlorethyl sulfide poisoning. Schweizerische Medizinische Wochenschrift 75:996-997. [In German]

Kucharik J, Telbisz A. 1947. Role of glutathione in poisoning with dichloroethyl sulfide. Archives Internationales de Pharmacodynamie et de Therapie 75:157-161.

Kulka RA, Schlenger WE, Fairbank JA, Hough RL, Jordan BK, Marmar CR, Weiss DS. 1990. Trauma and the Vietnam War Generation. New York: Brunner Mazel.

Kulling P. 1992. New antidotes for poisoning and mustard gas exposure are being introduced. Lakartidningen 89:548. [In Swedish]

Kumar PVN, Tabei SZ, Sotoodeh M. 1988. Aplastic anemia in chemical war injury patients. In: Abstracts of the First International Medical Congress on Chemical Warfare Agents in Iran. June 13-16, 1988. Mashhad University of Medical Sciences. Mashhad, Iran. No. 11.

Kurozumi S, Harada Y, Sugimoto Y, Sasaki H. 1977. Airway malignancy in poisonous gas workers. Journal of Laryngology and Otology 91:217-225.

Kuznetzovsky NY. 1930. Morphologic examination of vascular changes from Lewisite (dichlorvinyl arsine). Voy Med J 1:1930.

Kyle RA, Pierce RV, Bayrd ED. 1970. Multiple myeloma and acute myelomonocytic leukemia. New England Journal of Medicine 283:1121.

Lacotte B, Ceuterick M, Geerts ML, Willems J. 1989. Cutaneous burns due to mustard gas (yperite): two recent cases. Journal de Chirurgie 126:315-318. [In French]

Lainee P, Robineau P, Douget D, Zelter M. 1991. Pulmonary edema after endotracheal administration of sulfur mustard in dogs: improvement by mechanical ventilation (abstract). American Review of Respiratory Disease 143:A570.

Lammintausta K, Maibach HI. 1988. Exogenous and endogenous factors in skin irritation. International Journal of Dermatology 27:213-222.

Lancet. 1919. Mustard gas: its brief but inglorious career. 1:471-472.

Landis HR. 1923. Late effects of gassing. Progressive Medicine. 3:194.

Landrigan PJ. 1981. Arsenic: state of the art. American Journal of Industrial Medicine 2:5-14.

Lang. 1937. On the effect of Lost vesicant on protein. Deutsch Militararzt 2:480. [In German]

Lanks KW, Turnbull JD, Aloyo VJ, Dorwin J, Papirmeister B. 1975. Sulfur mustards induce neurite extension and acetylcholinesterase synthesis in cultured neuroblastoma cells. Experimental Cell Research 93:355-362.

Lannois, Sargnon. 1919. Stricture of the larynx and trachea following yperite exposure. Revue de Laryngologie, Otologie, Rhinologie 40:409-415. [In French]

Lawley PD. 1966. Mechanism of action of alkylating agents: comparisons with other cytotoxic, mutagenic, and carcinogenic agents. Colloquim-Gesellschaft fur Physiologische Chemie 17:126-141.

Lawley PD. 1976. Carcinogenesis by alkylating agents. In: Searle CE, ed. Chemical Carcinogens. American Chemical Series Monograph 173. Washington, DC: American Chemical Society. 83-244.

Lawley PD, Brookes P. 1965. Molecular mechanism of the cytotoxic action of difunctional alkylating agents and of resistance to this action. Nature 206:480-483.

Lawley PD, Brookes P. 1967. Interstrand cross-linking of DNA by difunctional alkylating agents. Journal of Molecular Biology 25:143-160.

Lawley PD, Brookes P. 1968. Cytotoxicity of alkylating agents towards sensitive and resistant strains of *Escherichia coli* in relation to extent and mode of alkylation of cellular macromolecules and repair of alkylation lesions in deoxyribonucleic acids. Biochemical Journal 109:433-447.

Lawley PD, Lethbridge JH, Edwards PA, Shooter KV. 1967. Inactivation of T7 and fX-174 bacteriophages by mono- and difunctional sulfur mustards. Quantitative correlations with phage deoxyribonucleic acid alkylation (abstract). Biochemical Journal 105:52P.

Lawley PD, Lethbridge JH, Edwards PA, Shooter KV. 1969. Inactivation of bacteriophage T7 by mono and difunctional sulphur mustards in relation to cross linking and depurination of bacteriophage DNA. Journal of Molecular Biology 39:181-198.

Lawson WE, Reid EE. 1925. Reactions of mustard gas with amino compounds. Journal of the American Chemical Society 47:2821-2836.

Lazenby JM. 1918. The treatment of irritant gas poisoning. British Medical Journal 2:342.

LeBeau MM, Albain KS, Larson RA, Vardiman JW, Davis EM, Blough RR, Golomb HM, Rowley JD. 1986. Clinical and cytogenetic correlations in 63 patients with therapy-related myelodysplastic syndromes and acute nonlymphocytic leukemia: further evidence for characteristic abnormalities of chromosomes no. 5 and 7. Journal of Clinical Oncology 4:325-345.

Lebedev DV. 1948. Mustard gas and mutations. Priroda (Moscow) 37:66-67. [In Russian]

Lecaplain J. 1919. Study of the lung affected by yperite. Normandie Medicale 31:286. [In French]

Leclerc J, Boez L. 1919. Yperite intoxication: early diagnosis and prognosis. Annales d'Hygiene Publique et de Medecine Legale 32:193-206. [In French]

Leduc P. 1919. Acute catarrhal laryngitis from toxic gases (dichloroethylsulfide and chlorodiphenylarsine). Revue de Laryngologie, Otologie, Rhinologie 11:73-78. [In French]

Lee AM, Fraumeni JF. 1969. Arsenic and respiratory cancer in man: an occupational study. Journal of the National Cancer Institute 42:1045-1052.

Lee TC, Tzeng YJ, Chang WJ, Lin YC, Jan KY. 1986. Posttreatments with sodium arsenite during G_2 enhance the frequency of chromosomal aberrations induced by S-dependent clastogens. Mutation Research 163:263-269.

Lee TC, Lee KC, Tzeng YJ, Huang RY, Jan KY. 1986. Sodium arsenite potentiates the clastogenicity and mutagenicity of DNA cross-linking agents. Environmental Mutagenesis 9:119-128.

Lee TC, Tanaka N, Lamb PW, Gilmer TM, Barrett JC. 1988. Induction of gene amplification by arsenic. Science 241:79-81.

Lee WR. 1975. Comparison of the mutagenic effects of chemicals and ionizing radiation using *Drosophila melanogaster* test systems. In: Nygaard OF, Adler HI, Sinclair WK, eds. Radiation Research: Biomedical, Chemical, and Physical Perspectives. New York: Academic Press. 976-983.

Lee WR, Abrahamson S, Valencia R, Von Halle ES, Wurgler FE, Zimmering S. 1983. The sex linked recessive lethal test for mutagenesis in *Drosophila melanogaster:* a report of the U.S. Environmental Protection Agency Gene-Tox Program. Mutation Research 123:183-279.

Legler F. 1978. Genesis of cancer through noxious substances in the environment, drugs, and food habits. Offentliche Gesundheitswesen 40:653-662. [In German]

Leipner N, Dewes W, Schuller H, Helm U. 1987. Changes in the lung parenchyma following mustard gas poisoning (dichlorodiethyl sulphide). Roefo, Fortschritte auf dem Gebiet der Roentgenstrahlen und der Nuklearmedizin 147:152-155. [In German]

Lemen R. 1986. Occupationally induced lung cancer epidemiology. In: Merchant JA, ed. Occupational Respiratory Diseases. NIOSH Publication No. 86-102. Washington, DC: U.S. Department of Health and Human Services. 629-656.

Lemesch C, Naim L, Donagi A. 1983. Occupational cancer in Israel. Israel Journal of Medical Sciences 19:655-56.

Lemoine JM, Fauvet J. 1953. On the subject of bronchial lesions from war gas. Journal Francais de Medecine et Chirurgie Thoraciques 7:388-392. [In French]

Leonard A, Lauwerys RR. 1980. Carcinogenicity, teratogenicity and mutagenicity of arsenic. Mutation Research 75:49-62.

Leopold IH, Lieberman TW. 1971. Chemical injuries of the cornea. Federation Proceedings 30:92-95.

LeQuesne PM, McLeod JG. 1977. Peripheral neuropathy following a single exposure to arsenic. Journal of Neurological Sciences 32:437-451.

Lereboullet, Lelong. 1925-1927. Clinical and therapeutic study of the major respiratory sequelae of war gas intoxications. Bruxelles Medical 7:476-477. [In French]

Leroux-Robert J, Worms G. 1935. Major esophageal dilations secondary to intoxications by war gases. Pr Med 42:337. [In French]

Levie LH. 1940. The effect of mustard gas on rats. Acta Brevia Neerlandica de Physiologia, Pharmacologia, Microbiologia E.A. 10:199-202. [In Dutch]

Levy M. 1965. The effect of mustard gases upon a property of nucleated erythrocytes and their ghosts. Cancer Research 25:752-759.

Lewin L. 1962. [Poisons and Poisonings] Gifte und Vergiftungen. 5th ed. Ulm: K.F. Haug Verlag. [In German]

Lewis WL, Perkins GA. 1923. The b-chlorovinylchloroarsines. Industrial and Engineering Chemistry 15:290.

Lewis WL, Stiegler HW. 1925. The b-chlorovinyl chloroarsines and their derivatives. Journal of the American Chemical Society 47:2546-2556.

Li F-X, Feng X-W, Cheng J-P. 1989. The comparison of sensitivity of bone marrow CFU-GM from human and various species to sulfur mustard. Chinese Journal of Pharmacology and Toxicology 3:149-153.

Lieske CN, Klopcic RS, Gross CL, Clark JH, Dolzine TW, Logan TP, Meyer HG. 1992. Development of an antibody that binds sulfur mustard. Immunology Letters 31:117-122.

Ligezinski A. 1974. Cytologic examination of nasal mucosa in rats in early stages of poisoning with mustard gas with reference to histologic findings. Otolaryngolgia 28:501. [In Polish]

Lillie RS, Clowes GHA, Chambers R. 1919. Action of mustard on the cells of marine organisms. Science 49:383-385.

Lillie RS, Clowes GHA, Chambers R. 1919-1920. On the penetration of dichloroethylsulphide (mustard gas) into marine organisms and the mechanism of its destructive action on protoplasm. Journal of Pharmacology and Experimental Therapeutics 14:75-120.

Lindenfelser R. 1970. Cancer of the urinary bladder as a late effect of exposure to chemical war agents. Zeitschrift fur Urologie und Nephrologie 63:175-177. [In German]

Lippman M, ed. 1992. Environmental Toxicants, Human Exposures and Their Health Effects. New York: Van Nostrand Reinhold.

Lisitsin MS, Zvirbul OD. 1931. Experimental study of wounds of joints and extremities in relation to dichlorethyl sulfide. Voy Med J 2:516-517.

List AF, Doll DC, Greco FA. 1985. Lung cancer in Hodgkin's disease: association with previous radiotherapy. Journal of Clinical Oncology 3:215-221.

Lister W, McPherson WG. 1922. History of the Great War Based on Official Documents. Vol. 1, Medical Services. London: His Majesty's Stationery Office.

Litz BT, Keane TM, Fisher L, Marx B, Monaco V. 1992. Physical health complaints in combat-related post-traumatic stress disorder: a preliminary report. Journal of Traumatic Stress 5:131-140.

Livingstone PC, Walker HM. 1940. Effects of liquid mustard gas upon eyes of rabbits and of certain methods of treatment. British Journal of Ophthalmology 24:67-97.

Lobanok EV, Pan'Shina EF, Fomichev YK 1979. Morphological changes in *Escherichia coli* cells damaged with nitrous yperite. Mikrobiologiya 48:307-310.

Lobbecke EA. 1967. Induction of somatic mutations in *Ephestia* by injections of mustard gas solutions. Molecular and General Genetics 99:115-124. [In German]

Lobbecke EA. 1967. Reaction of mustard gas and DNA. Molecular mechanisms and reaction of somatic cells. Proceedings of the 5th International Congress of Chemotherapy 2:167-170. [In German]

Locey BJ. 1988. Impact of bis(b-chloroethyl) Sulfide on Keratin Protein and Intermediate Filaments in Cultured Keratinocytes as Indicated by Monoclonal Antibody Binding (dissertation). University of Michigan.

Loeper M. 1919. Pulmonary sclerosis following yperite intoxication. Progres Medical 34:89. [In French]

Loeper M. 1920. Chronic dyspepsia from gases. Progres Medicale 35:113. [In French]

Lohs K. 1974. Synthetische Gifte [Synthetic Poisons]. 4th ed. Stockholm International Peace Research Institute Monograph. Berlin: Militarverlag der Deutschen Demokratischen Republik. 126-131, 272-279. [In German]

Lohs K. 1975. Delayed Toxic Effects of Chemical Warfare Agents. Stockholm International Peace Research Institute Monograph. Stockholm: Almqvist & Wilksell International.

Lomova T, Shakhova I, Antoshechkin A. 1978. Repair of DNA-protein cross-links in cultivated Chinese hamster cells. Doklady Akademii Nauk SSSR 240:1472-1474. [In Russian]

Lopez S, Miyashita Y, Simons SS. 1990. Structural based selective interaction with arsenite with steroid receptors. Journal of Biological Chemistry 265:16039-16042.

Loveless A. 1951. Qualitative aspects of the chemistry and biology of radiomimetic (mutagenic) substances. Nature 167:338-342.

Loveless A. 1966. Genetic and Allied Effects of Alkylating Agents. London: Butterworths.

Loveless A, Stock JC. 1959. Influence of radiomimetic substances on deoxyribonucleic acid synthesis and function studied in *Escherichia coli* phage systems. 1. Nature of the inactivation of T2 phage *in vitro* by certain alkylating agents. Proceedings of the Royal Society of London, Series B: Biological Sciences 150:423-445.

Loveless A, Cook J, Wheatley P. 1965. Recovery from the lethal effects of cross-linking alkylation. Nature 205:980-983.

Lucam F. 1939. Cutaneous lesions caused by war gas. An experimental study. Annales d'Anatomie Pathologique et d'Anatomie Normale Medico-Chirurgicale 16:705-732. [In French]

Ludewig S, Chanutin A. 1946. Chemical changes in rat adrenals after injection of b-chloroethyl vesicants. Endocrinology 38:376-384.

Ludlum DB, Papirmeister B. 1986. DNA modification by sulfur mustards and nitrosoureas and repair of these lesions. Basic Life Sciences 38:119-125.

Ludlum DB, Metha JR, Steiner RF, DeWitt J. 1978. Physical properties of poly(3,N4ethenocytidylic acid). Biophysical Chemistry 7:339-346.

Ludlum DB, Tong WP, Mehta JR, Kirk MC, Papirmeister B. 1984. Formation of O^6 ethylthioethyldeoxyguanosine from the reaction of chloroethylethyl sulfide with deoxyguanosine. Cancer Research 44:5698-5701.

Ludlum DB, Kent S, Mehta JR. 1986. Formation of O^6-ethylthioethylguanine in DNA by reaction with the sulfur mustard, chloroethyl sulfide, and its apparent lack of repair by O^6-alkylguanine DNA alkyltransferase. Carcinogenesis 7:1203-1206.

Luego G, Cassady G, Palmisano P. 1969. Acute maternal arsenic intoxication with neonatal death. American Journal of Diseases of Children 117:328-330.

Luening KG. 1951. Mustard gas and gynandromorph production in *Drosophila melanogaster.* Hereditas 37:488-500.

Luening KG. 1952. Studies on the origin of apparent gene mutations in *Drosophila melanogaster.* Acta Zoologica 33:193.

Lukas S, Voiculet N. 1968. Aspects of the interaction between deoxyribonucleic acid and bifunctional alkylating agents. I. The kinetics of the bifunctional alkylating agents-DNA complexing reaction. Oncologia si Radiologia 7:409-413. [In Romanian]

Lundin SJ, ed. 1991. Verification of Dual-Use Chemicals Under the Chemical Weapons Convention: The Case of Thiodiglycol. SIPRI Chemical & Biological Warfare Studies, No. 13. Oxford: Oxford UniversityPress .

Lunghetti B. 1926. Contributions to the study of anatomical changes by the study of war gases, especially mustard gas. Giornale de Medicina Militare 74. [In Italian]

Lustig A. 1917. Summary relationship of the effects of mustard gas and its means of defense. Circolare del Comando Supremo 32440. [In Italian]

Lustig A. 1918. On the effect of dichloroethyl sulfide (yperite). Sperimentale 72:119-128. [In Italian]

Lustig A. 1931. Medical therapy of dichlorethyl sulfide poisoning. Giornale de Medicina Militare 79:596-599. [In Italian]

Lustig A. 1932. Effects of war gases, hydrogen arsenide and dichlorvinyl arsine on skin. Sperimentale 86:155-173. [In Italian]

Lustig A. 1935. Mustard gas. Riforma Medica 51:239-244. [In Italian]

Lyle DJ. 1925. Mustard gas burn of cornea. American Journal of Ophthalmology 8:779-781.

Lynch V. 1918. The individual variation in susceptibility to mustard gas (with method of determining same). Bureau of Mines Report 22:71.

Lynch V, Smith HW, Marshall EK Jr. 1918. On dichloroethyl sulfide (mustard gas). The systemic effects and mechanisms of action. Journal of Pharmacology and Experimental Therapeutics 12:265-290.

Lyons JA, Keane TM. 1992. Keane PTSD scale: MMPI and MMPI-2 update. Journal of Traumatic Stress 5:111-117.

Machata G, Vycudilik W. 1984. Detection of mustard gas in biological material. Archives Belges (Supplement):53-55.

Mackenzie GM. 1918. Plates illustrating the pathological effects produced by some of the better known gases used by the Germans. U.S. Naval Medical Bulletin 12:173-183.

Macleod AD. 1991. Posttraumatic stress disorder in World War II veterans. New Zealand Medical Journal 104:285-288.

Mader R. 1938. Poison war gases and their physiological action. Militarwissenschaftliche Mitteilungen 69:241-248. [In German]

Madoc-Jones H, Mauro F. 1974. Site of action of cytotoxic agents in the cell life cycle. Handbuch der Experimentellen Pharmakologie 38:205-219.

Magee PN. 1977. Extrapolation of cellular and molecular level studies to the human situation. Journal of Toxicology and Environmental Health 2:1415-1424.

Maier G. 1938. General effects of mustard gas poisoning. Zeitschrift fur die Gesamte Experimentelle Medizin 103:458-478. [In German]

Maier H, De Vries N, Snow GB. 1991. Occupational factors in the aetiology of head and neck cancer. Clinical Otolaryngology 16:406-412.

Maire GLO. 1938. Present status of sequels of respiratory tract due to toxic action of war gases in World War veterans. Revista Sanitara Militara 109:745-789. [In Romanian]

Maisin H. 1966. Effect of the combination of cytotoxins and x-rays on bone marrow. Comptes Rendus des Seances de la Societe de Biologie et de Filiales 160:1104-1108. [In French]

Malbon RM, Parish JH. 1971. Fractions of RNA and ribonucleoproteinfrom bacterial polysomes . Part 2. Reactions with sulphur mustard. Biochimica et Biophysica Acta 246:542-552.

Malizia E, Smeriglio M, Milletti M, Borge S, Giomo E, Piccioni L, Silfen RE. 1991. Acute intoxication from chemical weapons: clinical observations in a group of patients and 7-year follow-up. Rivista di Tossicologia Sperimentale Clin 21:9-39. [In Italian]

Mallie H. 1920. Pulmonary sequelae of yperite. Journal de Medecine de Bordeaux et du Sud-Ouest 1:9-13. [In French]

Mallie H. 1930. Bronchopulmonary effects of dichlorethyl sulphide. Archives MedicoChirurgicales de l'Appareil Respiratoire 5:233-266. [In French]

Manage K, Shigenobu F, Okimoto T, Usui K. 1957. X-ray findings of the lungs of gas poisoned men at Okunoshima. Journal of the Hiroshima Medical Association 10:252-254. [In Japanese]

Mandel M, Gibson WS. 1917. Clinical manifestations and treatment of gas poisoning. Journal of the American Medical Association 69:1970-1971.
Mandl H, Freilinger G. 1984. First report on victims of chemical warfare in the Gulf War treated in Vienna. Archives Belges (Supplement):330-340.
Mangun GH. 1947. Toxicity Laboratory, University of Chicago. Chemical Corps Journal 1:25-26, 49-50.
Manieri A, Morelli E. 1929. Effects of war gases on epinephrine content of suprarenal glands. Sperimentale 83:45-62. [In Italian]
Manieri A, Rovida. 1928. Complications and posthumous results from victims of war gas. Giomale de Medicina Militare 76:397-426.
Mann I. 1942. Mustard gas lesions of the eyes. British Medical Journal 1:353-354.
Mann I. 1944. A study of eighty-four cases of delayed mustard gas keratitis fitted with contact lenses. British Journal of Ophthalmology 28:441-447.
Mann I, Pullinger BD. 1940. Experiments on the effect of ascorbicacid in mustard gas burns of the eye . British Journal of Ophthalmology 24:444-451.
Mann I, Pullinger BD. 1941. A study of mustard gas lesions on the eyes of rabbits and men. Proceedings of the Royal Society of Medicine 35:229-244.
Mann I, Pullinger BD. 1942. The pathology of cholestrin and fat deposition in mustard gas injuries of the cornea. British Journal of Ophthalmology 26:503-507.
Mann I, Pullinger BD. 1944. A study of mustard gas lesions of the eyes of rabbits and men. American Journal of Ophthalmology 26:1253-1277.
Mann I, Pirie A, Pullinger BD. 1946. Study of Lewisite lesions of the eyes of rabbits. American Journal of Ophthalmology 29:1215-1227.
Mann I, Pirie A, Pullinger BD. 1948. An experimental and clinical study of the reaction of the anterior segment of the eye to chemical injury, with special references to chemical warfare agents. British Journal of Ophthalmology Monograph Supplement XIII.
Manning KP, Skegg DCG, Stell PM, Doll R. 1981. Cancer of the larynx and other occupational hazards of mustard gas workers. Clinical Otolaryngologyand Allied Sciences 6:165-170.
Manolov KG, Boyadzhiev SB. 1977. Notes on some common aspects and differences in hematopoiesis impairment due to ionizing radiation and alkylation agents (radiomimetics). I. Ionizing radiation and sulfur yperite. Mediko-Biologichni Problemni 5:225-233. [In Bulgarian]
Marlow DD, Mershon MM, Mitcheltree LW, Petrali JP, Jaax GP. 1989. Evaluation of euthymic hairless guinea pigs. In: Proceedings of the Medical Defense Bioscience Review, Aberdeen Proving Ground, MD. 561-568.
Marlow DD, Mershon MM, Mitcheltree LW, Petrali JP, Jaax GP. 1990. Sulfur mustard-induced skin injury in hairless guinea pigs. Journal of Toxicology—Cutaneous and Ocular Toxicology 9:179-192.
Maronpot RR, Shimkin MB, Smith LH, Cline JM. 1986. Strain A mouse pulmonary tumor test results for chemicals previously tested in the National Cancer Institute carcinogenicity tests. Journal of the National Cancer Institute 76:1101-1112.
Marshak A. 1946. Effect of mustard gas (dichloroethyl sulfide) on mitosis and ^{32}P (radioactive phosphorus) uptake in regenerating liver. Proceedings of the Society for Experimental Biology and Medicine 63:118-120.
Marshall EK. 1918. Individual variation in susceptibility to mustard gas. Pharm Research Report No. 320
Marshall EK. 1919. Mustard gas. Journal of the American Medical Association 73:684-686.
Marshall EK, Williams JW. 1921. The toxicity and skin irritant effects of certain derivatives of dichloroethylsulfide. Journal of Pharmacology and Experimental Therapeutics 16:259-272.
Marshall EK, Lynch V, Smith HW. 1918. On dichloroethylsulfide (mustard gas). II.

Variations in susceptibility of the skin to dichlorethylsulphide. Journal of Pharmacology and Experimental Therapeutics 12:291-301.

Martens ME. 1991. Glucose metabolism and NAD^+ content in cultured human epidermal keratinocytes exposed to sulfur mustard (abstract). FASEB Journal 5:A823.

Martin CVF, Higgins JE. 1976. Byssinosis and other respiratory ailments. Journal of Occupational Medicine 16:455-462.

Martin F. 1920. Analytical characteristics of dichloroethyl sulfide. Journal de Pharmacie et de Chimie 22:161-165. [In French]

Martin H. 1925. Skin injury caused by yellow cross. Zentralblatt fur die Gesamte Hygiene mit Einschluss der Bakteriologe und Immunitatslehre 2:307.

Martinius J. 1958. On the toxic stability of arsenic-containing vesicants. Archiv fur Toxikologie 17:210-213. [In German]

Mason JW, Giller EL, Kosten TR, Ostroff RB, Podd L. 1986. Urinary free-cortisol levels in posttraumatic stress disorder patients. Journal of Nervous and Mental Disease 174:145-148.

Mason JW, Giller EL, Kosten TR, Ostroff RB, Harkness L. 1988. Elevation of urinary norepinephrine/cortisol ratio in posttraumatic stress disorder. Journal of Nervous and Mental Disease 176:498-502.

Mastrorilli A. 1958. Remote results of lesions from vesicant gases: clinical and statistical review of 102 cases. Giornale de Medicina Militare 108:349-361. [In Italian]

Mathias CGT. 1987. Clinical and experimental aspects of cutaneous irritation. In: Marzulli FN, Maibach HI, eds. Dermatotoxicology. 3rd ed. Washington, DC: Hemisphere.

Matos E. 1979. Occupational cancer. Medicina 39:536-539. [In Portuguese]

Matsumoto K. 1948. Zoological observation at Okuno-jima Island. Journal of the Hiroshima Medical Association 1:14-17. [In Japanese]

Maugh TH. 1978. Chemical carcinogens: the scientific basis for regulation. Science 201:1200-1205.

Maumenee AE, Scholz RO. 1948. The histopathology of the ocular lesions produced by sulfur and nitrogen mustards. Bulletin of Johns Hopkins Hospital 82:121-147.

Mauro F, Elkind MM. 1967. Sulfur mustard and x-rays: differences in expression of lethal damage. Science 155:1561-1563.

Mauro F, Elkind MM. 1968. Comparison of repair of sublethal damage in cultured Chinese hamster cells exposed to sulfur mustard and x-rays. Cancer Research 28:1156-1161.

Mauro F, Elkind MM. 1968. Differences in survival variations duringthe growth cycle of cultured Chinese hamster cells treated with sulfur mustard and x-rays . Cancer Research 28:1150-1155.

Mauro F, Madoc-Jones H. 1970. Age responses of cultured mammalian cells to cytotoxic drugs. Cancer Research 30:1397-1408.

Mauro F, Elkind MM, Sakamoto K. 1968. Comparison of survival kinetics and modifications of lethal damage in cultured Chinese hamster cells exposed to sulfur mustard and x-rays. Cancer Research 28:1143-1149.

Mauro F, Madoc-Jones H, Sakamoto K, Elkind M. 1969. Variations in sensitivity to cytotoxic drugs during the mitotic cycle of cultured mammalian cells. Idengaku Zasshi 44:31-32.

Maximum concentrations at the workplace and biological tolerance values for working materials, 1987. Weinheim, Federal Republic of Germany: VCH Publishers.

Mayer A. 1920. Mode of action of war gases through the course of the war. Comptes Rendus Hebdomadaires des Seances de l'Academie des Sciences 170:1073-1075. [In French]

Mayer A, Magne H, Plantefol L. 1920. Toxic action of bis(b-chloroethyl) sulfide. Comptes Rendus Hebdomadaires des Seances de l'Academie desSciences 170:1625-1628. [In French]

Maynard RL, Meredith TJ, Marrs TC, Vale JA. 1991. Management of chemical warfare injuries (letter). Lancet 337:122.

McAdams AJ Jr. 1956. A study of mustard vesication. Journal of Investigative Dermatology 26:317-326.

McClurkin T. 1939. Physiological effects of toxic gases. J Instn Petrol Tech 25:382-391.

McDiarmid MA. 1991. Occupational exposure in lung cancer patients: contribution of remote past work. American Journal of Preventive Medicine 7:348-351.

McDonagh JER. 1918. The treatment of mustard gas poisoning. Medical Press 106:365.

McDonagh JER. 1920. The relationship between arsine poisoning and yperite. British Journal of Dermatology and Syphilis 32:188-194

McDonagh JER. 1924. The action, prevention, and treatment of HS poisoning. British Journal of Dermatology and Syphilis 36:64-71.

McGavack TH. 1942. The symptoms, prevention and treatment of the effects of vesicants. Bulletin of the New York Medical College 5:85-93.

McGown EL, Van Ravenswaay T, Dumlao CR. 1987. Histologic changes in nude mouse skin and human skin xenografts following exposure to sulfhydryl reagents: arsenicals. Toxicologic Pathology 15:149-156.

McKellar JH. 1920. Recurring kerato-conjunctivitis following exposure to dichloroethylsulfide. American Journal of Ophthalmology 3:209-210.

McLean AEM. 1971. Environmental factors and toxic cell injury (abstract). Journal of Pathology and Bacteriology 103:Pxv-Pxvi.

McNaught PR. 1923. Effects, immediate and remote, of irritant gas (dichlorethyl sulphide) poisoning on the respiratory tract. Tubercle 4:345-348.

Meade RH. 1922. The late effects of war gas on the lungs and its relation to pulmonary tuberculosis. Journal of the Missouri State Medical Association 19:385-387.

Medema J. 1986. Mustard gas: the science of H. Nuclear, Biological, and Chemical Defense and Technology International 1:66-71.

Medical Bulletin (US Veterans Administration). 1928. Preliminary report of a board of medical officers on residual effects of warfare gassing. 4:681-683.

Mehtab M. 1953. Chromosomal re-arrangements in progeny of *Drosophila* males treated with mustard gas. Nature 171:262.

Mehzad M. 1988. Pathological study of skin lesions in chemical casualties. In: Abstracts of the First International Medical Congress on Chemical Warfare Agents in Iran. June 13-16, 1988. Mashhad University of Medical Sciences. Mashhad, Iran. No. 78.

Meier HL. 1991. The kinetics of 2,2'-dichlorodiethyl sulfide (sulfur mustard, HD) dependent cell death (abstract). FASEB Journal 5:553.

Meier HL, Johnson JB. 1992. The determination and prevention of cytotoxic effects induced in human lymphocytes by the alkylating agent 2,2'-dichlorodiethyl sulfide (sulfur mustard, HD). Toxicology and Applied Pharmacology 113:234-239.

Meier HL, Gross CL, Papirmeister B, Daszkiewicz J. 1984. The use of human leukocytes as a model for studying the biochemical effects of chemical warfare (CW) agents. In: Proceedings of the 4th Chemical Defense Bioscience Review, Aberdeen Proving Ground, MD.

Meier HL, Gross CL, Papirmeister B. 1987. 2,2'-Dichlorodiethyl sulfide (sulfur mustard) decreases NAD^+ levels in human leukocytes. Toxicology Letters 39:109-122.

Meier HL, Gross CL, Graham LM, Lusco CT, Johnson JB. 1987. The prevention of 2,2-dichlorodiethyl sulfide (sulfur mustard, HD) cytotoxicity in human lymphocytes by inhibitors of poly(ADP-ribose) polymerase. In: Proceedings of the 6th Chemical Defense Bioscience Review, Frederick, MD. 313-315.

Mellett LB. 1974. The constancy of the product of concentration and time. Handbuch der Experimentellen Pharmakologie 38:330-340.

Mellor SG, Rice P, Cooper GJ. 1991. Vesicant burns. British Journal of Plastic Surgery 44:434-437.

Menetrier P, Coyon A. 1921. A study of the pathological lesions in the respiratory passages caused by the action of yperite. Annales de Medecine 9:409-418. [In French]

Menezo JM. 1936. Action of war gases on the eye. Rev Med Mil (Madrid) 26:214-223. [In Spanish]

Menzel DB, Amdur MO. 1986. Toxicology responses of the respiratory system. In: Klaassen CD, Amdur MO, Doull J, eds. Casarett and Doull's Toxicology. 3rd ed. New York: Macmillan.

Menzel W, Willhoeft J. 1951. Hyperplastic late reaction of hemopoietic organs following dichloroethyl sulfide poisoning. Folia Haematologica 70:278-285.

Merkel H. 1951. Effect on dichloroethyl sulfide on blood and haemopoietic tissues: experimental studies. Zentralblatt der Allgemeinen Pathologie 87:76.

Mershon MM, Brinkley FB, Clark OE, Michie MW, Forster JS, Harrington DG. 1987. Quantitative test for drug efficacy in deducing vascular leakage after vesicant vapor exposure. In: Proceedings of the 6th Chemical Defense Bioscience Review, Frederick, MD. 329-331.

Mershon MM, Mitcheltree LW, Petrali JP, Braue EH, Wade JV. 1990. Hairless guinea pig bioassay model for vesicant vapor exposures. Fundamental and Applied Toxicology 15:622-630.

Meselson MS. 1979. Chemicals and Cancer. Boulder, CO:Colorado Associated University Press.

Metelmann J. 1938. On the treatment of skin wounds induced by yellow cross vesicant with chamomile. Archiv fur Experimentale Pathologie und Pharmakologie 191:263-265. [In German]

Meunier J. [1984]. Yperite: "war gas" yesterday . . . chemical warfare agent today. In: Hommage au Professeur Rene Truhaut: Jubile Scientifique du Professeur Rene Truhart. Publisher and place of publication unknown. Available at National Library of Medicine: 747-751. [In French]

Meyer E, Heubner W. 1929. Observations on arsine poisoning. Biochemische Zeitschrift 206:212-222. [In German]

Meyer HG, Lieske CN, Gross CL, Clark JH, Dolzine TW, Logan TP, Swift AT. 1990. Development of an antibody that binds sulfur mustard. Abstracts of the 200th American Chemical Society National Meeting, Washington, DC. Anyl 152.

Meyer V. 1886. Compounds of thiodiglycol. Chemischte Berichte 19:3259-3266. [In German]

Meyn RE, Murray D. 1984. Cell cycle effects of alkylating agents. Pharmacology and Therapeutics 24:147-163.

Michael WH. 1920. Mustard gas and the cardiovascular system. U.S. Naval Medical Bulletin 14:268-269.

Michailescu CN. 1936. Physiopathology of war gas poisoning. Revista Sanitara Militara 35:1331-1368. [In Romanian]

Miller AB. 1980. Identification of adults at high risk of lung cancer. Canadian Medical Association Journal 122:985-987.

Miller J. 1977. Cancer clues from chemical structures. Science News 111:362-363.

Miller RW. 1973. Etiology of cancer: epidemiological studies. In: Anfinsen CB, Potter M, Schechter AN, eds. Current Research in Oncology—1972. New York: Academic Press. 1-14.

Miller RW. 1977. Relationship between human teratogens and carcinogens (editorial). Journal of the National Cancer Institute 58:471-474.

Miller, Rainy. 1917. A note on the blood changes in gas poisoning. Lancet 1:19.

Minden H, Zschunke E. 1958. Extensive skin damage caused by mustard gas. Deutsche Gesundheitswensen 13:426-428. [In German]

Minkowski O. 1921. Diseases caused by poisonous gases. In:Handbuch der Arztlichen Erfahrungen im Weltkriege. III. [Internal Medicine]. Leipzig: Barth. 340-383.

Minton J. 1949. Occupational Eye Diseases and Injuries. New York: Grune & Stratton.

Mirhosaini ME, Banaei M, Balali M. 1988. Report of a case with muscular fibrosis and syphosis following sulfur mustard poisoning. In: Abstracts of the First International Medical Congress on Chemical Warfare Agents in Iran. June 13-16, 1988. Mashhad University of Medical Sciences. Mashhad, Iran. No. 69.

Mish Mast NG, Semnanian S, Hosseini ESA, Motamedi F. 1988. The analysis of spirometric data of chemically injured patients and their comparison with normal values. In: Abstracts of the First International Medical Congress on Chemical Warfare Agents in Iran. June 13-16, 1988. Mashhad University of Medical Sciences. Mashhad, Iran. No. 55.

Mitcheltree LW, Mershon MM, Wall HG, Pulliam JD, Manthei JH. 1989. Microblister formation in vesicant-exposed pig skin. Journal of Toxicology—Cutaneous and Ocular Toxicology 8:309-319.

Mitrofanov YA, Voskanyan AZ. 1972. Analysis of the cytogenetic effect of nitrogen mustard gas in two cellular generations. Tsitologiya i Genetika 6:422-425. [In Russian]

Miyaji T. Skin cancer in Japan: a nationwide 5 year survey. In: Urbach F, ed. Conference on Biology of Cutaneous Cancer. National Cancer Institute Monograph No. 10. 55-70.

Moallem A, Nassiri Z, Kouzeh Kanani A, Balali M. 1988. Determination of serum theophylline concentration in combatants with sulfur mustard poisoning. In: Abstracts of the First International Medical Congress on Chemical Warfare Agents in Iran. June 13-16, 1988. Mashhad University of Medical Sciences. Mashhad, Iran. No. 61.

Mohamed-Ali H. 1992. Late lesions due to poison gas survivors of the Iraqi poison warfare against the Kurdish people. Wiener Medizinische Wochenschrift 142:8-15. [In German]

Mol MAE, Van De Ruit ABC, Kluivers AW. 1989. NAD^+ levels and glucose uptake of cultured human epidermal cells exposed to sulfur mustard. Toxicology and Applied Pharmacology 98:159-165.

Mol MAE, de Vries R, Kluivers AW. 1991. Effects of nicotinamide on biochemical changes and microblistering induced by sulfur mustard in human skin organ cultures. Toxicology and Applied Pharmacology 107:439-449.

Momeni AZ, Enshaeih S, Meghadi M, Amindjavaheri M. 1992. Skin manifestations of mustard gas. A clinical study of 535 patients exposed to mustard gas. Archives of Dermatology 128:775-580.

Mongelli-Sciannameo N. 1960. Accident from dichlordiethyl sulfide in a group of fishermen. Rassegna di Medicina Industriale e di Igiene del Lavoro 29:441-454. [In Italian]

Moniteur Scientifique. 1919. Yperite and gaz toxiques. 86:147-153. [In French]

Monks TJ, Anders MW, Dekant W, Stevens JL, Lau SS, Van Bladeren PJ. 1990. Contemporary issues in toxicology: glutathione conjugate mediated toxicities. Toxicology and Applied Pharmacology 106:1-19.

Montesano R. 1979. Carcinogenic risk of chemicals: experimental evidence. Collection de Medecine Legale et de Toxicologie Medicale 111:29-33.

Montgomery H, Waismann M. 1941. Epithelioma attributable to arsenic. Journal of Investigative Dermatology 4:365-383.

Moore AM, Rockman JB. 1950. Study of human hypersensitivity to compounds of mustard gas (dichloroethyl) sulfide type. Canadian Journal of Research. Section E: Medical Sciences 28:169-176.

Moore KG, Schofield BH, Higuchi K, Kajiki A, Au KW, Pula P, Bassett DP, Dannenberg AM Jr. 1986. Two sensitive *in vitro* monitors of chemical toxicity to human and animal skin (in short-term organ culture): I. Paranuclear vacuolization in glycol methacrylate tissue sections. II. Interference with (^{14}C)leucine incorporation. Journal of Toxicology—Cutaneous and Ocular Toxicology 5:285-302.

Moore RF, Heckford F. 1929. Delayed corneal ulceration from mustard gas. British Medical Journal 1:497-498.

Moore S, Stein WH, Fruton JS. 1946. Chemical reactions of mustard gas and related compounds. II. The reaction of mustard gas with carboxyl groups and with the amino groups of amino acids and peptides. Journal of Organic Chemistry 11:675-680.

Moorhead TG. 1919. The clinical results of poisoning by mustard gas. Dublin Journal of Medical Sciences 147:1-7.

Moreaux R. 1919. Contribution to the study of vesicant gas action on the upper respiratory tract. Paris Medical 33:146-150. [In French]

Morgenstern P, Koss FR, Alexander WW. 1947. Residual mustard gas bronchitis: effects of prolonged exposure to low concentrations of mustard gas. Annals of Internal Medicine 26:27-40.

Morris HP, Laug EP, Morris HJ, Grant RL. 1938. The growth and reproduction of rats fed diets containing lead acetate and arsenic trioxide. Journal of Pharmacology and Experimental Therapeutics 64:420-445.

Morris RS. 1919. Clinical observations on the late pulmonary effects of gassing. In: Contributions to Medical and Biological Research Dedicated to Sir William Osler. Vol. 2. New York: P.B. Hoeber. 1138-1142.

Morrow LA, Ryan CM, Hodgson MJ, Robin N. 1991. Risk factors associated with persistence of neuropsychological deficits in persons with organic solvent exposure. Journal of Nervous and Mental Disease 179:540-545.

Motamedi F, Hosseini ESA, MishMast NG, Semnanian S. 1988. Study of the acute and chronic effects of chemical warfare on respiratory volumes and capacities and comparison of these parameters in injured patients in different confrontations. In: Abstracts of the First International Medical Congress on Chemical Warfare Agents in Iran. June 13-16, 1988. Mashhad University of Medical Sciences. Mashhad, Iran. No. 9.

Motekallem MH. 1988. Clinical evaluation and follow-up of seventeen victims who came in contact with mustard gas used by Iraq. In: Abstracts of the First International Medical Congress on Chemical Warfare Agents in Iran. June 13-16, 1988. Mashhad University of Medical Sciences. Mashhad, Iran. No. 21.

Mousavy Tadi SH, Eslami MB, Mahmoudi M, Keyhani A, Eftekhar B. 1988. Study of cellular immunity in Iranian combatants poisoned with mustard gas. In: Abstracts of the First International Medical Congress on Chemical Warfare Agents in Iran. June 13-16, 1988. Mashhad University of Medical Sciences. Mashhad, Iran. No. 60.

Mozier NM, Hoffman JL. 1990. Biosynthesis and urinary excretion ofmethylsulfonium derivatives of the sulfur mustard analog, 2-chloroethyl ethyl sulfide, and other thioethers . FASEB Journal 4:3329-333.

Muggia FM. 1983. Pulmonary toxicity of anti-tumor agents. Cancer Treatment Reviews 10:221-243.

Muller U. 1971. Properties and Mechanism of the Action of Mustard Gas (dissertation). University of Erlangen-Nurnberg. [In German]

Muller-Kiel U. 1933. The Chemical Weapon in the World War and Today. Dichlorodiethyl Sulphide (Yellow Cross). Berlin: Verlag Chemie, GMBH.

Munro N, Watson A, Ambrose K, Griffin G. 1990. Treating exposure to chemical warfare agents: implications for health care providers and community emergency planning. Environmental Health Perspectives 89:205-215.

Muntsch O. 1934. Blood changes following poisoning with chemical warfare agents. Gasschutz und Luftschutz 4:165-166. [In German]

Muntsch O. 1934. The changes of the blood in blister gas diseases as a diagnostic aid. Klinische Wochenschrift 13:482-485, 529-532. [In German]

Muntsch O. 1936. On the attempts to use a chloramide ointment as detoxicant against the effect of dichlorodiethylsulfide on the skin. Gasschutz und Luftschutz 6:50-51. [In German]

Muntsch O. 1937. Problem of chronic poisoning with war gases. Medizinische Welt 11:928-931. [In German]

Muntsch O. 1941. [Manual on the Pathology and Therapy of Blister Gas Diseases] Leitfaden der Pathologie und Therapie der Kampfgaserkrankungen. Leipzig: Georg Thieme. [In German]

Murnane JP, Byfield JE. 1981. Irrepairable DNA cross-links and mammalian cell lethality with bifunctional alkylating agents.Chemico-Biological Interactions 38:75-76.

Murphy ML, Del Moro A, Lacon C. 1958. The comparative effects of five polyfunctional alkylating agents on the rat fetus, with additional notes on the chick embryo. Annals of the New York Academy of Sciences 68:762-782.

Murray V, Volans G. 1991. Management of injuries due to chemical weapons (editorial). British Medical Journal 302:129-130.

Musajo Somma A, Tanzarella M, Petruzzelli C. 1983. Skin burns and chemical compounds. Rivista Italiana di Chirurgia Plastica 15:35-43. [In Italian]

Nagai M. 1958. Study of chronic poison gas effects. Kyosai Iho 7:386-398. [In Japanese]

Nagy SM, Golumbic C, Stein WH, Fruton JS, Bergmann M. 1946. The penetration of vesicant vapors into human skin. Journal of General Physiology 29:441-469.

Nakamura M, Rikimaru T, Yano T, Moore KG, Pula PJ, Schofield BH, Dannenberg AM Jr. 1990. Full-thickness human skin explants for testing the toxicity of topically applied chemicals. Journal of Investigative Dermatology 95:325-332.

Nakamura T. 1956. Studies on the warfare gas injury in Japan. Report I: On the general condition of the poison gas island. Gencho Hiroshima Igaku 4:1141-1149. [In Japanese]

Nakamura T. 1956. Studies on the warfare gas injury in Japan. Report II. On the desoxynucleodepolymerase in serum. Gencho Hiroshima Igaku 4:1150-1156. [In Japanese]

Nakamura T. 1956. Studies on the warfare gas injury in Japan. Report III. On the labour hygiene. Gencho Hiroshima Igaku 4:1157-1171. [In Japanese]

Nakamura T. 1956. Studies on the warfare gas injury in Japan. Report IV: On the x-ray view, especially on the view of bronchography. Gencho Hiroshima Igaku 4:1172-1181. [In Japanese]

Nakamura T. 1956. Studies on warfare gas-injury in Japan. Report V. On the investigations on the cause of death of the workers. Gencho Hiroshima Igaku 4:1182-1189. [In Japanese]

Nasrat GE. 1953. Dose-dependence relationship of mustard gas induced lethals and rearrangements in *D.* melanogaster (abstract). Proceedings of the Ninth International Congress of Genetics 9 (Part 2):694.

Nasrat GE. 1954. Some cytological observations on the delayed effect of mustard gas. Nature 174:968-969.

Nasrat GE, Kaplan WD, Auerbach C. 1954. A quantitative study of mustard gas induced chromosome breaks and rearrangements in *Drosophila melanogaster.* Zeitschrift fur Induktive Abstammungs-und Vererburgslehre 86:249-262.

National Commission for the Protection of Human Subjects of Biomedical and Behavioral Research. 1978. The Belmont Report: Ethical Principles and Guidelines for the Protection of Human Subjects of Research. Washington, DC: U.S. Government Printing Office.

National Research Council. Division of Medical Sciences. Committee on Treatment of Gas Casualties. 1942. Bulletin of the Committee on Treatment of War Gas Casualties. Minutes, March 18, 1942. Available at the National Research Council Archives, Washington, DC.

National Research Council. Division of Medical Sciences. Committee on Treatment of Gas Casualties. [1945]. Bibliography of the Medical Aspects of Chemical Warfare: Published Literature. Unpublished, available at the National Research Council Archives, Washington, DC.

National Research Council. Division of Medical Sciences. Committee on Treatment of Gas Casualties. 1945. Fasciculus on Chemical Warfare Medicine. 3 vols. Washington, DC: Prepared for the Committee on Medical Research of the Office of Scientific Research and Development. Available at the National Research Council Archives, Washington, DC.

National Research Council. 1979. Airborne Particles. Baltimore: University Park Press.

National Research Council. Committee on Chemical Environmental Mutagens. 1983. Identifying and Estimating the Genetic Impact of Chemical Mutagens. Washington, DC: National Academy Press.

National Research Council. Committee on Chemical Environmental Mutagens. 1983. Quantitative Relationship Between Mutagenic and Carcinogenic Potencies: A Feasibility Study. Washington, DC: National Academy Press.

National Research Council. Division of Medical Sciences. 1977. Arsenic. Medical and Biologic Effects of Environmental Pollutants. Washington, DC: National Academy of Sciences.

National Research Council. Committee on Toxicology. 1985. Possible Long-term Health Effects of Short-term Exposure to Chemical Agents. 3 vols. Washington, DC: National Academy Press.

Navarre M. 1939. Lesions and disorders due to war gases. Monde Medical 49:564-570. [In French]

Neame H. 1928. Late result after mustard gas burn of the eyes. Proceedings of the Royal Society of Medicine 22:25.

Needham DM. 1948. The action of mustard gas on enzymes *in vitro* and in tissues. In: Williams RT, ed. The Biochemical Reactions of Chemical Warfare Agents. Biochemical Society Symposia No. 2. Cambridge, England: University Press. 16-27.

Needham DM, Cohen JA, Barrett AM. 1947. The mechanism of damage to the bone marrow in systemic poisoning with mustard gas. BiochemicalJournal 41:631-639.

Neilands JB. 1973. Survey of chemical and related weapons of war. Naturwissenschaften 60:177-183.

Nelson N. 1977. Inhalation carcinogenesis. In:Science for Better Environment. Proceedings of the International Congress on the Human Environment. Oxford: Pergamon Press. 505-510.

Nemiah JC. 1980. Anxiety state. In: Kaplan HI, Freedman AM, Saddock BJ, eds. Comprehensive Textbook of Psychiatry. Baltimore: Williams and Wilkins. 1483-1493.

Nesnow S, Argus M, Bergman H, Chu K, Frith C, Helmes T, McGaughy R, Ray V, Slaga TJ, Tennant R, Weisburger E. 1987. Chemical carcinogens. A review and analysis of the literature of selected chemicals and the establishment of the Gene-Tox Carcinogen data base. Mutation Research 185:1-195.

Neubrauer O. 1947. Arsenical cancer: a review. British Journal of Cancer 1:192-244.

Neumann E, Schwank R. 1960. Multiple malignant and benign epidermal and dermal tumors following arsenic. Acta Dermato-Venereologica 40:400-409.

New York Medical Journal. 1918. Mustard gas and its effect upon the skin (editorial). 108:119.

Newman-Taylor A, Morris A. 1990. Experience with mustard gas casualties (letter). Lancet 337:242.

Nicholson W. 1988. IARC evaluations in the light of limitations of human epidemiologic data. Annals of the New York Academy of Sciences 534:44-54.

Niemann A. 1860. The action of brown sulfur chloride on ethylene gas. Annalen der Chemie und Pharmazie 113:288. [In German]

Nishimoto R. 1961. Clinical observations of former employees of the Okuno-jima poison gas factory. Journal of the Hiroshima Medical Association, Special Series 14:671-690. [In Japanese]

Nishimoto R. 1961. Historical review of Okuno-jima poison gas factory and its working environment. Journal of the Hiroshima Medical Association, Special Series 14:635-641. [In Japanese]

Nishimoto R. 1961. Review of literatures concerning the poison gas injuries at Okuno-jima Island. Journal of the Hiroshima Medical Association, Special Series 14:643-651. [In Japanese]

Nishimoto R. 1961. Studies on the cause of death of former employees of the Okuno-jima poison gas factory. Journal of the Hiroshima Medical Association, Special Series 14:653-669. [In Japanese]

Nishimoto Y, Katsuta S. 1958. Radiological observation of the chest in former workers of Okuno-jima poison gas factory, with special reference to relation between bronchography and pulmonary function study. Kekkaku 33:226. [In Japanese]

Nishimoto Y, Burrows B, Miyanishi M, Katsuta S, Shigenobu T, Kettel LJ. 1970. Chronic obstructive lung disease in Japanese poison gas workers. American Review of Respiratory Disease 102:173-179.

Nishimoto Y, Yamakido M, Ishioka S, Shigenobu T, Yukutake M. 1988. Epidemiological studies of lung cancer in Japanese mustard gas workers. International Symposium of the Princess Takamatsu Cancer Research Fund.Vol. 18, Unusual Occurrences as Clues to Cancer Etiology. 95-101.

Nishimoto Y, Yamakido M, Shigenobu T, Onari K, Yukutake M. 1983. Long-term observation of poison gas workers with special reference to respiratory cancers. Sangyo Ika Daigaku Zasshi 5(Suppl):89-94.

Nishimoto Y, Yamakido M, Shigenobu T, Yukutake M, Matsusaka S. 1986. Cancer of the respiratory tract observed in retired workers from a poison gas factory. Gan To Kagaku Ryoho 13:1144-1148. [In Japanese]

Noorbakhsh K, Balali M. 1988. Evaluation of gastrointestinal effects of sulfur mustard poisoning. In: Abstracts of the First International Medical Congress on Chemical Warfare Agents in Iran. June 13-16, 1988. Mashhad University of Medical Sciences. Mashhad, Iran. No. 33.

Noori SMH, Chitnis PS. 1988. Histological and histochemical studies on the affected tissues of integumentary system of Iranian combatants exposed to chemical warfare agents. In: Abstracts of the First International Medical Congress on Chemical Warfare Agents in Iran. June 13-16, 1988. Mashhad University of Medical Sciences. Mashhad, Iran. No. 79.

Nordenson I, Beckman G, Beckman L, Nordstrom S. 1978. Occupational and environmental risks in and around a smelter in northern Sweden. II. Chromosomal aberrations in workers exposed to arsenic. Hereditas 88:47-50.

Norman JE Jr. 1975. Lung cancer mortality in World War I veterans with mustard gas injury, 1919-1965. Journal of the National Cancer Institute 54:311-318.

Norouzi MR, Keyhani A, Eslami MB, Mahmoudi M, Eftekar B. 1988. Study of the humoral immunity in Iranian combatants poisoned with mustard gas. In: Abstracts of the First International Medical Congress on Chemical Warfare Agents in Iran. June 13-16, 1988. Mashhad University of Medical Sciences. Mashhad, Iran. No. 59.

Norris GW. 1918. Toxic gases in modern warfare with special reference to diagnosis and treatment. Journal of the American Medical Association 71:1822-1825.

Nourmahammadi I, Keshavarz A, Ghaemmaghamy M. 1988. Evaluation of zinc blood levels and related parameters in patients wounded by chemical warfare agents. In: Abstracts of the First International Medical Congress on Chemical Warfare Agents in Iran. June 13-16, 1988. Mashhad University of Medical Sciences. Mashhad, Iran. No. 16.

Novick M, Gard DH, Hardy SB, Spira M. 1977. Burn scar carcinoma: a review and analysis of 46 cases. Journal of Trauma 17:809-817.

Nowrouzbaigi Y. 1988. Disorders of ears, nose, throat and larynx of combatants with chemical war gas poisoning. In: Abstracts of the First International Medical Congress on Chemical Warfare Agents in Iran. June 13-16, 1988. Mashhad University of Medical Sciences. Mashhad, Iran. No. 34.

Ochoa MJ. 1963. Alkylating agents in clinical cancer chemotherapy. Annals of the New York Academy of Sciences 163:921-930.

Ogsten AG. 1948. The chemical reactions of mustard gas in aqueous solution. In: Williams RT, ed. The Biochemical Reactions of Chemical Warfare Agents. Biochemical Society Symposia No. 2. Cambridge, England: Cambridge University Press. 2-7.

Ohmine H, Fujita M, Goriki K, Asakawa J, Satoh C, Yamakido M, Nishimoto Y. 1984. A study of the genetic effects of occupational exposure to mustard gas. Jinrui Idengaku Zasshi 29:237-238.

Ollila O. 1939. [The chemical structure of war gases and their physiological effect]. Finska Kemustsamfundets Medd 48:72-86.

Olmer D. 1919. Hematologic research on fresh cases of yperite intoxication. Comptes Rendus des Seances de la Societe de Biologie et de Filiales 62:1292-1294. [In French]

Omaye ST, Elsayed NM, Klain GJ, Inase JL, Korte DW Jr. 1988. Biochemical response of mouse kidney to subcutaneous injection of butyl 2-chloroethyl sulfide (abstract). FASEB Journal 2:A795.

Omaye ST, Elsayed NM, Klain GJ, Korte DW Jr. 1991. Metabolic changes in the mouse kidney after subcutaneous injection of butyl 2-chloroethyl sulfide. Journal of Toxicology and Environmental Health 33:19-27.

Ormsbee RA, Henriques FC Jr, Ball EG. 1949. Reaction of mustard gas with skin proteins. Archives of Biochemistry 21:301-312.

Oster II. 1955. Nuclear changes in *Drosophila melanogaster* after combination treatment with various physical and chemical agents. Cancer Research Campaign, Annual Report. 415-417.

Osterchrist W. 1941. Clinical aspects of 4 cases of dichloroethylsulfide poisoning in children . Bruns Beitrage zur Klinischen Chirurgie 172:240-246. [In German]

Oswald A. 1920. Collapse of the double-sides closure of the central artery as a consequence of blister gas poisoning. Klinische Monatsblatter fur Augenheilkunde 64:381-387. [In German]

Owen BA, Jones TD. 1990. Hazard evaluation for complex mixtures: relative comparisons to improve regulatory consistency. Regulatory Toxicology and Pharmacology 11:132-148.

Paoletti C. 1956. Dichloroethyl toxicity: effect on erythropoietic cells. Revue de Medecine Navale 11:241-251. [In French]

Paoletti C, Boiron M, Truhaut R, Tubiania. 1956. Inhibition of the hemoglobin metabolism in reticulocytes by b-chloroethyl compounds. Comptes Rendus des Seances de la Societe de Biologie et de Filiales 150:1084-1088. [In French]

Papirmeister B, Davison CL. 1964. Elimination of sulfur mustard-induced products from DNA of *Escherichia coli.* Biochemical and Biophysical Research Communications 17:608-617.

Papirmeister B, Davison CL. 1965. Unbalanced growth and latent killing of *Escherichia coli* following exposure to sulfur mustard. Biochimica et Biophysica Acta 103:70-92.

Papirmeister B, Davison CL, Gross CL. 1968. Protection and reversal of lethal mustard damage resulting in recovery of cell viability. In: Proceedings of the 1968 Army Science Conference, West Point, NY.

Papirmeister B, Dorsey JK, Davison CL, Gross CL. 1970. Sensitization of DNA to endonuclease by adenine alkylation and its biological significance (abstract). Federation Proceedings 29:726A.

Papirmeister B, Gross CL, Petrali JP, Brinkley FB, Meier HL. 1983. A biochemical model for the cutaneous mustard injury. In: Proceedings of the Third Chemical Defense Bioscience Review, Aberdeen Proving Ground, MD.

Papirmeister B, Gross CL, Petrali JP, Hixson CJ. 1984. Pathology produced by sulfur mustard in human skin grafts on athymic nude mice. 1. Gross and light microscopic changes. Journal of Toxicology—Cutaneous and Ocular Toxicology 3:371-392.

Papirmeister B, Gross CL, Petrali JP, Meier HL. 1984. Pathology producedby sulfur mustard in human skin grafts on athymic nude mice . 2. Ultrastructural changes. Journal of Toxicology —Cutaneous and Ocular Toxicology 3:393-408.

Papirmeister B, Gross CL, Meier HL, Petrali JP, Johnson JB. 1985. Molecular basis for mustard-induced vesication. Fundamental and Applied Toxicology 5:S134-S149.

Papirmeister B, Gross CL, Petrali JP, Hixson CJ, Meier HL, Brinkley FB. 1985. Nude mice with human skin grafts used to study mustard gas-induced injury (abstract). Food and Chemical Toxicology 23:326-327.

Papirmeister B, Feister AJ, Robinson SI, Ford RD. 1991. Medical Defense Against Mustard Gas: Toxic Mechanisms and Pharmacological Implications. Boca Raton, FL: CRC Press.

Pappenheimer AM. 1918-1919. The effects of intravenous injections of dichlorethylsulphide in rabbits. Proceedings of the Society for Experimental Biology and Medicine 16:92.

Pappenheimer AM. 1919. The pathology of the poisonous gases used in the war. Proceedings of the New York Pathological Society 19:97-112.

Pappenheimer AM, Vance M. 1920. The effect of intravenous injections of dichlorethylsulphide in rabbits, with special reference to its leucotoxic action. Journal of Experimental Medicine 31:72.

Parisot J, Lecaplain. 1918. Delayed sequelae of vesicant gas exposure. Establishment of a file of gas victims. Trauvaux-Ambulance de l'Ocean de Panne (September). [In French]

Parker RM, Bucci TJ, Denny KH, Dacre JC. 1991. Negative dominant lethal study of Lewisite in CD-rats (abstract). Toxicologist 247.

Parlange JA. 1924. Ocular sequelae from war gases. Archives d'Ophtalmologie 41:278-283. [In French]

Paulet G. 1952. Cellular metabolism and cutaneous action of dichloroethyl sulfide. Comptes Rendus des Seances de la Societe de Biologie et de Filiales 146:925-928. [In French]

Paulet G. 1954. Further studies on cutaneous action. Biologie et Medecine 43:76-128.

Paulino O, Oliveira CJ, Monteiro J, de Carvalho Iervolino LA, Passos Pereira M, Ceplovitz M, de Almeida M, Matano VN, Giannico R, Saburo YR, Parise V, Carollo dos Santos W. 1980. Occupational cancer. Revista Brasileira de Saude Ocupacional 31:32-42. [In Portuguese]

Pauser G, Aloy A, Carvana M, Graninger W, Havel M, Koller W, Mutz N. 1984. Lethal intoxication by wargases on Iranian soldiers. Therapeutic interventions on survivors of mustard gas and mycotoxin immersion. Archives Belges (Supplement):341-351.

Pearce RG. 1919-1920. Notes on some respiratory studies made on late stages of gas poisoning. Journal of Laboratory and Clinical medicine 5:411-417.

Pedersen-Bjergaard J, Nissen NI, Sorensen HM, Hou-Jensen K, Larsen MS, Ernst P, Ersbol J, Knudtzon S, Rose C. 1980. Acute nonlymphocytic leukemia in patients with ovarian carcinoma following long-term treatment with Treosulfan (=dihydroxybusulfan). Cancer 45:19-29.

Pedersen-Bjergaard J, Philip P, Pedersen NT, Hou-Jensen K, Svejgaard A, Jensen G, Nissen NI. 1984. Acute nonlymphocytic leukemia, preleukemia, and acute antileukemic chemotherapy, and survival in a total series of 55 patients. Cancer 54:452-462.

Pedersen-Bjergaard J, Osterlind K, Hansen M, Philip P, Pedersen AG, Hansen HH. 1985.

Acute nonlymphocytic leukemia, preleukemia, and solid tumors following intensive chemotherapy of small cell carcinoma of the lung. Blood 66:1393-1397.

Pedersen-Bjergaard J, Ersboll J, Sorenson HM, Keiding N, Larsen SO, Philip P, Larsen MS, Schultz H, Nissen NI. 1985. Risk of acute nonlymphocytic leukemia and preleukemia in patients treated with cyclophosphamide for non-Hodgkin's lymphomas. Annals of Internal Medicine 103:195-200.

Pedersen-Bjergaard J, Specht L, Larsen SO, Ersboll J, Struck J, Hansen MM, Hansen HH, Nissen NI. 1987. Risk of therapy-related leukemia and preleukemia after Hodgkin's disease: relation to age, cumulative dose of alkylating agents and time from therapy. Lancet 2:83-88.

Pedersen-Bjergaard J, Ersboll J, Hansen VL, Sorensen BL, Christoffersen K, Hou-Jensen K, Nissen NI, Knudsen JB, Hansen MM. 1988. Carcinoma of the urinary bladder after treatment with cyclophosphamide for non-Hodgkin's lymphoma. New England Journal of Medicine 318:1028-1031.

Peeters H. 1939. Cases of poisoning by war gases. Nederlands Tijdschrift voor Geneeskunde 83:5646-5655. [In Dutch]

Penn I. 1976. Second malignant neoplasms associated with immunosuppressive medications. Cancer 37:1024-1032.

Penn I, Brunson ME. 1988. Cancers after cyclosporine therapy. Transplantation Proceedings 20 (Suppl 3):885-892.

Pentschew A. 1958. Intoxication. In: Handbuch der Speziellen Pathologischen Anatomie und Histologie. Vol. 13, Part 2. Berlin: Springer-Verlag. 1907-2502. [In German]

Perera J. 1985. Lewisite: new gas weapon in the Gulf war. New Scientist 105(1450):8.

Perera J, Thomas A. 1986. Britain's victims of mustard gas disaster. New Scientist 109:26-27.

Peronnet M. 1936. The discovery of yperite. Journal de Pharmacie et de Chimie 23:29-32. [In French]

Perpere. 1927. Current respiratory sequelae of gas victims. Pr Med 1:785-787. [In French]

Perry SW, Cella DF, Falkenberg J, Heidrich G, Goodwin C. 1987. Pain perception in burn patients with stress disorders. Journal of Pain and Symptom Management 2:29-33.

Perry SW, Cella DF, Falkenberg J, Heidrich G, Goodwin C. 1987. Painperception vs. pain response in burn patients . American Journal of Nursing 87:698-699.

Pershagen G. 1981. The carcinogenicity of arsenic. Environmental Health Perspectives 40:93-100.

Pershagen G, Lind B, Bjorklund N. 1982. Lung retention and toxicity of some inorganic arsenic compounds. Environmental Research 29:425-434.

Persson SA, Sellstrom A. 1991. Chemical warfare agents: current review on protection and treatment. Lakartidningen 88:3477-3481. [In Swedish]

Pesme P. 1924. Corneal fistula following yperite burns. Archives d'Ophtalmologie 41:278-283. [In French]

Peters CS. 1920. Gas poisoning: phosgene and mustard. Canadian Medical Association Journal 10:773-776.

Peters JM, Wegman DH. 1975. Epidemiology of toluene diisocyanate (TDI) induced respiratory disease. Environmental Health Perspectives 11:97-100.

Peters JM, Murphy RLH, Pagnotto LD, Van Ganse WF. 1968. Acute respiratory effects in workers exposed to low levels of toluene diisocyanate (TDI). Archives of Environmental Health 16:642-647.

Peters RA. 1936. Effects of dichlor-diethyl-sulfone on brain respiration. Nature 138:327-328.

Peters RA. 1947. Biochemical research at Oxford upon mustard gas. Nature 159:149-153.

Peters RA. 1947. Development and theoretical significance of British anti-Lewisite (BAL). British Medical Bulletin 5:313-318.

Peters RA. 1955. Biochemistry of toxic agents: I. Present status of knowledge of biochemical lesions induced by trivalent arsenical poisoning. Bulletin of Johns Hopkins Hospital 97:1-20.

Peters RA. 1963. The biochemical lesion in mustard gas poisoning. In: Biochemical Lesions and Lethal Synthesis. New York: Macmillan. 40-73.

Peters RA, Wakelin RW. 1949. Toxicity of vesicants and some other compounds to the pyruvate oxidase system (brain). British Journal of Pharmacology 4:51-62.

Peters RA, Walker E. 1923. Rate of liberation of acid by b,b-dichlorodiethylsulphide and its analogues in its relation to the acid theory of skin vesication. Biochemical Journal 17:260-276.

Peters RA, Sinclair HM, Thompson RHS. 1946. An analysis of the inhibition of pyruvate oxidation by arsenicals in relation to the enzyme theory of vesication. Biochemical Journal 40:516-524.

Petracek E. 1937. Chemical warfare and skin. Ceskoslovenska Dermatologie 17:7, 37. [In Czech]

Petrali JP, Oglesby SB, Mills KR. 1990. Ultrastructural correlates of sulfur mustard toxicity. Journal of Toxicology—Cutaneous and Ocular Toxicology 9:193-214.

Petrali JP, Oglesby SB, Meier HL. 1990. Ultrastructural correlates of the protection afforded by niacinamide against sulfur mustard-induced cytotoxicity of human lymphocytes *in vitro.* Ultrastructural Pathology 14:253-262.

Petrali JP, Oglesby SB, Justus TA. 1991. Morphologic effects of sulfur mustard on a human skin equivalent. Journal of Toxicology—Cutaneousand Ocular Toxicology 10:315-324.

Petri E. 1930. Pathological anatomy and histology of poisonings. In: Handbuch der Speziellen Pathologischen Anatomie und Histologie. Vol. 10. Berlin: Springer-Verlag. 114-130.

Pfister RR. 1986. The biology of persistent epithelial defects of the cornea. In: Brightbill, FS, ed. Corneal Surgery: Theory, Technique, and Tissue. Papers from the First International Cornea and Eye Banking Symposium. St. Louis: Mosby. 582-593.

Pfister RR, Paterson CA. 1980. Ascorbic acid in the treatment of alkali burns of the eye. Ophthalmology 87:1050-1057.

Pfister RR, Nicloaro ML, Paterson CA. 1981. Sodium citrate reduces the incidence of corneal ulcerations and perforations in extreme alkali burned eyes: acetylcysteine and ascorbate have no favorable effect. Investigative Ophthalmology 21:486-490.

Pfister RR, Haddox J, Paterson CA. 1982. The efficacy of sodium citrate in the treatment of severe alkali burns of the eye is influenced by the route of administration. Cornea 1:205-211.

Pfister RR, Haddox J, Barr D. 1991. The combined effect of citrate/ascorbate treatment in alkali-injured rabbit eyes. Cornea 10:100-104.

Phalen RF, Prasad SB. 1988. Morphology of the respiratory tract. In: McClellan RO, Henderson RF, eds. Concepts in Inhalation Toxicology. New York: Hemisphere.

Philips FS. 1950. Recent contributions to the pharmacology of bis(2-haloethyl) amines and sulfides. Pharmacological Reviews 2:281-323.

Phillips TJ. 1940. The delayed action of mustard gas and the treatment. Proceedings of the Royal Society of Medicine 33:229-232.

Physicians for Human Rights. 1989. Winds of Death: Iraq's Use of Poison Gas Against Its Kurdish Population. Somerville, MA: Physicians for Human Rights.

Physicians for Social Responsibility. The Physicians Task Force on the Health Risks of Nuclear Weapons Production. 1992. Dead Reckoning. A Critical Review of the Department of Energy's Epidemiologic Research. Washington, DC: Physicians for Social Responsibility.

Pickard HL. 1919. Ocular action of dichlorethylsulphide (mustard gas). American Journal of Ophthalmology 3:136.

Pierard GE, Dowlati A. 1989. The harmful effects of war blister gases. Revue Medicale de Liege 44:133-137. [In French]

Pierard GE, Dowlati A, Dowlati Y, Pierard-Franchimont C, Hermanns-Le T, Letot B. 1990. Chemical warfare casualties and yperite-induced xerodermoid. American Journal of Dermatopathology 12:565-570.

Pindborg S. 1955. Experiments on volunteers on the efficacy of anti-gas preparations on mustard gas. Militaerlaegen 61:111-114. [In Danish]

Pindborg S. 1961. Gas powder experiments with animals and humans. Tidskrift I Militar Halsovard 86:267-272. [In Danish]

Pinto SS, Nelson KW. 1946. Arsenic toxicology and industrial exposure. Annual Review of Pharmacology and Toxicology 16:95-100.

Pirie A. 1947. The action of mustard gas on ox cornea collagen. Biochemical Journal 41:185-190.

Pissarello C. 1918. On the morbid effects of shells containing yperite. Giornale de Medicina Militare 66:128-134. [In Italian]

Pitot HC. 1980. Relationships of bioassay data on chemicals to their toxic and carcinogenic risk for humans. Journal of Environmental Pathology and Toxicology 3:431-450.

Plant JE, Roberts JJ. 1971. Extension of the pre-DNA synthetic phase of the cell cycle as a consequence of DNA alkylation in Chinese hamster cells: a possible mechanism of DNA repair. Chemico-Biological Interactions 3:343-351.

Pochedly C. 1987. The discovery of the first drug effective in treating acute lymphoid leukemia in children. New York State Journal of Medicine 87:500-503.

Popa NE. 1969. Dynamics and spectrum of structural chromosome mutations of *Vicia faba* L. var-minor under the effect of mustard gas. Tsitologiya i Genetika 3:136-141. [In Russian]

Pope WJ. 1919. Mustard gas. Journal of the Society of Chemical Industry 38:344-345.

Potter D. 1989. Ida Mann: her wartime career, 1939-1949. Australian and New Zealand Journal of Ophthalmology 17:95-101.

Prakash UB. 1991. Chemical warfare and bronchoscopy (editorial). Chest 100:1486.

Preininer T. 1938. Unusual cutaneous lesions sustained by doctors and nurses while caring for patients suffering from dichloroethyl sulfide poisoning. Archiv fur Dermatologie und Syphilis 176:508-514. [In German]

Prentiss AM. 1937. Vesicant agents. In: Chemicals in Warfare: A Treatise on Chemical Warfare. New York: McGraw-Hill. 177-300.

President's Commission for the Study of Ethical Problems in Medicine and Biomedical and Behavioral Research. 1983. Summing Up: Final Report on Studies of the Ethical and Legal Problems in Medicine and Biomedical and Behavioral Research. Washington, DC: President's Commission.

Price CC, Gaucher GM, Konero P, Shibakowa R, Sowa JR, Yamaguchi M. 1968. Relative reactivities for monofunctional nitrogen mustard alkylation of nucleic acid components. Acta Biochimica et Biophysica Hungarica 166:327-359.

Price GB. 1938. After results of gassing. Medical Press 196:218-222.

Prokes J, Svoboda V, Hynie I, Proksova M, Kacl K. 1968. The influenceof x-radiation and mustard gas on methionin ^{35}S incorporation in erythrocytes . Neoplasma 15:393-398.

Prokulevich VA, Fomichev YK. 1975. Recombination process in *Escherichia coli* K-12. II. Effect of treating donor cells with yperite and uv light on recombination frequency in a merozygote. In: Rokitskii PF, Khotyleva LV, eds. [Studies on Theoretical and Applied Genetics] Issledovanieiia po Teoreticheskoei i Prikladnoei Genetike. Minsk: Nauka i Teknika. 209-214. [In Russian]

Prytkova TN, Gol'Dat SY. 1973. Effect of 2 nitrogenous analogues of mustard gas on variation of actinomyces-streptomycini producing streptomycin. Antibiotiki (Moscow) 18:590-595. [In Russian]

Przystasz T. 1985. The behavior of erythrocyte, leukocyte, and platelet counts in rabbits after contamination of a skin-muscle wound with a,b-dichlorodiethyl sulfide and b-chlorovinyldichloroarsine. Lekarz Wojskowy 61:596-602. [In Polish]

Przystasz T. 1986. Healing of cutaneomuscular wounds in rabbit after contamination with a,b-dichlorodiethyl sulfide and b-chlorovinyldichloroarsine. Lekarz Wojskowy 62:157-166. [In Polish]

Przystasz T. 1986. Serum levels of bilirubin, AIAT, and AspAT in rabbits after contamination of musculocutaneous wound with certain irritating and burning substances. Lekarz Wojskowy 62:311-316. [In Polish]

Pui C-H, Behm FG, Raimondi SC, Dodge RK, George SL, Rivera GK, Mirro JJ, Kalwinsky DK, Dahl GV, Murphy SB, Crist WM, Williams DL. 1989. Secondary acute myeloid leukemia in children treated for acute lymphoid leukemia. New England Journal of Medicine 321:136-142.

Pui C-H, Hancock ML, Raimondi SC, Head DR, Thompson E, Wiliams J, Kun LE, Bowman LC, Crist WM, Pratt CB. 1990. Myeloid neoplasia in children treated for solid tumors. Lancet 336:417-421.

Pullinger BD. 1947. Character of coagulation necrosis due to mustard gas (dichloroethyl sulfide). Journal of Pathology and Bacteriology 59:255-259.

Purko J. 1968. Biological action of sulfur mustard in cultured mammalian cells and in early teleost development (abstract). Proceedings of the Twelfth International Congress of Genetics 12:152.

Quiring K. 1979. Toxicity of bis(2-chloroethyl) sulfide (sulfur mustard gas). AMI-Berichte 1:32-40. [In German]

Railliet G. 1918. Parotitis in yperite victims. Bulletins et Memoires de la Societe Medicale des Hopitaux de Paris 3:1168-1171.[In French]

Rakhmanov VA, Inyakhina AV. 1969. Some aspects of current psoriasis therapy. Vestnik Dermatologii i Venerologii 43:16-29. [In Russian]

Rall DP. 1979. The role of laboratory animal studies in estimating carcinogenic risks for man. In: Davis W, Rosenfeld C, eds. Carcinogenic Risks: Strategies for Intervention. IARC Scientific Publication No. 25. Lyon: International Agency for Research on Cancer. 179-189.

Rall DP. 1988. Laboratory animal toxicity and carcinogenesis testing. Underlying concepts, advantages, and constraints. Annals of the New York Academy of Sciences 534:78-83.

Rall DP. 1990. Carcinogens in our environment. In: Vainio H, Sorsa M, McMichael AJ, eds. Complex Mixtures and Cancer Risk. IARC Scientific Publication No. 104. Lyon: International Agency for Research on Cancer. 233-239.

Rankin CJ, Jacobsen MK, Mitchell VA, Busbee DL. 1980. Reduction of nicotinamide adenine dinucleotide levels by ultimate carcinogens in human lymphocytes. Cancer Research 40:1803-1807.

Ratain MJ, Kaminer LS, Bitran JD, Larson RA, LeBeau MM, Skosey C, Purl S, Hoffman PC, Wade J, Vardiman JW, Daly K, Rowley JD, Golomb HM. 1987. Acute nonlymphocytic leukemia following etoposide and cisplatin combination chemotherapy for advanced non-small-cell carcinoma of the lung. Blood 70:1412-1417.

Ratnakar RP. 1919. A case of nystagmus caused by mustard gas. Lancet 1:423-424.

Razavi SM, Mahmoudi M, Keyhani A, Eslami MB, Eftekhar B. 1988. Study of some components of complement system in Iranian combatants poisoned with mustard gas. In: Abstracts of the First International Medical Congress on Chemical Warfare Agents in Iran. June 13-16, 1988. MashhadUniversity of Medical Sciences . Mashhad, Iran. No. 58.

Reed CI. 1920. The minimum concentration of dichlorethylsulphide (mustard gas) effective for the eyes of man. Journal of Pharmacology and Experimental Therapeutics 15:77-80.

Rees J, Harper P, Ellis F, Mitchell D. 1991. Mustard gas casualties (letter). Lancet 337:430.

Rehbein M. 1921. Observations on the action of dichloroethylsulfide on the skin: animal experiments. Dermatologische Wochenschrift 73:865-869. [In German]

Reiche F. 1930. Sequels of gas poisoning in warfare. Munchener Medizinische Wochenschrift 77:792-793. [In German]

Reid BD, Walker IG. 1966. Resistance to sulfur mustard: a comparison of some properties of strain L cells and a resistant subline. Cancer Research 26:1801-1805.

Reid BD, Walker IG. 1969. The response of mammalian cells to alkylating agents. II. On the mechanism of the removal of sulfur mustard-induced cross links. Biochimica et Biophysica Acta 179:179-188.

Reid EE. 1958. Mustard gas. In: Organic Chemistry of Bivalent Sulphur. Vol. 2. New York: Chemical Publishing. 237-451.

Reimer RR, Hoover R, Fraumeni JF Jr, Young RC. 1977. Acute leukemia after aklylatingagent therapy of ovarian cancer. New England Journal of Medicine 297:177-181.

Reimer RR, Hoover R, Fraumeni JF Jr, Young RC. 1978. Second primary neoplasms following ovarian cancer. Journal of the National Cancer Institute 61:1195-1197.

Rendu R. 1918. Lesions of the upper respiratory tract due to the new German vesicant gases. Lyon Medical 127:108-116. [In French]

Rendu R, Martin J.F. 1921. Histology of membranes in yperite victims. Lyon Medical 130:528-532.

Renga G. 1979. Bronchopulmonary tumors: epidemiology. Minerva Medica 70:173-194. [In Italian]

Rennie TAC, Small SM. 1943. Psychological Aspects of Chemical Warfare. Review Series, Vol. 1, No. 1. New York: Josiah Macy, Jr. Foundation.

Renshaw B. 1947. Observations on the role of water in the susceptibility of human skin to vesicant injury. Journal of Investigative Dermatology 9:75-85.

Requena L, Requena C, Sanchez M, Jaqueti G, Aguilar A, Sanchez-Yus E, Hernandez Moro B. 1988. Chemical warfare: cutaneous lesions from mustard gas. Journal of the American Academy of Dermatology 19:529-536.

Rhoads CP. 1947. Edward Gamaliel Janeway Lecture: the sword and the ploughshare (dichloroethyl sulfide poisoning at Bari, 1943 and work of Chemical Warfare Service, especially on nitrogen mustards or chloroethylamines). Journal of Mt. Sinai Hospital 13:299-309.

Ribiero PL. 1988. The Repair of bis-(b-chloroethyl) Sulfide (BCES)-Induced DNA Alkylation Damage and Its Possible Role in the Survival of Exposed Primary Monolayer Cultures of Rat Cutaneous Keratinocytes (dissertation). University of Michigan.

Ribiero PL, Mitra RS, Bernstein IA. 1991. Assessment of the role of DNA damage and repair in the survival of primary cultures of rat cutaneous keratinocytes exposed to bis(2-chloroethyl) sulfide. Toxicology and Applied Pharmacology 111:342-351.

Richard A. 1940. Disorders involving the motor nerves of the nervous system in war gas intoxication. Semaine des Hopitaux de Paris 16:97-104. [In French]

Richardson GM. 1948. Onset of pneumonic influenza in 1918 in relation to wartime use of mustard gas. New Zealand Medical Journal 47:4-16.

Ridley H. 1936. Mustard gas keratitis. Proceedings of the Royal Society of Medicine 29:962-963.

Rikimaru T, Nakamura M, Yano T, Beck G, Habicht GS, Rennie LL, Widra M, Hirshman CA, Boulay MG. 1991. Mediators initiating the inflammatory response released in organ culture by full-thickness human skin explants exposed to the irritant sulfur mustard. Journal of Investigative Dermatology 96:888-897.

Ritlop B. 1939. Poisoning of foods contaminated by chemical warfare agents. Orvosi Hetilap 83:311-315. [In Hungarian]

Riviere JE, Monteiro-Riviere NA. 1991. The isolated perfused porcine skin flap as an *in vitro* model for percutaneous absorption and cutaneous toxicology. Critical Reviews in Toxicology 21:329-344.

Roberts JJ. 1975. Inactivation and repair of the DNA template by cytotoxic alkylating agents. Excerpta Medica International Cancer Congress Series No. 353. 330-333.

Roberts JJ. 1978. The repair of DNA modified by cytotoxic, mutagenic, and carcinogenic chemicals. Advances in Radiation Biology 9:211-435.

Roberts JJ, Kotsaki-Kovatsi VP. 1986. Potentiation of sulfur mustard or cisplatin-induced toxicity by caffeine in Chinese hamster cells correlates with formation of DNA double-strand breaks during replication on a damaged template. Mutation Research 165:207-220.

Roberts JJ, Pascoe JM. 1971. Evidence for cross-linking of complementary strands of DNA in cultured mammalian cells by platinum compounds and mustard gas. In: Semonsky M, Hejzlar M, Masak S, eds. Advances in Antimicrobial and Antineoplastic Chemotherapy: Progress in Research and Clinical Application. Vol. II, Antineoplastic Chemotherapy Radioprotectives. Baltimore: University Park Press. 249-251.

Roberts JJ, Ward KN. 1973. Inhibition of post-replication repair of alkylated DNA by caffeine in Chinese hamster cells but not HeLa cells. Chemico-Biological Interactions 7:241-264.

Roberts JJ, Warwick GP. 1963. Studies of the mode of action of alkylating agents. 6. The metabolism of bis-2-chloroethylsulphide (mustard gas) and related compounds. Biochemical Pharmacology 12:1329-1334.

Roberts JJ, Brent TP, Crathorn AR. 1968. Mechanism of the cytotoxic action of alkylating agents on mammalian cells. In: Campbell PN, ed. Interaction of Drugs and Subcellular Components in Animal Cells. Boston: Little, Brown. 5-27.

Roberts JJ, Crathorn AR, Brent TP. 1968. Repair of alkylated DNA in mammalian cells. Nature 218:970-972.

Roberts JJ, Brent TP, Crathorn AR. 1971. Evidence for the inactivation and repair of the mammalian DNA template after alkylation by mustard gas and half mustard gas. European Journal of Cancer 7:515-524.

Roberts JJ, Pascoe JM, Plant JE, Sturrock JE, Crathorn AR. 1971. Quantitative aspects of the repair of alkylated DNA in cultured mammalian cells. Part 1. The effect of HeLa and Chinese hamster cell survival of alkylation of cellular macromolecules. Chemico-Biological Interactions 3:29-47.

Roberts JJ, Pascoe JM, Smith BA, Crathorn AR. 1971. Quantitative aspects of the repair of alkylated DNA in cultured mammalian cells. Part 2. Non-semiconservative DNA synthesis (repair synthesis) in HeLa and Chinese hamster cells following treatment with alkylating agents. Chemico-Biological Interactions 3:49-68.

Roberts JJ, Friedlos F, Scott D, Ormerod MG, Rawlings CJ. 1986. The unique sensitivity of Walker rat tumour cells to difunctional agents is associated with a failure to recover from inhibition of DNA synthesis and increased chromosome damage. Mutation Research 166:169-181.

Robinson JP. 1967. Chemical warfare. Science Journal 4:33-40.

Roche L, Grunwald E, Rouanet J. 1957. Occupational emphysema caused by yperite. Archiv des Maladies Professionelles de Medecine du Travail et de Securite Sociale 18:339-342. [In French]

Rodriques-Roisin R, Picado C, Roca J, Arrigo S, Agusti-Vidal A. 1986. Early lung function changes after short heavy exposure to chrysotile asbestos in non-smoking women. Bulletin Europeen de Physiopathologie Respiratoire 22:225-229.

Roese FH. 1933. Ocular lesions from war gases: review. Zentralblatt fur die Gesamte Ophthalmologie und Ihre Grenzgebiete 28:513-528. [In German]

Rogers EH, Chernoff N, Kavlock RJ. 1981. The teratogenic potential of cacodylic acid in the rat and mouse. Drug and Chemical Toxicology 4:49-61.

Rohrschneider W. 1937. Late ocular lesions (keratitis) following yellow cross (dichloroethylsulfide) poisoning. Klinische Monatsblatter fur Augenheilkunde 99:447-455. [In German]

Roland G. 1940. Review on yperite. Journal de Pharmacie de Belgique 22:199-204. [In French]

Romanowska-Gorecka B. 1971. Biochemical studies on the activity of lysosomal hydrolases in sulfur mustard poisoning. I. Effect of sulfur mustard on the activity of acid phosphatase in serum and isolated lysosome fraction of liver homogenates. Polskie Archiwum Medycyny Wewnetrznej 46:55-60. [In Polish]

Romanowska-Gorecka B. 1971. Biochemical studies on the activity of lysosomal hydrolases in sulfur mustard poisoning. II. Effect of sulfur mustard on the stability of lysosomal membranes and activity of b-glucuronidase in serum and in isolated lysosome fractions of liver homogenates. Polskie Archiwum Medycyny Wewnetrznej 46:61-67. [In Polish]

Rommereim RL, Hackett PL. 1986. Evaluation of the teratogenic potentialof orally administered sulfur mustard in rats and rabbits (abstract) . Teratology 70C.

Ronai ZA, Lambert ME, Weinstein IB. 1990. Inducible cellular responses to ultraviolet light irradiation and other mediators of DNA damage in mammalian cells. Cell Biology and Toxicology 6:105-126.

Rosenow EC, Wilson WR, Cockerill FR. 1985. Pulmonary disease in the immunocompromised host. Mayo Clinic Proceedings 60:473-487.

Rosenthal SM, Voegtlin C. 1930. Biological and chemical studies of the relationship between arsenic and crystalline glutathione. Journal of Pharmacology and Experimental Therapeutics 39:347-367.

Rosner F, Grunwald H. 1975. Multiple myeloma terminating in acute leukemia. American Journal of Medicine 58:339.

Ross PM, Carter DM. 1984. Cutaneous DNA repair mechanisms. In: Drill VA, Lazar P, eds. Cutaneous Toxicity. New York: Raven Press.

Ross WCJ. 1958. *In vitro* reaction of biological alkylating agents. Annals of the New York Academy of Sciences 68:669-681.

Ross WCJ. 1959. The tumour growth inhibitory activity of some sulphur mustard gas derivatives. Biochemical Pharmacology 2:215-220.

Ross WCJ. 1962. Biological Alkylating Agents: Fundamental Chemistry and Design of Compounds for Selective Toxicity. Butterworths: London.

Ross WD. 1966. Neuroses following trauma and their relation to compensation. In: Arieti S, ed. American Handbook of Psychiatry. New York: Basic Books. 131-147.

Rossman TG. 1981. Enhancement of UV-mutagenesis by low concentrations of arsenite in *E. coli.* Mutation Research 91:207-211.

Roth EF. 1974. Effects of nitrogen and sulfur mustards on hemoglobin S & on intact red cells. 15th Congress of the International Society of Hematology, Sept. 1-6, 1974, Jerusalem, Israel.

Roth F. 1956. Chronic arsenicism and cancers among vineyard workers in the Moselle Valley. Zeitschrift fur Krebsforschung und Klinische Onkologie 61:287-319.

Rothea F. 1933. Toxicology of war gases. Their pathogenic chemical action and therapeutic treatment. Bulletin des Biologistes Pharmaciens 20:5-14. [In French]

Rothlin E. 1942. On yperite experiences in man and animal. Schweizerische Medizinische Wochenschrift 72:385-388. [In German]

Rothlin E. 1943. Results of animal experiments concerning the progress of poisoning with phosgene and yperite. Schweizerische Medizinische Wochenschrift 73:1205-1210. [In German]

Roubal J, Pokorny F. 1939. Effect of aqueous solutions of dichlorovinylarsine and dichloroethylsulfide on eye of rabbits. Casopis Lekaru Ceskych 78:633-639. [In Czech]

Roueche H, Poirson. 1919. A contribution to the study of laryngeal symptoms in persons intoxicated by gas. Montpellier Medical 38:207-209. [In French]

Rovida G. 1929. Experimental research with Lewisite. III. Action on the human skin. Sperimentale 83:115-120. [In Italian]

Rovida G. 1929. Lewisite II. Action on the skin of animals. Sperimentale 83:101-113. [In Italian]

Rowland KM Jr, Murthy A. 1986. Hodgkin's disease: long-term effects of therapy. Medical and Pediatric Oncology 14:88-96.

Rozmiarek H, Capizzi RL, Papirmeister B, Fuhrman WH, Smith WJ. 1973. Mutagenic activity in somatic and germ cells following chronic inhalation of sulfur mustard (abstract). Mutation Research 21:13-14.

Rudge E. 1984. Chemists at war. Chemistry in Britain 20:138-141.

Rudolovitch M. 1931. Ocular lesions due to dichloroethylsulfide and their treatment. Vojno-Sanitetski Glasnik 2:627-639. [In Serbo-Croatian]

Ruhland A, Kircher M, Wilborn F, Brendel M. 1981. A yeast mutant specifically sensitive to bifunctional alkylation. Mutation Research 91:457-462.

Runne U, Gartmann H 1973. Vesicular lesion of the male genitalia due to an aminoalkyl compound (halogenated alkylamine). Berufsdermatosen 21:6-11. [In German]

Ryland A. 1919. The laryngeal changes induced by mustard gas. Journal of Laryngology, Rhinology and Otology 24:153-156.

Sabshin Zl. 1921. Chronic effects of war-gassing: notes on the examination of 1200 cases. New York Medical Journal 114:232.

Salzgeber B. 1967. On the experimental study of phocomelic genesis in the chick embryo. Comptes Rendus Hebdomadaires des Seances de l'Academie des Sciences. D: Sciences Naturelles 264:395-397. [In French]

Salzgeber B. 1968. Study of the distal malformations obtained after treatment with nitrogen yperite of ectodermal constituent of the extremity buds. Comptes Rendus Hebdomadaires des Seances de l'Academie des Sciences. D: Sciences Naturelles 267:90-92. [In French]

Salzgeber B. 1971. Concerning the development of the limb buds of chicken embryos treated with nitrogen yperite at the 48 hour stage of incubation. Comptes Rendus Hebdomadaires des Seances de l'Academie des Sciences. D: Sciences Naturelles 272:3083-3086. [In French]

Salzgeber B. 1975. The effect of nitrogen mustard on the incorporation of tritiated thymidine into chick embryo limb buds. Autoradiographic study. Comptes Rendus de l'Academie des Sciences 280:911-914. [InFrench]

Salzgeber B. 1976. Genesis of malformations of the distal parts of chick embryo extremities (adactylia, hemimelia). Comptes Rendus de l'Academie des Sciences 283:1241-1244. [In French]

Salzgeber B, Wolff E. 1966. Embryological explanation of experimental phocomelia in the chick embryo. Bulletin de L'Academie Nationale de Medecine 150:565-571. [In French]

Sampaolo A, Binetti R. 1989. Improvement of a practical method for priority selection and risk assessment among existing chemicals. Regulatory Toxicology and Pharmacology 10:183-195.

Sams WM, Lynch PJ. 1990. Principles and Practice of Dermatology. New York: Churchill Livingstone.

Sandall TE. 1922. The later effects of gas poisoning. Lancet 2:857-859.

Sandelowsky I, Simon GA, Bel P, Barak R, Vincze A. 1992. N1-(2-hydroxyethylthioethyl)4-methyl imidazole (4-met-1-imid-thiodigylcol) in plasma and urine: a novel metabolite following dermal exposure to sulphur mustard (1). Archives of Toxicology 66:296-297.

Sanders KM, Innace JK, Gross CL, Smith WJ. 1989. Flow cytometric analysis of peripheral blood lymphocytes for the assessment of toxicity by alkylating agents. Alternative Methods in Toxicology 7:255-263.

Sangailo AK. 1930. [Effect of methyldichlorarsine on cardiovascular system]. Jurnal Eksperimental'noy Biologii i Medetoiny Series A 14:76-80, 81-99.

Sanotskiy VA. 1932. Pathologic and clinical aspect of dichlorethyl sulfide poisoning in warfare. Klinicheskaia Meditsina 10:179-188. [In Russian]

Santoro A, Bonadonna G, Valagussa P, Zucali R, Viviani S, Villani F, Pagnoni AM, Bonfante V, Musumeci R, Crippa F, Tess JDT, Banfi A. 1987. Long-term results of combined chemotherapy-radiotherapy approach in Hodgkin's disease: superiority of ABVD plus radiotherapy versus MOPP plus radiotherapy. Journal of Clinical Oncology 5:27-37.

Sanyal MK, Kitchin KT, Dixon RL. 1981. Rat conceptus development *in vitro:* comparative effects of alkylating agents. Toxicology andApplied Pharmacology 57:14-19.

Saracci R. 1981. The IARC monograph program on the evaluation of the carcinogenic risk of chemicals to humans as a contribution to the identification of occupational carcinogens. In: Peto R, Schneiderman M, eds. Quantification of Occupational Cancer.Banbury Report, Vol. 9. Cold Spring Harbor, NY: Cold Spring Harbor Laboratory. 165-176.

Saroya SZ. 1989. The Effect of bis(b-chloroethyl) Sulfide on DNA Synthesis of a Stratified Epidermal Culture System (dissertation). University of Michigan.

Sartorelli E, Giubielo M, Bartalini E. 1957. Contribution to the study of chronic asthmatiform bronchitis with pulmonary emphysema as a consequence of occupational mustard gas poisoning.Medicine del Lavoro 48:336-346. [In Italian]

Sasser LB, Cushing JA, Buschbom RL, Dacre JC. 1989. Two generation reproduction study of sulfur mustard in rats (abstract). Toxicologist 274.

Sasser LB, Cushing JA, Dacre JC. 1990. Dominant lethal effect of sulfur mustard in rats (abstract). Toxicologist 225.

Sato T, Utsumi S, Kajikawa K, Ikeda H. 1967. A case of cancer ofthe larynx found in mustard gas poisoning . Nippon Jibiinkoka Gakkai Kaiho 70:1773-1778. [In Japanese]

Satterlee A. 1960. The arsenic poisoning epidemic of 1900. New England Journal of Medicine 263:676-684.

Sauerbrier I. 1981. Is the extent of teratogenic action determined by time of induction? Verhandlungen der Anatomischen Gesellschaft 75:465-468. [In German]

Savage JRK, Breckon G. 1981. Differential effects of sulfur mustard on S phase cells of primary fibroblast cultures from Syrian hamsters *Mesocricetus-auratus.* Mutation Research 84:375-387.

Sayers RR, Dudley HC. 1938. Toxicology of phenyldichloroarsine. II. Response of man to PDA-oil mixtures. Public Health Reports 53:1292-1301.

Scavarelli-Karantsavelos RM. 1989. Comparative Effect of bis(b-chloroethyl) sulfide on Basal and Differentiated Keratinocytes (dissertation). University of Michigan.

Scavarelli-Karantsavelos RM, Saroya SM, Vaughan FL, Bernstein IA. 1990. Pseudoepidermis, constructed *in vitro,* for use in toxicological and pharmacological studies. Skin Pharmacology 3:115-125.

Schasteen CS, Reed DJ. 1983. The hydrolysis and alkylation activities of S-2 halo ethyl-L cysteine analogs evidence for extended half-life. Toxicology and Applied Pharmacology 70:423-432.

Scherf D. 1942. Pathologic physiology of the respiration and circulation in war gas poisoning. Bulletin of the New York Medical College (June-October):52-58.

Scherling SS, Blondis RR. 1945. The effect of chemical warfare agents on the human eye. Military Surgeon 96:70-78.

Schilling A. 1962. Repetitive massive chemotherapy with sulfur mustard analogs and bone marrow protection by vascular occlusion. Cancer Chemotherapy Reports 16:527-530.

Schirren C. 1963. Impotentia coeundi as a vesicant-induced injury. Zeitschrift fur Haut und Geschlechtskrankheiten 34:189-195. [In German]

Schmahl D, Osswald H. 1970. Experimental studies on the carcinogenic effects of anticancer chemotherapeutic and immunosuppressive agents. Arzneimittelforschung 20:1461-1467.

Schmelzer H. 1933. Eye injuries due to blister gas (recognition of first aid and further treatment). Munchener Medizinische Wochenschrift 80:70-83. [In German]

Schmidt L, Schennetten F. 1956. On the clinical toxicology of nitrogen lost. Zeitschrift fur die Gesamte Innere Medizin und Ihre Grenzgebiete 11:477-480. [In German]

Schneider W. 1940. Yellow cross induced injury to the skin. Archiv fur Gewerbepathologie und Gewerbehygiene 10:370-377. [In German]

Schoene K, Bruckert HJ, Schreiber G, Wodtke G. 1989. A method for correlating skin exposure to S mustard vapor with skin damage. American Industrial Hygiene Association Journal 50:569-573.

Scholtyssek H. 1960. Late keratitis after mustard gas poisoning and the importance of its early diagnosis for therapy and expertise. Klinische Monatsblatter fur Augenheilkunde 136:243-254. [In German]

Scholz RO, Woods AC. 1947. Relapsing and chronic ocular lesions following mustard gas burns. Archives of Ophthalmology 37:137-148.

Schonwald S. 1992. Mustard gas. PSR Quarterly 2:92-97.

Schowing J. 1981. Application of a direct method in experimental embryology, the deposit of a teratogenic substance. Revue Medicale de la Suisse Romande 102:423-427. [In French]

Schrafl A. 1936. Protection against and prophylactic treatment of dichloroethylsulfide poisoning. Schweizerische Medizinische Wochenschrift 66:591-593. [In German]

Schrafl A. 1938. [The symptoms, prophylaxis and therapy of vesicant wounds on the skin caused by war gases]. Protar 4:96-98.

Schreiber G, Mueller OG, Baumert HP, Schoene K. 1981. Investigations Concerning the Protection and Decontamination of Skin from Highly Toxic Organophosphorus Compounds and Mustard Gas. U.S. Army Foreign Science and Technology Center Translation No. FSTC-HT-68-80. DTIC AD-B056-000. [In German]

Schwartz F. 1937. [Experimental studies on mustard action]. Protar 3:34-37.

Schwartz F. 1938. [Physiological sensitization to mustard gas]. Protar 4:17-18.

Scianowski J. 1977. The influence of b,b'-dichlordiethyl sulfide on lactate dehydrogenase (E.C.1.1.1.27) and its isoenzymes in the internal organs of rats. Acta Medica Polona 18:97-112.

Scianowski J. 1988. Alterations of liver superoxide dismutase and catalase activities in rats intoxicated with b,b-dichlorodiethyl sulfide. Bromatologia i Chemia Toksykologiczna 21:283-287. [In Polish]

Scianowski J, Smok W. 1990. Glutathione peroxidase activity and malonic dialdehyde content in blood serum and liver homogenates of rats intoxicated with sulfuric mustard yperite. Bromatologia i Chemia Toksykologiczna 23:57-61. [In Polish]

Scianowski J, Strzelczyk M, Strozynski H. 1988. Comparative evaluation of acute toxicity and mutagenicity of sulfur mustard gas and its structural homolog 2-chloroethyl-3chloropropyl sulfide. Lekarz Wojskowy 64:432-438. [In Polish]

Scott D. 1976. Action of anticancer drugs on the cell cycle. Chromosome damage and differential stage sensitivity. In: Hellmann K, Connors TA, eds. Chemotherapy. Proceedings of the 9th International Congress on Chemotherapy. Vol. 7. New York: Plenum Press. 95-103.

Scott D. 1977. Chromosome aberrations, DNA post-replication repair and lethality of tumour cells with a differential sensitivity to alkylating agents. Chromosomes Today 6:391-401.

Scott D, Bigger TRL. 1972. The induction of chromosomal aberrations by sulphur mustard in marsupial lymphocytes. Chromosomes Today 3:162-176.

Scott D, Fox M, Fox BW. 1974. The relationship between cell survival, chromosome aberrations and DNA repair in tumour cell lines of differential sensitivity to x-rays and sulphur mustard (abstract). British Journal of Cancer 29:99.

Scott D, Fox M, Fox BW. 1974. The relationship between chromosomal aberrations, survival, and DNA repair in tumour cell lines of differential sensitivity to x-rays and sulphur mustard. Mutation Research 22:207-221.

Scott D, Fox M, Fox BW. 1975. Differential induction of chromosome aberrations in mammalian cell lines (abstract). Mutation Research 29:201-202.

Scott D, Fox M, Marshall RR. 1975. Effect of caffeine on the survival of pairs of mammalian cell lines of differential sensitivity to radiation and alkylating agents (abstract). British Journal of Cancer 32:759.

Seaton A. 1984. Occupational pulmonary neoplasms. In: Morgan WK, Seaton A, eds. Occupational Lung Diseases. Philadelphia: W.B. Saunders. 657-675.

Seeley TD, Nowicke JW, Meselson M, Guillemin J, Akratanakul P. 1985. Yellow rain. Scientific American 253:128-137.

Segal D. 1987. The Soviet Union's mighty chemical warfare machine. Army 37:26-38.

Seidel R, Westphal HJ. 1973. Treatment and process of severe skin damages due to yperite. Deutsche Gesundheitswesen 28:2473-2476. [In German]

Seiler JP. 1975. Chemical mutagenesis. Chimia 29:8-17.

Seligman AM, Milden M, Friedman OM. 1949. Study of inhibition of tumor growth in mice and rats with 10-methyl-1,2-benzanthracene and derivatives related to nitrogen and sulfur b-chloroethyl vesicants. Cancer 2:701-706.

Sellei C. 1939. Gas metabolism studies with heteroplastic transplanted tumors. Zeitschrift fur Krebsforschung und Klinische Onkologie 48:520-531. [In German]

Semnanian S, Mishmast NG, Motmedi F, Hosseini ES A. 1988. Evaluation of the relationship between 30 different factors with 14 spirometric parameters in 200 Iranian patients injured with chemical warfare agents. In: Abstracts of the First International Medical Congress on Chemical Warfare Agents in Iran. June 13-16, 1988. Mashhad University of Medical Sciences. Mashhad, Iran. No. 53.

Semon HC. 1922-1923. Case of gas-burn scarring. Proceedings of the Royal Society of Medicine 16:95.

Sergent E. 1925. The respiratory sequelae of war-gas poisoning. Presse Medical 33:201-205. [In French]

Sergent E, Haas J. 1920. Pulmonary tuberculosis and the sequelae of intoxications by gas. Medecine 1:466, 469-471. [In French]

Shafiee A, Cheraghali A, Kebriaeizadeh A. 1989. Spectrophotometric and gas chromatographic methods for detection and essay of mustard gas. In: International Conference on Combatting the Use of Chemical and Biological Weapons. May 24-27, 1989, Geneva. 139-152.

Shelby MD. 1988. The genetic toxicity of human carcinogens and its implications. Mutation Research 204:3-15.

Shelby MD, Zeiger E. 1990. Activity of human carcinogens in the *Salmonella* and rodent bone-marrow cytogenetics tests. Mutation Research 234:257-261.

Shigenobu H. 1957. Capillary figures of the skin observed among the workers in the so-called poison gas factory at Okuno island. Journal of the Hiroshima Medical Association 10:465-466. [In Japanese]

Shigenobu T. 1973. Clinical observations on retired workers of Okuno-jima poison gas factory with special reference to chronic bronchitis. Gencho Hiroshima Igaku 21:409-499. [In Japanese]

Shigenobu T. 1980. Occupational cancer of the lungs: cancer of the respiratory tract among

workers manufacturing poisonous gases. Nippon Kyobu Shikkan Gakkai Zasshi 18:880-885. [In Japanese]
Shigenobu T, Nishimoto Y. 1985. Examples of health disorders due to chemical substances: health disorders due to sulfur mustard gas on Okuno Island. Hoshasen Seibutsu Kenkyu 20:127-138. [In Japanese]
Shimkin MB, Weisburger JH, Weisburger EK. 1966. Bioassay of 29 alkylating chemicals by the pulmonary-tumor response in strain A mice. Journal of the National Cancer Institute 36:915-935.
Shimura K, Nomura H, Niimoto M, Hattori T, Kamitsuna A, Tokuoka S. 1978. Pathohistological studies on four cases of gastric carcinoma in Okuno-jima poison gas workers. Journal of the Hiroshima Medical Association 31:277-281. [In Japanese]
Shirazi SF, Balali M. 1988. Comparison of early and late toxic effects of sulfur mustard poisoning in two-year periods. In: Abstracts of the First International Medical Congress on Chemical Warfare Agents in Iran. June 13-16, 1988. Mashhad University of Medical Sciences. Mashhad, Iran. No. 73.
Shire TH. 1938. Gas: an introduction to reality. Guys Hospital Gazette 52:385-390.
Shooter KV. 1972. Some aspects of the interaction of carcinogenic and mutagenic agents with purines in nucleic acids. Jerusalem Symposia on Quantum Chemistry and Biochemistry 4:509-518.
Shooter KV, Edwards PA, Lawley PD. 1971. The action of mono- and di-functional sulphur mustards on the ribonucleic acid-containing bacteriophage μ-2. Biochemical Journal 125:829-840.
Shufflebotham F. 1919. Influenza among poison gas workers. British Medical Journal 1:478-479.
Silva C. 1919. Gangrene from gassing. Riforma Medica 35:28-30. [In Italian]
Sim VM. 1976. Chemicals used as weapons of war. In: DiPalma J, ed. Drill's Pharmacology in Medicine. 4th ed. New York: McGraw-Hill. 1232-1248.
Simons SS, Chakraborti KK, Cavanaugh AH. 1990. Arsenite and cadmium(II) as probes of glucocorticoid receptor structure and function. Journal of Biological Chemistry 265:1938-1945.
Simsa J, Mraz J. 1974. Changes in the contents of some rat serum proteins under pathologic conditions. 2. The influence of some further noxious factors in comparison with the effect of ionizing radiation. Radiobiologia Radiotherapia 15:509-519.
Sinclair DC. 1948. The clinical features of mustard gas poisoning in man. British Medical Journal (Part 2):290-294.
Sinclair DC. 1949. The clinical reaction of the skin to mustard gas vapour. British Journal of Dermatology and Syphilis 61:113-125.
Sinclair DC. 1949. Treatment of skin lesions caused by mustard gas. British Medical Journal 1:476.
Sinclair DC. 1950. Disability produced by exposure of skin to mustard gas vapour. British Medical Journal (Part 1):346-349.
Singh R. 1984. Triazenes as transport form of sulfur mustard synthesis of 3-S-2 chloroethylthioethylaryltriazenes and study of their reactions in aqueous and nonaqueous solutions. Indian Journal of Chemistry, Section B, Organic Chemistry, Including Medicinal Chemistry 23:1088-1097.
Skegg DCG, Manning K, Doll R, Stell PM. 1980. Mustard gas and laryngeal cancer (abstract). Clinical Otolaryngology 5:78.
Skipper PL, Tannenbaum SR. 1990. Protein adducts in the molecular dosimetry of chemical carcinogens. Carcinogenesis 11:507-518.
Skot-Hansen JP. 1934. Dichloroethylsulfide injuries in war. Ugeskrift for Laeger 96:685-691. [In Danish]

Skurnik Y, Baniel J, Shemer J. 1985. Effects of mustard gas in chemicalwarfare . Harefuah 109:240-242. [In Hebrew]
Slizynska H. 1968. Triplications and the problem of nonhomologous crossing-over. Genetics Research 11:201-208.
Smith G, Walker J. 1990. The histopathology of pulmonary reactions to drugs. Clinics in Chest Medicine 11:95-117.
Smith HW, Clowes GA, Marshall JV. 1919. On dichloroethylsulfide (mustard gas). IV. The mechanism of absorption by the skin. Journal of Pharmacology and Experimental Therapeutics 13:1-30.
Smith PG, Doll R. 1978. Age and time-dependent changes in the rates of radiation-induced cancers in patients with anklylosing spondylitis following a single course of x-ray treatment. In: Late Biological Effects of Ionizing Radiation. Vol. 1. Vienna: International Atomic Energy Agency. 205-218.
Smith PK, Nadkarni MV, Trams EG, Davison C. 1958. Distribution and fate of alkylating agents. Annals of the New York Academy of Sciences 68:834-852.
Smith WJ, Dunn MA. 1991. Medical defense against blistering chemical warfare agents. Archives of Dermatology 127:1207-1213.
Smith WJ, Gross CL, Chan P, Meier HL. 1990. The use of human epidermal keratinocytes in culture as a model for studying the biochemical mechanisms of sulfur mustard toxicity. Cell Biology and Toxicology 6:285-292.
Smith WJ, Sanders KM, Gales YA, Gross CL. 1991. Flow cytometric analysis of toxicity by vesicating agents in human cells *in vitro.* Cutaneous and Ocular Toxicology 10:33-42.
Smith WJ, Cowan FM, Broomfield CA. 1991. Increased proteolytic activity in human epithelial cells following exposure to sulfur mustard (abstract). FASEB Journal 5:A828.
Snider TH, Joiner RL, Feder PI, Kiser RC, Keys WB, Fisher GL. 1987. Tissue distribution of arsenic in the rabbit following subcutaneous administration of Lewisite with or without British anti-Lewisite therapy. In: Proceedings of the 6th Chemical Defense Bioscience Review. Frederick, MD. 743-746.
Snider TH, Hobson DW, Olson CT, Dill GS, Joiner RL. 1989. Evaluation of hematologic endpoints in response to sulfur mustard exposure in rats and rabbits (abstract). Toxicologist 287.
Snider TH, Wientjes M, Joiner R, Fisher G. 1990. Arsenic distribution in rabbits after Lewisite administration and treatment with British anti-Lewisite (BAL). Fundamental and Applied Toxicology 14:262-272.
Sobels FH. 1962. Rates of forward and reverse mutation in *Drosophila* after exposure to mustard gas and x-rays. Genetica 33:31-44.
Sobels FH. 1975. A comparison of the mutagenic effects of chemicals and ionizing radiation. In: Nygaard OF, Adler HI, Sinclair WK, eds. Radiation Research: Biomedical, Chemical and Physical Perspectives. New York: Academic Press. 958-965.
Sobels FH, Van Steenis H. 1957. Chemical induction of crossing-over in *Drosophila* males. Nature 179:29-31.
Sohier R. 1939. Treatment of lesions due to war gases. Presse Medicale 47:1064-1069. [In French]
Sohrabpour H. 1987. Observation and clinical manifestations of patients injured with mustard gas. Medical Journal of the Islamic Republic of Iran 1:32-37.
Sohrabpour H, Masjedi MR, Bahadori M. 1988. Pulmonary complications in chemically injured patients. In: Abstracts of the First International Medical Congress on Chemical Warfare Agents in Iran. June 13-16, 1988. Mashhad University of Medical Sciences. Mashhad, Iran. No. 22.
Sohrabpour H, Masjedi MR, Bahadori M. 1989. Late complications of sulfur mustard in

respiratory system. In: International Conference on Combatting the Use of Chemical and Biological Weapons. May 24-27, 1989. Geneva. 108-116.
Sollman T. 1918. Action of HS. Journal of Pharmacology and Experimental Therapeutics 11:229.
Sollman T. 1919. Dichloroethylsulfide (mustard gas). I. The influence of solvents, adsorbents, and chemical antidotes on the severity of the human skin lesions. Journal of Pharmacology and Experimental Therapeutics 12:303-318.
Sollman T. 1919. Dichloroethylsulfide (mustard gas). II. The question of induced hypersusceptibility of the skin. Journal of Pharmacology and Experimental Therapeutics 12:319-321.
Sollmann MT. 1957. A Manual of Pharmacology and Its Application to Therapeutics and Toxicology. 8th ed. Philadelphia: W.B. Saunders. 192-201.
Soloman Z. 1989. Characteristic psychiatric symptomatology of post-traumatic stress disorder in veterans: a three year follow-up. Psychological Medicine 19:927-936.
Solomon Z, Mikulincer M, Benbenishty R. 1989. Locus of control and combat-related post-traumatic stress disorder: the intervening role of battle intensity, threat appraisal and coping. British Journal of Clinical Psychology 28:131-144.
Soloway AH, Brumbauch RJ, Witiak DT. 1983. Carbinolamines and related structures: potential alkylating metabolites of clinically active anticancer drugs. Journal of Theoretical Biology 102:361-374.
Somani SM, ed. 1992. Chemical Warfare Agents. San Diego:Academic Press.
Somani SM, Babu SR. 1989. Toxicodynamics of sulfur mustard. International Journal of Clinical Pharmacology, Therapy and Toxicology 27:419-435.
Sommer. 1938. Late eye injuries following dichlorodiethylsulfide erosion. Deutsch Militararzt 3:519-521. [In German]
Sommers S, McManus RG. 1953. Multiple arsenical cancers of skin and internal organs. Cancer 6:347-359.
Sonbati EM, Auerbach C. 1960. The brood pattern for intragenic and intergenic changes after mustard gas treatment of *Drosophila* males. Zeitschrift fur Vererbungslehre 91:253-258.
Sorscher D, Conolly R. 1989. Pretreatment of primary rat cutaneous epidermal keratinocyte culture with a low concentration of MNNG: effect on DNA cross-linking measured *in situ* after challenge with bis-2-chloroethyl sulfide. Journal of Toxicology and Environmental Health 27:367-379.
Sourdille GP. 1936. Lesions of cornea, especially marginal vascular lesions by yperite. Bulletin de la Societe d'Ophtalmologie de Paris. 799-801. [In French]
Sourdille GP. 1949. Late keratitis due to dichloroethyl sulfide: treatment by conjunctival graft. Ophthalmologica 118:893-908. [In French]
Speer FD. 1942. Pathology of the vesicants and pulmonary irritants. Bulletin of the New York Medical College 5:59-63.
Spehl PM. 1924. The sequelae of war gases. Archives Medicales Belges 77:1-15.
Spiegelberg U. 1961. Psychopathological-neurological damages following the effect of synthetic poisons. In: Wehrdienst und Gesundheit. Vol. 3. Darmstadt: Wehr und Wissen Verlagsgesellschaft. [In German]
Spiegelberg U. 1963. Psychopathological-neurological late and permanent damage following occupational intoxication due to phosphate ester. Excerpta Medica International Cancer Congress Series No. 62. 1778-1780. [In German]
Spiers EM. 1986. Chemical Warfare. Urbana: University of Illinois Press.
Spiers EM. 1989. Chemical Weaponry: A Continuing Challenge. New York: St. Martins Press.
Spiker R. 1980. The Stoltzenberg syndrome: West Germany's biggest environmental scandal continues. Ambio 9:256-257.

Spitz S. 1948. The histological effects of nitrogen mustards on human tumors and tissues. Cancer 1:383-398.

Spurr CL. 1947. Influence of nitrogen mustards on the antibody response. Proceedings of the Society for Experimental Biology and Medicine 64:259.

Squibb KS, Fowler BA. 1983. The toxicity of arsenic and its compounds. In: Fowler BA, ed. Biological and environmental affects of arsenic. Vol. 6. Amsterdam: Elsevier Science. 233-269.

Srytr F. 1930. Yperite ocular injuries. Oftal Sborn 5:215-222. [In German]

Stack HE, Coombs CF, Rolfe R. 1919-1920. Poisoning by mustard gas. Bristol MedicoChirurgical Journal 37:151-162.

Staehelin R. 1920. Late effects on the respiratory organs of poisoning by the war gases. Jahreskurse fur Arztliche Fortbildung (February). [In German]

Stampfer JF. 1982. Respirator canister evaluation for nine selected organic vapors. American International Hygiene Association Journal 43:319-328.

Stedman's Medical Dictionary. 1976. 23rd ed. Baltimore: Williams & Wilkins.

Stein WH, Fruton JS. 1946. Chemical reactions of mustard gas and related compounds. IV. Chemical reactions of b-chloroethyl-b-hydroxyethylsulfide. Journal of Organic Chemistry 11:686-691.

Stein WH, Moore S, Bergmann M. 1946. Chemical reactions of mustard gas and related compounds. I. The transformation of mustard gas in water. Formation and properties of sulfonium salts derived from mustard gas. Journal of Organic Chemistry 11:664-674.

Stenhouse HA. 1923. Cholecystitis of chemical origin in man following inhalations of poison gas. U.S. Naval Medical Bulletin 19:291-296.

Sterlin RN, Emeleyanov VI, Zimin VI. 1973. Vesicant toxic agents. In: Civil Defense of the USSR. 2nd ed. Moscow: U.S. Army Foreign Science and Technology Center Translation No. FSTC-HT-23-1000-73. DTIC AD-770 735.

Stevens CM, Mylroie A. 1950. Mutagenic activity of b-chloroalkyl amines and sulphides. Nature 166:1019.

Stewart I. 1948. Organizing Scientific Research for War. Science in World War II: Office of Scientific Research and Development. Boston: Little, Brown.

Stiefler G. 1923. Striated symptom complex of a gas poisoning suffered in the field. Zentralblatt fur die Gesamte Neurologie und Psychiatrie 81:142-157. [In German]

Stiegler HW. The b-chlorvinyl-arsines and their derivatives. Easton, PA: Eschenbach Printing.

Stockholm International Peace Research Institute. 1971. The Problem of Chemical and Biological Warfare: A Study of the Historical, Technical, Military, Legal, and Political Aspects of Chemical and Biological Warfare and Possible Disarmament Measures. Vol. 1, The Rise of Chemical and Biological Weapons. Stockholm: Almqvist & Wiksell.

Stott H, Fox W, Girling DJ, Stephens RJ, Galton DA G. 1977. Acute leukemia after busulphan. British Medical Journal 2:1513-1517.

Strauss BS. 1971. Physical-chemical methods for the detection of the effect of mutagens on DNA. Chemical Mutagens: Principles and Methods for Their Detection. 145-174.

Strauss BS. 1976. Repair of DNA adducts produced by alkylation. In: Smith KC, ed. Aging, Carcinogenesis, and Radiation Biology: The Role of Nucleic Acid Addition Reactions. New York: Plenum Press. 287-314.

Strong LC, Herson J, Osborne BM, Sutow WW. 1979. Risk of radiation-related subsequent malignant tumors in survivors of Ewing's sarcoma. Journal of the National Cancer Institute 62:1401-1406.

Strozynski H, Strzelczyk M, Scianowski J. 1987. Studies on mutagenicity of 3-chloroprophylsulfide derivatives, potential antipsoriatic agents. Polish Journal of Pharmacology and Pharmacy 39:405-410.

Strudhold H. 1923. The effect of the vesicants diphenylchloroarsine (blue cross) and

ethyl dichloroarsine on the human skin. Zeitschrift fur Biologie 78:195-231. [In German]
Stuttgen G. 1980. The pathophysiology of blistering. Dermatologica 160:145-160. [In German]
Sulzberger MB, Baer RL, Kanof A, Lowenberg C. 1947. Skin sensitization to vesicant agents of chemical warfare. Journal of Investigative Dermatology 8:365-393.
Symonds M. 1980. The "second injury" to victims. Evaluation and Change (Special Issue):36-38.
Szarejko R. 1974. Assessment of changes induced by mustard gas in the upper respiratory passages and trial of early treatment in rats. Otolaryngologia Polska 28:617-618. [In Polish]
Szczygiel A, Just J. 1937. Influence of yperite (mustard gas) on hair growth in rabbits. Archives Internationales de Medecine Experimentale 12:626-631.
Szigeti A, Kramer M, Petranyi G. 1980. Clinical observations in cutaneous lymphomas. Orvosi Hetilap 121:2249-2252. [In Hungarian]
Tabarestani M, Farhoudi M, Balali M. 1988. Stem cell and erythroid precursors disorders in three patients with sulfur mustard poisoning. In: Abstracts of the First International Medical Congress on Chemical Warfare Agents in Iran. June 13-16, 1988. Mashhad University of Medical Sciences. Mashhad, Iran. No. 10.
Tabatabai SM, Rizabi N, Kamkardel MH, Balali M. 1988. Study of psychiatric complications of poisonings with chemical warfare agents. In: Abstracts of the First International Medical Congress on Chemical Warfare Agents in Iran. June 13-16, 1988. Mashhad University of Medical Sciences. Mashhad, Iran. No. 66.
Tahara E. 1968. Pathologic studies on carcinomas of the upper respiratory tract in mustard gas workers. Gencho Hiroshima Igaku 16:133-172. [In Japanese]
Taher AA. 1992. Cleft lip and palate in Tehran. Cleft Palate Craniofacial Journal 29:15-16.
Tainter ML. 1942. Medical aspects of war gases and chemicals: summary. Stanford Medical Bulletin 1:6-8.
Takahashi A. 1959. Studies on function of lungs and circulation observed in cases exposed to poison gases. Report I: Rentgenological study on chest affected by poison gas exposure. Journal of the Hiroshima Medical Association, Special Series 12:882-891. [In Japanese]
Takahashi A. 1959. Studies on functions of lungs and circulation observed in cases exposed to poison gases. Report II: Ventilatory function affected by poison gas exposure. Journal of the Hiroshima Medical Association, Special Series 12:892-906. [In Japanese]
Takahashi A. 1959. Studies on functions of lungs and circulation, observed in cases exposed to poison gases. Report III: Circulatory function tests affected by poison gas exposure. Journal of the Hiroshima Medical Association, Special Series 12:1077-1094. [In Japanese]
Takahashi A, Ishida S, Takemoto S, Takashina S, Mukai M, Origuchi K. 1959. Influence of changing body position upon pulmonary ventilation in cases exposed to poison gases. Journal of the Hiroshima Medical Association, Special Series 12:1193-1196. [In Japanese]
Takeshima Y, Inai K, Kobuke T, Tokuoka S. 1990. Histopathological study of hyperplastic lesions in the tracheobronchial epithelium of former workers engaged in mustard gas factory. Hiroshima Daigaku Igaku Zasshi 38:853-864. [In Japanese]
Tanaka K, Miura N, Satokata I, Miyamoto I, Yoshida MC, Satoh Y, Kondo S, Yasui A, Okayama H, Okada Y. 1990. Analysis of a human DNA excision repair gene involved in group A xeroderma pigmentosum and containing a zinc-finger domain. Nature 348:73-76.

Tanaka Y. 1988. Poison gas: the story Japan would like to forget. Bulletin of the Atomic Scientists 44:10-19.

Tarbell DS, Tarbell AT. 1981. Roger Adams: Scientist and Statesman. Washington, DC: American Chemical Society.

Tari AS, Rahimi F, Lashay AR. 1988. Ocular lesions of chemical vesicatory weapons. In: Abstracts of the First International Medical Congresson Chemical Warfare Agents in Iran . June 13-16, 1988. Mashhad University of Medical Sciences. Mashhad, Iran. No. 35.

Tarras-Wahlberg B. 1939. Gas injuries from dermatologic point of view. Svenska Lakartidningen 36:1079-1087. [In Swedish]

Telbisz A, Kucharik J. 1939. Mechanism of action of mustard gas. Orvosi Hetilap 390-393. [In Hungarian]

Telbisz A, Kucharik J. 1940. Loss of weight following administration of dichlorethylsulphide. Biochemische Zeitschrift 307:82-96. [In German]

Telbisz A, Kucharik J. 1940. Mustard gas. Magyar Orvosi Archivum 41:261. [In Hungarian]

Teulieres M. 1918. Ocular-palpebral lesions from the new gases. Journal de Medecine de Bordeaux et du Sud-Ouest 48:37-39. [In French]

Thalhammer O, Heller-Szoelloesy E. 1955. Exogenous malformations (defects) due to mustard gas injection into the pregnant mouse. Zeitschriftfur Kinderheilkunde 76:351-365. [In German]

Thatcher CJ, Walker IG. 1969. Sensitivity of confluent and cycling embryonic hamster cells to sulfur mustard, 1,3-bis(2-chloroethyl)-1-nitrosourea, and actinomycin D. Journal of the National Cancer Institute 42:363-68.

Thomas CB, Kohn KW, Bonner WM. 1972. Characterization of DNA-protein cross-links formed by treatment of L1210 cells and nuclei with bis(2-chloroethyl) methylamine (nitrogen mustard). Biochemistry 17:3954-3958.

Thomas IJ, Gough J. 1945. Dichloroethyl sulfide burns. Lancet 2:496-498.

Thompson RHS. 1946. The effect of arsenical vesicants on the respiration of the skin. Biochemical Journal 40:525-529.

Thompson RHS. 1947. The action of chemical vesicants on cholinesterase. Journal of Physiology 105:370-381.

Thompson RHS. 1948. The reaction of arsenicals in living tissues. In: Williams RT, ed. The Biochemical Reactions of Chemical Warfare Agents. Biochemical Society Symposia No. 2. Cambridge: Cambridge University Press.

Timko IM. 1943. Organization of treatment and care of victims of vesicant war gases. Soviet Med 7:35-37.

Tokuoka S. 1960. Patho-anatomical studies on mice treated with nitrogen mustard N-oxide, particularly on cases that developed leukemia and related conditions. Gencho Hiroshima Igaku 8:479-518. [In Japanese]

Tokuoka S. 1985. Early cancer and related changes in the bronchial epithelium of former mustard gas workers. Gan To Kagaku Ryoho 12:708-713. [In Japanese]

Tokuoka S, Inai K. 1988. Lung cancer and related environmental factors. Yakugaku Zasshi 108:1013-1022. [In Japanese]

Tokuoka S, Naito F. 1954. An autopsy case of atypical pneumonia found in a male following occupational mustard gas poisoning. Nippon Byori Gakkai Kaishi 43:115-117. [In Japanese]

Tokuoka S, Hayashi Y, Inai K, Egawa H, Aoki Y, Akamizu H, Eto R, Nishida T, Ohe K, Kobuke T, Nambu S, Takemoto T, Kou E, Nishina H, Fujihara M, Yonehara S, Tsuya T, Suehiro S, Horiuchi K. 1986. Early cancer and related lesions in the bronchial epithelium in former workers of mustard gas factory. Acta Pathologica Japonica 36:533-542.

Tomatis L. 1979. The predictive value of rodent carcinogenicity tests in the evaluation of human risks. Annual Review of Pharmacology and Toxicology 19:511-530.

Tomatis L. 1988. The contribution of the IARC monographs program to the identification of cancer risk factors. Annals of the New York Academy of Sciences 534:31-38.

Tomatis L. 1990. Cancer: Causes, Occurrence and Control. IARC Scientific Publication No. 100. Lyon: International Agency for Research on Cancer.

Tomatis L, Bartsch H. 1990. The contribution of experimental studies to risk assessment of carcinogenic agents in humans. Experimental Pathology 40:251-266.

Tomatis L, Agthe C, Bartsch H, Huff JE, Montesano R, Saracci R, WalkerE , Wilbourn J. 1978. Evaluation of the carcinogenicity of chemicals: a review of the monograph program of the International Agency for Research on Cancer (1971 to 1977). Cancer Research 38:877-885.

Tomatis L, Aitio A, Wilbourn J, Shuker L. 1989. Human carcinogens so far identified. Japanese Journal of Cancer Research 80:795-807.

Torchi M. 1939. Lupus carcinoma developed on burns due to war gas (dichloroethyl sulfide): case. Dermosilfilografo 14:545-553. [In Italian]

Trams EG, Nadkarni MV, Smith PK. 1961. On the mechanism of action of the alkylating agents. 1. Interaction of alkylating agents with nucleic acids. Cancer Research 21:560-566.

Trendelenburg F, Mall W. 1970. Epidemiology and detection of bronchial carcinoma. Internist 11:303-317. [In German]

Treves N, Pack GT. 1930. Development of cancer in burn scars: analysis and report of 34 cases. Surgery, Gynecology, and Obstetrics 51:749-782.

Trimble, MR. 1985. Post-traumatic stress disorder: history of a concept. In: Figley CR, ed. Trauma and Its Wake. New York: Brunner Mazel. 5-14.

Trotot R. 1935. War lesions due to toxic gases and their therapy. Tunisie Medicale 29:293-304. [In French]

Truhaut R, Deysson G. 1959. Comparative studies on the antimitotic properties of bis(2-chloroethyl) and its sulfur (yperite) and nitrogen (nitrogen mustard) homologs. Comptes Rendus Hebdomadaires des Seances de l'Academie des Sciences 248:732-734. [In French]

Tsuchiya K. 1985. Occupational lung cancer in Japan. Kokyu 4:582-593. [In Japanese]

Tsuchiya K. 1988. Occupational health and epidemiology. Asian Medical Journal 31:491-500.

Tucker MA, Meadows AT, Boice JD Jr, Hoover RN, Fraumeni JF Jr. 1984. Cancer risk following treatment of childhood cancer. In: Boice JD Jr, Fraumeni JF Jr, eds. Radiation Carcinogenesis: Epidemiology andBiological Significance . New York: Raven Press. 211-224.

Tucker MA, Misfeldt D, Coleman N, Clark WH Jr, Rosenberg SA. 1985. Cutaneous malignant melanoma after Hodgkin's disease. Annals of Internal Medicine 102:37-41.

Tucker MA, D'Angio GJ, Boice JD Jr, Strong LC, Li FP, Stovall M, Stone BJ, Green DM, Lombardi F, Newton W, Hoover RN, Fraumeni JF Jr. 1987. Bone sarcomas linked to radiotherapy and chemotherapy in children. New England Journal of Medicine 317:588-593.

Tucker MA, Meadows AT, Boice JD Jr, Stovall M, Oberlin O, Stone BJ, Birch J, Voute PA, Hoover RN, Fraumeni JF Jr. 1987. Leukemia after therapy with alkylating agents for childhood cancer. Journal of the National Cancer Institute 78:459-464.

Tucker MA, Coleman CN, Cox RS. 1988. Risk of second cancers after treatment for Hodgkin's disease. New England Journal of Medicine 318:76-81.

Tucker MA, Caganna M, Kelsey K, Coleman CN. In press. Secondary cancers. Cancer Medicine.

Turanow NM, Trofimonwa LJ, Bolschakowa GM, Chapilowa WI. 1977. Experience report from the USSR on the management of psoriasis patients using "Psoriasin." Zeitschrift fur Hautkrankheiten 52:1045-1049. [In German]

Turnbull JD, Aloyo VJ, Papirmeister B. 1973. Induction of axon growth and increased acetylcholinesterase production in mouse neuroblastoma by sulfur mustards. Federation Proceedings.

Twomey JJ, ed. 1982. The Pathophysiology of Human Immunologic Disorders. Baltimore: Urban & Schwarzenberg.

Uchida K, Morita T, Sato T, Ogura T, Yamashita R, Noguchi S, Suzuki H, Nyunoya H, Miwa M, Sugimura T. 1987. Nucleotide sequence of a full-length cDNA for human fibroblast poly (ADP-ribose) polymerase. Biochemical and Biophysical Research Communications 148:617-622.

Uhde GJ. 1946. Burns of upper respiratory tract and eyes from nitrogen mustard (b-chloroethylamine) gas: clinical study. Archives of Otolaryngology 44:701-709.

Uhde GJ. 1946. Mustard gas (dichloroethyl sulfide) burns of human eyes in World War II. American Journal of Ophthalmology 29:929.

Uhde GJ. 1946. Personal experience in treatment of mustard gas burns of upper respiratory tract during World War II. Military Surgeon 98:483-487.

Uhrig HT. 1962. Some medical aspects of chemical agents. Journal of the Medical Association of the State of Alabama 32:144-150.

UICC Technical Report Series. 1977. Workshop on lung cancer. II. Epidemiology. 25:3-41.

Underhill FP. 1919. Poisoning with the lethal war gases. Journal of the American Medical Association 73:686-689.

Underhill FP. 1920. The Lethal War Gases: Physiology and Experimental Treatment. New Haven, CT: Yale University Press.

Underhill FP. 1936. Toxicology or the Effect of Poisons. Philadelphia: Blakiston.

Underhill FP, Amatruda FG. 1923. The transmission of arsenic from mother to fetus. Journal of the American Medical Association 81:2009-2015.

United Nations. 1969. Chemical and Bacteriological (Biological) Weapons and the Effects of Their Possible Use. New York: United Nations.

United Nations Security Council. 1984. Report of the Specialists Appointed by the Secretary-General to Investigate Allegations by the Islamic Republic of Iran Concerning the Use of Chemical Weapons. March 26, 1984. S/16433. New York: United Nations.

United Nations Security Council. 1986. Report of the Mission Dispatched by the Secretary-General to Investigate Allegations of the Use of Chemical Weapons in the Conflict Between the Islamic Republic of Iran and Iraq. March 12, 1986. S/17911, S/17911/ Addendum 1, and S/17911/Addendum 2. New York: United Nations.

United Nations Security Council. 1987. Report of the Mission Dispatched by the Secretary-General to Investigate Allegations of the Use of Chemical Weapons in the Conflict Between the Islamic Republic of Iran and Iraq. May 8, 1987. S/18852 and S/18852/ Addendum 1. New York: United Nations.

United Nations Security Council. 1988. Report of the Mission Dispatched by the Secretary-General to Investigate Allegations of the Use of Chemical Weapons in the Conflict Between the Islamic Republic of Iran and Iraq. April 25, 1988. S/19823 and S/19823/ Addendum 1. New York: United Nations.

United Nations Security Council. 1988. Report of the Mission Dispatched by the Secretary-General to Investigate Allegations of the Use of Chemical Weapons in the Conflict Between the Islamic Republic of Iran and Iraq. July 20, 1988. S/20060 and S/20060/ Addendum 1. New York: United Nations.

United Nations Security Council. 1988. Report of the Mission Dispatched by the Secretary-General to Investigate Allegations of the Use of Chemical Weapons in the Conflict Between the Islamic Republic of Iran and Iraq. July 25, 1988. S/20063 and S/20063/ Addendum 1. New York: United Nations.

United Nations Security Council. 1988. Report of the Mission Dispatched by the Secretary-General to Investigate Allegations of the Use of Chemical Weapons in the Conflict Between the Islamic Republic of Iran and Iraq. August 19, 1988. S/20134. New York: United Nations.

Urbanetti JS. 1988. Battlefield chemical injury. In: Loke J, ed. Pathophysiology and Treatment of Inhalation Injuries. New York: Marcel Dekker. 281-348.

Ursano RJ, ed. 1987. Groups and Organizations in War, Disasters, and Trauma. Bethesda, MD: Uniformed Services University of the Health Sciences.

Ursano RJ, ed. 1987. Individual and Group Behavior in Toxic and Contained Environments. A Conference to Explore the Psychological Effects of Chemical and Biological Warfare. December 12-14, 1986, Airlie, VA. Bethesda, MD: Uniformed Services University of the Health Sciences.

Ursano RJ. 1988. Combat stress in the chemical and biological warfare environment. Aviation, Space, and Environmental Medicine 59:1123-1124.

Ursano RJ, ed. 1988. Performance and Operations in Toxic Environments. 2nd ed. Bethesda, MD: Uniformed Services University of the Health Sciences.

Ursano RJ, ed. 1988. Psychological and Behavioral Responses to a Chemical and Biological Warfare Environment: Final Recommendations. Bethesda, MD: Uniformed Services University of the Health Sciences.

Ursano RJ, ed. 1988. Training for the Psychological and Behavioral Effects of the CBW Environment: A Conference to Explore Training for Operational and Medical Personnel for Coping, Adaptation and Performance in the High Stress Environment of Chemical and Biological Warfare. November 6-8, 1987, Airlie, VA. Bethesda, MD: Uniformed Services University of the Health Sciences.

Ursano RJ, Holloway HC. 1985. Military psychiatry. In: Kaplan HI and Saddock BJ, eds. Comprehensive Textbook of Psychiatry. 4th ed. Baltimore: Williams Wilkins.

Ursano RJ, Boydstun JA, Wheatley RD. 1981. Psychiatric illness in U.S. Air Force Vietnam prisoners of war: a 5 year follow-up. American Journal of Psychiatry 138:310-314.

U.S. Department of Health and Human Services. 1991. Arsenic and certain arsenic compounds. In: Sixth Annual Report on Carcinogens. Summary 1991. Washington, DC: DHHS. 22-27.

U.S. Department of Health and Human Services. 1991. Mustard gas. In: Sixth Annual Report on Carcinogens. Summary 1991. Washington, DC: DHHS. 66-67.

U.S. Department of Health, Education, and Welfare. Office of the Secretary. Task Force on the Compensation of Injured Research Subjects. 1977. Report, Secretary's Task Force on the Compensation of Injured Research Subjects. OS-77-003. Bethesda, MD: National Institutes of Health.

U.S. Department of Health, Education, and Welfare. Office of the Secretary. Task Force on the Compensation of Injured Research Subjects. 1977. Secretary's Task Force on the Compensation of Injured Research Subjects. Appendix B. OS-77-005. Bethesda, MD: National Institutes of Health.

U.S. Environmental Protection Agency. 1985. Lewisite. In: EPA Chemical Profiles. Revised 1987. Washington, DC: EPA.

U.S. Environmental Protection Agency. 1985. Mustard gas. In: EPA Chemical Profiles. Revised 1987. Washington, DC: EPA.

Vahedi P. 1988. Pathophysiology of respiratory effects of chemical warfare agents. In: Abstracts of the First International Medical Congress on Chemical Warfare Agents in Iran. June 13-16, 1988. Mashhad University of Medical Sciences. Mashhad, Iran. No. 6.

Valagussa P. 1990. Second malignancies: the experience of the Milan Cancer Institute (abstract). Journal of Cancer Research and Clinical Oncology 116 (Supplement):982.

Valagussa P, Santoro A, Fossati-Bellani F, Banfi A, Bonadonna G. 1986. Second acute leukemia and other malignancies following treatment for Hodgkin's disease. Journal of Clinical Oncology 4:830-837.

Valagussa P, Tancini G, Bonadonna G. 1987. Second malignancies after CMF for resectable breast cancer. Journal of Clinical Oncology 5:1138-1142.

Van Delft JH, Van Weert EJ, Schellekens MM, Claassen E, Baan RA. 1991. The isolation of monoclonal antibodies selected for the detection of imidazole ring-opened N7-ethylguanine in purified DNA and in cells *in situ.* Crossreaction with methyl, 2-hydroxyethyl and sulphur mustard adducts. Carcinogenesis 12:1041-1049.

Van der Kolk BA, ed. 1984. Post-Traumatic Stress Disorder: Psychologicaland Biological Sequelae . Washington, DC: American Psychiatric Press.

Van der Steen J, Van Wichen BH, Benckhuysen C. 1982. Effects of alkylating agents on the DNA replication of cultured Yoshida sarcoma cells. Chemico-Biological Interactions 39:191-204.

Van Duuren BL. 1976. Tumor-promoting and co-carcinogenic agents in chemical carcinogenesis. In: Searle CE, ed. Chemical Carcinogens. American Chemical Series Monograph 173. Washington, DC: American Chemical Society. 24-51.

Van Duuren BL. 1980. Prediction of carcinogenicity based on structure, chemical reactivity and possible metabolic pathways. Journal of Environmental Pathology and Toxicology 3:11-34.

Van Duuren BL, Segal A. 1976. Inhibition of two-stage carcinogenesis in mouse skin with bis(2-chloroethyl)sulfide. Cancer Research 36:1023-1025.

Van Duuren BL, Witz G, Sivak A. 1974. Chemical carcinogenesis. In: Homburger F, ed. The Physiopathology of Cancer. 3rd ed. Vol. 1, Biology and Biochemistry. Basel: S. Karger. 1-63.

Van Genderen J, Wolthuis OL. 1986. New models for testing skin toxicity. In: Marks R, Plewig G, eds. Skin Models: Models to Study Function and Disease of Skin. Berlin: Springer-Verlag. 85-93.

Van Genderen J, Mol M, Wolthuis O. 1985. On the development of skin models for toxicity testing. Fundamental and Applied Toxicology 5:s98-s111.

Van Hooidonk C, Ceulen BI, Bock J, Van Genderen J. 1983. CW agents and the skin. Penetration and decontamination. FOA Reports. 153-160.

Van Leeuwen FE, Somers R, Taal BG, Van Heerde P, Coster B, Dozeman T, Huisman SJ, Hart AAM. 1989. Increased risk of lung cancer, non-Hodgkin's lymphoma, and leukemia following Hodgkin's disease. Journal of Clinical Oncology 7:1046-1058.

Van Melckebeke F. 1932. Yperite. Journal de Pharmacie de Belgique 14:93-99. [In French]

Vaughan FL, Zaman S, Scavarelli R, Bernstein IA. 1988. Macromolecular metabolism of a differentiated rat keratinocyte culture system following exposure to sulfur mustard. Journal of Toxicology and Environmental Health 23:507-518.

Vedder EB. 1925. The Medical Aspects of Chemical Warfare. Baltimore: Williams & Wilkins.

Veil WH, Sturm A. 1942. [The Pathology of the Brainstem] Die Pathologie des Stammhirns. Jena: Gustav Fischer Verlag.

Venitt S. 1968. Interstrand cross-links in the DNA of *Escherichia coli* B/r and B_{s-1} and their removal by the resistant strain. Biochemical and Biophysical Research Communications 31:355-360.

Venitt S, Brookes P, Lawley PD. 1968. Effects of alkylating agents on the induced synthesis of b-galactosidase by *Escherichia coli* B_{s-1} . Biochimica et Biophysica Acta 155:521-535.

Viale G. 1918. Treatment of cutaneous lesions from yperite with silver nitrate. Policlin (Rome) 25:1061-1063. [In Italian]

Vijayaraghavan R, Sugendran K, Pant SC, Husain K, Malhotra RC. 1991. Dermal intoxication of mice with bis(2-chloroethyl)sulphide and theprotective effect of flavonoids . Toxicology 69:35-42.

Villard H, DeJean C. 1933. Deep central bilateral keratitis following dichloroethylsulfide poisoning: case. Archives de la Societe des Sciences Medicales et Biologiques de Montpellier et du Languedoc Mediterraneen 14:16-19. [In French]

Vincent. 1919. The otorhinolaryngologist's view of persons intoxicated by the new war gases. Revue de Laryngologie, Otologie, Rhinologie 40:313-328. [In French]

Vineis P, Thomas T, Hayes RB, Blot WJ, Mason TJ, Williams PL, Correa P, Fontham ETH, Schoenberg J. 1988. Proportion of lung cancers in males due to occupation in different areas of the USA. International Journal of Cancer 42:851-856.

Vinsonneau G, Bron J, Putot R. 1920. Ocular lesions from gas poisoning. Archives d'Ophthtalmologie 37:475. [In French]

Visser J, Thvos JJ. 1935. On the influence of mustard gas on malignanttumors . Geneeskundig Tijdschrift voor Nederlansch-Indie 75:1363-1377. [In Dutch]

Vocci FJ, Ballard TA, Yevich P, Punte CL. 1963. Inhalation toxicity studies with aerosols of sesqui-mustard. Toxicology and Applied Pharmacology 5:677-684.

Vogel E, Natarajan AT. 1982. The relation between reaction kinetics and mutagenic action of monofunctional alkylating agents in higher eukaryotic systems: interspecies comparisons. Chemical Mutagens: Principles and Methods for Their Detection 7:295-336.

Vogel F, Rohrborn R, eds. 1970. Chemical Mutagenesis in Mammals and Man. New York: Springer-Verlag.

Vogel F, Rohrborn G, Schleiermacher E, Schroder TM. 1967. Mutations due to chemical effects in mammals and man. Deutsche Medizinische Wochenschrift 92:2249-2254. [In German]

Vogt RF Jr. 1983. Skin Injury by Sulfur Mustard: A Model for Acute Inflammation Accompanying Slow Cell Death (dissertation). Baltimore: Johns Hopkins University.

Vogt RF Jr, Dannenberg AM Jr, Castracane S, Papirmeister B. 1978. Basophil participation in non-allergic inflammatory responses (abstract). Journal of the Reticuloendothelial Society 24:70a.

Vogt RF Jr, Dannenberg AM Jr, Schofield BH, Hynes NA, Papirmeister B. 1984. Pathogenesis of skin lesions caused by sulfur mustard. Fundamental and Applied Toxicology 4:S71-S83.

Voiculet N, Niculescu-Duvaz I. 1968. Interaction of carcinogenic and antitumor agents with nucleic acids. Oncologia si Radiologia 7:205-218. [In Romanian]

Voiculet N, Sbenghe M, Lukas S, Barbu C, Sandulescu T, Tudor M. 1975. [The production and repair of DNA lesions induced by ionizing radiations and cytostatic alkylating agents]. Onkologiia 41:115-130.

Voina I. 1926. On the mechanism of the toxic effect of yperite. Technika (Moscow) 83:334. [In Russian]

Voivenel P, Martin P. 1919. Pulmonary problems due to recent yperite intoxication. Progres Medicale 3:99-103. [In French]

Vojvodic V. 1982. Toxicology of War Gases. Belgrade, Yugoslavia: Literature Research Co.

Vojvodic V, Milosabljevic Z, Boskovic B, Bojanic N. 1985. The protective effect of different drugs in rats poisoned by sulfur and nitrogen mustards. Fundamental and Applied Toxicology 5:S160-S168.

Voljanski IV. 1934. Vesicants. Arhiv za Hemiju i Farmaciju 8:93-99. [In Serbo-Croatian]

Von den Velden R. 1921. War-gas poisoning. X. Clinical picture of dichloroethylsulfide poisoning. Zeitschrift fur die Gesamte Experimentelle Medizin 14:1-27. [In German]

von Hess H. 1937. On the question concerning the causes of yellow cross induced cachexia and its treatment. Medizinische Welt 11:1568-1569. [In German]

Vos JG. 1977. Immune suppression as related to toxicology. CRC Critical Reviews in Toxicology 5:67-101.

Vycudilik W. 1985. Detection of mustard gas bis(2-chloroethyl)-sulfide in urine. Forensic Science International 28:131-136.

Vycudilik W. 1987. Detection of bis(2-chlorethyl)-sulfide (yperite) in urine by high resolution gas chromatography-mass spectrometry. Forensic Science International 35:67-71.

Vyner HM. 1988. Invisible Trauma: The Psychosocial Effects of Invisible Environmental Contaminants. Lexington, MA: Lexington Books.

Vyner HM. 1988. The psychological dimensions of health care for patients exposed to radiation and the other invisible environmental contaminants. Social Science in Medicine 27:1097-1103.

Wachtel C. 1941. Chemical Warfare. Brooklyn, NY: Chemical Publishing.

Wada S, Miyanishi M. 1954. A case report of bronchogenic carcinoma seen following chronic mustard gas exposure. Shindan to Chiryo 42:791-794. [In Japanese]

Wada S, Nishimoto Y, Miyanishi M. 1958. Clinical studies on the lungs in poison gas injury. Transactions of the Fifth International Congress on Diseases of the Chest. 169-170.

Wada S, Nishimoto Y, Miyanishi M, Katsuta S, Nishiki M, Yamada A, Tokuoka S, Umisa H, Nagai M. 1962. Malignant respiratory tract neoplasms related to poison gas exposure. Hiroshima Journal of Medical Sciences 11:81-91.

Wada S, Yamada A, Nagai M, Nishimoto Y, Miyanshi M. 1962. Respiratory diseases related to poison gas exposure, with particular comment on malignant neoplasms. Nippon Kyobu Rinsho 21:160-166. [In Japanese]

Wada S, Nishimoto Y, Miyanishi M, Katsuta S, Nishiki M. 1962. Review of Okuno-Jima poison gas factory regarding occupational environment. Hiroshima Journal of Medical Sciences 11:75-80.

Wada S, Nishimoto Y, Miyanishi M. 1962. Studies on former employees of the poison war-gas institute at Okuno Island, Japan (first report). Studies on the cause of death of former employees. Japanese Journal of Medicine 1:250-251.

Wada S, Nishimoto Y, Miyanishi M, Katsuta S. 1962. Studies on the late effects of poison gas exposure observed in former employees in Okuno-jima Island (second report). Nippon Naika Gakkai Zasshi 50:45-46. [In Japanese]

Wada S, Yamada A, Nishimoto Y, Tokuoka S, Miyanishi M, Katsuta S, Umisa H. 1963. Neoplasms of the respiratory tract among poison gas workers. Journal of the Hiroshima Medical Association 16:728-745.

Wada S, Nishimoto Y, Miyanishi M, Kambe S, Miller RW. 1968. Mustard gas as a cause of respiratory neoplasia in man. Lancet 2:1161-1163.

Waitt AH. 1942. Gas Warfare: The Chemical Weapon, Its Use, and Protection Against It. New York: Duell, Sloan, Pierce.

Waitt AH. 1942. War gas cases: first aid treatment. American Journal of Nursing 42:489-498.

Walker IG. 1966. Sulfur mustard: reaction with L-cells treated with 5-fluorodeoxyuridine. Science 151:99-101.

Walker IG. 1971. Intrastrand bifunctional alkylation of DNA in mammalian cells treated with mustard gas. Canadian Journal of Biochemistry 49:332-336.

Walker IG, Reid BD. 1971. Some properties of substrains of L-cells with a decreased sensitivity to bis(2-chloroethyl)sulfide. Cancer Research 31:510-515.

Walker IG, Smith JF. 1969. Protection of L-cells by thiols and against the toxicity of sulfur mustard. Canadian Journal of Physiology and Pharmacology 47:143-151.

Walker IG, Thatcher CJ. 1968. Studies on the lethal effects of sulfur mustard on dividing mammalian cells. Radiation Research 34:110-127.

Walker IG, Watson WJ. 1961. The reaction of mustard gas (b,b-dichloroethyl sulphide) with purines and pyrimidines. Canadian Journal of Biochemistry and Physiology 39:377-393.

Wall RL, Clausen KP. 1975. Carcinoma of the urinary bladder in patients receiving cyclophosphamide. New England Journal of Medicine 293:271-273.

Walpole AL. 1958. Carcinogenic action of alkylating agents. Annals of the New York Academy of Sciences 68:750-761.

Walter F. 1935. On the mechanism of action of mustard gas on theskin of animals . Bulletin International de l'Academie Polonaise des Sciences et des Lettres. Classe de Medicine. [In French]

Ward JR, Seider RP 1984. Activation energy for the hydrolysis of bis(2-chloroethyl) sulfide. Thermochimica Acta 81:343-348.

Warthin AS, Weller CV. 1917-1918. The pathology of skin lesions produced by mustard gas (dichlorethylsulphide). Journal of Laboratory and Clinical Medicine 3:447-479.

Warthin AS, Weller CV. 1918. The toxic action of dichlorethylsulphide (mustard gas). Proceedings of the Society for Experimental Biology and Medicine 16:143-147.

Warthin AS, Weller CV. 1918-1919. Researches on the pathology of mustard gas (dichloroethylsulphide) poisoning. Journal of Laboratory and Clinical Medicine 4:265-306.

Warthin AS, Weller CV. 1918-1919. The lesions of the respiratory and gastrointestinal tract produced by mustard gas (dichlorethyl sulphide). Journal of Laboratory and Clinical Medicine 4:229-264.

Warthin AS, Weller CV. 1918-1919. The pathology of dichlorethylsulphide (mustard gas). Proceedings of the Society of Experimental Biology and Medicine 16:147-151.

Warthin AS, Weller CV. 1918-1919. The treatment of dichlorethylsulphide (mustard gas) injuries. Journal of Laboratory and Clinical Medicine 4:833-848.

Warthin AS, Weller CV. 1919. The Medical Aspects of Mustard Gas Poisoning. St. Louis: C.V. Mosby.

Warthin AS, Weller CV, Herrmann GR. 1918-1919. The ocular lesions produced by dichlorethylsulphide (mustard gas). Journal of Laboratory and Clinical Medicine 4:785-832.

Waters MD, Garrett NE, Covone-de Serres CM, Howard BE, Stack HF. 1983. Genetic bioassay data on some known or suspected human carcinogens. In: Castellani A, ed. The Use of Human Cells for the Evaluation of Risk from Physical and Chemical Agents. NATO Advance Science Institutes Series. Series A, Life Sciences. New York:Plenum Press. 495-570.

Waters MD, Garrett NE, Covone-de Serres CM, Howard BE, Stack HF. 1983. Genetic toxicology of some known or suspected human carcinogens. Chemical Mutagens: Principles and Methods for Their Detection 8:261-341.

Waters MD, Garrett NE, Covone-de Serres CM, Howard BE, Stack HF. 1984. Genetic toxicology of 14 agents causally associated with cancer in humans. Environmental Science Research 31:625-687.

Waters WA, Williams JH. 1950. Hydrolyses and derivatives of some vesicant arsencials. Journal of the Chemical Society (London) 18-22.

Watjen J. 1940. Pathological-anatomic effects of poison gases employed in war. Medizinische Welt 14:341-343. [In German]

Watry. 1921. Dental sequelae of war gas intoxication. Revue Belge de Stomatologie 19:1021. [In French]

Watson AP, Griffin GD. In press. Sulfur mustard. In: Handbook of Hazardous Materials. New York: Academic Press.

Watson AP, Griffin GD. In press. Toxicity of vesicant agents scheduled for destruction by the Chemical Stockpile Disposal Program. Environmental Health Perspectives.

Watson AP, Ambrose KR, Griffin GD, Leffingwell SS, Munro NB, WatersLC . 1989. Health effects of warfare agent exposure: implications for stockpile disposal. Environmental Professional 11:335-353.

Watson AP, Jones TD, Griffin GD. 1989. Sulfur mustard as a carcinogen: application of

relative potency analysis to the chemical warfare agents H, HD, and HT. Regulatory Toxicology and Pharmacology 10:1-25.
Watson-Williams E. 1930. Granuloma of larynx following exposure to mustard gas. Journal of Laryngology, Rhinology and Otology 45:717-722.
Weger N. 1975. Mustard gas intoxication by inhalation. Therapiewoche 25:5908.
Weidner H. 1937. Contributions to the studies on resorptive action of dichlorodiethylsulfide. Deutsch Militararzt 2:247-251. [In German]
Weigand DA, Haygood C, Gaylor G. 1974. Cell layers and density of Negro and Caucasian stratum corneum. Journal of Investigative Dermatology 62:563-568.
Weill G. 1939. Corneal ulcers after dichloroethylsulfide poisoning: case. Bulletin de la Societe d'Ophtalmologie de Paris 51:281-282. [In French]
Weinstein IB. 1978. Current concepts on mechanisms of chemical carcinogenesis. Bulletin of the New York Academy of Medicine 54:366-383.
Weiss A. 1958. Blister gases as carcinogenic substances. Kongressbericht, Nordwestdeutsche Gesellschaft fur Innere Medizin 51:23-24. [In German]
Weiss A. 1992. From trench warfare to war on cancer. The development of chemotherapy for malignant disease. Hospital Practice (Office Edition) 27:141-143, 147-148.
Weiss A, Weiss B. 1975. Carcinogenesis from exposure to mustard gas in man, an important sign for therapy with alkylating agents. Deutsche Medizinische Wochenschrift 100:919-923. [In German]
Weiss A, Weiss B. 1975. Malignant tumors and leukemia: abnormally frequent cause of death in former blister gas workers. Comparative study. Kongressbericht, Nordwestdeutsche Gesellschaft fur Innere Medizin 84:27-28. [In German]
Weiss A, Weiss B. 1976. Glioblastoma and neurofibroma following the action of mustard gas. Munchener Medizinische Wochenschrift 118:875-878. [In German]
Weiss RB, Muggia FM. 1980. Cytotoxic drug-induced pulmonary disease: update 1980. American Journal of Medicine 68:259-266.
Weiss W. 1979. Lung cancer due to chemicals. Comprehensive Therapy 5:18-23.
Weiss W. 1981. Lung cancer and occupational lung disease. Clinics in Chest Medicine 2:289-300.
Weiss W. 1984. The epidemiology of endocrine tumors of the lung. In: Becker KL, Gazdar AF, eds. The Endocrine Lung in Health and Disease. Philadelphia: W.B. Saunders. 373-388.
Weissberg JB, Herion JC, Walker RI, Palmer JG. 1978. Effect of cycloheximide on the bone marrow toxicity of nitrogen mustard. Cancer Research 38:1523-1527.
Weller CV. 1918. The pathology of gassing. Bulletin of the International Association of Medical Museums 7:126-147.
Welsh GA. 1920-1921. The effect of the inhalation of gases. Journal of Industrial Hygiene 2:328-332.
Wen J-F. 1984. Biological detection of chemical warfare agents. Archives Belges (Supplement):74-80.
Wessely. 1935. On the effect of dichloroethylsulfide on the eyes. Klinische Monatsblatter fur Augenheilkunde 94:100. [In German]
West CJ. 1919. History of mustard gas. Science 49:412-417.
Wester DH. 1939. Vesicants. Pharmaceutisch Weekblad 76:1317-1324. [In Dutch]
Wester DH. 1940. [A new point of view regarding the toxicity and the possibility of use of mustard gas]. NVL Stud Org Luchtbeschermingsvaagstuk 1:326-329.
Wheeler GP. 1962. Studies related to the mechanisms of action of cytotoxic alkylating agents: a review. Cancer Research 22:651-688.
Wheeler GP. 1967. Some biochemical effects of alkylating agents. Proceedings of the Society for Experimental Biology and Medicine 26:885-892.
Wheeler GP. 1971. Pharmacological considerations of alkylating agents. Recent Results in Cancer Research 36:137-146.

Whitaker PH. 1946. Radiologic appearance of chests of workers engaged in production of toxic gases. British Journal of Radiology 19:158-164.
White SA. 1938. Some medical aspects of chemical warfare. Chemical Warfare Bulletin 24:135-144.
Whiting MH. 1940. Discussion on gas injuries to the eye. Proceedings of the Royal Society of Medicine 33:225-236.
Wickstrom G. 1972. Arsenic in the ecosystem of man: a review. Work, Environment, Health 9:2-8.
Wiedling S. 1940. Internal injuries caused by war gases and their therapy. Svenska Lakartidningen 37:2068-2072. [In Swedish]
Wiencke JK, Yager JW. 1992. Specificity of arsenite in potentiating cytogenetic damage induced by the DNA cross-linking agent diepoxybutane. Environmental and Molecular Mutagenesis 19:195-200.
Wiencke JK, Christiani DC, Kelsey KT. 1991. Bimodal distribution of sensitivity to SCE induction by diepoxybutane in human lymphocytes. I. Correlation with chromosomal aberrations. Mutation Research 248:17-26.
Wilbourn J, Haroun L, Vainio H, Montesano R. 1984. Identification of chemicals carcinogenic to man. Toxicologic Pathology 12:397-399.
Wilbourn J, Haroun L, Heseltine E, Kaldor J, Partensky C, Vainio H. 1986. Response of experimental animals to human carcinogens: an analysis based upon the IARC Monographs Programme. Carcinogenesis 7:1853-1863.
Willems JL. 1988. Controversies regarding diagnosis and treatment in Iranian chemical war casualties brought to Europe. In: Abstracts of the First International Medical Congress on Chemical Warfare Agents in Iran. June 13-16, 1988. Mashhad University of Medical Sciences. Mashhad, Iran. No. 94.
Willems JL. 1989. Clinical management of mustard gas casualties. Annales Medicinae Militaris Belgicae 3:S1-61.
Willems JL, De Leenheer AP, De Bisschop HC. 1991. Management of chemicalwarfare injuries (letter) . Lancet 337:121-122.
Willhite CC. 1981. Arsenic-induced axial skeletal (dysraphic) disorders. Experimental and Molecular Pathology 34:145-158.
Williams GM. 1985. Genotoxic and epigenetic carcinogens. In: Homburger F, ed. Safety Evaluation and Regulation of Chemicals. 2. Impact of Regulations: Improvement of Methods. Basel: S. Karger. 251-256.
Williams JM. 1919. Protective ointments against mustard gas. Journal of the American Pharmaceutical Association 8:824-829.
Williams RN, Bhattacherjee P. 1984. Inhibition of the acute ocular responses to nitrogen mustard by colchicine. Experimental Eye Research 39:721-729.
Williamson CE, Seligman AM, Witten B. 1972. Extracellular toxic reactions of some sulfur and nitrogen mustards. Journal of Pharmacology and Experimental Therapeutics 182:77-82.
Wils ERJ, Hulst AG, De Jong AL, Verweij A, Boter HL. 1985. Analysisof thiodiglycol in urine of victims of an alleged attack with mustard gas . Journal of Analytical Toxicology 9:254-257.
Wils ERJ, Hulst AG, Van Laar J. 1988. Analysis of thiodiglycol in urine of victims of an alleged attack with mustard gas. Part II. Journal of Analytical Toxicology 12:15-19.
Wilson CM, Mackintosh JM. 1920. Mustard gas poisoning. Quarterly Journal of Medicine 13:201-240.
Winternitz MC. 1919. Anatomical changes in the respiratory tract initiated by irritating gases. Military Surgeon 44:476-493.
Winternitz MC. 1919. Chronic lesions of the respiratory tract. Journal of the American Medical Association 73:689-691.

Winternitz MC. 1920. Collected Studies on the Pathology of War Gas Poisoning. New Haven, CT: Yale University Press.

Woessner JF JR, Dannenberg AM Jr, Pula PJ, Selzer MG, Ruppert CL, Higuchi K, Kajiki A, Nakamura M, Dahms NM, Kerr JS, Hart GW. 1990. Extracellular collagenase, proteoglycanase and products of their activity, released in organ culture by intact dermal inflammatory lesions produced by sulfur mustard. Journal of Investigative Dermatology 95:717-726.

Woitowitz HJ. 1986. Occupational noxae in etiology of lung cancer. Atemswegs und Lungenkrankheiten 12:482-484. [In German]

Wolf ME, Mosnaim AD, eds. 1990. Posttraumatic Stress Disorder: Etiology, Phenomenology and Treatment. Washington, DC: American Psychiatric Press.

Wolthers K, Reumert T. 1970. Mustard gas lesions in Bornholm in 1969. Ugeskrift for Laeger 132:835-836. [In Danish]

Wood JR. 1944. Chemical warfare: a chemical and toxicological review. American Journal of Public Health 34:455-460.

Working PK, Smith-Oliver T, White RD, Butterworth BE. 1986. Induction of DNA repair in rat spermatocytes and hepatocytes by 1,2-dibromoethane: the role of glutathione conjugation. Carcinogenesis 7:467-472.

World Health Organization. 1970. Health Aspects of Chemical and Biological Weapons: Report of a WHO Group of Consultants. Geneva: WHO.

Wormall A. 1947. Mustard gas (dichloroethyl sulfide). St. Bartholomews Hospital Journal 51:6-7.

Worms G, Bolotte. 1926. Ocular sequelae due to vesicant gas lesions: problems of the cornea. Soc de Med Mil Franc Bull Mems (Paris) 20:229-235. [In French]

Worms G, Leroux-Robert J. 1934. Extensive secondary dilations caused by dichloroethyl sulfide poisoning: study of pathogenesis of mega-esophagus. Annales d'Oto-Laryngologie et de Chirurgie Cervico Faciale 1:669-681. [In French]

Wormser U. 1991. Toxicology of mustard gas. Trends in Pharmacological Sciences 12:164-167.

Wulf HC. 1979. Mutagenicity assessed by the incidence of sister chromatid exchanges. Ugeskrift for Laeger 141:2371-2373. [In Danish]

Wulf HC, Aasted A, Darre E, Niebuhr E. 1985. Sister chromatid exchanges in fishermen exposed to leaking mustard gas shells (letter). Lancet 1:690-691.

Wurgler FE, Ramel C, Moustacchi E, Carere A. 1986. Assays for genetic activity in *Drosophila melanogaster.* In: Montesano R, ed. Long-Term and Short-Term Assays for Carcinogens: A Critical Appraisal. IARC Scientific Publication No. 83. Lyon: International Agency for Research on Cancer. 351-394.

Wynder E. 1982. The etiology, epidemiology, and prevention of lung cancer. Seminars in Respiratory Medicine 3:135-139.

Wynn-Williams N. 1953. Bronchiectasis caused by mustard gas. British Journal of Tuberculosis 47:35-38.

Wyzga R. 1988. The role of epidemiology in risk assessments of carcinogens. In: Cothersn CR, Mehlman MA, Marcus WL, eds. Risk Assessment and Risk Management of Industrial and Environmental Chemicals. Princeton, NJ: Princeton Scientific Publishing. 189-208.

Yamada A. 1954. Bronchial carcinoma found in persons who succumbed to occupational mustard gas poisoning. Transactiones Societatis Pathologicae Japonicae 43:13-14. [In Japanese]

Yamada A. 1955. Mustard gas (yperite): its relation to lung cancer. Sogo Igaku 12:355-368. [In Japanese]

Yamada A. 1956. Hematological and histological studies on rabbits following prolonged administration with nitrogen mustard N-oxide (abstract). Nippon Haematologica Japonica 18:248-249. [In Japanese]

Yamada A. 1957. An autopsy case of generalized amyloidosis with renal calculi which was found in a person who succumbed to occupational mustard gas poisoning (abstract). Transactiones Societatis Pathologicae Japonicae 46:394. [In Japanese]

Yamada A. 1959. A case of occupational cancer of the larynx. Gencho Hiroshima Igaku 7:2711-2715. [In Japanese]

Yamada A. 1959. Patho-anatomical studies on respiratory cancers developed in workers with occupational exposure to mustard gas. Gencho Hiroshima Igaku 7:719-761. [In Japanese]

Yamada A. 1962. Occupational lung cancer in Japan. Hiroshima Daigaku Igaku Zasshi 10:1-7. [In Japanese]

Yamada A. 1963. On the late injuries following occupational inhalation of mustard gas, with special references to carcinoma of the respiratory tract. Acta Pathologica Japonica 13:131-155.

Yamada A. 1972. Pathology related to poison gas. Nippon Kyobu Shikkan Gakkai Zasshi 10:483-485. [In Japanese]

Yamada A. 1974. Pathoanatomical studies on the late injuries following occupational inhalation of mustard gas. Transactiones Societatis Pathologicae Japonicae 63:17-61. [In Japanese]

Yamada A, Matsubayashi Y. 1962. On occupational mustard gas poisoning observed among employees of the Japanese Army war gas manufacturing plant on the island of Okuno-jima, with special references to its sequelae (abstract). Transactiones Societatis Pathologicae Japonicae 51:368. [In Japanese]

Yamada A, Nagai M. 1960. A case of cancer of the pharynx developed in a worker suffering from occupational mustard gas and Lewisite poisoning. Gencho Hiroshima Igaku 8:1107-1112. [In Japanese]

Yamada A, Tamura H. 1961. A case of carcinoma of the hypopharynx found in a male following prolonged occupational exposure to mustard gas. Gencho Hiroshima Igaku 9:1427-1432. [In Japanese]

Yamada A, Hirose F, Miyanishi M. 1953. An autopsy case of bronchial carcinoma found in a patient succumbed to occupational mustard gas poisoning. Gann 44:216-218. [In Japanese]

Yamada A, Hirose F, Nakamura T, Nagai M. 1956. An autopsy case of esophagus cancer with extension into the left side first bronchus found in a person who succumbed to occupational mustard gas poisoning. Gann 47:696-698. [In Japanese]

Yamada A, Hirose F, Nagai M, Nakamura T. 1957. Five cases of cancer of the larynx found in persons who suffered from occupational mustard gas poisoning. Gann 48:366-368.

Yamada A, Tokuoka S, Maeda T, Nagai M. 1962. Report of four autopsied cases with bronchogenic carcinoma in individuals exposed to poison gases under occupational circumstances. Journal of the Hiroshima Medical Association, Special Series 15:157-158. [In Japanese]

Yamakido M, Onari K, Ishioka S, Yukutake M, Nishimoto Y. 1983. Study on prevention of carcinogenesis among poison gas sufferers by immunological method (abstract). Japanese Journal of Medicine 22:286.

Yamakido M, Yanagida J, Ishioka S, Matsuzaka S, Hozawa S, Takaishi M, Akiyama M, Nishimoto Y. 1985. Effects of *Nocardia rubra* cell wall skeleton on lymphocyte subsets in former poison gas factory workers. Japanese Journal of Cancer Research 76:1091-1098.

Yamakido M, Matsuzaka S, Yanagida J. 1985. Function of monocytes in the retired workers of the Okuno-jima poison gas factory. Hiroshima Journal of Medical Sciences 34:425-430.

Yamakido M, Nishimoto Y, Shigenobu T, Onari K, Satoh C, Goriki K, Fujita M. 1985. Study of genetic effects of sulphur mustard gas on former workers of Okuno-jima poison gas factory and their offspring. Hiroshima Journal of Medical Sciences 34:311-322.

Yamakido M, Shigenobu T, Kamitsuna A, Nishimoto Y, Yukutake M. 1985. The causes of death in the retired workers of Okuno-jima poison gas factory (abstract). Japanese Journal of Medicine 24:394.

Yamakido M, Yanagida J, Ishioka S, Matsuzaka S, Hozawa S, Takaishi M, Inamizu T, Akiyama M, Nishimoto Y. 1986. Immune functions of former poison gas workers. I. Mitogenic response of lymphocytes and serum factors. Hiroshima Journal of Medical Sciences 35:117-126.

Yamakido M, Yanagida J, Ishioka S, Matsuzaka S, Hozawa S, Takaishi M, Inamizu T, Akiyama M, Nishimoto Y. 1986. Immune functions of former poison gas workers. II. Lymphocytic subsets and interleukin 2 production. Hiroshima Journal of Medical Sciences 35:127-134.

Yamakido M, Ishioka S, Hozawa S, Matsuzaka S, Yanagida J, Shigenobu T, Otake M, Nishimoto Y. 1992. Effect of *Nocardia rubra* cell-wall skeleton on cancer prevention in humans. Cancer Immunology, Immunotherapy 34:389-392.

Yamamoto T, Yamada A. 1955. Pathological studies in rats injected with methyl bis(b-chloroethyl) amine N-oxide hydrochloride (nitromin) in various ways. Gann 46:376-377. [In Japanese]

Yamasaki H, Wilbourn JD, Haroun L. 1982. Use of data from short-term tests in the evaluation of the carcinogenicity of environmental chemicals to humans. In: Sorsa M, Vainio H, eds. Progress in Clinical and Biological Research. Vol. 109, Mutagens in Our Environment. New York: Alan R. Liss. 169-180.

Yanagida J, Hozawa S, Ishioka S, Maeda H, Takahashi K, Oyama T, Takahashi M, Hakoda M, Akiyama M, Yamakido M. 1988. Somatic mutation in peripheral lymphocytes of former workers at the Okuno-jima poison gas factory. Japanese Journal of Cancer Research 79:1276-1283.

Yang YC, Szafraniec LL, Beaudry WT, Ward JR. 1988. Kinetics and mechanisms of the hydrolysis of 2-chloroethyl sulfides. Journal of Organic Chemistry 53:3293-3297.

Yang YC, Szanfraniec LL, Beaudry WT. 1990. A comparison of the oxidative reactivities of mustard (2,2'-dichlorodiethyl sulfide) and bivalent sulfides. Journal of Organic Chemistry 55:3664-3666.

Yang ZX. 1983. Cutaneous injuries by mustard gas in 48 cases. ChungHua Wai Ko Tsa Chih 21:110-112. [In Chinese]

Yeh S, Now SW, Lin CS. 1968. Arsenical cancer of skin. Histologic study with special reference to Bowen's disease. Cancer 21:312-339.

Yokoro K, Ohara I, Oshita A, Yamada A, Nagai M, Ueoka R. 1958. Two cases of mustard gas cancer of the respiratory tract. Gann 49:351-353. [In Japanese]

Young L. 1947. Observations on the effects of mustard gas on the rat. Canadian Journal of Research. Section E: Medical Sciences 141 151.

Yourick JJ, Clark CR, Mitcheltree LW. 1991. Niacinamide pretreatment reduces microvesicle formation in hairless guinea pigs cutaneously exposed to sulfur mustard. Fundamental and Applied Toxicology 17:533-542.

Yourick JJ, Clark CR, Mitcheltree LW. 1991. Niacinamide pretreatment reduces microvesicle formation in sulfur mustard cutaneously exposed hairless guinea pigs (abstract). Toxicologist 284.

Zachariae H. 1988. Mustard gas should not be investigated as a treatment protocol. Sygeplejersken 88:24. [In Danish]

Zackheim HS, Smuckler EA. 1980. Tumorigenic effect of topical mechlorethamine, BCNU and CCNU in mice. Experientia 36:1211-1212.

Zagari. 1918. Gastric ulcerations as effects of the toxic gases of war. Pensiero Medica 8:91-94. [In Italian]
Zakharov IA, Fedorova IV. 1980. Induced mutational process in radio sensitive yeast mutants and prospects for using *Saccharomyces cerevisiae* yeast as objects for testing mutagenic factors. Izvestiya Akademii Nauk Tadzhikskoi Ssr Otdelenie Biologicheskikh Nauk. 3-10. [In Russian]
Zandieh T, Marzban S, Tarabadi F, Ansari H. 1990. Defects of cell-mediated immunity in mustard gas injury after years (abstract). Scandinavian Journal of Immunology 32:423.
Zeehuisen H. 1922. The action of certain gases and toxic vapors on guinea pigs and white rats. Archives Neerlandaises de Physiologie de l'Homme et des Animaux 7:146.
Zettl W. 1940. Symptomatology and therapy of late injuries of eyescaused by mustard gas . Klinische Monatsblatter fur Augenheilkunde 104:217-222. [In German]
Zhang B-Z, Wu Y. 1987. Toxicokinetics of sulfur mustard. Chinese Journal of Pharmacology and Toxicology 1:188-194.
Zimmermann FK. 1971. Genetic aspects of carcinogenesis. Biochemical Pharmacology 20:985-995.
Zunz E. 1917. Research on blood pressure during gas intoxications. Trauvaux-Ambulance de l'Ocean de Panne 1:257-284. [In French]
Zunz E. 1919. Changes to the blood during exposure to yperite or the arsines. TrauvauxAmbulance de l'Ocean de Panne 2:427-444. [In French]
Zunz E. 1919. On blood pressure during exposure to yperite and the arsines. TrauvauxAmbulance de l'Ocean de Panne 2:411-425. [In French]
Zunz E, de Harven J. 1919. Histological evidence from mortal cases of yperite intoxication in man. Trauvaux-Ambulance de l'Ocean de Panne 2:475-484. [In French]

MILITARY REPORTS UNITED STATES

U.S. Army Reports

1923-1946

Adler FH. 1944. Report of consultant in ophthalmology on 6 cases of H vapor burns occurring at Bushnell Field Installation on April 20, 1944. Memo to Colonel C.P. Rhoads, dated May 1, 1944. Available at the National Archives, Suitland Reference Branch, Suitland, MD. Record Group 175, Group 4B, Folder 319.1.
Chemical Warfare Service (CWS). 1944. Report of Chemical Warfare Service Conference. October 10-13, 1944. Available at the National Archives, Suitland Reference Branch, Suitland, MD. Record Group 175, Group 4B, Folder 337.
DPGMR (Dugway Proving Ground Memorandum Report) No. 23. The assessment of the M47A2 and M70 bombs filled with Levinstein mustard. Dugway Proving Ground Mobile Unit, Bushnell, FL. 1945.
EAL 539. A preliminary report on the ocular action of dichlorethyl sulfide (mustard gas) in man as seen at Edgewood Arsenal. Otto CE. 1946.
EAMRD (Edgewood Arsenal Medical Research Division Report) 9. Blistering concentration of mustard gas vapors for exposures from five minutes to three hours. Temple JW. 1923.
EAMRD 11. Toxicity of certain compounds for mice and vesicant action on man. Wells WJ. 1923.

EAMRD 12. Digest of the toxic effects of mustard gas. 1/31/23.
EAMRD 27. Report on pathology of animals recovered from gassing with phosgene, mustard, Lewisite, chlorine, chlorpicrin and methyldichlorarsine. Koontz AR, Witherspoon MG, Allen MS. 1925.
EAMRD 39. Persistency of mustard in living tissues. 6/16/25.
EAMRD 89. On the incidence of bacteria in gassed lungs and in pneumonia produced by war gases: a study of the bacteriology of chemical pneumonia. Koontz AR, Allen MS. 1928.
EATR (Edgewood Arsenal Technical Report) 78. Constants and physiological action of chemical warfare agents. 7/19/32.
EATR 285. Lewisite (M-1). Summary of physiologic and toxicologic data. Wardell EL. 3/15/41.
EATR 292. Pathology of heat and mustard burns: a comparison. 1/10/39.
EATR 293. Arsine as a potential chemical warfare agent. Summary of available data. 1/31/39.
MD (EA) Memorandum Report (Medical Division, Edgewood Arsenal) 22. The determination of biological methods of the length of time irritant substances remain active in the skin of living goats following application of mustard. Project MR1.1-1. 1941.
MD(EA) Memorandum Report No. 38. Treatment of conjunctivitis and laryngitis due to exposure to mustard vapor. 10/27/41.
MD(EA) Memorandum Report No. 42. Treatment of chemical casualties: eye and respiratory tract injuries due to mustard vapor. Gilpin BB. 1/19/42.
MD(EA) Memorandum Report No. 78. Effects of mustard on muscle tissue. 1/9/43.
MDR (Medical Division, Edgewood Arsenal) Report No. 22. The relation of time to the dose to produce a given physiological effect. 2/3/45.
MDR No. 36. Construction and operation of a gassing chamber for human tests. 6/28/45.
MRL(DPG) Report (Medical Research Laboratory, Dugway Proving Ground) No. 4. The effectiveness of low altitude airplane spray of Levenstein and thickened mustard (HV) against men with and without protective clothing. 1946.
MRL (DPG) Report No. 5. A study of the vesicant power of falling drops of thickened and unthickened Levinstein mustard. 1946.
MRL(EA) Report (Medical Research Laboratory, Edgewood Arsenal) No. 3. The pathology of mustard burns of human skin. Ginzler AM, Davis MI. 9/30/43.
MRL(EA) Report No. 9. Continued exposure of human eyes to H vapor (MIT subjects). 1/1/44.
MRL(EA) Report No. 18. Eye examination of factory workers handling H, CN, and CG. Laughlin RC. 4/19/44.
MRL(EA) Report No. 20. Pathological changes in tissues of victims of the Bari incident. Rich AR, Ginzler AM. 5/18/44.
MRL(EA) Report No. 23. Correlation of eye changes in rabbits with CT exposure to H. 6/12/44.
SJPR (San Jose Project Report) No. 24. Relative sensitivity to liquid mustard gas of continental U.S. troops and Puerto Rican troops in a tropical climate. 10/27/44.
SJPR No. 82. The effectiveness of standard anti-gas training as applied to jungle warfare, and the effects produced on troops by wearing impregnated clothing and gas masks in the jungle (Exercise Sandfly). 9/1/45.
TDMR (Technical Division, Edgewood Arsenal, Memorandum Report) 457. A mixture of 50% Lewisite, 50% mustard. The median lethal dosage by skin application. 10/20/42.
TDMR 462. Mustard gas: its effects on syphilis serology. 11/2/42.
TDMR 468. Distribution and rate of penetration of Lewisite in skin. 5/1/43.
TDMR 490. Memorandum report. The effect of relative humidity and temperature on the vesicant action of liquid mustard. Breazle EL, Hunt CM. 1942.

TDMR 491. Precipitation produced by the action of mustard on human serum. 12/3/42.
TDMR 512. The comparative vesicant action of, and penetration of impregnated cloth by mustard and Lewisite airplane spray mixtures. 12/19/42.
TDMR 515. The effect of heat and humidity on the vesicant action of Lewisite and mustard. 12/24/42.
TDMR 528. The effect of mustard on the complement titers of rabbit sera. Breazle EL, Lowry JJ, Lankford E, Hunt CM. 1943.
TDMR 567. Complement fixation test for mustard sensitivity. 2/11/43.
TDMR 575. H and HT review of British and U.S. literature. Cone NM, Rouiller CA. 1943.
TDMR 632. Memorandum report. Local sensitization of human skin to HS by means of sensitive serum. Breazle EL, Lankford E. 1943.
TDMR 639. The effect of mustard and Lewisite on the colloidal gold curve of spinal fluid. 5/4/43.
TDMR 648. Distribution and rate of penetration of Lewisite in skin. 5/1/43.
TDMR 666. Effects of mustard on peritoneal tissue. 5/28/43.
TDMR 668. Destruction of immune antibodies by mustard vapors. 6/4/43.
TDMR 710. Studies of the mechanism of physiological action of mustard: effect on calcium ion liberation. 8/30/43.
TDMR 720. Field tests with airplane mustard spray to compare protection afforded by three types of permeable protective clothing. 8/27/43.
TDMR 731. The value of permeable protective shorts as a means of reducing the number of casualties from exposure to H vapor. Gas chamber tests. 9/9/43.
TDMR 860. Protection afforded by single-layer protective outfits against a single mustard vapor exposure. 8/19/44.
TDMR 930. The H vapor protection afforded by one and one-half layer protective outfits worn by the same men in successive exposures. Gas chamber tests. Part I. 3/1/45.
TDMR 970. Investigation of the physiological effects, protective life span, and protection afforded by protective clothing at Edgewood Arsenal. 2/8/45.
TDMR 994. Preliminary gas chamber tests of protection afforded by single layer protective outfits against successive mustard vapor exposures. 3/12/45.
TDMR 1009. Preliminary gas chamber tests of protection afforded by double-layer protective outfits against a single mustard vapor exposure. 7/4/45.
TDMR 1012. The H vapor protection afforded by various protective outfits by the same men in successive exposures. 6/27/45.
TDMR 1018. Protection afforded by one and one-half layer protective outfits against a single exposure to mustard vapor. 7/9/45.
TDMR 1042. Protection afforded by one and one-half layer protective outfits against successive exposures to H vapor. Gas chamber tests. Part V. 5/22/45.
TDMR 1062. Exposures of permeable protective fabrics on men's arms to H vapor. Preliminary gas chamber tests. 8/21/45.
TRLR (Toxicological Research Laboratory, Edgewood Arsenal, Report) 1. Lewisite. Determination of vesicant action on man by use of a continuous flow chamber. Silver SD. Project A10.3. 8/21/43.
TRLR 18. L, HN-1, H and HQ: effects of 0.1 mg drops on eyes of rabbits. Wallen LJ, Horton RG, Ferguson RL. Project A10.3. 12/20/43.
TRLR 47. Gassing chamber for human tests: construction and operation. 10/25/44.

1947-1992

Capizzi BL, Papirmeister B. The pre-clinical detection of chemical mutagens and their evaluation as genetic health hazards. Edgewood Arsenal, MD: Biomedical Laboratory. 1972.

Chan P, ed. Proceedings of the Vesicant Workshop, February 1987. Aberdeen Proving Ground, MD: Medical Research Institute of Chemical Defense. AD-A188 222. 1987.

Christensen M. Optimization of a polymer-based system which reduces agent penetration into the skin. Contract No. DAMD17-85-C-5097, Final Report. Aberdeen Proving Ground, MD: Medical Research Institute of Chemical Defense. DTIC AD-A210-157. 1988.

Cochrane RC. Medical research in chemical warfare. Available through the U.S. Army Chemical Defense Research, Development, and Engineering Center, Aberdeen Proving Ground, MD. [1946].

Cochrane RC. U.S. Army Chemical Corps Historical Studies. Gas warfare in World War I: The 78th Division at the Kriemhilde Stellung, October 1918. AD-A955 211/8/XAB. 1957.

Cochrane RC. U.S. Army Chemical Corps Historical Studies. Gas warfare in World War I: The 89th Division in the Bois de Bantheville, October, 1918. AD-A955 210/0/XAB. 1960.

Dorsey JK, Davison CL, Gross CL, Turnbull JD, Papirmeister B. Sensitization of deoxyribonucleic acid to endonuclease by adenine alkylation. EATR 4650. Edgewood Arsenal, MD: Medical Research Laboratory. AD-743 446. 1972.

Forster JS, Jacinto B, Gadson C, Mershon MM. Development of a screening procedure for use in the evaluation of topical protectants after challenge with vesicating agents. Aberdeen Proving Ground, MD: U.S. Army Medical Research Institute of Chemical Defense. AD-A234-9470XSP. 1991.

Hassett CC. Study of long-term human and ecological effects of chemical weapons systems. Chemical Research and Development Laboratories Special Publication 2-52. Edgewood Arsenal, MD: Chemical Research and Development Laboratories. 1963.

Jaeger J, Schmid P. Vesicant studies. 2. Volatility determinations of butyl 2-chloroethyl sulfide on various surfaces, including pig skin. San Francisco: Letterman Army Institute of Research. AD-175-251-8. 1986.

Joiner RL. Multiple animal studies for medical chemical defense program in soldier/patient decontamination and drug development. Contract No. DAMD17-83-C-3129, Task No. 85-01, Final Report. Fort Detrick, MD: Medical Research and Development Command. 1987.

Joiner RL. Assessment of subcutaneous lethality of mustard and efficacy of enzyme inhibitors against mustard lethality in the mouse. Contract No. DAMD17-83-C-3129, Task No. 84-8, Final Report. Fort Detrick, MD: Medical Research and Development Command. 1988.

Koon WS, Swan DJ, Simmonds JS, Oberst FW. The protective effectiveness of the E13R16 protective mask with the E33R4 hood worn by men in mustard vapor. EATM 113-3. Edgewood Arsenal, MD: Medical Research Laboratory. 1966.

Lieberman MW. Development of a system to study gene activation in mammalian cells treated with bis (2-chloroethyl)sulfide. Contract No. DAMD17-86-C-6066. Annual report. Aberdeen Proving Ground, MD: Medical Research Institute of Chemical Defense. AD-A217-974. 1986.

Maas JM, McAdams AJ, McShane WP. Mustard, Lewisite and phosgene oxime burns on the depilated skin of a heifer to determine the animal's vesicating potential. Army Chemical Center, MD: Chemical Corps Medical Labs. AD-061 556/7. 1955.

McGown EL, Van Ravenswaay T, Damlao CR. Histologic changes caused by application of Lewisite analogs to mouse skin and human skin xenografts. San Francisco: Letterman Army Institute of Research. AD-A159 554. 1985.

McGrath FP, Koon WS, Billups NB, Barry MC. Protection afforded against mustard vapor during wearing trials of permeable protective clothing. Technical Report CWLR 2173. Army Chemical Center, MD: U.S. Army Chemical Corps Research and Development Command. 1957.

McKinley MD, McKinley FR, McGown EL. Thiosulphate as an antidote to mustard poisoning. A review of the literature. Institute Report No. 127. San Francisco: Letterman Army Institute of Research. 1982.

McNamara BP. Medical aspects of chemical warfare. Project No. 4C08-02-023. Edgewood Arsenal, MD: Chemical Research and Development Laboratories. 1960.

McNamara BP, Owens EJ, Christensen MK, Vocci FJ, Ford DF, Rozimarek H. Toxicological basis for controlling levels of mustard in the environment. Edgewood Arsenal Special Publication EB-SP-74030. Aberdeen Proving Ground, MD: U.S. Army Armament Command, Edgewood Arsenal Biomedical Laboratory. 1975.

Musselman NP, Fencel RM, Koon WS, Oberst FW. Protection afforded by combination of sateen fatigues with chloramide-treated protective liner worn by men exposed to mustard vapor. Technical Report CRDLR 3225. Edgewood Arsenal, MD: Chemical Research and Development Laboratories. 1964.

Oberst FW, Koon WS, Musselman NP, Wilinski FT, Fencel RM. Mustard vapor tests on several chloramide-treated clothing materials worn by men in chamber tests. Technical Report CRDLR 3210. Edgewood Arsenal, MD: Chemical Research and Development Laboratories. 1964.

Oberst FW, Musselman NP, Graf CH, Trapp GA, Dawson PB. Protection afforded by experimental XXCC3-impregnated Navy work/combat clothing worn by men exposed to mustard vapor. Technical Report CRDLR 3254. Edgewood Arsenal, MD: Chemical Research and Development Laboratories. 1965.

Oberst FW, Swan DJ, Koon WS, Musselman NP, Billups NB, Drew DA, Simmonds JS, Vancil ME. Protection afforded by an experimental zippered hood (E33R4) and the standard M6A1 hood worn by men in mustard vapor. EATR 4031. Edgewood Arsenal, MD: Medical Research Laboratory. 1966.

O'Connor RJ, McGown EL, Dill K. Interaction of phenyldichloroarsine with biological molecules. San Francisco: Letterman Army Institute of Research. AD-A175 296. 1986.

O'Connor RJ, McGown EL, Adams ER, Dill K. Nuclear magnetic resonance study of the interaction of phenyldichloroarsine with calf thymus DNA. San Francisco: Letterman Army Institute of Research. AD-A203 463. 1988.

Papirmeister B. On the mechanism of inhibition of T2 bacteriophage by mustard gas. EA Special Publication No. 2-45. Edgewood Arsenal, MD. DTIC AD-267 365. 1961.

Papirmeister B, Davison CL. Unbalanced growth and latent killing of *Escherichia coli* following exposure to sulfur mustard. Technical Report CRDLR 3257. Edgewood Arsenal, MD: Chemical Research and Development Laboratories. DTIC AD-614-284. 1965.

Papirmeister B, Westling AW, Schroer J. Mustard: the relevance of DNA damage to the development of the skin lesion. EATR 4294. Edgewood Arsenal, MD: Medical Research Laboratory. DTIC 688-866. 1969.

Papirmeister B, Gilbert RM, Rowland S, Davison CL. Involvement of separate pathways in the repair of mutational and lethal lesions induced by a monofunctional sulfur mustard. B-TR-74052. AD-787 315/1. 10/74.

Penski E. An expanded model for the hydrolysis of mustard and its applications. Technical Report No. ARCLS-TR-83021. Aberdeen Proving Ground, MD: U.S. Army Chemical Systems Laboratory. AD-B075-118. 1983.

Petrali JP, Oglesby SB, Mills KR. Ultrastructural correlates of sulfur mustard toxicity. USAMRICD-TR-89-15. Fort Detrick, MD: U.S. Army Medical Research and Development Command. 1989.

Rosenblatt DH, Miller TA, Dacre JC, Muul I, Cogley DR, eds. Preliminary assessment of ecological hazards and toxicology of environmental pollutants at Rocky Mountain Arsenal. Fort Detrick, MD: U.S. Army Medical Bioengineering Research and Development Laboratory. 1975.

Rosenblatt DH, Miller TA, Dacre JC, Muul I, Cogley DR. Problem definition studies on potential environmental pollutants. II. Physical, chemical, toxicological, and biological properties of 16 substances. TR-7509. Fort Detrick, MD: U.S. Army Medical Bioengineering Research and Development Laboratory. AD-A030. 1975.

Rudin DO. A review of the biological action of mustard as a basis for therapy. Special Report No. 21. Army Chemical Center, MD: U.S. Army Chemical Corps Medical Laboratories. DTIC AD-15 565. 1953.

Sidell FR. Clinical notes on chemical casualty care. USAMRICD Technical Memorandum 90-1. Aberdeen Proving Ground, MD: Medical Research Institute of Chemical Defense. 1990.

Sidell FR. Clinical notes on chemical casualty care. USAMRICD Technical Memorandum 90-2. Aberdeen Proving Ground, MD: Medical Research Institute of Chemical Defense. 1990.

Smith GP, Schumchyk MJ. Performance evaluation of gas filters. June 1971-March 1976. Aberdeen Proving Ground, MD: Chemical Systems Laboratory. ARCSL-SP-78004. 1978.

Taylor JR, Johnson WN. Research report concerning the use of volunteers in chemical agent research. DAIG-IN 21-75. Washington, DC: Department of the Army, Office of the Inspector General and Auditor General. 1975.

U.S. Army Chemical Research, Development, and Engineering Center. Lewisite. Material safety data sheet. Aberdeen Proving Ground, MD: U.S. Army Chemical Research, Development, and Engineering Center. 1988.

U.S. Army Chemical Research, Development, and Engineering Center. HD and THD. Material safety data sheet. HCSDS No. 20058A. Aberdeen Proving Ground, MD: U.S. Army Chemical Research, Development and Engineering Center. 1990.

U.S. Department of the Army. Chemical agent data sheets, Vol. 1. Technical Report EO-SR-74001. Edgewood Arsenal, Maryland. 1974.

U.S. Department of the Army. Final Programmatic Environmental Impact Statement for the Chemical Stockpile Disposal Program. Aberdeen Proving Ground, MD. 1988.

Williamson CE, Witten B. Reactivity of some nitrogen and sulfur mustards in biologic media. EATR 4587. Edgewood Arsenal, MD: Medical Research Laboratory. AD-734 828. 1971.

Yaverbaum S. 14C-sulfur mustard adducts of calf thymus DNA. USAMRICD-TR-91-02. Aberdeen Proving Ground, MD: Medical Research Institute of Chemical Defense. AD-A233 313/6/ XAB. 1991.

U.S. Navy Reports

Naval Research Laboratory, Anacostia Station, Washington DC

Report No. P-1898. Prophylaxis and treatment of burns caused by chemical warfare agents. I. Treatment of mustard burns with S-461 ointment in a series of controlled experiments on human subjects. 4/24/42.

Report No. P-1899. Prophylaxis and treatment of burns caused by chemical warfare agents. II. Prophylaxis, as applied to prevention of burns by liquid mustard with S-461 ointment. 5/26/42.

Report No. P-2208. Report on chamber tests with human subjects. I. Design and operation of chamber. II. Initial tests of Navy issue protective clothing against H vapor. Taylor WH, Carhart HW, Daily LE. AD 223752. 12/22/43.

Report No. P-2219. Chamber tests with human subjects. III. Design, operation and calibration of a chamber for exposing forearms to H vapor. 1/22/44.

Report No. P-2239. Chamber tests with human subjects. IV. Tests of carbon clothing against H vapor. Carhart HW, Taylor WH, Daily LE. 2/25/44.

Report No. P-2322. The evaluation of activated carbon as an anti-vesicant agent in protective clothing. Billica HR, Gantz GM. AD B-953304L. 7/3/44.

Report No. P-2364. A controlled laboratory experiment to compare lesions resulting from application of mustard, Lewisite and nitrogen mustards to the skin of the forearms of humans 9/1/44.

Report No. P-2406. Report on wearing trials of protective clothing at Camp Lejeune, North Carolina. 11/13/44.

Report No. P-2464. Chamber tests with human subjects. V. Arm chamber exposures to HN vapors. Carhart HW, Taylor WH. AD B953306L. 3/45.

Report No. P-2483. Chamber tests with human subjects. VI. Arm chamber exposures to L vapor. Taylor WH, Carhart HW. AD B-968917. 5/31/45.

Report No. P-2528. Chamber tests with human subjects. VII. The effect of concentration of H vapor and time of exposure on the protection afforded by CC-2 impregnated clothing. Taylor WH, Carhart HW, Heinen JH. AD 471548. 7/5/45.

Report No. P-2579. Chamber tests with human subjects. IX. Basic tests with H vapor. Heinen JH, Carhart HW, Taylor WH, Stolp BN, Conner JC Jr, Clausen NM. AD 396275L. 8/14/45.

Report No. P-2590. Chamber tests with human subjects. X. Protection afforded by CC-2 impregnated clothing under various conditions of exposure. Taylor WH, Carhart HW, Heinen JH. 9/7/45.

Report No. P-2597. Chamber tests with human subjects. VIII. Evaluation of worn CC-2 impregnated clothing. Taylor WH, Carhart HW, Heinen JH. AD B-968912. 9/5/45.

Report No. P-2602. Chamber tests with human subjects. XI. Evaluation of modified aqueous CC-2 impregnation systems. Taylor WH, Carhart HW, Heinen JH. 8/18/45.

Report No. P-2603. Chamber tests with human subjects. XII. Suit and man "breaks" with CC-2 impregnated clothing. Carhart HW, Taylor WH, Heinen JH. 8/31/45.

Report No. P-2604. Chamber tests with human subjects. XIII. Special tests of CC-2 and carbon protective clothing. Taylor WH, Carhart HW, Heinen JH. AD 223227. 8/18/45.

Report No. P-2682. Second wearing trial of protective clothing at Camp Lejeune, North Carolina. 11/26/45.

Report No. P-2688. Chamber tests with human subjects. XVII. Supplementary tests of CC-2 protective clothing. Taylor WH, Carhart HW, Heinen JH. 11/15/45.

Report No. P-2695. A summary of wearing trials of permeable protective clothing. 11/26/45.

Report No. P-2701. Chamber tests with human subjects. XIV. Tests of new carbon clothing. Carhart HW, Taylor WH, Heinen JH. AD 223534. 12/3/45.

Report No. P-2702. Chamber tests with human subjects. XV. Tests of worn carbon clothing. Carhart HW, Taylor WH, Heinen JH. AD 223228. 1/16/46.

Report No. P-2703. Chamber tests with human subjects. XVI. Tests of regenerated and contaminated carbon clothing.

Report No. P-2729. Chamber tests with human subjects. XIX. Studies of clothing design. Gantz GM, Taylor WH, Carhart HW. 12/10/45.

Report No. P-2734. Chamber tests with human subjects. XVIII. Tests with HN vapors. Heinen JH, Taylor WH, Stolp BN, Conner JC Jr, Clausen NM. 1/9/46.

Report No. P-2760. Chamber tests with human subjects. XX. Hypersensitivity to H as demonstrated by patch tests before and after chamber exposure to H vapor. Heinen JH, Carhart HW, Taylor WH, Stolp BN, Conner JC Jr, Clausen NM. 5/15/46.

Other U.S. Military Reports

Augerson WS, Sivak A, Marley WS. Chemical casualty treatment protocol developmenttreatment approaches: mustards. Brooks Air Force Base, TX: U.S. Air Force Systems Command. DTIC AD-B112-915. 1986.

U.S. Army and U.S. Air Force. Military chemistry and chemical compounds. Field Manual No. FM3-9, Regulation No. AFR 355-7. 1975.

U.S. Department of the Army, the Navy, and the Air Force. Treatment of Chemical Warfare Casualties. Technical Manual No. 8-285; NAVMED P-5041; Air Force Manual No. 160-12. Washington, DC: U.S. Government Printing Office. 1956.

UNITED KINGDOM

Chemical Defence Experimental Station, Porton Down

Porton Report No. 483. The effects of the vapor of H on the human skin. Goodwin GS. 5/31/27.

Porton Report No. 930. Sensitivity to mustard gas. 7/22/31.

Porton Report No. 948. Further report on sensitivity to mustard gas. 9/23/31.

Porton Report No. 2150. Lewisite shock. 12/21/40.

Porton Report No. 2201. Assessment of danger of systemic poisoning by Lewisite. 4/29/41.

Porton Report No. 2246. Systemic effects induced by mustard gas poisoning. First report. 7/30/41.

Porton Report No. 2297. Effect of mustard gas vapour on the eyes. 11/8/41.

Porton Report No. 2320. Systemic effects produced by mustard gas poisoning. Sedimentation rate, corpuscular fragility, and coagulation time of blood after mustard gas poisoning, with a note on Lewisite. 12/26/41.

Porton Report No. 2343. Medical report on casualties produced by airburst mustard gas shells. 3/10/42.

Porton Report No. 2354. The protective value of carbon impregnated fabrics against H vapour. 5/20/42.

Porton Report No. 2374. The protection of vulnerable areas of the body against H vapour in hot climates. 5/22/42.

Porton Report No. 2383. Third report on mustard gas. 6/11/42.

Porton Report No. 2398. Fourth report on mustard gas. General systemic effects of mustard gas applied to the skin. 8/7/42.

Porton Report No. 2429. The relative insensitivity to mustard gas of the skin of the hand. 9/21/42.

Porton Report No. 2460. Blood changes due to S derivatives. 12/1/42.

Porton Report No. 2483. Chemical studies on the mode of action of mustard gas on proteins. Part 1. The nature of the combination of H with proteins. 2/19/43.

Porton Report No. 2508. Lymph drainage in Lewisite poisoning. 6/2/43.

Porton Report No. 2526. The behavior *in vitro* of serum and bone marrow from H poisoned animals. 8/14/43

Porton Report No. 2543. Capillary permeability factors related to the action of CW agents. Part I. 9/21/43.

Porton Report No. 2553. Effect of Lewisite vapor on small animals and on man. 10/29/43.

Porton Report No. 2560. Some further studies on the treatment of mustard gas blisters and a comparison of the healing of mustard gas and Lewisite burns. 11/10/43.

Porton Report No. 2587. Treatment of systemic effects of H. Part I. A preliminary survey of therapeutic agents of high competition factor. 3/10/44.

Porton Report No. 2590. Bone marrow cultures as test objects for the treatment of injury due to CW agents. 1/29/44.
Porton Report No. 2596. Toxicity of H in droplet form. 2/3/44.
Porton Report No. 2626. The treatment of the leucopenia of vesicant poisoning. 6/20/44.
Porton Report No. 2635. Capillary permeability factors related to the action of CW agents. Part II. 7/21/44.
Porton Report No. 2638. Anti-gas ointments for use in the tropics. Part 1. Exploratory experiments. 8/14/44.
Porton Report No. 2647. The mode of penetration of the skin by mustard gas. 9/8/44.
Porton Report No. 2658. The treatment of the systemic effects of H. Part II. Dithiocarbamates as therapeutic agents. Burgess F, Call F, Cameron GR, Rydon HN. 1944.
Porton Report No. 2680. The effect of mustard gas upon binary fistula animals. 5/29/45.
Porton Departmental Report No. 76. Report on the skin burning power of various vesicant mixtures through service dress.
Ptn. (Porton Letters) 1200 (R.12035 and R.14727). Comparative vesicant properties of vesicant compounds. 12/19/41.
Ptn. 1601A (V.1358). Effects of H vapour on man. 2/7/45.
Ptn. 2800 (T.12752 and T.12908). The biochemical mode of action of mustard gas. A review. 10/2/43.
Ptn. 2809 (U.672). A clinicopathological report of human eyes splashed with mustard gas. 2/3/44.
Ptn. 2820 (T.9506A). Comparison of the effects of H and Lewisite on living tissues. 7/10/43.
CDE Technical Paper No. 522. The reaction between sulphur and nitrogen mustards and collagen. Goodlad GAJ. DTIC AD-70 757. 1956.
CDE Technical Paper No. 573. Cross-linking of collagen by S- and N-mustards. Goodlad GAJ. DTIC AD-119 934. 1956.
CDE Technical Paper No. 575. Esterification of amino acids by S-mustards. Goodlad GAJ. DTIC AD-123 274. 1956.
CDE Technical Note No. 840. A literature review on the toxicology, mechanism of action and treatment of sulphur and nitrogen mustard poisoning. Whitfield D. 1987.

Other British Military Reports

Great Britain War Office. Medical Manual of Chemical Warfare. Admiralty BR/235/1954, War Office Code No. 10562, Air Ministry AP3326. London: Her Majesty's Stationery Office. 1955.

AUSTRALIA

CD (Chemical Defence, Australia) Note No. 21. The effects of exposure to mustard vapor on the blood coagulation time in man. 5/23/44.
CD (Australia) Note No. 41. The effect of high and low mustard vapour concentrations on the dosage required to penetrate impregnated clothing. 5/16/45.
CD (Australia) Note No. 50. Exposure of subjects to mustard vapor dosages of 50, 120, 125, and 220 mg. min./m under tropical conditions. 1945.
CD (Australia) Note No. 56. The effect of cumulative exposures to mustard vapour. 7/2/45.
CD (Australia) Report No. 15. The differential white count in man following exposure to mustard vapor. 1/24/44.
CD (Australia) Report No. 20. The differential white count in man following exposure to mustard vapor. Second report. 3/29/44.

CD (Australia) Report No. 30. An attack on a small island with the M.47 bomb charged Levinstein mustard. Final report. 4/1/44.
CD (Australia) Report No. 37. The systemic effects of mustard gas under tropical conditions as seen in four cases of liquid contamination. 5/8/44.
CD (Australia) Report No. 40. Effect of varying humidity and temperature upon the sensitivity of human skin to mustard vapor. 5/22/44.
CD (Australia) Report No. 41. Effect of varying activity of subjects during exposure on the sensitivity of human skin to mustard vapor. 5/22/44.
CD (Australia) Report No. 55. A clinical analysis of a series of mustard burns occurring under tropical conditions. 10/18/44.
CD (Australia) Report No. 78. A further comparison between physiological effects of mustard vapour in the chamber and in the field. 5/25/45.
CD (Australia) Report No. 85. Estimation of the dose-lesion relationship for human exposure to mustard vapour. 2/28/46.
Gillis RG. Australian field tests with mustard gas 1942-1945. Department of Defence, Australia. 1985.
Gray PJ. A literature review on the mechanism of action of sulphur and nitrogen mustard. MRL-TR-89-24. Maribyrong, Australia: DSTO Materials Research Laboratory. ADA125 070. 1989.

CANADA

Experimental Station, Suffield, Alberta

Suffield Field Report No. 133. The physiological effects of mustard vapour at low temperatures. 8/9/45.
Suffield Technical Minute No. 72. A comparison of the erythema and vesicle producing capacity of HT/MM and HT. 9/15/44.
Suffield Technical Minute No. 92. Vapour damage from gross mustard contamination (Field Experiment 141). 10/8/43.
Suffield Technical Minute No. 103. Effectiveness of given Ct's of mustard gas vapour for long exposure periods under tropical conditions. 8/13/45.

Chemical Warfare Laboratories, Ottawa

Physiological Section Report No. 27. Observations on the treatment of experimentally produced mustard lesions. 12/21/43.
Physiological Section Report No. 35. Studies on the immunological behavior of H. II. The production of immune bodies by the use of H-serum protein complexes. 2/22/44.

INDIA

CDRE (Chemical Defence Research Establishment, Rawalpindi, India) Report No. 245. The effect of mustard gas vapour on eyes under Indian hot weather conditions. 11/9/42.
CDRE (India) Report No. 255. The appearances and treatment of mustard gas burns of the skin under Indian conditions. 7/15/43.
CDRE (India) Report No. 285. Report on two cases of severe skin burns from mustard gas vapor under tropical conditions in India. 11/29/44.

OTHER MILITARY REPORTS

Alexander SF. Final report of Bari mustard casualties. Allied Force Headquarters. Director of Medical Services. Surgeon NATO. 1944.
Project Coordination Staff. Part 1. Technical aspects of chemical warfare in the field. 1946.
Project Coordination Staff. Part 2. Discussion of experimental data. Technical aspects of chemical warfare in the field. 1946.

TECHNICAL REPORTS

1942-1946

Office of Scientific Research and Development

OSRD 451. The study of the mechanism of the physiological action of mustard by means of radioactive mustard. Moritz AR, Henriques FC Jr, Schneider WG, Halford RS. Harvard University. 3/16/42.
OSRD 527. The toxicity of mustard (redistilled Levinstein). Coon JM, Last JH, Lushbaugh CC, Rotariu GC. University of Chicago Toxicity Laboratory. 4/21/42.
OSRD 823. Report on toxic effects of various arsine derivatives. Geiling EMK, McLean FC. University of Chicago. 8/24/42.
OSRD 1199. Comparison of the lethal effects of several dichloroarsines on mice exposed to the vapors at low relative humidity. Hutchens JO, Doyle WL, Merrill R, Glass HG, Lushbaugh CC. University of Chicago Toxicity Laboratory. 2/16/43.
OSRD 1248. The inactivation of enzymes by mustard gas. Cannan RK. New York University. 3/10/43.
OSRD 1253. Toxic effects of various arsine derivatives. Hutchens JO, Glass HG, Gurney HW, Merrill R. University of Chicago Toxicity Laboratory. 3/11/43.
OSRD 1313. A study of the hematological changes following exposure to certain war gases. Lusbaugh CC. University of Chicago Toxicity Laboratory. 4/3/43.
OSRD 1439. Some reactions of mustard gas with proteins and proteolytic enzymes. Bergmann M, Fruton JS, Irving GW, Moore S, Stein WH. Rockefeller Institute. 5/21/43.
OSRD 1630. A study of the reaction between mustard gas and proteins. Davis SB, Ross WF, Ball EG. Harvard University. 7/22/43.
OSRD 1717. Review of the literature on the systemic action of mustard gas to August 1, 1943. Smith HW. New York University. 8/16/43.
OSRD 1824. The action of mustard gas on certain enzymatic reactions. McKee RW, Marston EL, Ormsbee RA, Ball EG. Harvard University. 9/21/43.
OSRD 1825. A study of the "fixed mustard" in skin tissues. Ormsbee RA, Henriques FC Jr. Harvard University. 9/21/43.
OSRD 1855. The reactions of amine mustards with chemical constituents of biological systems. 9/28/43.
OSRD 1911. Determination of the distribution of H and L in skin and eye tissues by radio-autographic techniques. Hamilton G, Axelrod D. University of California. 10/13/43.
OSRD 3249. Preparation and testing of substances as neutralizing or therapeutic agents for H burns. Folkers K, Phillips RF, Shunk CH, Molitor H, Kuna S. Merck and Company. 2/15/44.
OSRD 3269. Induced hypersensitivity to bis(b-chloroethyl)sulphide (mustard gas) and 2,3-dimercaptopropanol (BAL) in guinea pigs. Kidd JG, Landsteiner K. Rockefeller Institute. 3/4/44.

OSRD 3366. A study of the ability of compounds with high competition factors to counteract the injurious effects of mustard gas. Buchanan JM, Doering WE, Marston EL, McKee RW, Ormsbee RA. Harvard University. 3/16/44.

OSRD 3386. Tests of chloroamide-containing ointments for protection and decontamination of human skin against vesicants. Savit J, Thomson JF, Goldwasser E, Debruyn P, Bloom MA. University of Chicago Toxicity Laboratory. 3/21/44.

OSRD 3437. A. Investigation of detoxicants or decontaminants for mustard gas. B. Enzyme studies and other chemical studies bearing upon the action of certain vesicants. Hellerman L, Porter CC, Irvin JL, Presta UA, Lindsay A. Johns Hopkins University. 4/4/44.

OSRD 3467. Studies on the cause of death after systemic intoxication with the b-chloroethyl vesicants. Smith HW, Crawford B, Houck CR. New York University. 4/12/44.

OSRD 3620. The mechanism of cutaneous injury by mustard gas. An experimental study using mustard prepared with radioactive sulfur. Henriques FC, Moritz AR, Breyfogle HS, Patterson LA. Harvard University. 5/9/44.

OSRD 3653. Reactions of H with enzymes and proteins. Herriott RM, Anson ML, Northrop JH. Rockefeller Institute. 5/20/44.

OSRD 3942. Protective clothes. I. Irritancy of vapor-contaminated samples on human skin. II. Penetration of vapor and liquid vesicants. University of Chicago. 8/30/44.

OSRD 3943. Analysis of variations in size of blister after applications of H. Wright S. University of Chicago Toxicity Laboratory. 7/27/44.

OSRD 3944. A vapor-train study of the comparative vesicancy of mustard and several related amines and sulfides on human skin. Black S, DuBois KP, Lipton MA. University of Chicago Toxicity Laboratory. 8/30/44.

OSRD 4176. Toxicity and vesicant tests of compounds referred to the University of Chicago Toxicity Laboratory. Young HD. 10/3/44.

OSRD 4638. Tests for decontamination of mustard and nitrogen mustard on human skin. Goldwasser E, DeBruyn PPH, Thomson JF, Savit J. University of Chicago Toxicity Laboratory. 12/12/44.

OSRD 4841. Biochemistry of the action of sulfur containing vesicants. du Vigneaud V, Carpenter FH, McDuffie HF, McKennis H, Melville OB, Rachele JR, Stevens CM, Wood LJ. Cornell University. 3/20/45.

OSRD 4852. I. The necrotizing action of certain substances related to mustard gas, H, or to the nitrogen mustards. II. A comparison of the vesicant action exerted on human skin by mustard gas, H, and by mixtures of H with wetting agents or solvents. Hogeboom GH, McMaster PD, Sulzberger MB, Baer RL, Kanof A. Rockefeller Institute. 3/25/45.

OSRD 4853. The development of methods for testing the abilities of agents to combat the effects of mustard gas, H, and other vesicants upon the skin. McMaster PD, Hogeboom GH, Sulzberger MB, Baer RL, Kanof A. Rockefeller Institute. 3/24/45.

OSRD 4854. A search for decontaminating and treatment agents for skin exposed to mustard gas. McMaster PD, Hogeboom GH, Sulzberger MB, Baer RL, Kanof A. Rockefeller Institute. 3/24/45.

OSRD 4855. The penetration of vesicant vapors into human skin. Bergmann M, Fruton JS, Golumbic C, Nagy SM, Stahmann MA, Stein WH. Rockefeller Institute. 3/24/45.

OSRD 5000. The effect of flow rate on the toxicities of H, Q, HN1, HN3 and L by inhalation, by total exposure, and by body exposure. Albaum HG, Benedict D, DuBois KP, Glass HS, Henderson JH, Hutchens JO. University of Chicago Toxicity Laboratory. 4/28/45.

OSRD 5026. Changes in the circulation and in the permeability of vessels within and about mustard gas and Lewisite lesions of rabbit skin. McMaster PD, Hogeboom GH. Rockefeller Institute. 5/3/45.

OSRD 5027. The inhibition of vesiculation in mustard gas, H, lesions of human skin by

BAL. McMaster PD, Hogeboom G, Sulzberger MB, Baer RL, Kanof A. Rockefeller Institute. 5/3/45.
OSRD 5032. A formal analysis of the action of liquid vesicants on bare skin. Landahl HD. University of Chicago Toxicity Laboratory. 5/5/45.
OSRD 5169. Observations on the role of water in the susceptibility of human skin to vesicant vapors. Renshaw B. 6/1/45.
OSRD 5180. The comparative systemic effects of mustard and the nitrogen mustards (HN1, HN2, HN3) in rats. Graef I, Jager VB, Karnofsky DA. New York University. 5/1/45.
OSRD 5181. The penetration of vesicant vapors into human skin (Supplement to OSRD No. 4855). Bergmann M, Fruton JS, Nagy SM, Stein WH. Rockefeller Institute. 6/6/45.
OSRD 5194. Tests for vesicancy on human skin. Thomson JF, Young HD, Savit J, Goldwasser E, Murray RG, DeBruyn P. University of Chicago Toxicity Laboratory. 6/1/45.
OSRD 5245. Effects of bis(b-chloroethyl) sulfide (H) and bis(b-chloroethyl) methylamine (HN2) on enzymes *in vitro* and *in vivo*. Cori CF, Colowick SP, Berger L, Slein MW. Washington University School of Medicine. 6/20/45.
OSRD 5527. Final technical report, University of Chicago Toxicity Laboratory. Cannan RK, Geiling EMK. 9/11/45.
OSRD 5979. Protective and therapeutic agents for war gases: therapeutic agents for mustard and nitrogen mustards II. Salzberg PL, Lazier WA, Pavlic AA, Rigby GW, Vinton WH. 1/10/46.
OSRD 6084. The NDRC method for laboratory evaluation of permeable protective fabrics. Guss WI, Adkins H, University of Wisconsin. 10/17/45.
OSRD 6325. Summary report on the systemic pharmacology and pathology of the sulfur and nitrogen mustards. Anslow WP, Houck CR. New York University. 10/1/45.
OSRD 6664. The effects of vesicants on cell division in *Arbacia punctulata*. Cannan RK, Levy M. New York University. 6/1/46.
Office of Scientific Research and Development. National Defense Research Committee. Chemical Warfare Agents, and Related Chemical Problems. 2 vols. Summary Technical Report of Division 9, NDRC. Washington, DC: NDRC. AD-234 249. 1946.

1947-1992

Anders MW. Glutathione-dependent toxicity: biosynthesis and bioactivation of cytotoxic S-conjugates. Rochester, NY: Rochester University. School of Medicine and Dentistry. AD-A247-1126XSP. 1988.
Aposhian HV. Prevention and treatment of vesication and poisoning caused by arsenicals. Annual summary report, Feb. 1980-Jan. 1981. Tucson: Arizona University. AD-A1203850XSP. 1981.
Back KC, Thomas AA, MacEwen JD. 1972. Reclassification of material listed as transportation health hazards. Office of Hazardous Materials of the Assistant Secretary for Safety and Consumer Affairs, U.S. Department of Transportation. TSA-2072-3. PB-214 270/1.
Benschop HP. Verification, dosimetry and biomonitoring of mustard gas exposure via immunochemical detection of mustard gas adducts to DNA and proteins. Midterm report 15 May 1988-14 Nov. 1989. Rijswijk, Netherlands: Prins Maurits Lab. AD-A240471-3XSP. 1990.
Bernstein IA, Brabec MJ, Conolly RC, Gray RH, Kulkarni A, Mitra R, Vaughan FL. Chemical blistering: cellular and macromolecular components. AD-A190 313. Ann Arbor: University of Michigan. 1985.
Dannenberg AM Jr. Pathogenesis and treatment of skin lesions caused by sulfur mustard:

inflammatory mediators and modulators released from organ-cultured inflammatory lesions produced *in vivo* in rabbit skin by sulfur mustard. Final report. Baltimore, MD: Johns Hopkins University. AD-A207 976. 1987.

Dannenberg AM Jr. Sulfur mustard (SM) lesions in organ-cultured human skin: markers of injury and inflammatory mediators. Final report. DTIC AD-B134-651. 1990.

Dannenberg AM Jr, Vogt RF Jr. Pathogenesis and treatment of skin lesions caused by sulfur mustard. Inflammatory mediators and modulators released in organ culture from rabbit skin lesions produced *in vivo* by sulfur mustard. Annual report. AD-A182 786. Baltimore, MD: Johns Hopkins University. 1986.

Feister A, Papirmeister B, Robinson S, Kiebzak G, McNally R, Ford R, Baggett J, Gottlieb J, Bareis D. Sulfur mustard and Lewisite: current perspectives and future directions. Prepared by Science Applications International Corporation for U.S. Army Medical Research Institute of Chemical Defense. Unpublished. 1989.

Hackett PL, Rommereim RL, Burton FG, Buschbom RL, Sasser LB. Teratology studies on Lewisite and sulfur mustard agents: effects of sulfur mustard in rats and rabbits. Richland, WA: Pacific Northwest Laboratory. AD-A187 495. 1987.

Hackett PL, Rommereim RL, Burton FG, Buschbom RL, Sasser LB. Teratology studies on Lewisite and sulfur mustard agents. effects of sulfur mustard in rats and rabbits. Part 2. Appendices. Richland, WA: Pacific Northwest Laboratory. NTIS AD-A189-161. 1987.

Hackett PL, Sasser LB, Rommereim RL, Cushing JA, Buschbom RL, Kalkwarf DR. Teratology studies of Lewisite and sulfur mustard agents: effects of Lewisite in rats and rabbits. PNL-6408, Pt. 1. Richland, WA: Pacific Northwest Laboratory. 1987.

Jostes RF Jr, Sasser LB, Rausch RJ. Toxicology studies on Lewisite and sulfur mustard agents: genetic toxicity of sulfur mustard (HD) in Chinese hamster ovary cells. PNL-6922. Richland, WA: Pacific Northwest Laboratory. AD-A215 475. 1989.

Koppikar A, McGaughy R, Rhomberg L. Upper-bound quantitative cancer risk estimates for populations adjacent to sulfur mustard incineration facilities. EPA 600891053, OHEAC291. Washington, DC: U.S. Environmental Protection Agency. PB 921-372-07XSP. 1991.

Landry L. Develop a biomedical database on the medical aspects of chemical defense. Final report. Washington, DC: Associate Consultants. NTIS-AD-A206 941-7. 1986.

Ludlum D. Protection against the acute and delayed toxicity of mustards and mustard-like compounds. Final Report. Sept. 1, 1982-February 28, 1987. Albany Medical College, NY. AD-A183-573-5. 1987.

Monteiro-Riviere NA, King JR, Riviere JE. Cutaneous toxicity of mustard and Lewisite on the isolated perfused porcine skin flap. Raleigh: North Carolina State University, School of Veterinary Medicine. AD-A229-9220XSP. 1990.

Opresko DM. Occupational criteria documents for chemical agents. Review of methodologies used for establishing occupational health criteria with particular reference to chemical agents. ORNL-6387. Oak Ridge, TN: Oak Ridge National Laboratory. AD-A233 858/0/XAB. 1988.

Sasser LB, Cushing JA, Kalkwarf DR, Buschbom RL. Toxicology studies on Lewisite and sulfur mustard agents: modified dominant lethal study of sulfur mustard in rats. Richland, WA: Pacific Northwest Laboratory. AD-A214 556. 1989.

Sasser LB, Miller RA, Kalkwarf DR, Buschbom RL, Cushing JA. Toxicology studies on Lewisite and sulfur mustard agents: subchronic toxicity of sulfur mustard (HD) in rats. Richland, WA: Pacific Northwest Laboratory. AD-A214-555. 1989.

Sasser LB, Cushing JA, Kalkwarf DR, Mellick PW, Buschbom RL. Toxicology studies on Lewisite and sulfur mustard agents: subchronic toxicity study of Lewisite in rats. Richland, WA: Pacific Northwest Laboratory. AD-A217 886. 1989.

Sasser LB, Cushing JA, Kalkwarf DR, Mellick PW, Buschbom RL. Toxicology studies on

Lewisite and sulfur mustard agents: two-generation reproduction study of Lewisite in rats. Richland, WA: Pacific Northwest Laboratory. AD-A214 311. 1989.

Sasser LB, Miller RA, Kalkwarf DR, Buschbom RL, Cushing JA. Toxicology studies on Lewisite and sulfur mustard agents: two-generation reproduction study of sulfur mustard (HD) in rats. Richland, WA: Pacific Northwest Laboratory. AD-A216 423. 1989.

Smulson ME. Molecular biology basis for the response of poly(ADP-Rib) polymerase and NAD metabolism to DNA damage caused by mustard alkylating agents. Washington, DC: Georgetown University. AD-A235 314/2/XAB. 1991.

Stewart DL, Sass EJ, Fritz LK, Sasser LB. Toxicology studies on Lewisite and sulfur mustard agents: mutagenicity of Lewisite in the *Salmonella* histidine reversion assay. Richland, WA: Pacific Northwest Laboratory. AD-A213-146. 1989.

Stewart DL, Sass EJ, Fritz LK, Sasser LB. Toxicology studies on Lewisite and sulfur mustard agents: mutagenicity of sulfur mustard in the *Salmonella* histidine reversion assay. Richland, WA: Pacific Northwest Laboratory. AD-A213 102. 1989.

Watson AP, Munro NB. Reentry planning: the technical basis for offsite recovery following warfare agent contamination. ORNL-6628. Oak Ridge, TN: Oak Ridge National Laboratory. 1990.

Watson AP, Adams JD, Cerar RJ, Hess TL, Kistner SL, Leffingwell SS, MacIntosh RG, Ward JR. Estimated general population control limits for unitary agents in drinking water, milk, soil, and unprocessed food items. ORNL/TM-12035. Oak Ridge, TN: Oak Ridge National Laboratory. 1992.

AVAILABILITY INFORMATION

The following addresses and information are provided for those persons interested in obtaining copies of the military and technical reports listed in the bibliography. Consult the listed repositories for fees and availability.

OFFICE OF SCIENTIFIC AND RESEARCH DEVELOPMENT (OSRD) REPORTS:

Including National Defense Research Committee (NDRC) and Committee on Medical Research (CMR) reports.

Civil Reference Branch
National Archives and Records Administration
Washington, DC 20408
Records are stored in Record Group 227, Series 29.

COMMITTEE ON TREATMENT OF GAS CASUALTIES (CTGC) REPORTS:

Office of Archives and Information Services
National Academy of Sciences
2101 Constitution Avenue, N.W. Washington, DC 20418

NATIONAL TECHNICAL INFORMATION SERVICE (NTIS):

For unclassified reports listing an AD number:
National Technical Information Service
5285 Port Royal Road Springfield, VA 22161-0001

NAVAL RESEARCH LABORATORY REPORTS:

Reports with AD numbers can be ordered through the National Technical Information Service. Other reports can be requested from the NRL Public Affairs Office.

Public Affairs Office
Naval Research Laboratory
4555 Overlook Avenue, S.W.
Washington, DC 20375-5000

Chemical Warfare Service reports:

a) Administrative records and correspondence:
   National Archives and Records Administration
   Suitland Reference Branch
   Washington, DC 20409
   Records are stored in Record Group 175 at the Suitland, MD facility.
b) Reports can be requested through the Freedom of Information Act from: Commander U.S. Army Chemical Research, Development, and Engineering Center
   Aberdeen Proving Ground, MD 21010-5423

BRITISH REPORTS:

Chemical and Biological Defence Establishment
Porton Down
Salisbury, Wiltshire
United Kingdom SP4 OJQ

A subject bibliography, entitled *Health Effects of Mustard Gas and Lewisite: Subject Bibliography,* is available through the National Technical Information Service.

APPENDIXES

A

Scientific and Background Presentations Made to the Committee

January 6, 1992

David Saumweber, Director, Office of Archives and Information Services, National Academy of Sciences. Historical Background: The Role of the National Academy of Sciences During World War II.

Robert Cook-Deegan, Institute of Medicine; Linda Rosenstock, University of Washington; Bailus Walker, Jr., University of Oklahoma. Poison Gas: A Continuing Threat, Experiences from Korea and Iraq.

April 15-16, 1992

Bruno Papirmeister, Science Applications International Corporation. Current Research into the Biological Mechanisms of Mustard Gas Toxicity.

Frederick Sidell, United States Army Medical Research Institute of Chemical Defense. Clinical Aspects of Mustard Gas.

Howard Skipper, Southern Research Institute (retired). Chemical Warfare Research During World War II.

Robert Ursano, Uniformed Services University of the Health Sciences. Psychological Aspects of Chemical Warfare Environments.

Karen Freeman, Pennsylvania State University. Researching the World War II Testing Programs.

Annetta Watson, Oak Ridge National Laboratory. Outline of Concurrent Activities Dealing with Chemical Weapons Disposal and Risk Assessment.

JUNE 11-12, 1992

Robert L. Dedrick, National Cancer Institute. Toxicology Lessons from Cancer Chemotherapy.

David H. Wegman, University of Massachusetts. Examination of the Effects of Certain Acute Environmental Exposures on Future Respiratory Health Consequences.

O. Michael Colvin, Johns Hopkins University School of Medicine. Nitrogen Mustard Therapy and Second Cancers.

James M. Melius, State of New York Department of Health. The Bhopal Disaster.

Han K. Kang, Department of Veterans Affairs. Feasibility of Developing a Cohort of Veterans Exposed to Mustard Gas During WWII Testing Programs.

Richard Solana, United States Army Medical Research Institute of Chemical Defense. Toxicology of Lewisite.

Peter Sandman, Rutgers University. Communicating Risk.

Jay Katz, Yale University. Informed Consent: History, Development, and Application to Special Populations.

B

Excerpt from The Residual Effects of Warfare Gases (1933)

Harry L. Gilchrist, M.D., and Philip B. Matz, M.D.

SUMMARY AND CONCLUSIONS

1. Mustard gas as used during the World War affected particularly the skin, mucous membrane of the upper portion of the respiratory tract, and the eyes and their appendages. Secondary bronchopneumonia was a frequent complication and cause of death following mustard gas poisoning.
2. The principal residua noted in the series studied were: Chronic bronchitis, emphysema, bronchial asthma, and certain chronic disabilities of the eyes, such as conjunctivitis, blepharitis, keratitis, and corneal opacities.
3. A group of 89 living and 53 deceased ex-service men who had been gassed with mustard during the World War was selected for the study of residual effects. The standard of selection was on the basis of complete clinical histories, severity of gassing and aftereffects. A number of cases with a doubtful relationship of the disabilities to mustard gassing were also investigated.
4. Of the 89 living cases included in the study it was found that 27 gave evidence of definite anatomic or symptomatic residua which could

From: Gilchrist HL, Matz PB. 1933. The Residual Effects of Warfare Gases. Washington, DC: U.S. Government Printing Office.

be attributed to mustard gassing. In one instance the relationship of gassing to present disabilities was questionable.

5. Of the 53 deceased beneficiaries with a history of mustard gassing, in 11 instances death was an immediate result and in 4 instances a late result of mustard gassing. In 38 of the deceased cases death was not considered a result of mustard gassing.
6. The principal residual disabilities of the positive cases noted 9 to 10 years after mustard gassing were: Chronic bronchitis and emphysema, bronchial asthma, chronic conjunctivitis, and corneal opacities.
7.

   (a) Cases with chronic conjunctivitis gave a history of an acute inflammation of the conjunctivae from the date of gassing, followed by continual symptoms referable to the eyes and the development of chronic conjunctivitis, blepharitis, keratitis, or corneal ulcerations and opacities.

   (b) Emphysema was frequently found in combination with bronchitis. It usually appeared immediately after gassing and was compensatory in character, due to the extensive atelectasis found following gassing with mustard. It may also have been due to the obstruction by the exudate or false membrane in the bronchi or bronchioli which brought about an imprisonment of the inspired air in the pulmonary alveoli and resulted in their distention.

   (c) Bronchial asthma as a residual disability following mustard gassing may be due to the following:

(1) Hypersusceptibility of the asthmatic subject to mustard gas.
(2) Irritation of upper or lower portions of the respiratory tract resulting in spasm of the bronchi or bronchioli and swelling of the mucous membrane. Such spasm and swelling interfere with normal inspiration and expiration.
(3) Anoxemia or prolonged shortage of oxygen and an increase of carbon dioxide due to interference with normal respiration may cause a degeneration of the myocardium. the latter in turn may result in abnormal heart action and abnormal respiratory effort.

8. Of the 27 positive cases [only] 2 gave evidence of residual pulmonary tuberculosis following mustard gassing. In these two cases the histories indicate that a quiescent tuberculosis was present at the time of gassing and that, as a result of severe traumatism of the respiratory parenchyma by mustard gas and a lowered resistance, the latent tuberculosis foci became reactivated and a reinfection of the lungs followed.
9. In the four cases in which death was due to a residual disability following mustard gassing, in one instance pneumonia was the cause of

death and in three instances pulmonary tuberculosis. It is believed that gassing reactivated quiescent tuberculosis foci which ultimately resulted in an activity and extension of the tuberculosis disease and later resulted in death of the veterans.

10. Of 38 men who had been gassed by mustard and subsequently died from various disease, in no way related to gassing, tuberculosis was the cause of death in 8 instances.
11. One must not generalize on the relationship of mustard gassing to the development of tuberculosis. Each case must receive individual study for the purpose of establishing the part played by mustard gas in the inception of tuberculosis.

C

Involvement of the National Academy of Sciences Complex in World War II Research Programs: A Summary

Chartered in 1863 during Abraham Lincoln's presidency, the National Academy of Sciences (NAS) serves as an independent scientific advisory organization to Congress and government agencies. Three other components of the Academy complex were added later under the Academy's original charter, the National Research Council (NRC) in 1916, the National Academy of Engineering in 1964, and the Institute of Medicine in 1970. Each component of this complex works through committees of scientific experts who gather to discuss, share their expertise, and make recommendations on scientific policy and issues. A key aspect of all such committees is their independence from outside pressure and influence. Through a variety of mechanisms, each committee represents a group of individuals who operate with consideration for all known positions or views on a specific subject, but who also operate with complete independence from any of the involved parties, including those providing the funds for the study.

As the United States mobilized for the Second World War, the level of involvement of NRC committees in the actual research enterprise was dramatically changed. Of interest to the present study is the fact that certain NRC committees acted directly for the government in the supervision of war-related research, including animal studies of the effects of mustard gas on the body and tests with human volunteers of ointments to protect against mustard gas burns. The specific agencies and relationships involved in this collaboration of the NRC with the government are described in other sources, but the important aspect is that the strict, distinct boundary lines we see today between the role of the NRC committees and the role of the government were absent (Andrus et al., 1948; Stewart, 1948).

In 1941, President Roosevelt established two branches and an Advisory Council under the Office of Scientific Research and Development to initiate and supervise war-related research. The first of these branches was the Committee on Medical Research (CMR), charged with studying the medical effects of various warfare agents and situations and with developing protocols of treatment for everything from malaria to mustard gas burns. CMR came directly to the NRC and took advantage of already standing NRC committees to form, among others, the Committee on Treatment of Gas Casualties. It was from this group that the first requests for human volunteers were made in order to test protective ointments then under development.

The other branch, the National Defense Research Committee (NDRC), was charged with the development of protective clothing, gas masks, and other equipment-type items. Thus, the NDRC was responsible for the chamber tests of protective clothing against mustard gas, such as the tests done at the Naval Research Laboratory. Although seemingly separate, CMR (and NRC's Committee on Treatment of Gas Casualties) and NDRC dealt with many overlapping issues and, thus, were in relatively constant communication with each other. The overlap among the groups can be seen today in the reports of the testing programs, some of which list NDRC as sponsor, some of which list CMR, and some of which list those two along with NRC's Committee on Treatment of Gas Casualties.

The specific role of the Committee on Treatment of Gas Casualties was to review and supervise the 23 grants dealing with chemical warfare agents. This supervisory role for NRC committees ended in 1944, when CMR expanded its staff and reorganized in order to coordinate and supervise all medical research contracts. By 1947, many of the responsibilities and functions of the CMR were incorporated into newly formed government agencies, such as the National Science Foundation.

REFERENCES

Andrus EC, Bronk DW, Carden GA Jr, Keefer CS, Lockwood JS, Wearn JT, Winternitz MC, eds. 1948. Advances in Military Medicine. Vols. I and II. Boston: Little, Brown.

Stewart I. 1948. Organizing Scientific Research for War. Boston: Little, Brown.

D

Excerpts from Chamber Tests with Human Subjects I, II, and IX. Naval Research Laboratory Reports Nos. P-2208 and P-2579

UNCL

CONFIDENTIAL SERIAL No. 33

~~22 December 1943~~ (5) 627 400 (18) (19)

NRL ~~Report No~~. P-2208

NAVY DEPARTMENT

~~Report on~~

(6) Chamber Tests with Human Subjects
I. Design and Operation of Chamber
II. Initial Tests of Navy Issue Protective Clothing Against H Vapor.

(7 thru 9) NA

NAVAL RESEARCH LABORATORY
ANACOSTIA STATION
Washington, D. C.

Number of Pages: Text - 39 Tables - 22 Plates - 30

~~Authorization~~ (16) ~~Project~~ #547/41, "~~Maintenance, Bureau of Ships~~", ~~dated 16 December 1940; Bureau of Ships letter S-S77-3(Dz), Serial 811 of 17 December 1940.~~

Date of Tests: August 1943 - November 1943

~~Submitted~~ (10) by:

William E. Taylor, Jr., ~~Associate Chemist~~

(11) 22 Dec. 43

Homer W. Carhart, ~~Chemist~~ and

(12) 37 p.

(13 thru 15) NA

L. Eugene Daily, ~~Lt. Comdr. (MC), USN~~

(17) NA

Reviewed by:

W. C. Lanning, Senior Chemist

(20) U

(21) NA

P. Borgstrom, Head Chemist, Superintendent, Chemistry Division

Approved by:

A. H. Van Keuren, Rear Admiral, USN, Director

Distribution: BuShips (30)
BuM&S (4)

56 19712

vc

CONFIDENTIAL

UNCL

TABLE OF CONTENTS

CONFIDENTIAL

ABSTRACT

This report ~~is divided into two sections. The first~~ deals with the design, calibration and operation of a chamber for the exposure of human volunteers to the vapors of chemical warfare agents. The construction of the chamber is such that the temperature, relative humidity and concentration of vapor of the chemical warfare agent can be controlled closely over a wide range of conditions.

The second part deals with the testing of Navy issue S-145 impregnated Arnzen protective clothing, protective ointments and masks. Men dressed in water suspension, solvent and solvent + ZnO impregnated clothing have been exposed to H vapor at CT's ranging from 200 to 2500. A series of tests is in progress in which men dressed in the three types of suits have been exposed repeatedly to H vapor at a CT of 1200. No significant difference has been found in the protection afforded by these three types of suits. The effects of leakage of H through the suits : discussed.

The irritancy of S-461 and S-330 Protective Ointment when applied to the face, ears and neck of the men before exposure has been compared. S-330 is far less irritating than S-461.

The rubber of the gas mask face-pieces and connecting tubes absorbed enough H after 12 to 15 exposures to cause conjunctivitis, laryngitis and erythema of the face. The connecting hoses have been encased in impregnated cloth sleeves, and no break has been observed after 16 exposures.

A screening test ~~has been~~ run on the CT's required to cause burns of different degrees of severity on the bare skin of the arm.

AUTHORIZATION

1. This work was authorized under Project 547/41, "Maintenance, Bureau of Ships," dated 16 December 1940. The problems which were proposed for study were given in Bureau of Ships letter S-S77-2 (Dz), Serial 811 of 17 December 1940.

STATEMENT OF PROBLEM

2. This investigation was undertaken to design, calibrate and study the operation of a gas chamber for the exposure of human volunteers to the vapors of chemical warfare agents, and to evaluate Navy Issue Impregnated Protective Clothing and Masks when exposed to H vapor, and test the irritancy of Protective Ointments.

KNOWN FACTS BEARING ON PROBLEM

3. At present the Navy is issuing single layer protective clothing which requires suitable testing against vesicant vapors on human beings. Newer developments in protective devices also require extensive testing before they can be adopted. Therefore, it is essential to test such items as clothing, masks, ointments, etc. under carefully controlled conditions so that proper evaluation can be made of existing protecting measures, and to test newer developments still in the experimental stages.

4. TDMR #731 from CWS, Edgewood Arsenal, Md. describes chamber tests on subjects protected only by impregnated shorts. Complete protection against H vapor was afforded to the scrotal area by the impregnated shorts whereas burns of casualty severity resulted on other areas of the body from exposure to 315 to 600 mg. min./m^3 (CT).

THEORETICAL CONSIDERATIONS

5. The use of a properly constructed gas chamber for testing protective equipment against chemical warfare agent vapors is the best available method which will most closely simulate actual field trials and yet be operated under conditions which can be controlled critically. The whole body or, by suitable use of proper protection, any area of the body can be used for testing. The temperature, humidity, concentration of vesicant vapor and length of exposure can be varied at will in the chamber so that any type of condition can be achieved. Relatively high temperatures and humidities have been used in the tests actually carried out so far since the human skin is more sensitive to H vapor under these conditions. It can be assumed that if protective devices, such as clothing, prove to be adequate in these tests they will also be adequate under more temperate conditions.

PREVIOUS WORK DONE AT THIS LABORATORY

6. No gas chamber work has been done previously at this Laboratory.

EXPERIMENTAL WORK

I. GAS CHAMBER DESIGN, CALIBRATION AND OPERATION

GENERAL DESCRIPTION

7. The NRL gas chamber consists of a lead-lined room built as an addition to the laboratory building. It is designed as a static chamber, i.e., no air is passed through the chamber during a test, but the air in the chamber is continually circulated and volatilized agent is added as required to maintain the desired concentration. The volume of the chamber is such as to conveniently accommodate a maximum of ten subjects engaged in moderate activity, and construction is according to the following general specifications.

8. Size: Inside dimensions are 10 ft. by 15 ft. and 12 ft. high, giving a volume of 1,800 ft.^{3} or 50 m.^{3}.

9. Construction: The chamber is of transite covered frame construction insulated with rock wool. The floor is concrete and is provided with a center drain. The ceiling and walls are lined with lead, all joints being soldered.

10. Entrance: Entrance to the chamber is made through an antechamber approximately 5 ft. by 3 ft. and 7 ft. high. Doors of both the inner chamber and the antechamber are 2'6" by 6'8", open outward, and are weatherstripped and gas proof.

11. Observation Window: This window, approximately 12" by 18", is located near the entrance to the antechamber. It is a single pane, double window with a dead air space between.

12. Porch: An open porch of frame construction is built on to the gas chamber and the laboratory as an approach to the chamber entrance. The roof contains two skylight windows for lighting, and an exhaust fan, General Electric Spec. 272905-1, is mounted in the roof near the antechamber door for ventilation.

13. Exhaust System: An exhaust blower, Buffalo Limit Load Conoidal Fan, size #2, single width, Type LL, clockwise, with direct connected 1/2 H.P. 220-volt motor, is mounted in a gas proof compartment in one corner of the chamber. This compartment is approximately 42" by 30" and 36" high, with a door to the outside for access to the blower. A 12" diameter sheet metal duct extends through the compartment wall 2" into the chamber. The duct opening is equipped with a removable sheet metal cover.

Summary

Relative Humidity

Average = 67% Maximum = 71% Minimum = 63%

CT

Actual CT = (60 x 20.4) = 1224

II. INITIAL TESTS OF NAVY ISSUE PROTECTIVE CLOTHING AGAINST H VAPOR

SUBJECTS

75. There has never been any difficulty in getting volunteers for the experiments despite the fact that only two inducements were offered: i.e., leave and liberty—change of scenery. However, these facts definitely support the assumption that leave and new surroundings are still uppermost on the average sailor's priority list. Financial remunerations, which seem to play an important part in the rewards offered to volunteers in other countries, i.e., England, Canada, Australia, etc., have never been considered by us nor asked for by the men.

76. It has been impressed on the men that they are not "guinea pigs". They are told that they are expected to use their heads as well as their bodies; and if they do not understand anything to ask questions, these questions being answered in a simple and non-technical language.

77. During their stay at this activity, which varies from one to four weeks, the men pick up an amazing amount of gas warfare fundamentals and, if this is supplemented by a moderate amount of instruction, they leave with a basic amount of knowledge of defensive gas warfare which should be sufficient for the duties required of an enlisted man in the Navy Defensive Gas Warfare Program. The fact that has been most obvious throughout these experiments is that when the men first begin the work they should not be told too much. If they are, it sets up a fear reaction that remains for varying lengths of time and definitely affects their "virgin" runs in the chamber, and, occasionally, requires a removal from the chamber before the run is completed. However, after the first two runs in the chamber, the men become veterans and can be told almost anything without affecting their morale.

78. The men take any resulting casualty extremely well. Even the hospital cases, who, on a few occasions, were incapacitated for a month or so, were not upset and even volunteered for further trials.

79. Occasionally there have been individuals or groups who did not cooperate fully. A short explanatory talk, and, if necessary, a slight verbal "dressing down" has always proven successful. There has not been a single instance in which a man has refused to enter the gas chamber. Our opinion is that the men who have come through this program are much better equipped both mentally and physically to withstand gas warfare if and when it comes.

Physical Examination and Requirements

80. Emphasis must be placed on physical fitness. If not, the experiments are doomed to failure due to inability of the man to remain in the suits and masks and perform effectively when exposed to the high temperature and humidity of the chamber. The so-called false positive readings, due to physical unfitness, such as conjunctivitis, laryngitis, nausea, shock, etc., can easily be mistaken as gas manifestations. Another common symptom, headache, may be attributed to the tight mask straps, etc., when it is actually due to a systemic condition not caused by the chamber. In this connection, it may also be said that it is impossible to give the men liberty during a regular series of experiments and expect them to be in good physical condition the next morning; there always are a few that imbibe too freely and stay ashore too late to be in good condition for the experiments the next day. Because of the above conditions, a thorough physical examination is performed by the Medical Officer, particular attention being paid to the parts of the body most liable to be affected by the gas, i.e., the skin, eyes, genitalia, throat, etc. Many abnormalities are noted and also brought to the man's attention before he enters the chamber. This prevents false interpretations by both the examiner and the men.

81. As a supplement to the actual physical examination complete blood counts, urinalysis, and temperatures are taken; the work being done by qualified laboratory technicians. Blood counts are repeated after a cumulative CT of 4800. The history of each man is briefly checked by the Medical Officer, emphasis being placed on asthma, allergy, hay fever, skin diseases, etc. At this time, a quick psychological impression is also obtained.

82. Upper respiratory infections are the most common disabling factors, and if objective symptons are present, the man is not sent into the gas chamber. Immediate treatment is instituted and it is usually possible to use the man in a later experiment. This procedure also applied to any other minor physical disability.

83. No man is sent into the chamber without the Medical Officer's approval. Occasionally, at this point, malingerers and psychoneurotics are discovered. These cases have all been handled so far by minimizing their symptoms and then sending them into the chamber.

GENERAL PROCEDURE FOR CHAMBER TESTS

84. Each man exposed in the chamber was equipped with the following:

a) Navy diaphragm masks, Mark III.
b) Impregnated Arnzen protective suits.
c) Standard Navy underwear (unimpregnated).
d) Impregnated cotton socks and impregnated elbow length wool gloves.
e) Overshoes (Arctics).
f) Protective ointment for face and neck.
g) Impregnated undershorts for exposure to CT's above 1000. (Heavy cotton rib-knit underwear cut off at knee and rolled to give gas-tight fit.)

85. The impregnated Arnzen protective suits used in these tests were of three types.

a) Water suspension - Impregnated in a Navy Portable Plant with a water suspension at room temperature using the following formula: 100 S-145/75 CP/25 ZnO/3.75 PVA/0.75 Daxad 11/0.15 Duponol ME/9 Pigment, with enough water added to give a bath containing approximately 10% S-145.

b) Solvent - Impregnated in a Navy M-1 Plant with a solution of S-145 in tetrachloroethane at 55°C.

c) Solvent + ZnO - Impregnated in a Navy M-1 Plant with a solution of S-145 in tetrachloroethane containing 15% ZnO based on the weight of S-145 at 35°C.

86. The physically fit men chosen for a given test were instructed in the use of the gas mask and then checked with masks on in an atmosphere containing a high concentration of a lachrymator (CN). This was done to make sure the masks fitted properly without leakage. The men wore dungarees in this test to avoid subsequent contamination of the chamber atmosphere.

87. The men then dressed in protective clothing under close supervision to insure gas-tight seals at waist, face, ankles and wrists. Unimpregnated underwear was put on first, then impregnated shorts, followed by suit, socks, arctics, ointment, gloves and mask. Protective ointment was applied to the neck and face extending just inside the edge of the mask facepiece. A final inspection was made of masks and clothing just before the men entered the chamber.

88. Before each chamber test, qualified persons were required to sign a log attesting to the satisfactory condition of the following: (a) canisters, (b) active chlorine content of clothing, (c) concentration of agent in the chamber, (d) physical condition of the men, (e) proper adjustment of protective clothing and masks.

89. The men entered the chamber through the antechamber in groups of five. The chamber was operated under conditions considered average for the tropics, namely, 90°F, 65% Relative Humidity (R.H.).

90. Continuous visual and audio communication was maintained between the officer in charge and the men in the chamber. Every five minutes each man was required to move to a position on the opposite side of the chamber, otherwise they were permitted to move about at will. The time of each exposure was one hour, after which the men left the chamber and remained in the open five minutes to aerate their clothing and then removed their masks and gloves. The clothing was worn an additional four hours, outdoors in the shade on warm days and in a room at 75–80°F on cold days. During this time the men were not exercised but were allowed to move about freely.

91. Clothing was removed and the men were examined immediately and at subsequent twenty-four hour intervals, the areas most vulnerable to H vapor being closely checked. The face and neck were examined for evidences of ointment irritation.

EXPERIMENTAL RESULTS

Test No. 1 - Irritancy of Impregnated Arnzen Suits

92. In order to determine the irritancy of the Protective Clothing under severe conditions, ten men with full equipment were subjected to a temperature of 96°F and 81% R.H. in the chamber for one hour. Five of the men wore Arnzen suits impregnated with S-145 by the water suspension process, and the other five men wore Arnzen suits impregnated by the solvent process (without ZnO). S-461 Protective Ointment (15% Cl^+) was applied to the neck and face at the edges of the mask before the test. The results are summarized in Table XII. The outside temperature was 90°F and the R.H. was 37%.

SUMMARY AND CONCLUSIONS

1. A chamber for the exposure of human volunteers to the vapors of chemical warfare agents has been built and is in operation. The construction is such that the temperature, relative humidity and concentration of vapor of the chemical warfare agent can be controlled closely over wide ranges. The design, calibration and operation of the chamber are described in detail.

2. The subjects used in the tests are volunteer enlisted personnel who have just completed their basic training at the N.T.S., Bainbridge, Maryland. The men are given a thorough physical and laboratory examination before exposure in the chamber, and only those found physically fit are used for the tests.

3. A preliminary test was run to determine the irritancy of S-145 impregnated Arnzen protective clothing. The subjects developed transient erythema of the body and limbs, when exposed to 96°F and 81% R.H. for one hour, in both the water suspension and solvent type of impregnated suits. There was no significant difference between the two types, the erythema being due to the irritant qualities of the clothing and/or the high temperature during exposure.

4. A series of tests was run on the water suspension, solvent and solvent + ZnO S-145 impregnated Arnzen suits in which fully clothed men were exposed to H vapor at CT's ranging from 200 to 2500. At low CT's (up to 1200), positive reactions consisting of mild erythema of back, shoulders or arms occurred only if a man received more than one exposure. At a CT above 1200 some positive reactions occurred on a single exposure but none were severe even at a CT of 2500. It was concluded that new suits gave adequate protection for single exposures at a CT up to 2500 and there was no significant difference in the three types. The loss of active chlorine in any of these exposures was too small to measure its relationship to H vapor concentration.

5. A series of tests is in progress on the three types of suits in which fully clothed men have been given repeated exposures to H vapor at a CT of 1200 during one hour, followed by four hours wear of the clothing. The same suits, without laundering, are being used throughout the series, and the men are replaced after an average of three to four exposures. The water suspension suits have been exposed nineteen times at a total CT of 22,980, the solvent suits twenty-one times at a total CT of 25,460, and the solvent + ZnO suits nine times at a total CT of 11,050.

6. The tests have shown that there is a definite leakage of H vapor through all the suits. When a man is exposed a sufficient number of times with a short interval between exposures, a positive reaction results from the cumulative effect of this leakage. The leakage shown by these three types of suits increases with the number of exposures.

7. For the most part the burns varied in degree from the mildest type to moderate erythema with only occasional severe erythema and vesication. The majority of these burns reached their peak in 48-72 hours, but there have been a few cases of delayed reactions in which the peak occurred 5-8 days after the last exposure. Burns of the dorsal thorax, shoulders and arms predominated. The more sensitive areas of the body did not show the degree of erythema that was anticipated.

8. S-330 (10% Cl+) Protective Ointment was far less irritating on single and repeated applications than S-461 (10% Cl+).

9. After an average of 12-15 exposures in the chamber at a CT of 1200 each, the gas masks absorbed enough H to cause erythema and pigmentation of the face and moderate conjunctivitis and laryngitis. Other masks having the hose connecting tubes covered by impregnated cloth sleeves have not shown signs of a break after 16 exposures.

RECOMMENDATIONS

1. None. The chamber tests are being continued at high priority.

AD-396275

CONFIDENTIAL

UNCLASSIFIED

SERIAL NO. 41

Navy Department - Office of Research and Inventions

NAVAL RESEARCH LABORATORY
WASHINGTON 20, D. C.

DECLASSIFIED: By authority of DOD DIR 5200.9 ... Date 1970
Entered by ... NRL Code

CHEMISTRY DIVISION - PROTECTIVE CHEMISTRY SECTION

14 August, 1945

CHAMBER TESTS WITH HUMAN SUBJECTS
IX. BASIC TESTS WITH H VAPOR

J. H. Heinen, Lt MC(S) USNR
H. W. Carhart
W. H. Taylor
B. N. Stolp, Lt. USNR
J. C. Conner, Jr., Lt(jg) USNR
N. M. Clausen, Lt(jg) MC USNR

~~Report P-2579~~ NRL 2579

~~CONFIDENTIAL~~

Approved by:

Dr. W. C. Lanning, Head, Protective Chemistry Section

Dr. P. Borgstrom
Superintendent, Chemistry Division

Rear Adm. A. H. Van Keuren, USN (Ret.)
Director, Naval Research Laboratory

Preliminary Pages . . . a - d
Numbered Pages 28
Plates 17
Distribution List . . . e

NRL Problem P-60-S

ABSTRACT

This report describes the results of exposures to H vapor of men wearing ordinary clothing and unprotected except for masks and, in some cases, protective shorts, over a wide range of exposure conditions. Various methods for the evaluation of the results obtained are presented and discussed.

The severity and locations of burns from a given CT of H vapor were markedly influenced by the temperature of exposure. At low temperatures (70° F.,) active sweat secretion and H vapor burns were predominantly in the axillary and genital regions. At high temperatures (90° F.), both sweating and H vapor burns were generalized. The threshold temperature for generalized sweating, and consequent increased susceptibility to H vapor, was approximately 85° F. for lightly clothed, resting men. Variation in relative humidity had the most pronounced effect on susceptibility to H vapor at 85° F.

Conditioning of the men before exposure, either artificially or because of climatic conditions, had a significant effect on the reactions produced from exposure to H vapor. Suppression of sweating by application of aluminum chloride to the axillae prior to exposure, reduced the severity of the resulting H burns. The application of lanolin to the skin prior to exposure had no effect on the resulting H burns, whereas wetting of the skin with artificial sweat increased the severity of the burns.

The scrotal region was the most vulnerable area of the body to H vapor and would be the most important area in the production of casualties. It was found that ulcerated and crusted lesions of the penoscrotal region required from three to four weeks to heal with the men at bed rest.

KEY WORDS

Aluminum Chloride, Antiperspiants, Chamber tests, Human subjects H vapor, Mustard gas, Chemical warfare, Chlothing, Erythema, Exposure, Humidity, Persistance, Contamination, HN vapor, Impregnated chlothing, Lanolin, Levenstein H, Nitrogen mustard, Physiological effects, Toxic agents, Vesicants.

TABLE OF CONTENTS

TABLE OF CONTENTS

INTRODUCTION

A. Authorization.

1. This work was authorized under Project No. 547/41, "Maintenance, Bureau of Ships" dated 16 December 1940. The problems which were proposed for study were given in BuShips letter S-S77-2(Dz), serial 811 of 17 December 1940.

2. Participation of volunteer Naval personnel in tests for the study of vesicant gases was approved by the Secretary of the Navy (Acting Sec. Nav. ltr. to OSRD dated 8 May 1942). Performance of such tests at the Naval Research Laboratory was approved by the Chief of the Bureau of Medicine and Surgery (BuMed ltr Serial No. 446 X:OA A11/EN10(430320)(SC) dated 20 March 1943).

B. Statement of Problem

3. This investigation was undertaken to determine the effect of H vapor at different concentrations, and at various conditions of temperature and relative humidity, on men wearing gas masks and ordinary clothing. The results obtained would serve two purposes, namely, a more complete knowledge of the effects which would be expected from an H attack in the field, and also, by comparison with other chamber data, a more adequate evaluation of the protection afforded by different types of protective clothing under diverse conditions. During the course of the investigation, it became evident that it would be worth-while also to determine the extent of sweating of different areas of the body under various conditions of temperature and relative humidity and to correlate the results with those obtained by exposure to H vapor under the same conditions.

C. Known Facts Bearing on the Problem.

4. It has generally been recognized that when men are exposed to H vapor:

(1) Ordinary clothing offers very little protection;
(2) Increase in CT, by increasing either the concentration or the agent or the time of exposure, increases the severity of the skin reaction;
(3) At a given concentration, more severe reactions occur at higher temperatures and humidities, in summer, and during exercise;
(4) In any individual, marked differences exist in the susceptibilities of many of the body regions; and
(5) There is also considerable variation in the susceptibilities of different individuals.

5. Little precise information pertinent to the above statements or the mechanisms involved has been available. However, it is suggested in much of the available literature that, directly or indirectly, the susceptibility of the skin to H vapor is closely related to the sweating process.

6. In C.D. (Australia) Report No. 40 dated 22 May 1944, it is stated that men exposed in the chamber to H vapor at CT 150-200, 90° F., and 62-85% RH, received severe generalized erythema. The men wore regular clothing and were sweating profusely during exposure so that their clothing was saturated with sweat. Exposure at similar CTs at 70° F., 66-95% RH, affected

mainly the neck, axillae, and scrotum, but two-thirds of the men showed generalized erythema. The skin of these latter men was described as damp and the clothing as dry. All subjects performed moderate exercise while in the chamber. The writers concluded that temperature played an important part in determining the severity of the lesions but that, within the range tested, humidity played no significant role. In C.D. (Australia) Report No. 41, dated 19 May 1944, it is stated that exercise of the subjects, during field or chamber exposure to H vapor, plays an important part in increasing the severity of the lesions.

7. Experiments carried out at Porton indicated that the important factor in increasing the susceptibility is the moisture present on the skin surface, but no correlation was found between skin temperature and physiological effects.

8. By means of exposures of human forearms on the vapor train at UCTL, Dr. Simon Black, et al, have demonstrated that H vapor in moist air caused much more severe reactions than H vapor in dry air at any given CT. Sweating skin was found to be somewhat more susceptible than non-sweating, but the difference was slight compared to the increased susceptibility when the H vapor was in moist air rather than in dry (OSRD Report No. 3944, dated 30 August 1944).

9. In C.D.R.E. (India) Report No. 285, dated 29 November 1944, it is reported that two men, wearing tropical battledress suits and CC-2 impregnated shorts, were exposed in a chamber to H vapor at an integrated CT of 750 (T-16 minutes), 87° F., and 84% RH. The men who had been exercising so that their clothing was sweat-saturated at the time of entry into the chamber, sustained generalized second degree burns with edema and vesication and were both incapable of further military duties for four hours to fourteen days after exposure.

10. At the University of Chicago Toxicity Laboratory, studies have been made of the vesicancy of HN3 smoke on the forearm held in the wind tunnel (NDRC, Division 9, Informal Monthly Progress Report No. 9-4-1-23 dated 10 December 1944). Since active secretion of sweat lowers electrical resistance of the skin, skin resistance measurements were taken. "In any single run it was possible to predict which man would develop more severe reactions on the basis of skin resistance measurements. The predictions are much less reliable between runs and do not appear to be useful between profusely sweating observers and those at the incipient sweating level."

11. In Porton 1601 A(V.1358) dated 7 February 1945, chamber tests on resting men (presumably not showing generalized sweating) at even higher CTs [CT 1190 (T-85 minutes), 58° F., 59% RH and CT 1860 (T- 60 minutes, 47° F., 57% RH] are reported. Intense reactions were observed on the necks and axillae, and edema and vesication of the penis occurred in most of the men. Reactions on the rest of the body were much milder. Similar results were obtained with CT 1170 (T- 6 1/2 Minutes), 58° F., 59% RH, and it was concluded "that the rule CT-constant had been demonstrated with sufficient accuracy" for men wearing plain clothing.

12. In OSRD Report No. 5169, "Observations on the Role of Water in the Susceptibility of Human Skin to Vesicant Vapors", dated 12 April 1945, Renshaw has reviewed the literature pertinent to the mechanism by which sweating skin is more susceptible to H vapor and has performed some original experiments. He found that the skin of the arms of men who were not visibly sweating suffered a greater injury from H vapor (applied by vapor cups) if the skin were wet with a film of distilled water at the time of vapor exposure than if the skin were "unwetted" or "dry". Increased susceptibility was not observed if the wetted skin were allowed to dry before exposure to H vapor. Wetting the skin after H vapor exposure did "not markedly influence the severity of the injuries which subsequently developed". No marked difference was found if the skin were wetted with 4% NaCl instead of distilled water. It was concluded that the presence of water on and in the surface layers is, to a considerable degree, responsible for the well known heightened susceptibility of hot, sweating skin to H vapor.

13. Recent man-chamber tests conducted by Lt. Comdr. J. F. Truxel, MC. USN(Ret.), at UCTL (Informal Report No. N.S. 1, dated 15 April 1945), which are especially significant because of the large numbers of subjects employed, demonstrated that, at any given temperature, an increase in relative humidity resulted in an increased severity of reaction to H vapor at CT 100. Conversely, at any constant relative humidity, a rise in temperature also resulted in an increased severity. The greatest increase in severity for a 10° F. rise in temperature occurred between 80° F. and 90° F.

I. Theoretical Considerations

14. Recording and evaluation of data. Almost every group studying the effects of vesicants on human volunteers employs a different system of nomenclature, recording, and evaluation of data. While a more uniform method would be desirable, it is also evident that particular problems involve different considerations, e.g. studies on unprotected men as compared with those in which protective clothing is worn and studies on small skin areas as compared with those in which total man exposure is employed.

15. This problem has been cogently discussed by the Physiology Section of A.F.E.S. under Lt. Col. F. Gorrill in Phys/C/5, an interim discussion "on Reporting Results of Human Exposure to Vesicant Vapor", dated 25 January 1945. In this discussion, it is recommended that the body surface be divided into a given number of anatomical regions and scoring be done "on the basis of maximum lesion sustained by the subject in any given area, irrespective of the time at which it occurs".

16. In C.D. Australia 55, by Lt. Sinclair dated 18 October 1944, the effects under tropical conditions of H vapor from CT 60 to 765 on 98 men under varying conditions of exposure time, temperature, and RH have been considered. Lesions for each area have been evaluated by a coefficient which combines incidence and severity (number of men affected in that region x the mean score per lesion). This method was used for a given region, but was not used to assess the total severity of reaction of a group of men under specific conditions of temperature and RH.

17. A similar system has been in use at NRL which combines incidence and severity for a region. The average sum of this product for all the regions represents the total damage index for a given set of conditions. Recently it has been found desirable to increase the number of body subdivisions described in taking readings although those may be condensed later to facilitate comparison with older data.

18. The index for a given region should probably take the following factors into account:

(1) Intensity of reaction
(2) Area of skin represented by a region
(3) Casualty effect
(4) Regional susceptibility
(5) Individual variation

19. Intensity of reaction. The scale of readings used at this Laboratory has developed gradually and while the terminology is somewhat clumsy, it has been found satisfactory for subsequent translation into numerical values. Since most of the studies in this report involve no damage greater than a severe erythema (except for the axillae and genital region) it is evident that the description of severity is quite different than for the more severe effects observed by the Australian group. It would be ideal to incorporate the time to reach maximum severity and the healing time into an index, but the complexity introduced indicates it is better to treat these factors descriptively.

20. Area of skin represented by a region. It is well known that the total skin area involved is highly important in thermal burns (and probably for chemical burns) in determining the degree of shock produced. The striking differences in area involved in reactions to mustard vapor at high and low temperatures made it evident that it would be desirable (for special studies) to introduce a factor weighting areas so that regions of totally different areas like the thigh and axilla would not be weighted equally in the index. The factors used in the present report are highly arbitrary, but have been helpful in expressing the area effect. It is felt that the maximum reading for any area is adequate and that it would be impractical to attempt to delimit the readings any further.

21. Casualty effect. Casualty effect would be a desirable and most practical factor in an index. It has been impossible to evaluate this in the present work except on a descriptive basis. Most of the experiments in this series did not result in true casualties, except those which caused lesions of the axillae and scrotum which will be discussed below. Outstanding studies in the evaluation of casualties with great practical importance have been carried out by the Australian group.

22. Regional susceptibility. It would be ideal if the threshold CT required to produce erythema and vesication were accurately known for each region under a given set of conditions. It would be of interest to know if the ratio of erythema level to vesication level is constant for different regions. Adequate data are not available to ascertain this.

23. Individual variation. A simple test to predict how a given individual will react to H vapor has not been developed. The influence of this variable must still be minimized by using as many men as possible in each test.

E. Previous Work Done at this Laboratory.

24. NRL Report No. P-2208 dated 22 December 1943 contains a description of the design and operation of a chamber for exposing human subjects to known concentrations of H vapor under controlled conditions of temperature and humidity. NRL letters to BuShips, C-S77-2(459-BWC), C-459-604 dated 20 October 1944 and C-S77-2(459-BWC/JHH), C-459-119/45 dated 10 March 1945, include a preliminary report on the effects of CT, temperature, and relative humidity on the reactions of men exposed to H vapor when wearing ordinary clothing. In the present report, all the basic tests with H vapor carried out at this Laboratory to May 1945, are summarized.

EXPERIMENTAL PART

Part I - Procedure.

A. Basic Tests with H Vapor.

25. Basic tests, as defined at this Laboratory, are tests with vesicant vapors carried out on men wearing ordinary clothing and unprotected except for masks and, in some cases, protective shorts.

(1) General Procedure for Basic Tests.

26. Test subjects. The men used in these tests were volunteer Naval personnel from USNTC, Bainbridge, Maryland, and were usually seamen second class, from eighteen to twenty years of age who had just completed their "boot" training. Most of their homes were in the Atlantic Seaboard States, both north and south of Washington, D. C. All men received a routine physical and laboratory examination (blood and urine) and only those approved by a medical officer participated in the experiments. At the end of the tests, the men were granted special leave and an entry was made in their service records attesting their attendance at this activity. Recently, authorization has been granted for the commanding officer to give commendation to especially deserving individuals.

27. Clothing. During chamber exposure, the men wore standard issue skivvy shirts and Nainsook shorts, watch caps, blue denim shirts, dungaree pants, standard socks and shoes. Shirt collars were buttoned and shirt sleeves were buttoned at the wrist. All men wore ND Mark III or IV masks. In some of the tests, the Nainsook shorts were replaced by CC-2 impregnated shorts. These latter were of the rib-knit type, impregnated by the aqueous process, and contained about 0.5 mg. Cl^{+}/cm^{2}. In all tests since 1 January 1945, the men have worn suspenders made of carbon coated cloth (August model). The protection afforded by these suspenders causes a subjacent area of relatively normal skin which contrasts with the erythematous areas and facilitates observations (Fig. 35, Plate 10). Subjects dressed for a basic test are illustrated in Fig. 1, Plate I.

28. The chamber. The chamber has been described in detail in NRL Report No. P-2208. Briefly, it is 10 feet by 15 feet by 12 feet high and has a volume of 1800 ft.^{3}. It is operated as a static chamber, no air being passed through during a test, but the air present is kept in motion (average velocity = 2.0 m.p.h.) by suitable fans.

29. Concentration of agent. Volatilized redistilled thiodiglycol mustard was introduced into the chamber as needed to establish and maintain the desired concentration on the basis of Northrop titrimeter analyses (bromine method) which were made every five minutes. The average concentrations of H vapor in this series of tests varied from 1.67 to 11.7 micrograms H/liter.

30. Time of exposure. All basic tests were single exposure tests. In all cases the time of exposure was 60 (± 2) minutes except the two tests at CT 50 in which the exposure time was thirty minutes.

31. CT. CT represents the product of the concentration of the agent and the time of exposure; and where the units are not expressed, is understood to be in microgram minutes per liter. In this series of tests, the CT employed was varied from 50 to 700. A complete list of these CTs is presented in Table III.

32. Temperature and relative humidity (RH). The chamber temperature was elevated by electric heaters, and was lowered by means of ice. Humidification was accomplished by the introduction of steam; dehumidification, by the use of ice. Both temperature and humidity were regulated and recorded by a two-point Brown recording controlling potentiometer which operated through wet and dry bulb thermocouples. All temperatures given are dry bulb temperatures of the ambient air expressed in degrees Fahrenheit; Measurements of radiant energy effects have not been made.

33. Activity in the chamber. The men stood at ease in the chamber, but were required to change positions about every five minutes. No tests on the effects of exercise during chamber exposure are included in this series.

34. Activity before and after chamber exposure. In general, before and after chamber exposure, the men led a relatively sedentary existence with occasional mild athletics. In none of the tests in this series were the men assigned to strenuous work after chamber exposure.

35. Season and climate. The majority of the tests were performed as listed in Table III (Page 13), when the weather was relatively cool. Tests 2, 4, 8, 13 and 16 were carried out under the hot summer weather conditions of Washington, D. C. The chamber exposures were usually performed between the hours of 1000 and 1500, i.e. during the warmer part of the day. When the weather was fair, the men were allowed to be out of doors before and after chamber tests; when the weather was cold or inclement, the men were kept indoors at the conditions prevailing in the Laboratory.

36. Daily readings and the recording of data. The men were inspected daily by a medical officer for four to eight days or longer after exposure. To facilitate recording and subsequent use of data, subdivisions of the body surface were listed as ordinates on graph paper (one-quarter inch squares) and daily intensity readings for those areas were recorded as abscissas. A list of the body regions is given in Table I (page 10).

37. Photography. Kodachromes were taken of many of the groups of men used in the basic tests. These were usually taken on the fourth day after exposure; subsequent pictures were taken when deemed necessary. More satisfactory pictures were obtained after the skin had started to pigment since mild degrees of erythema did not show enough contrast on the pictures. It must be emphasized that kodachromes alone are not adequate records and that the daily readings are a more reliable reference for following the intensity of reaction. Inadequate illumination may lead to under-exposure of certain skin areas with a resultant apparent erythema which is an artefact. The prints which are included in the appendix were prepared by the Ansco "Printon" process. The larger ones were prepared from 4 x 5 inch transparencies; the smaller, from 35 mm. transparencies.

(2) Special Basic Tests.

38. Effect of environmental temperature immediately prior to chamber exposure. During the summer, the men were necessarily warm and often sweating at the time of entry into the chamber. When the weather was cool or cold, the men were usually indoors at a comfortable room temperature for one to two hours (or more) prior to exposure. Tests 24 and 25 (Table III) were carried out to see if precooling had any effect on the reactions from chamber exposure. Five men were exposed nude to a temperature of 55° to 60° F., approximately 65% RH, for over two hours prior to entering the chamber. Five other men remained clothed and at a comfortable room temperature, approximately 75° F. and 50% RH, for a similar period before exposure. Both groups donned fresh clothing, including CC-2 impregnated shorts, and were exposed in the chamber simultaneously to H vapor at CT 300, 90° F., 65% RH.

39. Use of aluminum chloride. To suppress axillary sweating, in Test 28 (Table III), a 25% aqueous solution of aluminum chloride was applied with a cotton pledget once daily to the left axilla of each man in the group for three successive days before he entered the chamber.

40. Use of lanolin. Since H is lipoid soluble, it was considered possible that the presence of sebum on the surface of the skin might influence the susceptibility of the skin to H vapor. To simulate sebum, lanolin (hydrated approximately 50%) was used. A thin film was applied shortly before entry to the chamber over an area of about thirty square centimeters on the forearm, posterior neck, and posterior shoulder of the men in Test 28, who were exposed to CT 500, 70° F., 48% RH. In a later test (Test 24), in which the men were exposed to CT 300, 90° F., 65% RH, lanolin was applied to the forearm, posterior shoulder, posterior neck, and left axilla. No men took showers for at least four hours after exposure in the tests in which lanolin was used.

41. Artificial wetting of the skin. On the basis of Dr. Renshaw's informal report on the effects of H vapor on simulated wet, sweating skin, the following experiment was performed. Standard skivvy shirts were treated with paraffin wax so that a waxed strip about two inches wide extended vertically the length of the shirt in the midline front and rear.

CONFIDENTIAL

One-half of the shirt was thoroughly moistened with artificial sweat,* the other half was left dry. These skivvies were then donned by six men who also wore denim shirts and the usual clothing for basic tests (including CC-2 impregnated shorts). The environmental temperature was sufficiently low so that the men did not show generalized sweating prior to entry into the chamber. These non-sweating men were exposed to H vapor at CT 300, 70° F., 45% RH on 2 April 1945.

(3) Evaluation of Data.

42. Since it is impractical to present a detailed description of each subject in every test, an attempt has been made to present the data quantitatively so that the results may be more readily visualized. Each degree of reaction was given an arbitrary numerical value as follows:

0 = no reaction
1 = mild erythema
2 = moderate erythema
3 = intense erythema
4 = a., Erythema with edema
b. Maceration of axillary skin
c. Dry scaling of scrotum
5 = a. Vesicle
b. Numerous pinpoint vesicles
c. Crusting or ulceration of scrotum or axilla.

43. From these numerical values, three quantitative methods of treating data were devised which are considered in this report; (1) Maximum severity; (2) Total damage index; and (3) Percentage of exposed area affected. Special cases, such as lesions of the scrotum, are discussed separately.

---o0o---

FOOTNOTE:

* The synthetic sweat solution was prepared according to a formula supplied by Dr. Dana Burks and is approximately five times as concentrated as that secreted by the glands.

Constituent	Concentration gms.or cc./100 cc.
Sodium chloride	3.65
Ammonium acetate	0.33
Urea	0.47
Dextrose	0.065
Potassium chloride	0.48
Magnesium chloride	0.036
Potassium dihydrogen phosphate	0.045
Calcium carbonate	0.065
Lactic acid (85%)	0.1

Five drops each of formic acid, acetic acid, butyric acid, propionic acid, methylamine and trimethylamine.

44. Maximum severity. This method has been used at this Laboratory and has been described in previous letter reports (NRL letter C-S77-2(459-BWC) C-459-604 dated 20 October 1944). The average maximum severity recorded over a period of several days after exposure, regardless of the region of the body affected, was taken to represent the effect of the agent on the body under a given set of conditions. A satisfactory relationship between CT and intensity of reaction could be demonstrated using this method as long as all tests were conducted at 90° F., 65% RH. However, when the temperature and RH were varied, it was found that the maximum severity method was no longer applicable since men with an obvious difference in reaction to the agent might, nevertheless, have the same maximum reading. For example, one man might have as a maximum reading an intense erythema of the axillae with essentially no burns elsewhere and another man might have an intense erythema over most of his body. By the maximum severity method of evaluation, both men would be classified as the same. Figure 4, Plate 3 gives an illustration of this since the men who do not show generalized erythema did have intense erythema of the axillae.

45. Total damage index. Since it was observed that often men exposed to H vapor at various conditions of temperature and RH differed in their reactions mainly in the areas of skin affected rather than in the intensities of reaction, and since it is common knowledge that the total area of skin affected is a highly important factor in determining systemic reactions to thermal burns, it was felt desirable to obtain an approximation of the percentage represented by a given body region of the total area exposed in a test. In order to facilitate use of older data, the areas were combined in the manner shown in Table I. These areas were then marked off with ink in arbitrary fashion on ten men of various weights and statures. The regions were measured and their areas were calculated according to the simplest geometric form represented. Their sums represent the total area considered in the basic tests, and each region has been represented as a percentage of this sum. These percentages are called "area factors" and are listed in Table I. It is of interest that for any given man, this sum represents 73 ± 6% of his theoretical total body area as obtained from standard height-weight nomograms. Although these area factors represent only a crude approximation and are highly arbitrary, it is felt that they are useful in evaluating the data obtained from the basic tests.

46. In calculating the total damage index, the intensity scale mentioned in paragraph 42 was used. The intensity factor for a given area in a test represented the average maximum reading for that area for that group of men. The product of the intensity factor and the area factor described above gives the total damage for that area for the group under consideration. The sum of these products for the eighteen areas considered represents the total damage index. An example of the method employed is given in Appendix B. In order to compare data on men who wore protective shorts with data on those who did not, it was necessary to subtract from the final index the figures for those areas on the unprotected men which would have been covered had shorts been worn.

TABLE I

Body Regions Described in Daily Readings

Complete List	Abbreviation	Summarized List	"Area Factor" Percentage of total exposed skin represented by each region*
Anterior neck	ane		
Posterior neck	pne	ne	2.4%
Lateral neck	lne		
Anterior shoulder	ash	sh	5.9%
Posterior shoulder	psh		
Anterior axillary fold	aaf		
Axillae	ax	ax	2.8%
Posterior axillary fold	paf		
Scapulae	sc	sc	3.3%
7th cervical vertebrae	C7		
Ventral thorax	vth	vth	6.0%
Lateral thorax	lth	lth	3.5%
Dorsal thorax	dth	dth	5.3%
Upper abdomen	uab		
Lower abdomen	lab		
Iliac	il	abd	11.1%
Lumbar	lum		
Inguinal region	ing	ing	2.4%
Penis	pen	pen	1.1%
Scrotum	scr	scr	1.1%
Buttocks	bt	bt	6.3%
Anterior arms	aar		
Forearms	fa		
Posterior arms	par	ar	14.6%
Elbows	el		
Cubital fossae	cf	cf	0.9%
Dorsal hands and wrists	wr	wr	1.4%
Anterior thighs	athi		
Knees	kn	thi	22.4%
Posterior thighs	pthi		
Popliteal fossae	pop	pop	1.7%
Anterior leg	ale	le	7.7%
Posterior leg	ple		
			99.9%

* See paragraph 45.

47. Percentage of exposed area affected. This expression was developed primarily to emphasize the difference between exposures at high and low temperatures. The number of men in any one group showing a moderate erythema* or greater on any particular area was multiplied by the corresponding area factor. The sum of these products for the eighteen areas divided by the number of men in the group for any given test represents the average percentage of the total skin surface exposed showing moderate erythema or worse. This expression likewise must be corrected for comparison with those groups which did not wear protective shorts. It should be emphasized that since the area covered by the shorts represents about 11% of the exposed surface, the maximum figure possible after correction is 89% and not 100%. An example of the method employed in this calculation is given in Appendix C.

B. Sweat Tests.

48. The starch-iodine method of Minor was used to demonstrate sweating at the various conditions of temperature and relative humidity used in the tests. An iodine solution (15 g. iodine, 15 g. potassium iodide, 100 ml. castor oil and 95% ethanol to make one liter) was applied to the entire body with a cotton pledget. After the iodine solution had dried, soluble starch powder was applied with a powder duster or cotton puff. Immediately after the application of the iodine and starch the men donned clothing and masks and remained in the chamber for one hour at the prescribed conditions. If sweat secretion occurred, the starch and iodine dissolved and reacted, changing color of the skin from a light brown to a dark purple. Except for the absence of H vapor in the chamber, the test conditions were identical with those of the basic tests described above. Since the men wore dungarees during the exposure to H vapor, it was considered a more comparable test to have them wear dungarees for the sweat test, although clothing did tend to rub off some of the starch-iodine complex, especially at the pressure points, and make the subsequent photographs less spectacular. It should be mentioned that this starch-iodine mixture washes off readily with soap and water and causes no inconvenience except for a temporary smarting of the scrotum at the time of application of the iodine. No cases of hypersensitivity to iodine were observed. Each group consisted of ten men and the tests performed are listed in Table II.

TABLE II

Sweat Tests Performed

Group	Date	Temperature	RH	Sweating**
I	2/5/45	70° F.	51%	Minimal
II	2/5/45	79° F.	86%	Minimal
I	2/6/45	85° F.	36%	Minimal
II	2/6/45	85° F.	75%	Moderate
I	2/7/45	97° F.	42%	Profuse
II	2/7/45	90° F.	62%	Profuse

* Burns of moderate erythema or greater were used in the calculations since it was considered that they represent significant H burns whereas readings of mild erythema may not always be due to H and their use would detract rather than add to the significance of the results.

** On body regions other than axillary and genital.

TABLE IX

Average Percentage of Skin Area per Man showing Moderate Erythema or Worse

Test No.	1	2	3	4	5	6	7	8	9	10	11	12	13	14	15	16
CT	50	50	100	100	100	100	150	150	200	200	200	200	200	200	250	250
Temp.	90	90	90	90	100	100	90	90	70	80	80	90	90	100	90	90
RH	65	65	65	65	35	67	65	65	53	46	86	65	65	66	65	65
Month	Mar	Jul	Mar	Jul	Jan	Jan	Apr	Jul	Nov	Jan	Jan	Apr	Jul	Jan	Apr	Jul
No. Men	6	5	6	5	6	6	10	6	7	6	5	10	6	4	15	6
neck		4.8	4.8	9.6	7.2	4.8	19.2	14.4	4.8	7.2	7.2	24.0	14.4	9.6	33.6	14.4
ax		2.8	2.8	8.4			2.8	16.8	5.6			14.0	14.0	11.2	30.8	16.8
sh				5.9	5.9	5.9	17.7	11.8	5.9			47.2	29.5	23.6	59.0	29.5
se			6.6	3.3		3.3	3.3	19.8				16.5	19.8	13.2	26.4	19.8
vth			6.0	24.0			18.0	36.0				24.0	36.0	18.0	54.0	36.0
lth			7.0				3.5	14.0				7.0	14.0	14.0	31.5	17.5
dth		5.3	10.6	5.3		5.3	21.2	31.8	5.3			37.1	31.8	21.2	68.9	31.8
abd		11.1	11.1	44.4			11.1	66.6				33.3	66.6	33.3	66.6	66.6
ing								2.4						2.4	4.8	9.6
pen			1.1										1.1		7.7	4.4
scr			2.2	1.1	1.1			4.4		2.2		6.6	6.6	4.4	14.3	6.6
bt							6.3	18.9				12.6	31.5	25.2	37.8	37.8
arms			29.2	14.6		14.6	43.8	73.0				87.6	87.6	58.4	160.6	87.6
ef		0.9	1.8	2.7			3.6	5.4	0.9	0.9	0.9	9.0	5.4	3.6	10.8	5.4
hands			1.4			1.4		1.4		2.8		5.6	2.8	5.6	11.2	4.2
thi			44.8				22.4	44.8				44.8	67.2	89.6	89.6	89.6
pop								6.8					3.4	5.1	10.2	8.5
leg							7.7	15.4				15.4	7.7	15.4	38.5	30.8
Tot. Av.	0.0	5.0	21.6	23.9	2.4	5.9	18.1	64.0	3.2	2.2	1.6	38.5	73.2	88.5	50.4	86.2
Tot. Av.*	0.0	5.0	21.0	23.6	2.2	5.9	17.4	59.7	3.2	1.8	1.6	36.6	66.7	80.5	46.1	76.4

* Corrected to exclude area covered by protective shorts.

Average Percentage of Skin Area per Man showing Moderate Erythema or Worse

Test No.	17	18	19	20	21	22	23	24	25	26	27	28	29	30	31	32	33
CT	300	300	300	300	300	300	300	300	300	400	500	500	600	600	600	600	700
Temp.	70	80	85	85	90	90	90	90	90	70	60	70	60	60	65	70	60
RH	60	85	36	77	37	65	65	65	65	62	60	46	57	62	69	52	67
Month	Nov	Jan	Feb	Feb	Mar	Apr	Nov	Mar	Mar	Dec	Jan	Jan	Jan	Jan	Mar	Jan	Jan
No. Men	6	4	8	8	7	5	6	5	5	6	8	8	8	8	7	8	6
neck	9.6	9.6	7.2	19.2	16.8	4.8	14.4	12.0	12.0	12.0	19.2	19.2	19.2	19.2	16.8	19.2	14.4
ax	16.8	11.2	16.8	16.8	19.6	11.2	16.8	14.0	11.2	16.8	22.4	22.4	22.4	22.4	16.8	19.6	22.4
sh	5.9	17.7	11.8	41.3	29.5	23.6	35.4	29.5	17.7	23.6	17.7	29.5	35.4	23.6	29.5	29.5	17.7
se		6.6	9.9	23.1	16.5	13.2	19.8	16.5	9.9	16.5	13.2	13.2	16.5	9.9	16.5	13.2	13.2
vth	6.0	6.0		12.0	18.0	24.0	36.0	30.0	24.0		18.0	24.0	30.0	30.0	24.0	24.0	18.0
lth	3.5	7.0	7.5	14.0	10.5	14.0	21.0	17.5	10.5		10.5	17.5	14.0	21.0	21.0	17.5	10.5
dth		10.6	15.9	37.1	26.5	21.2	31.8	26.5	21.2	15.9	10.6	10.6	21.2	15.9	26.5	21.2	15.9
abd		33.3		22.2	44.4	44.4	66.6	55.5	33.3	11.1	33.3	22.2	44.4	55.5	44.4	44.4	33.3
ing	2.4	2.4									9.6	14.4		12.0		2.4	
pen	1.1		1.1								5.5	8.8		8.8			
scr	3.5		2.2	1.1		1.1				3.3	8.8	8.8	1.1	8.8		1.1	
bt		12.6		6.3	12.6	6.3				12.6	31.5	25.2		25.2	6.3		6.3
arms		14.6	14.6	43.8	58.4	58.4	50.4	73.0	43.8	58.4	43.8	73.0	29.2	14.6	58.4	29.2	
cf		2.7		6.3	3.6	3.6	5.4	4.5	4.5	1.8	5.4	6.3	4.5	5.4	4.5	6.3	4.5
hands		1.4		2.8	7.0	1.4	1.4	7.0	5.6	2.8	5.6	9.8	2.8	2.8	5.6	5.6	
thi	67.2		22.4	134.4	67.2	67.2	134.4	112.0	89.6	44.8	89.6	44.8	89.6	89.6	44.8		
pop	5.1		1.7	8.5	1.7	6.8	6.8	8.5	6.8	6.8	10.2	11.9		6.8	1.7	3.4	1.7
leg	15.4		7.7	38.5	15.4	30.8	7.7	38.5	30.8		15.4	7.7			30.8		
Tot. Av.	22.7	33.9	14.4	53.4	49.7	60.4	76.0	89.0	64.2	37.7	46.3	51.9	35.7	46.4	56.1	35.2	19.7
Tot. Av.*	21.6	30.7	13.9	52.5	47.9	64.9	76.0	89.0	64.2	35.1	39.4	44.6	35.6	39.6	55.[illegible]	34.7	19.0

* Corrected to exclude area covered by protective shorts.

APPENDIX B
Calculation of the Total Damage Index

The following is the method used in calculating the total damage index, using Test No. 9 as an example:

It will be seen from Table VII (Test No. 9) that out of seven (7) men, three (3) sustained a mild erythema of the neck (numerical value = 1), two (2) sustained a moderate erythema (numerical value = 2) and two (2) sustained no burn. On a numerical basis for the group, therefore, the average is 1.0 (3 X 1 + 2X2 = 1.0). On the axillae, one (1) had no burn (= 0), four (4) had mild erythema (4X1 = 4) and two (2) had severe erythema (2X3 = 6) giving an average of 1.4 (6+4 = 10; 7 = 1.4). The method is continued until averages for all areas involved are obtained.

The average intensities are then multiplied by the "area factors" given in Table I. Thus, for the neck, the intensity times the "area factor" is 1.0 X 2.4 = 2.4, for the axillae, 1.4 X 2.8 = 3.9 etc. These values (the damage indices for each area) are listed in Table VIII. The sum of the products, then, gives the total damage index for any one test.

APPENDIX C
Calculation of the Percentage of Exposed Area Affected

The following is the method used in calculating the percentage of exposed area affected, using Test No. 9 as an example:

It will be seen from Table VII (Test No. 9) that only two (2) out of seven (7) men sustained burns of the neck of moderate erythema or worse. The two (2) is multiplied by the "area factor" given in Table I to give a value of 4.8. Similarly for the axillae 2 X 2.8 = 5.6, for the shoulder 1 X 5.9 = 5.9, etc. These values are listed in Table IX. The sum of these values, then, gives the total area involved for the group, which, divided by the number of men in the group gives the average percentage of area involved per man.

E

Interim Report and Addendum Han K. Kang, Dr. P.H.*

Feasibility of Developing a Cohort of Veterans Exposed to Mustard Gas During WWII Testing Programs

I. Background

In January of 1992 Dr. Susan Mather, ACMD for Environmental Medicine and Public Health, requested that my office investigate whether or not a cohort of WWII veterans who had known exposure to mustard gas could be identified (Attachment 1).

Mustard gas was not known to be used for warfare during WWII but a significant number of US military volunteers were exposed to the vesicant chemical during laboratory experiments on protective clothing, ointments and equipment. Soldiers were also exposed to the chemical during the course of field tests to determine the value of protective clothing, masks and ointments while traversing tropical and sub-tropical terrains contaminated with mustard gas. There is no central roster of sailors and soldiers who volunteered for either the laboratory experiments or the field tests. Therefore the actual number of veterans exposed to the vesicant is unknown.

---

* Director, Environmental Epidemiology Service, Veterans Health Administration, Department of Veterans Affairs, Interim report, June 1992, Addendum, August 1992.

II. Naval Research Laboratory (NRL) Tests

A. NRL Lab Log

On February 10, 1992, Mr. Larry Stockmoe of my office and I visited the NRL and met with Mr. Dean Bundy, Navy archivist, and Ms. Maria Lloyd, to review and discuss the available records concerning the WWII mustard gas experiments at the NRL. They made all relevant documents including 10 volumes of laboratory notebooks available to us for our examination. The laboratory notebooks contain meticulous records of each experiment with each subject's name listed as a part of the record. However, the volunteers were mainly identified by only their last names and occasionally their initials (Attachment 2). I reviewed all 10 volumes along with other documents and could not find any full names of participants entered in the materials. The NRL staff has abstracted the names for each experiment and also prepared a combined listing in alphabetical order. These rosters were later given to us by the VA Compensation and Pension Service.

B. Suitland Federal Records Center

During our discussions, Mr. Bundy mentioned that 18 boxes of materials pertaining to the tests were also archived at the Suitland Federal Records Center. He told us it would be highly unlikely that any stored documents would contain personal identification. It was also said that the documents still contain classified information and would therefore be accessible only by individuals with security clearances. We arranged for an inspection of the archived materials by the US Army and Joint Services Environmental Support Group (Attachment 3). During the months of March and April, 1992, Mr. Don Hakenson, Director, ESG, and three members of his staff reviewed the documents and came to the conclusion that none of the materials contain the names of test participants (Attachment 4).

C. National Personnel Record Center (NPRC), St. Louis, MO

One of the documents that I reviewed was a copy of memorandum from Lt. Commander J.H. Heinen, Jr. (MC) USNR to the Chief of the Bureau of Medicine and Surgery dated May 20, 1946. The memo stated that the tests at the NRL for the study of vesicant gases with Navy volunteers were approved by the Secretary of the Navy. Furthermore, the memo went on to describe a file card index of all the men who volunteered to participate in the tests. These men were sent to the NRL from the U.S. Navy Training Center, Bainbridge, Maryland (Attachment

5). I contacted Mr. Jan Herman, Naval historian at the Navy Bureau of Medicine and Surgery and asked him the whereabouts of the 3" x 5" file cards that were mentioned in the memo. He thought the cards were sent to the NPRC in St. Louis and put into each individual's military personnel record folder. Neither the Navy Bureau of Medicine and Surgery nor the NPRC kept a master list of these volunteers before the cards were sent to the NPRC or before the cards were inserted into the individual personnel folders. This account was corroborated by Mr. Bundy, the NRL archivist.

To test the existence of the 3" x 5" card in the personnel folder, we searched the personnel folders of five veterans who contacted the NRL and whose last names also appeared on the list of last names compiled by the NRL. The personnel folders were located for all five veterans. There were 3" x 5" cards or other supporting documents in the folders of four of the five veterans (Attachment 6). It appears very likely that once we identify putative participants from a military source, they can be easily validated by the examination of their personnel records.

D. Bainbridge, MD, Navy Training Center

Because it is not feasible or practical to search personnel records stored at the NPRC by last name only, we need to somehow narrow down the potential search list. The Navy records indicate that all volunteers for the NRL tests came from the Bainbridge Training Center. Therefore, there is a fair chance that the records of the individuals who were transferred to the NRL for temporary duty exist at the Bainbridge Training Center. I contacted the Military Reference Branch of the National Archives. Indeed, there are 29 rolls of microfilm which comprise the muster rolls of the Naval Training Center at Bainbridge from January 1, 1943, to December 31, 1945. I made arrangements for purchasing the muster rolls. I expect to receive the rolls by June 1992. If there is an entry of temporary duty assignment to the NRL for each volunteer in the muster rolls, the entire 29 rolls of microfilm can be reviewed and their names and service numbers and other relevant information can be abstracted. These individuals' personnel records can then be searched at the NPRC for the 3" x 5" cards and any other corroborating information.

If the muster rolls do not contain the records of temporary assignments, we can still search the muster rolls for potential matches by the last names listed in the NRL lab notebook and appropriate time periods. For example, Walker listed in Book #4211, January 1, 1944, to April 29, 1944, full-body chamber tests will be searched among the Bainbridge trainee muster rolls for the appropriate calendar period. There may be only one trainee with the last name Walker or there may be 5, 10, 15 and etc. These individual's personnel records can be searched at the NPRC for the 3" x 5" card and other supporting documents.

E. Conclusions

1. There is no central roster of the NRL test participants with their full names and service numbers.
2. Personnel records archived at the NPRC in St. Louis will contain supporting documents for those who participated in the NRL tests.
3. The muster rolls from the Bainbridge Naval Training Center from January 1, 1943, to December 31, 1945, may contain names of those volunteers or assist us in narrowing down the potential search list.
4. The muster rolls have been purchased and they will be reviewed shortly.

III. The U.S. Army Chemical Warfare Service Tests

A. Special Orders No. 152

The Army also conducted numerous tests of protective clothing, equipment and ointments on volunteers in the laboratory and in the field. The test sites included Edgewood Arsenal (MD), Camp Sibert (AL), Bushnell (FL), Dugwood Proving Ground (UT), and San Jose Island (Panama Canal Zone).

Volunteers were mainly from the CWS units, but some field tests were carried out on infantry troops stationed at Camp Paraiso, Panama Canal Zone, and a company of the 94th Medical Battalion at Bushnell, Florida. During the review of the 18 boxes of documents stored at the Suitland Federal Records Center, the ESG team found a document which recorded that the 150th Infantry Regiment stationed in the Panama Canal Zone participated in a field test. The troops entered a contaminated jungle two hours after a test bombing of mustard gas. They remained there for 24 hours. Now a declassified military document, "Medical Research in Chemical Warfare" states that "between September 1943 and February 1945, 1002 enlisted men and officers voluntarily submitted to tests conducted by the Medical Division and were commended by the Chief of the Chemical Warfare Service for participating 'beyond the call of duty by subjecting themselves to pain, discomfort, and possible permanent injury for the advancement of research in protection of our armed forces.' " The actual number of ground troops who participated in the field test is unknown.

In reviewing documents submitted to VA for compensation claims by veterans, we came across a copy of Special Orders No. 152 issued on June 25, 1944, by the office of the Chief, Chemical Warfare Service

(Attachment 7). This must be the source document used for the above quoted statement. The special order lists participants by rank, full name, service number and date of participation.

We selected a sample of eleven veterans listed in the Special Order and requested the NPRC's assistance in locating and reviewing their military personnel records. The 1973 fire at the NPRC destroyed or damaged the army records from this period and the NPRC could find the personnel records for only four of the eleven veterans. For each of the four veterans there were official documents attesting to their participation in a chemical test (Attachment 8).

B. Conclusions

1. There is a roster of approximately 1,000 army volunteers who participated in tests for mustard gas by the Army Chemical Warfare Service during WWII.
2. For those who were listed in the roster, their personnel records archived at the NPRC will contain corroborating documents if the records were not destroyed by the fire.
3. The identity of ground troops who participated in various field tests is unknown. It may require a substantial effort to trace their military unit records and research them.

IV. Summary

1. Establishing a cohort of Navy volunteers who participated in experiments at the Naval Research Lab during WWII is feasible but may require substantial work.
2. A cohort of approximately 1,000 Army volunteers who participated in various tests for mustard gas during WWII has been identified and the fact of their participation in the tests can be documented.
3. Some ground troop units are known to have participated in field tests (e.g., 150th Infantry Regiment, 94th Medical Battalion), but it may require a substantial effort to trace these unit records.

ADDENDUM

AUGUST 1992

I. Naval Research Laboratory (NRL) Test Participants

We have reviewed all 29 rolls of microfilm which comprise the muster rolls of the trainees at the Bainbridge Navy Training Center, Maryland from January 1, 1943, to December 31, 1945. I observed the following:

1. Approximately 30,000 sailors were present at the center on any given time during this period.
2. The individuals who participated in the NRL experiments can be found on the muster rolls with their full names and military service numbers.
3. There is no record on the muster rolls of any of these NRL test participants having transferred to the NRL for temporary duty. For example, I found the following entries of activities from a particular veteran's personnel records and the muster rolls.

Source	Date	Activities
Muster Roll	1-2245	Received by the Bainbridge Training Center
Muster Roll	4-9-45	Change of rating from Apprentice Seaman to Seaman 2nd Class
3" x 5" Card & Personnel Records	4-2845 to 6-23-45	Temporary assignment to NRL
3" x 5" Card	6-4-45 to 6-1145	Leave
3" x 5" Card	5-18-45	Participated in Experiment #107 at NRL
Muster Roll	6-3045	Transferred to Navy Training Station, Norfolk, VA

The muster rolls did not indicate that the veteran was transferred to the NRL for the period from April 28 to June 23, 1945.

4. Consequently, for those veterans with common last names (e.g. Adams, Jones, Smith, Williams), numerous personnel records have to be reviewed at the National Personnel Records Center to identify the individuals who actually participated in the experiment.

II. U.S. Army Chemical Warfare Service Test Participants

On July 20, 1992, Mr. Don Hackenson, Director, US Army and Joint Services Environmental Support Group, Mr. Larry Stockmoe of my staff and I visited the Edgewood Arsenal in Maryland. Three of us reviewed volumes of documents, both classified and unclassified. Most of the documents deal with technical aspects of mustard gas experiments, the status of projects and progress reports of different projects. None of the documents contained names of individuals who participated in the tests. However, in reviewing the documents we came across some names of military units which were subject to field tests. These unit names and times of testing are as follow:

Unit	Period
2nd Platoon, Company E, 295th Infantry	May 8, 1945-May 22, 1945
94th Medical Gas Treatment Battalion	Jan-Feb 1944
71st Chemical Smoke Generator Company "(colored)"	July-October 1942
67th Chemical Smoke Generator Company	1944
68th Chemical Smoke Generator Company	1944
27th Chemical Decontamination Company (Puerto Rican)	1944
95th Chemical Company	1944
135th Station Hospital (less nurses)	1944
Company B, 295th Infantry	April 1945
Company L, 150th Infantry	April 1945
Company C, 295th Infantry	April 1945
2nd Battalion, 12th Infantry	October 11, 1945

I do not know whether the above list includes all of the units which participated in the field tests or to what extent this list is complete. In the meantime, I obtained three additional documents which list names and service numbers of test participants:

1) Special Order No. 61 issued on March 11, 1945, by the Chief of the Chemical Warfare Service, 78 veterans listed.
2) Special Order No. 130 issued on May 27, 1945, by the Chief of Chemical Warfare Service, 32 veterans listed.
3) A memo from Lt. Col. Thomas Thompson to Chief, Chemical Warfare Service dated on May 23, 1945, 42 veterans listed.

The total number of Army veterans with full names and service numbers that I have been able to identify to date is 617, which is still short of the number "1002" stated in the historical military document written by Raymond C. Cochrane. Further efforts are required to locate the military records of the units listed above and research them for identification of additional individuals who participated in the tests.

As suggested by Dr. Connie Pechura, NAS, I contacted Lt. Col. Richard Parry at Fort Detrick. He did not have or know of any documents that contain names of WWII test participants.

III. Conclusions

1. The muster rolls of the Bainbridge Navy Training Center will help narrow down the number of personnel record reviews necessary to

identify the sailors who participated in the NRL tests. However, it will require substantial manual work.

2. If the total number of Army veterans who participated in the test conducted by the Chemical Warfare Service is "1002" as stated in the historical military documents, over half of them have been identified with their full names and service numbers.
3. Further efforts are needed to locate the military records of the Army units that participated in the tests and to identify individuals who actually volunteered for the tests.
4. At least 12 Army units have been identified as having been involved with the tests.

F

Summary of the Department of the Army Report by James R. Taylor and William Johnson

USE OF VOLUNTEERS IN CHEMICAL AGENT RESEARCH (1975)

This report was the result of a request to the Inspector General and Auditor General of the Department of the Army (DoA) by the Secretary of the Army to research the use and treatment of human volunteers in chemical agent research. The request was prompted by congressional inquiries, during 1975 and 1976, by the Senate Select Committee on Intelligence Activities and joint hearings by the Senate's Labor and Public Welfare Committee and the Judiciary Committee, individual Members of Congress, private citizens, and the press regarding the use of human volunteers in testing of hallucinogenic substances in DoA chemical warfare research. Although the report focused largely on psychochemical testing programs and on testing programs from approximately 1950 to 1975, it also related certain specific aspects of the history of chemical warfare research in regards to treatment of human volunteers and general attitudes toward and compliance with the Nuremberg Codes of 1947. The major conclusion of these authors was that the secrecy, applied to the projects, to the overall research program, and even to the official guidelines governing use of human volunteers, left

Summarized from: Taylor JR, Johnson WN. 1975. Research Report Concerning the Use of Volunteers in Chemical Agent Research. DAIG-IN 21-75. Washington, DC: Department of the Army, Office of the Inspector General and Auditor General.

ample room for misinterpretation, lack of knowledge about, and outright disregard for established policies and guidelines.

The June 1942 request for human volunteers to the Secretary of War, and its subsequent approval, gave authority to the War Department to use human volunteers in the World War II sulfur mustard and Lewisite testing programs. The approval of this request became the basis on which such authority was retained by the Army and the Chemical Warfare Service to use human volunteers in all other World War II and later testing programs up to the mid 1950s. In July 1950, research was placed under the control of the Army by the Organization of the Army Act. Despite the establishment in 1947 of the Nuremberg Codes regarding the appropriate use and treatment of human subjects in research, Taylor and Johnson reported that no documentation could be found about whether the Army was explicitly bound by the Codes. By 1952, the Armed Forces Medical Policy Council filed a request to use human subjects and suggested that the Nuremberg Codes be used as guidelines.

The possible guidelines were discussed at a meeting at Edgewood Arsenal in March 1953. Recommendations were made at the conference that distinctions be made between hazardous and nonhazardous test situations so that nonhazardous procedures/tests would not require approval of the human-use research protocols. The examples given for nonhazardous situations were training exercises in which men, equipped with gas masks, went through gas chambers filled with high concentrations of sulfur mustard. A further proposal was that any human-use codes based on the Nuremberg Codes should only apply to biological warfare testing, not to chemical or radiological testing. This proposal was rejected.

Formalized guidelines were finally issued in June 1953 in a Chief of Staff Memo (MM 385). These guidelines represented an official adoption of the Nuremberg Codes (although somewhat modified) and were meant to apply to all types of chemical, radiological, and biological warfare testing. Further, they required all projects to be approved by the Secretary of the Army. However, no detailed descriptions of what types of experiments required this approval were included, and the report authors argue later that this was a "loophole" that permitted "selective compliance" with the guidelines. For example, in August 1953 seven research projects were sent for approval, one on vesicants and other agents, one on phosgene, and five on nerve agents. Not sent for approval was a research project labelled a "local field exercise" at Fort McClellan, Alabama (Operation TOP HAT). This operation involved use of Chemical Corps personnel in tests of decontamination methods for biological warfare agents, sulfur mustard, and nerve gases. These personnel were not informed and were not volunteers. The justification for the lack of a request for approval was that the project fell under the "line of duty" definition and was not subject to protocol approval.

Another example given by Taylor and Johnson pointed out that protocols were often submitted to test a class of drugs, rather than a specific drug at defined dosage levels. One project entitled, "Retention of Nerve Gas in the Human Respiratory Tract" was given only a cursory examination prior to approval, despite the fact that the specific nerve agents to be used were not listed in the proposal. By 1955, when research into psychochemicals began, approvals were still being given for general research types and not for specific protocols. In 1957, more potent nerve agents were being tested, but the protocols for this research were not sent for examination and approval on the justification that they were simply extensions of projects already approved years earlier. By 1959, Secretary of the Army Brucker gave blanket approval for all projects utilizing "non-lethal incapacitating agents," and the period between 1959 and 1975 was typified by great inconsistency in policy and practices relating to research with human volunteers. The situation became so bad, and the outcry from Congress, the press, and the citizenry so intense, that all research with human volunteers was suspended in 1975 by Acting Secretary of the Army Norman Augustine.

KEY ELEMENTS OF THE NUREMBERG CODE OF 1947

1. *The voluntary consent of the human subject is absolutely essential.* This means that the person involved should have legal capacity to give consent; should be so situated as to be able to exercise free power of choice, without the intervention of any element of force, fraud, deceit, duress, overreaching or other ulterior form of constraint or coercion; and should have sufficient knowledge and comprehension of the elements of the subject matter involved so as to enable him to make an understanding and enlightened decision. This latter element requires that before the acceptance of an affirmative decision by the subject there should be made known to him the means by which it is to be conducted; all inconveniences and hazards reasonably to be expected; and the effects upon his health or person which may possibly come from his participation in the experiment. The duty and responsibility for ascertaining the quality of the consent rests with each individual who initiates, directs or engages in the experiment. It is a personal duty and responsibility which may not be delegated to another with impunity.
2. The experiment should be such as to yield fruitful results for the good of society, unprocurable by other methods or means of study, and not random and unnecessary in nature.
3. The experiment should be so designed and based on the results of animal experimentation and a knowledge of the natural history of the disease or other

problems under study that the anticipated results will justify the performance of the experiment.

4. *The experiment should be so conducted as to avoid all unnecessary physical and mental suffering and injury.*
5. No experiment should be conducted where there is an a priori reason to believe that death or disabling injury will occur; except, perhaps, in those experiments where the experimental physicians will also serve as subjects.
6. The degree of risk to be taken should never exceed that determined by the humanitarian importance of the problem to be solved by the experiment.
7. Proper preparations should be made and adequate facilities provided to protect the experimental subject against even the remote possibilities of injury, disability or death.
8. *The experiment should be conducted only by scientifically qualified persons.* The highest degree of skill and care should be required through all stages of the experiment of those who conduct or engage in the experiment.
9. During the course of the experiment the human subject should be at liberty to bring the experiment to an end if he has reached the physical or mental state where continuation of the experiment seems to him to be impossible.
10. *During the course of the experiment the scientist in charge must be prepared to terminate the experiment at any stage, if he has probable cause to believe, in the exercise of the good faith, superior skill, and careful judgment required of him that a continuation of the experiment is likely to result in injury, disability, or death to the subject.*

G

Public Hearing Announcement

PUBLIC HEARING ANNOUNCEMENT

COMMITTEE TO SURVEY THE HEALTH EFFECTS OF MUSTARD GAS AND LEWISITE

Wednesday, April 15, 1992
1:00 PM until 4:00 PM
NATIONAL ACADEMY OF SCIENCES LECTURE ROOM
2101 Constitution Avenue, N.W.
Washington, D.C.

This committee will be holding a public hearing to encourage veterans who participated as subjects in mustard gas and Lewisite tests during their military service, or their families or representatives, to tell the committee their experiences in these tests and subsequent health problems. All information given to the committee will be considered, in addition to the published scientific literature, as part of the committee's overall task to:

I. Review the published literature on the long-term health effects of mustard gas and Lewisite
II. Summarize the strength of association between exposure to these chemicals and specific diseases
III. Identify gaps in the knowledge regarding the long-term health

effects associated with these chemicals

IV. Recommend ways to decrease the gaps in knowledge that may be found.

Those interested in giving a brief oral presentation to the committee *must* respond, giving their name, address, and telephone number, by *Monday, March 16th* to:

Constance M. Pechura, Ph.D. (Staff Director)
Institute of Medicine
F03036
2101 Constitution Avenue, N.W.
Washington, DC 20418
TEL: 202/334-3387
FAX: 202/334-2939

The committee also encourages submission of information in written form to the above address; telephone calls will also be accepted. **Attendance at the public hearing is *not required* for consideration of your experiences by the committee**. Please contact the staff director for additional information.

VETERANS WHO APPEARED AT THE PUBLIC HEARING

Glenn Jenkins, Nokomis, Florida
Johnnie H. Ross, Robersonville, North Carolina
Richard Snow, West Sunbury, Pennsylvania
Nathan J. Schnurman, Charles City, Virginia
Richard W. Rawls, Stone Mountain, Georgia
Dan Gentile, Scottsdale, Arizona
George Avery, Salem, Oregon
David D. Fallin, San Antonio, Texas
Charles Cavell, Midlothian, Virginia
Stanley Weintraub, Washington, D.C.
Bernard Klonowski, Arlington, Virginia
Frank Kozdras, Port Charlotte, Florida
Millard Scudder, Dillsboro, Indiana
Russell H. O'Berry, Richmond, Virginia
Joseph L. Butash, Scranton, Pennsylvania
R. B. Moore, Mechanicsville, Virginia
Victor R. Barnhardt, Concord, North Carolina
Victor LaBate, Jarrettsville, Maryland
Elmer L. Hood, Monroe, North Carolina
Walter Langston, Rectortown, Virginia

SUMMARY OF HEALTH PROBLEMS REPORTED BY VETERANS

The following information is included to inform readers of the general types of health problems that were reported to the committee by the veterans. It is important to note, however, that this information was not collected in the rigorous manner required for quantitative analysis. Thus, no conclusions were drawn from the distribution or frequencies of specific diseases listed here by the committee and, likewise, such conclusions should not be drawn by the readers.

The total number of respondents represents each individual who contacted the committee through letters, phone calls, or appearances at the public hearing. The numbers indicated with serious injuries represent those veterans who were hospitalized within days following their exposure, whether or not this exposure was due to accidental explosions, normal testing conditions, or, in a very few cases, combat injuries. Some of the veterans also participated in other types of tests. Most often the additional tests were patch or drop tests of liquid mustard gas, but some others included atomic bomb tests and drug tests. Finally, the number with scars still present and the number of veterans who reported no health problems that they attribute to their exposure are also listed.

Finally, it should be noted that the disease and health problem categories are arbitrary and in some cases a number of different specific diseases are grouped into one category. For example, heart attacks, congestive heart problems, and angina are all listed under heart problems. Respiratory problems encompass difficulty in breathing, chronic colds and infections, lung collapses, and chronic cough. Esophageal stricture includes complaints of difficulty in swallowing.

Total Responses (as of June 29, 1992)	
Total number of respondents	257
Number with serious injury	53
Number with scars	28
Number in other tests	21
Number reporting no health effects	12
Asthma	45
Chronic bronchitis	63
Laryngitis	25
Emphysema/lung disease	75
Conjunctivitis/opacities/keratitis	10
Skin rashes/blisters	55
Pneumonia (repeated)	16

Sinus problems	16
Other respiratory problems	42
Cancers	
Lung/laryngeal	6
Oral/nasal	1
Skin	20
Bladder	6
Prostate	6
Intestinal	1
Thyroid	1
Pancreas	1
Kidney	1
Leukemia/lymphoma	1
Unspecified	1
Tumors/polyps	
Laryngeal	5
Intestinal	3
Brain	1
Heart problems	63
Vascular disease/stroke	15
Blood pressure problems	18
Chest pain	2
Diabetes	11
Cataracts/eye problems	50
Hearing problem	14
Nausea/stomach ulcers	22
Esophageal stricture	11
Hiatal hernia	6
Headaches	19
Arthritis/bone disease	40
Neurological problems	32
Depression/anxiety/post-traumatic stress disorder	52
Chronic pain	3
Alcoholism	3
Allergies	8
Blood/lymphatic disorders	6
Prostate disease	10
Kidney disease	10
Tuberculosis	4
Hepatitis B	1
Liver disease	4
Muscle spasms	2
Hair/tooth loss	4
Impotence/sterility/sexual problems	7
Genital burns and scars	7

H

YALE LAW SCHOOL
401A YALE STATION
NEW HAVEN, CONNECTICUT 06520

Jay Katz
Elizabeth K. Dollard Professor
of Law, Medicine, and Psychiatry

June 16, 1992

Dr. David Rall, Director
National Institute of
Environmental Health Sciences
5302 Reno Road, N.W.
Washington, D.C. 20015

Dear David:

On my flight back to New Haven, I reviewed the remarks I had delivered to your Committee. They had been composed before benefitting from the opportunity to talk with some of you Thursday evening and listen to your reactions to Mr. Sandman's presentation. In light of what I had learned, I deviated somewhat from what I had intended to present to you but, in retrospect, not sufficiently so. Thus, this letter which I hope you will share with the members of the Committee. I shall not repeat what I said in my presentation which essentially reviewed the history of the regulation of human experimentation from its beginnings through the 1950's. Therefore, this letter may require amplification. If so, feel free to ask for it.

Let me begin by acknowledging that I had not read carefully the "Statement of Task" which IOM had prepared for you. I had glanced at it and it seemed so limited in scope that I assumed your Committee had been provided with more extensive instructions which had not been made available to me. I now have carefully reviewed this document and find it both too limited and somewhat confusing.

Apparently your primary assignment is to conduct a scientific review of the "long-term health effects of mustard gas" and identify "gaps of knowledge," etc. One omission already strikes me: You were not charged with making recommendations, on the basis of your study, on who is deserving of compensation for the ill-effects (physical and psychological) from participation in these experiments. You were also asked to hold a public hearing so that you could inform yourselves about "[the veterans'] experiences and concerns about exposure to mustard gas and lewisite." At the same time you were not instructed whether, and how, these "experiences and concerns" should be reflected in your report. Moreover, you were not charged with making a judgment about the ways in which

Dr. David Rall, Director
June 16, 1992
Page 2

these experiments were conducted and their impact on how these veterans now feel about having been manipulated, exploited and deceived. The psychological damage inflicted upon them could be considerable. Finally, you were not asked to make a judgment about the ethics of conducting such experiments in light of the opinion of the Nuremberg Tribunal which, as I tried to convey to you, also constituted a retrospective judgment about conduct by Nazi physician-scientists during the days of the second World War.

With all the restrictions imposed on you, I believe that your task is an unenviable one. Against the background of an unethically conducted series of experiments, you are charged to contemplate, on the basis of cold scientific knowledge, the relationship between exposure to mustard gas and subsequent physical illness. Should you do more? This to me is a crucial question. If you attend only to this limited scientific task, you may be abdicating your responsibility to condemn such studies for 1992 and beyond, even if you wish to excuse them for 1943-45 when supposedly our ethical sensitivities were less developed. Whatever your answer to my question, should you at least note in your final report that, since your assigned task was restricted, you do not wish to comment on these burning issues? Perhaps you may only wish to recommend that another IOM Committee consider these issues forthwith.

If you decide to do more, you of course could leave it by noting that in 1943-45 no policies had been promulgated for the conduct of research, either by the medical profession or by the State. You could also point to the fact that Federal courts, including the U.S. Supreme Court, when confronted with allegations about wrongful conduct by members of the Armed Forces, CIA, PHS, etc. in the pursuit of scientific knowledge, by-and-large concluded that "sovereign immunity" and the exigencies of wartime conditions required that no liability be imposed. (As you know, I argued that the dissenting opinions of Justices Blackmun and O'Connor in Stanley should have prevailed.) Therefore, you could conclude that in 1943-45 these experiments, though perhaps unfortunate, were legally licit.

You also have the choice to go beyond your charge. I shall briefly review some of the issues you could then raise, though not those that I already brought to your attention Friday morning. (If you intend to so proceed, I shall be glad to elaborate on them in a subsequent letter):

(1) The excuse that "military discipline" (which underlies the "sovereign immunity" doctrine) and wartime necessity may require such deceptive experimental studies is all too lightly made. To be sure, as someone mentioned at the meeting, the harm inflicted by these experiments pales in comparison with the death

Dr. David Rall, Director
June 16, 1992
Page 3

and injury our soldiers, sailors and marines suffered at Anzio or Iwo Jima. However, our fighting men knew what their fate might be. With regard to the mustard gas and lewisite experiments, the soldiers did all too often not know what they were volunteering for. No attempt was made to learn whether a sufficient number of volunteers could be obtained after full disclosure of what they were asked to do. As I argued before your Committee, I believe that volunteers would have come forward; I may be wrong, but at least the attempt should have been made. That the subjects of research were not carefully monitored after the war was over is of course also egregious, but it is a separate issue; for the experiments themselves were conducted in the spirit of cheap availability of human beings for these purposes, and with utter disregard to alternative ways of proceeding (not to speak of giving consideration to the minimum number of human subjects required for the conduct of such scientific studies, etc.) or without regard to the future welfare of the participants. The lack of any assigned continuing accountability, once the physician-investigators who designed and conducted the experiments had returned to civilian life, is most disturbing.

(2) The soldiers who "volunteered" for these experiments had every expectation that they would be treated fairly by their officers and surely by the physicians who superintended the experiments. As doctors we ask our patients to trust us and this trust was manipulated, exploited and betrayed. To be sure, no authoritative policies then existed for the conduct of research, and this was one of the defenses introduced by the Nazi physicians at the Nuremberg Trial. The Tribunal made short shrift of this defense: "Certain basic principles must be observed [in the conduct of research] to satisfy moral, ethical and legal concepts [and one of them is that]the voluntary consent of the human subject is absolutely essential." Andrew Ivy and Leo Alexander, on behalf of the prosecution, averred that obtaining consent is the customary practice in American research. Regulation or no regulation, our soldiers expected better of doctors, but instead they were treated as guinea pigs. That fact, "good" excuses notwithstanding, must remain deeply etched in your minds as you proceed with your deliberations.

(3) The question was raised during our discussion whether all subjects who participated in these experiments need be identified and examined. If you conclude that some of them may have been physically or psychologically damaged as the result of participation, then clearly you have no choice but to recommend that they be apprised of what had been done to them. Doing otherwise is an abdication of medical responsibility. If others who have suffered no ill effects will now become outraged and perhaps unduly concerned, this must be faced and handled in appropriate ways.

Dr. David Rall, Director
June 16, 1992
Page 4

(4) I said earlier that yours is an unenviable task if you stay within your mandate to evaluate only cold scientific facts; if you decide to evaluate only data and not the misuse of human beings. When I served on the DHEW Panel that investigated the Tuskegee Syphilis Study (1932-1972) we faced the same dilemma, and we failed to rise to the occasion by not refusing to restrict our Final Report to the evaluation of the data and not of the persons who conducted the Study and the USPHS officials who reviewed it during the forty years of its existence. In retrospect, this was a grievous error. I urge you to do better for many reasons:

(a) It is all too often asserted that much has changed since 1943, 1972 or even later. This assertion is only partially true. We still have not learned how to safeguard the dignitary interests (consent, self-determination) of subjects of research when these interests conflict with the interests of science (acquisition of knowledge) and society (wartime emergencies).

(b) We are reluctant to pass judgment on past transgressions because at least preconsciously we know that they are still with us in today's world. Santayana once said that "those who do not know the past are condemned to repeat it." Let me elaborate on his wisdom by reminding you that those who do not squarely face up to past transgressions contribute to their perpetuation in the present. At least I urge you to assert that what was done in 1943 was unconscionable and, if it still is being done today, that it must be judged henceforth unconscionable and subjected to sanctions.

If I can be of further assistance to your Committee, please be in touch. I would appreciate your circulating this letter to the members of your Committee because I do not believe that I had done justice to what I wanted to say when I appeared before them.

Cordially yours,

Jay Katz

JK/j

cc: Dr. Constance Pechura

I

Risk Assessment Considerations for Sulfur Mustard

William Nicholson and Annetta Watson

It would have been ideal to have both exposure and epidemiological data on the groups of individuals exposed to sulfur mustard, such as those exposed during manufacturing. Such information could be used to develop exposure-risk relationships and, *if exposures were known,* applied to individuals who participated in World War II testing programs or who were otherwise exposed. If epidemiologic studies of such exposed individuals had been conducted, the results would provide risk information of importance, even without knowledge of exposure. Absent these ideal data, we are forced to make risk estimations using other information.

Experimental animal data can be used to make estimates of carcinogenic potency in humans by using cancer risk models and standard interspecies extrapolation procedures. Unfortunately, quantitative human cancer risk estimates for sulfur mustard are impractical because the experimental data from animal studies have three large uncertainties:

- only a few experiments were conducted;
- many were in a mouse strain that exhibited a high genetic susceptibility to spontaneous pulmonary tumors;
- routes of administration tested and duration of follow-up observations are not comparable to the human exposures of concern.

Nevertheless, approximate estimates can be made of the carcinogenic potency of sulfur mustard relative to other carcinogenic agents for which we have knowledge and concern. These estimates allow us to place sulfur mustard on a scale of cancer potency that spans more than seven orders of magnitude between highly potent carcinogens, such as aflatoxin B_1, and carcinogens of very much lower potency, such as saccharin.

SULFUR MUSTARD

Two sets of experiments provide information on the carcinogenicity of sulfur mustard (see Chapter 6). One is the series of injection studies of Heston and colleagues (1949, 1950, 1953a) using strains A and C3H mice. These studies are difficult to interpret because of the extreme sensitivity of strain A mice for development of pulmonary tumors and the use of only subcutaneous injections in studies using the strain C3H mice. The second experiments are the chamber exposure studies of McNamara and colleagues (1975) in which Sprague-Dawley-Wistar rats were exposed to air concentrations of 0.001 and 0.1 mg/m^3 for various periods of time up to 52 weeks. These latter experiments are more useful because of the long durations of exposure and long subsequent follow-up periods.

A comparison of the results of the intravenous injection studies of Heston (1950) indicates that both sulfur mustard and nitrogen mustard (HN2) readily produce pulmonary tumors in strain A mice. However, because of the high genetic susceptibility for development of pulmonary tumors in this strain of mice, it is inappropriate to make quantitative estimates of risk from these data alone. Table 6-1 does indicate that sulfur and nitrogen mustards have a similar potential to produce pulmonary tumors in this strain. This conclusion follows from the finding that the number of nodules produced from comparable injections of sulfur or nitrogen mustard is similar at 16 weeks of follow-up. This similarity and the similarity of DNA alkylation action make HN2 data relevant to evaluating the carcinogenic potency of sulfur mustard. However, caution must be exercised, because the pharmacokinetics of the two compounds differ. Sulfur mustard is more rapidly metabolized and may not act at distant sites as readily as HN2. This is seen in the strain C3H and C3Hf subcutaneous injection studies of Heston (1953a) where injection site malignancies were similar for both sulfur mustard and HN2, yet only HN2 produced an excess of pulmonary tumors (see Table 6-3).

The single inhalation study by Heston (1953b) is also inappropriate to use for estimating either animal or human cancer potency: first, because strain A mice were used; and second, because of the extremely large uncertainty of the exposure concentration tested. In this study, sulfur mustard evaporated from a soaked filter paper and was distributed by a fan through an 8-liter desiccator. The actual concentration of sulfur mustard during a 15-minute inhalation exposure is unknown. The importance of this experiment is not that of a potency estimate, but that of clear evidence of lifetime risk of pulmonary tumors from a 15-minute inhalation exposure.

The inhalation chamber exposures of McNamara and colleagues (1975) present the best available experimental data on sulfur mustard carcinogenicity (Tables 6-5 and 6-6). Of the rats exposed to 0.1 mg/m^3 for 6.5 hours 5 days per week for 52 weeks, 10 of 23 developed agent-related tumors in a "carcinogenicity study" and 9 of 39 in a "toxicity study." Of the 19 tumors, 17 were either squamous cell or basal cell carcinoma of the skin. No carcinomas were present in 52 control mice observed for similar periods. In the atmospheric exposure experiments of McNamara and colleagues, it should be noted that the only excess malignancies in exposed groups were skin tumors. With the exception of exposed strain A/J mice (which developed pulmonary tumors, but not in excess), no pulmonary tumors in any exposed group were reported in these inhalation experiments. Further, at similar inhalation exposures, but inadequate follow-up time, McNamara and colleagues observed no increase in agent-related skin or other malignancies in experiments with dogs, guinea pigs, rabbits, and A/J mice. The skin malignancies, however, are noteworthy. First, their cumulative incidence in the high-exposure group approached 50 percent. Second, the absence of other tumors suggests that the skin malignancies were the result of external, rather than systemic, exposure. This suggests that similar or even lower cumulative air exposures may be of concern for human skin carcinogenicity, particularly in combination with high exposure to sunlight.

NITROGEN MUSTARD

Because of a similar structure and toxicity, malignancies caused by nitrogen mustard (HN2) are also of interest. The intravenous and subcutaneous experiments using HN2 by Heston (1950, 1953a) have been mentioned above.

Another study by Shimkin and colleagues (1966) reviewed early National Cancer Institute bioassay data for 29 alkylating chemicals tested in strain A mice. Of the 29 compounds, the potency of HN2-HCl in producing pulmonary nodules was exceeded only by uracil mustard. The data for the five most potent compounds and that for chlorambucil and cyclophosphamide are shown in Table I-1.

Gold and colleagues (1984) reported a TD_{50} for all malignancies of 22.8 µg (0.0456 mg/kg) for HN2, based on laboratory rat data from Schmähl and Osswald (1970). TD_{50} is the cumulative exposure that is expected to produce an excess cancer mortality of 50 percent in a two-year followup; a low TD_{50} indicates a high potency. The TD_{50} for HN2 translates into a cancer potency that is only one-fifth that of aflatoxin B1 for liver tumors (aflatoxin B_1 TD_{50} = 4.19 μg in male rats). Because that experimental cancer potencies can differ by as much as 10 million-fold, these data suggest similar potencies for HN2 and aflatoxin B_1.

TABLE I-1 Comparative Carcinogenicity of Selected Agents in Strain A Mice

Compound	Response Dose[a]	Relative Potency[b]
Uracil mustard	0.96	10,420
Nitrogen mustard	3.0	3,300
Melphalan	3.8	2,630
Aziridyl benzoquinone	12	830
Chloroquine mustard	18	560
Chlorambucil	60	170
Cyclophosphamide	360	28

[a] The total dose required to produce 1.0 tumor per mouse at 39 weeks.
[b] *10,000* divided by the response dose.
SOURCE: Shimkin et al., 1966.

In a study comparing the effects of HN2 and other radiomimetic chemicals with the effects of X-radiation, Conklin and colleagues (1965) found the incidence of thymic lymphoma from four injections of 2.4 mg/kg each to be less than that of four doses of 300 rads each of X-rays (21 percent lymphoma incidence for HN2, 33 percent for X-radiation, 10 percent for controls). Thus, the midlethal doses of HN2 were similar to midlethal doses of X-radiation.

CANCER RISKS FROM SIMILAR COMPOUNDS

A variety of mustard compounds are used in chemotherapy or treatment of other diseases. Many of these have also demonstrated a potential to produce malignancy during or following treatment. Sulfur mustard, HN2, and these other mustards are alkylating agents. It is believed that they act by producing interstrand and intrastrand DNA-DNA cross-links; this action may be related to carcinogenic effects (Colvin and Chabner, 1990). While the pharmacokinetics of the alkylation process differs for the different chemicals, the similarity of ultimate action suggests that data on the carcinogenic potential of these other mustard compounds may be relevant to the risks of developing cancer from exposure to sulfur mustard.

A malignancy commonly arising from treatment by alkylating agents is acute nonlymphocytic leukemia (ANL). Of the various medicinal mustard compounds, good data on human exposure-based risk of developing ANL are available for chlorambucil, cyclophosphamide, and melphalan. Kaldor and colleagues (1988) have calculated a leukemia potency index (10-year incidence per gram total dose) for each of these drugs and compared it with the corresponding rodent TD_{50}'s. Figure I-1 is a graphic representation of data from Kaldor and colleagues (1988) of the reciprocal of the TD_{50}'s for various antineoplastic drugs plotted against the relative human leukemogenic potency. No data exist that would allow a direct estimate to be made of the human leukemia potency index for HN2-HCl, but its TD_{50} in male rats (0.23 mg/kg per day) would suggest that its relative leukemogenic potency would be comparable to or exceed that of melphalan. This comparability is also suggested by comparing the pulmonary tumorigenic potency of HN2 and the other compounds in the strain A mice bioassay data of Shimkin

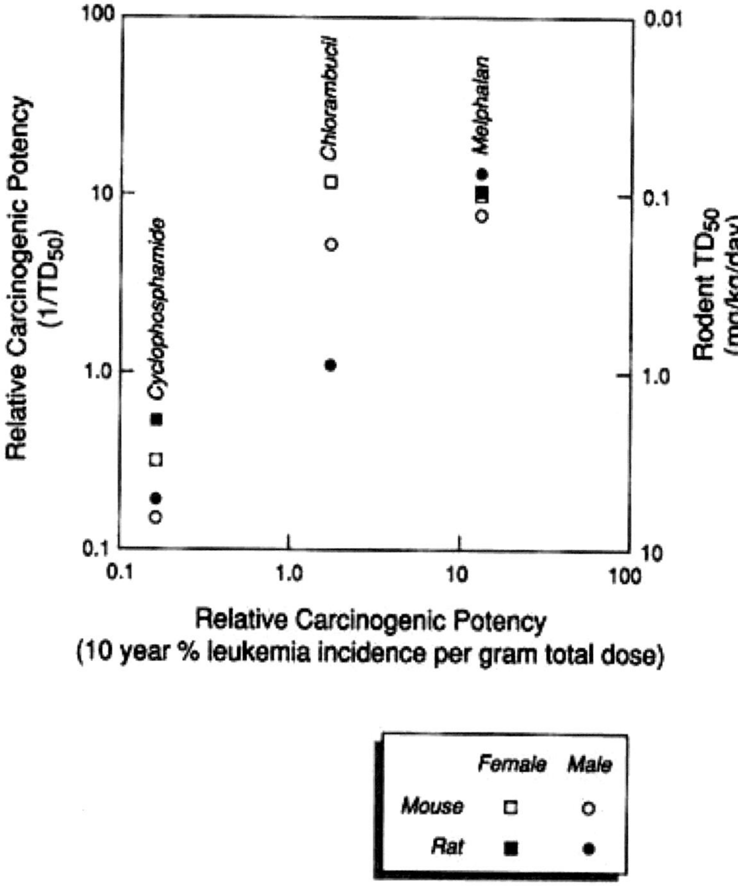

FIGURE I-1 Relative carcinogenic potency of three nitrogen derivative antineoplastic agents. SOURCE: Kaldor et al., 1988

and colleagues (1966) mentioned above.

Because of the similarity of sulfur mustard and HN2 in producing pulmonary tumors in strain A mice, we would expect sulfur mustard to carry a potential ANL cancer risk. However, the magnitude of the effect may differ from HN2, because a different fraction of sulfur mustard inhaled would be transported to the bone marrow.

Dedrick and Morrison (1992) have also compared the carcinogenic potency of chlorambucil, cyclophosphamide, and melphalan in humans and rodents, but based on the integrated exposure of the ultimate carcinogenic chemicals. This procedure changes the relative positions of the chemicals, but not the correlation between human and rodent data. Thus, comparisons between effects data derived from external exposure are considered relevant, as pharmacokinetic effects of these compounds appear to be similar in humans and rodents.

RAPID SCREENING OF HAZARD (RASH) APPROACH

Watson and colleagues (1989) have estimated the carcinogenic potency of sulfur mustard by the "rapid screening of hazard" (RASH) method. This approach compares exposures that produce documented toxic effects from an agent of interest to exposures of a reference chemical producing a similar effect (Jones et al., 1985, 1988). The results of these relative potency analyses are usually similar to those of the more traditional CAG (Carcinogen Assessment Group) and NTP (National Toxicology Program) carcinogenicity assessments in establishing exposure standards used by EPA, OSHA, and other regulatory agencies (Glass et al., 1991; Jones and Easterly, 1991; Jones et al., 1988; Owen and Jones, 1990). Watson and colleagues (1989) applied the RASH procedures to Heston's intravenous injection studies (1950) and subcutaneous injection experiments (1953a). Other available Heston data were not incorporated into the analysis because of Heston's incomplete characterization of exposure concentrations or high animal mortality induced by the experimental protocol. The experiments of McNamara and colleagues (1975) would have been desirable to use, but comparison exposures for primary or secondary standards were not available.

By considering all possible combinations of experiments and several reference compounds, sulfur mustard tumorigenicity was determined to be comparable with nitrogen mustard (HN2 and HN2-HCl) tumorigenicity in laboratory rodents. In the analysis of nitrogen mustard, data from the studies of Abell and colleagues (1965), Boyland and Horning (1949), Conklin and colleagues (1965), Heston (1949, 1950, 1953a), Schmähl and Osswald (1970), Shimkin and colleagues (1966), and Zackheim and Smuckler (1980) were used. Additional relative potency comparisons were made for the therapeutic nitrogen mustards melpha lan and chlorambucil, and the alkylating carcinogenic compound, bis (chloromethyl) ether. Comparisons of laboratory rodent data indicated that sulfur mustard and nitrogen mustard had tumorigenic potencies comparable with melphalan and bis(chloromethyl) ether; the tumorigenic potencies of sulfur and nitrogen mustard were possibly greater than that of chlorambucil.

RELEVANCE OF EXPERIMENTAL RESULTS TO HUMAN RISKS

The above observations and comparisons indicate that sulfur mustard is an animal carcinogen and, to the extent that its action is similar to HN2, a potent one. Both excess pulmonary tumors and skin malignancies were demonstrated to occur from sulfur mustard exposure in experimental studies. Such data are in agreement with

- excess lung cancer observed in groups of individuals occupationally exposed during sulfur mustard production; and
- skin cancer observed in patients undergoing topical treatment with therapeutic concentrations of nitrogen mustard.

Data from studies of effects of exposure to therapeutic nitrogen mustards are suggestive of risks of additional malignancies. Mustard compounds used as chemotherapeutic agents have demonstrated a high potential to generate acute nonlymphocytic leukemia. The similarity between the alkylating action of these compounds and sulfur mustard suggests that sulfur mustard exposure might also result in such malignancies. This is also suggested by the finding of an increased incidence of thymic lymphoma in female strain RF mice injected with HN2.

In summary, experimental studies establish that exposure to sulfur mustard produces a substantial risk of lung tumors in laboratory animals and also produces a risk of skin cancer from air exposure. For each pathological site, the cancer potency of sulfur mustard is high. Experimental data from exposures to HN2 suggest that sulfur mustard exposure may lead to an increased risk of developing thymic lymphoma, and perhaps acute nonlymphocytic leukemia, based on findings in humans treated with therapeutic alkylating agents.

HUMAN EXPOSURES

Mustard agents were positively associated with human respiratory tract cancer incidence by the International Agency for Research on Cancer (IARC) in 1975. By 1981, IARC had categorized "mustard gas" as a "Class 1" human carcinogen (Saracci, 1981). Exposure and dose-response data are not available that would allow precise risk estimates to be made for the specific exposure circumstances of veterans exposed during World War II or otherwise. Human epidemiological studies of workers employed in the production of sulfur mustard demonstrate a significant excess of cancers of the lung and larynx.

Lung cancer has been found to be doubled among British veterans exposed to mustard agents (Case and Lea, 1955) and increased by up to 50 percent in United States veterans (Beebe, 1960). It is suggested in Chapter 3 of this report that the cumulative exposures for some of the subjects in the WWII testing programs may have been similar to battlefield exposures. To the extent that they were comparable, a similar increased risk of lung cancer would be expected. It is not known how many individuals this finding describes.

REFERENCES

Abell CW, Falk HL, Shimkin MB. 1965. Uracil mustard: a potent inducer of lung tumors in mice.Science 147:1443-1445.

Beebe G. 1960. Lung cancer in World War I veterans: possible relation to mustard gas injury and 1918 influenza epidemic. Journal of the National Cancer Institute 25:12311252.

Boyland E, Horning ES. 1949. The induction of tumours with nitrogen mustards. British Journal of Cancer 3:118-123.

Case RM, Lea AJ. 1955. Mustard gas poisoning, chronic bronchitis, and lung cancer: an investigation into the possibility that poisoning by mustard gas in the 1914-1918 war might be a factor in the production of neoplasia. British Journal of Preventive and Social Medicine 9:62-72.

Colvin M, Chabner BA. 1990. Alkylating agents. In: Chabner BA, Collins JM, eds. Cancer Chemotherapy: Principles and Practice. Philadelphia: J.B. Lippincott.

Conklin JW, Upton AC, Christenberry KW. 1965. Further observations on late somatic effects of radiomimetic chemicals and X-rays in mice. Cancer Research 25:20-28.

Dedrick RL, Morrison PF. 1992. Carcinogenic potency of alkylating agents in rodents and humans. Cancer Research 52:2464-2467.

Glass LR, Easterly CE, Jones TD, Walsh PJ. 1991. Ranking of carcinogenic potency using a relative potency approach. Archives of Environmental Contamination and Toxicology 21:169-176.

Gold LS, Sawyer CB, Magaw R, Backman GM, de Veciana M, Levinson R, Hooper NK, Havender WR, Bernstein L, Peto R. 1984. A carcinogenic potency database of the standardized results of animal bioassays. Environmental Health Perspectives 58:9-319.

Heston WE. 1949. Induction of pulmonary tumors in strain A mice with methyl-bis(ßchloroethlyl) amine hydrochloride. Journal of the National Cancer Institute 10:125-130.

Heston WE. 1950. Carcinogenic action of the mustards. Journal of the National Cancer Institute 11:415-423.

Heston WE. 1953a. Occurrence of tumors in mice injected subcutaneously with sulfur mustard and nitrogen mustard. Journal of the National Cancer Institute 14:131-140.

Heston WE. 1953b. Pulmonary tumors in Strain A mice exposed to mustard gas. Proceedings of the Society for Experimental Biology and Medicine 82:457-460.

International Agency for Research on Cancer (IARC). 1975. IARC Monographs on the Evaluation of Carcinogenic Risk of Chemicals to Man. Vol. 9, Some Aziridines, N, S- & O- Mustards and Selenium. Lyon: IARC. 181-192.

Jones TD, Easterly CE. 1991. On the rodent bioassays currently being conducted on 44 chemicals: a RASH analysis to predict test results from the National Toxicology Program. Mutagenesis 6:507-514.

Jones TD, Walsh PJ, Zeighami EA. 1985. Permissible concentrations of chemicals in air and water derived RTECS entries: a "RASH" chemical scoring system. Toxicology and Industrial Health 1:213-234.

Jones TD, Walsh PJ, Watson AP, Owen BA, Barnthouse LW, Sanders DW. 1988. Chemical scoring by a Rapid Screening Hazard (RASH) method. Risk Analysis 8:99-118.

Kaldor JM, Day NE, Hemminki K. 1988. Quantifying the carcinogenicity of antineoplastic drugs. European Journal of Cancer and Clinical Oncology 24:703-711.

McNamara BP, Owens EJ, Christensen MK, Vocci FJ, Ford DF, Rozimarek H. 1975. Toxicological Basis for Controlling Levels of Mustard in the Environment. Edgewood Arsenal Special Publication EB-SP-74030. Aberdeen Proving Ground, Maryland: U.S. Army Armament Command. Edgewood Arsenal Biomedical Laboratory.

Owen BA, Jones TD. 1990. Hazard evaluation for complex mixtures: relative comparisons to improve regulatory consistency. Regulatory Toxicology and Pharmacology 11:132-148.

Saracci R. 1981. The IARC monograph program on the evaluation of the carcinogenic risk of chemicals to humans as a contribution to the identification of occupational carcinogens. In: Peto R, Schneiderman M, eds. Quantification of Occupational Cancer. Vol. 9. Banbury Report . Cold Spring Harbor, NY: Cold Spring Harbor Laboratory. 165-176.

Schmähl D, Osswald H. 1970. Experimental studies on the carcinogenic effects of anticancer chemotherapeutic and immunosuppressive agents. Arzneimittelforschung 20:1461-1467. [In German]

Shimkin MB, Weisburger JH, Weisburger EK. 1966. Bioassay of 29 alkylating chemicals by the pulmonary-tumor response in strain A mice. Journal of the National Cancer Institute 36:915-935.

Watson AP, Jones TD, Griffin GD. 1989. Sulfur mustard as a carcinogen: application of relative potency analysis to the chemical warfare agents H, HD, and HT. Regulatory Toxicology and Pharmacology 10:1-25.

Zackheim HS, Smuckler EA. 1980. Tumorigenic effect of topical mechlorethamine, BCNU and CCNU in mice. Experientia 36:1211-1212.

J

Examination of the Effects of Certain Acute Environmental Exposures on Future Respiratory Health Consequences

David H. Wegman, M.D., M.S.*

I have been asked to address two primary issues: Can human exposures to nitrogen mustard gases over relatively brief intervals eventually lead to chronic nonmalignant pulmonary effects? Can such effects be of a degree sufficiently severe to disable them? In both clinical and public health terms these issues can be restructured as follows: 1) Does the occurrence of an acute pulmonary reaction identify an individual at risk for long-term respiratory sequelae? 2) If so, can the probability or degree of damage be predicted from the magnitude of the acute response? 3) What assurance is there that the absence of an acute pulmonary reaction identifies individuals who will not develop long-term sequelae?

To begin, it is recognized that little structured study of the mustards has occurred among human populations acutely overexposed. Without attempting to estimate the relative importance of the different respiratory conditions or to estimate the dose levels necessary to cause each, a variety of conditions have been noted. Low dose inhalation exposure of mustard gas has been noted to result in chest tightness while higher doses have resulted in sneezing, lacrimation, rhinorrhea, nasal bleeding, sore throat, hoarseness and cough.[1] The WHO has reported the pulmonary effects in Iranian patients exposed during the Iran-Iraq war.[2],[3] Acute effects included: (2-4 hrs) chest tightness, sneezing, lacri

---

*University of Massachusetts at Lowell. Prepared for the Institute of Medicine's Committee to Survey the Health Effects of Mustard Gas and Lewisite.

mation, rhinorrhea, epistaxis, hacking cough; (4-16 hrs) sinus pain, severe epistaxis; (16-24 hrs) severe cough; (24-48 hrs) severe dyspnea, pulmonary edema; (48-72 hrs) bronchopneumonia. Delayed effects measured after two years included: chronic bronchitis, asthma, rhinopharyngitis, tracheobronchitis, laryngitis, recurrent pneumonia and bronchiectasis. In summary, then, overexposure to nitrogen mustards has caused both airway reactions and parenchymal damage with increased risk of pulmonary infection.

Since the respiratory conditions described in instances of severe overexposure to nitrogen mustards have been both acute and chronic, there is evidence that some level of exposure to these agents can cause both severe acute and chronic respiratory disease.

However, there is little evidence available on the effects of lower level acute exposure and none apparently on brief exposures which are not too excessive. In the absence of follow-up studies that directly address the question, much of the exploration must involve examining indirect evidence from the study of compounds which might behave similarly. In order to evaluate the respiratory health risk associated with repeated brief overexposures at lower levels, review of indirect evidence will focus on examining the link between acute and chronic respiratory responses to such agents.

One central difficulty in examining this question is the absence of knowledge about the early stages of environmentally related pulmonary diseases coupled with almost no knowledge on the natural (longitudinal) evolution of the clinical conditions. Thus there is little to guide us in what is the appropriate disease model and what pattern(s) of exposure are relevant to disease etiology.

REVIEW OF AGENTS CAUSING ACUTE AND CHRONIC RESPIRATORY EFFECTS

The following agents have been selected to illustrate the range of exposure-effect relationships that may be relevant to the effects of the mustards. It will be noted that the agents selected have been shown to result in (to cause) a mix of respiratory conditions: Exposures to irritant gases (chlorine, SO_2, combustion products); exposures to materials of plant origin (cotton); exposures to chemicals (isocyanates); and exposures to inorganic dusts (silica, beryllium, asbestos).

Chlorine

Since chlorine gas is such an irritating substance, there are a number of reports of overexposures with documented acute respiratory effects.

These provide clear evidence of severe acute respiratory damage with intensity commensurate with the exposure level and duration.[4,5]

Several investigators have attempted to follow subjects who have experienced acute symptomatic effects to determine whether these either persist or progress. To date, most of the reports of accidental overexposures suggest that individuals return to normal.[6,7] A few reports have documented persistent obstructive or restrictive effects in follow-up of 1, 3 and 12 years.[8,9,10] In one case report a previously healthy individual suffered onset of irreversible and debilitating asthma following a single overexposure.[11] In all instances there has been an absence of pre-accident health assessments so whatever residual effects are present at from 6 months to 13 years later cannot be attributed to the acute accidental exposure alone.

In contrast to these studies of single acute overexposure episodes, there are two populations of workers occupationally exposed to chlorine gas where the importance of accidental gassings has been evaluated. In a cross-sectional study of chlorine gas plant workers, 55 of 139 were reported to have experienced accidental overexposures against a background of less than 1 ppm.[12] Of those overexposed, only three showed significant impairment which might only represent statistical variability in the population. Overall there was, however, evidence of persistent minor airflow abnormalities.

In a more detailed study of pulpmill workers, evidence of an effect of gassings was presented.[13] Although several irritant gases were present in the pulpmill, chlorine was the dominant irritant exposure. Overall the pulpmill workers had more symptoms but no pulmonary function differences compared to a control working population. However, those with gassings had all the excess of symptoms and had more airflow obstruction compared with the other pulpmill workers. Evidence of healthy worker selection was also noted as likely to have decreased the strength of the findings.

This population had been studied seven years previously by the same investigators. Those workers who had experienced chlorine gassings once or twice during the study interval were compared to the remainder. [14] For those whose gassing resulted in their seeking first-aid assistance at the time, there was evidence that respiratory symptoms and airflow obstruction had progressed when 1988 measures were compared to ones taken in 1981. These effects were present after controlling for cigarette smoking.

In summary, there is clear evidence of acute respiratory effects from chlorine exposures which include both acute bronchitis and airway obstruction as well as inflammatory bronchoconstriction (RADS). There is also evidence for a chronic effect from exposure to chlorine. The relationship of acute respiratory effects to chronic ones is also present, although the evidence is less clear in the absence of continued very low level exposure to chlorine gas independent of the overexposure. In at least one case a single exposure in a previously asymptomatic individual resulted in a debilitating chronic condition. There are no studies that followed individuals acutely overexposed but not acutely symptomatic to determine whether chronic effects were likely to develop.

Sulfur Dioxide

Sulfur dioxide is an agent with irritant potential equal to that of chlorine. Both agents are moderately soluble in water, are likely to result in inflammatory bronchoconstriction with the amount of upper respiratory tract irritation dependent on exposure level. Sulfur dioxide also shows clear evidence of acute respiratory effects related to short exposures. These include evidence of bronchial constriction and chest tightness.[4]

Studies of chronic respiratory effects associated with sulfur dioxide exposures have yielded mixed results. However, those which included individual assessment or assignment of SO_2 exposure have demonstrated accelerated loss of pulmonary function overall as well as among those with poorer baseline function.[15,16]

There are, however, some important differences between the two irritants when considering the impact of acute overexposures. As with chlorine, the primary respiratory response seen as a result of accidental overexposure is obstructive and includes both pulmonary function and symptoms of productive cough, wheeze and chest tightness. However, whereas those who experience single accidental exposures to chlorine generally recover most of their pulmonary function losses, those similarly overexposed to sulfur dioxide almost always do not. In three studies where acute overexposed subjects were followed for 4 months to 4 years all but one showed evidence of functional abnormality (predominantly obstructive in nature).[17,18,19]

In summary then, there is clear evidence that sulfur dioxide can cause acute respiratory effects and it is likely that lower level exposures over time result in chronic obstructive abnormalities. Single overexposure episodes which produce acute respiratory symptoms appear to cause irreversible damage. No follow-up studies of those acutely overexposed but not symptomatic have been reported.

Miscellaneous Irritants

A variety of reports have been published where single accidents or episodes have resulted in overexposure to many different irritant agents. In several of these there has been an attempt to follow-up those who were adversely affected in order to determine the persistence of any immediate abnormalities.

Prominent among these has been the examination of firefighters who have been acutely exposed to a variety of combustion products.[20] As might be expected, in a number of the reports only transient pulmonary function changes have occurred.[21,22] In cases where fires involved polyvinyl chloride,[23] isocyanates[24] or "plastics"[25] there was evidence of persistent effects ranging from increased symptoms to progressive fatal asthma. A follow-up of survivors of a subway fire suggested persistence of small airway damage and respiratory symptoms at six months and two years.[26]

In 1985 Brooks, et al., described the clinical condition of reactive airways dysfunction syndrome (RADS).[27] The condition is asthma-like but differs from occupational asthma because of an absence of a preceding period for sensitization to occur and the onset of illness after a single overexposure. A single exposure to a glacial acetic acid spill was carefully studied and the relative odds of developing RADS was estimated to be as large as 10 for the highest exposure group. [28] Although no general estimate is available of the probability that RADS will follow from a single high irritant exposure, the number of agents which have been reported is quite varied and includes uranium hexafluoride, hydrazine, heated acids, perchlorethylene and toluene diisocyanate.[27,29,30] The severe disability associated with some of these episodes is disturbing.

In summary, a number of specific agents as well as poorly described irritant exposures have been shown to cause long-term disability and even death after a single severe overexposure. The suggestion is that any severe exposure to a wide variety of respiratory irritants has a reasonable likelihood of producing serious long-term effects on the respiratory system. The exposure circumstances described in this section did not lead to evaluation of asymptomatic exposed individuals. Therefore, the effects of exposures high enough to cause disability in those who were symptomatic do not directly address the question of chronic effects in the absence of acute illness.

Cotton Dust

In 1956, Schilling published a summary of descriptive studies which elegantly characterized the persisting risk of disease among cotton textile workers.[31] It was Schilling who also was responsible for the development of the symptom score which still best describes the acute symptom complex—ranging from occasional Monday morning chest tightness to chest tightness persisting throughout the workweek but remitting over a weekend.

Ample evidence exists to show the relationship of acute chest tightness, cough and dyspnea related to cotton dust exposure.[32,33,34] The classic Monday morning chest tightness has been shown to be dose related[35] and to increase with increasing duration of employment.[36] Reports of prevalence of the syndrome have varied to levels as high as 50 percent.[37] The relationship of these striking acute symptoms to pulmonary function changes has not been well established. Cotton textile workers do show dust-related cross-shift loss in pulmonary function but the relationship of this change to byssinosis has been variable.[38,39,40] As suggested in a recent report on the effects of exposures to endotoxin in cotton dust among human volunteers, this problem may be resolved as better characterization of the actual agent evolves.[41]

In the few efforts to examine the persistence of the acute symptoms in affected workers, studies have shown remission in some of those still exposed[42,43] and persistence among those who have been removed from exposure.[44]

To date there remains no consensus for the mechanism of the acute symptoms related to cotton dust exposures. Although the symptoms are similar to asthma for many of those affected, the dose-response relationship and the absence of clear evidence of immunogenesis have led this acute condition to be classified as a pharmacologic bronchoconstriction. [45]

The presence of a chronic effect of cotton dust on respiratory health has been debated for a number of years. There is, however, mounting evidence that such effects do occur. Early mortality studies31 which indicated excess respiratory deaths have not always been confirmed. [46,47] However, the earlier studies were among more highly exposed workers and mortality analysis is generally a very imprecise measure of chronic respiratory disease.

A number of studies, both cross-sectional and longitudinal, have examined the effect of cotton textile employment on pulmonary function. Many of the cross-sectional studies have shown lower function in cotton workers,[33,39,48,49,50] but inconsistent findings have been reported for the relationship of the lower function to duration of employment[35,36] and current dust exposures.[33,48] One study did show a relationship to a dust index.[32]

The longitudinal studies have consistently shown greater loss of FEV1, among cotton textile workers.[22,36,38,42,51] In only one of those with under three years follow-up was an association of function loss with duration noted.[9] In the three with five or more years, despite methodologic flaws there was not only lower function in the cotton workers but lower function in those who reported byssinosis.[16,49] In the most recent one a relationship to job category and dose was described. [51]

Although there remains controversy over the presence of a chronic respiratory effect from cotton dust exposures, the mounting evidence appears to lead to the acceptance of such an effect. If the evidence reviewed is accepted then there is substantial support for a relationship of the acute respiratory syndrome to the various chronic effects. These include the accelerated loss of function among byssinotics, [49] the increased prevalence of chronic bronchitis among byssinotics, [36,39,42,52] and the recent identification of a relationship between cross-shift and five year change in FEV1 in Chinese cotton textile workers.[53]

In summary, exposure to cotton dust or components of the dust has been noted to cause 1) acute effects represented by both a disabling acute respiratory syndrome (byssinosis) and a cross-shift decrement in FEV_1, and 2) chronic effects represented by both accelerated annual decrements in FEV1 and chronic bronchitis. The acute syndrome does not necessarily remit on removal from exposure and there is a growing body of evidence to suggest a relationship between the acute effects and chronic airflow limitation. There is evidence that short-term exposures result in acute effects[41,54] but none for whether such exposures, with or without an acute response, ultimately lead to chronic effects.

Isocyanates

These reactive chemical agents have been reported to cause a wide range of nonmalignant respiratory health effects although there are substantial differences in the proportion of the population at risk for each.[55]

In 1951, case reports of isocyanate-related pulmonary effects called attention to a new occupational asthma.[56] Sensitization was reported to occur as early as after only one high exposure[57] but onset could also be delayed for a number of years after first exposure.[58,59] Unlike with high molecular weight agents which cause occupational asthma, however, further study of isocyanate asthma cases showed that measurable IgE and airway hyperreactivity was not invariably an essential feature.[55,60,61] The course of isocyanate asthma includes the fact that as many as half of seriously affected workers who leave work do not recover.[62,63,64]

Although it was not initially expected, continued investigation of the effects of work exposure to isocyanates revealed evidence of a number of respiratory effects other than asthma which were important and a series of epidemiologic studies ensued. There is good evidence that high levels of exposure are associated with chemical bronchitis [65] and, recently, with reactive airway dysfunction syndrome.[66]

Independent of asthma or acute response to serious overexposures, there are well-documented acute and chronic non-specific airway effects from isocyanate exposures. Peters' studied workers exposed to very low levels of TDI. His work demonstrated significant change in FEV1 over the workshift.[67] A three year prospective study of these same workers demonstrated the losses were not just acute; exposed workers experienced accelerated losses over three years and there appeared to be a correlation between the magnitude of cross-workshift change and the accelerated functional loss.[68,69] My own work provided preliminary information on the relationship of acute and chronic losses to exposure level and that the effects persisted after controlling for cigarette smoking.[70,71] Epidemiology studies have also provided a preliminary estimate of a no-effect level[72].

Work by Weill and his colleagues, in a prospective study of a new plant confirmed the broad population dose-related effects of chronic TDI exposure while directly accounting for healthy worker selection. [73,74] Their work raised the possibility that peak exposures might be important not only for asthma[74] but for the non-asthma related chronic losses as well. These epidemiologic studies provided sufficient evidence that TDI was not a problem for only a small population of sensitized workers, but that the agent was a risk for all TDI workers.

Finally no epidemiologic studies have yet examined the importance of the case reports of hypersensitivity pneumonitis as an exposure consequence.[55] As these are being reported with increasing frequency their relative importance as a risk from this chemical exposure is an important priority to determine.

In summary, isocyanates (a series of small molecular weight chemical compounds) have been noted to cause 1) several acute conditions: chemical bronchitis, allergic bronchoconstriction (after short-term high exposures as well as following longer-term lower exposures), and to cause large dose-related cross-shift loss in FEV_1; and 2) chronic respiratory effects including accelerated loss in pulmonary function over several years (suggesting the development of chronic airway limitation) and irreversible asthma. There is also evidence that the acute effects are related to the chronic effects both as asthma that does not remit and as cross-shift loss which is related to the rate of subsequent annual loss. Finally, there is evidence that short-term exposures can cause acute responses that are irreversible and progressive, but there is no evidence, either way, as to whether short-term exposures without acute response result in irreversible respiratory effects.

Notes on Selected Inorganic Agents

Beryllium exposures have been associated with both acute and chronic pulmonary disease.[75] Both acute and chronic conditions have been shown to be disabling. Relevant to the questions being considered here is that exposures in one instance as brief as one week and in several instances occurring for less than 10 weeks have resulted in disease certified for inclusion in the Beryllium Registry established at the Massachusetts General Hospital.[76]

Silica exposures are generally considered important only for chronic disease. There is, however, the condition of acute silicosis caused by relatively short-term high intensity exposures to very fine silica particulate.[77] Exposures as brief as six months have been associated with acute silicosis. But even in the absence of such severe overexposures and acute clinical disease, there is evidence that even low level exposures not associated with clinical symptoms or chest X-ray abnormality can result in irreversible pathology.[78]

Asbestos exposures similarly are noted for causing chronic respiratory illness as well as cancer. Brief high exposures are not recognized to cause respiratory complaints. However, brief exposures (on average less than four years) have been associated with increased respiratory symptoms and decreased vital capacity.[79]

THE FUNDAMENTAL QUESTION

Narrowly Structured

For an acute exposure (intense but over only a short interval) to have a causal association with subsequent respiratory disease conditions, must the exposure cause acute irreversible and progressive damage evidence by acute clinical illness or, at minimum, objective changes in lung parameters or severe subjective symptoms?

Alternative Broadly Structured

For an acute exposure (intense but over only a short interval) to have a causal association with subsequent respiratory disease conditions, either the exposure must cause acute irreversible (and possibly progressive) damage with or without clinical evidence of the damage, or it must result in or lead to an alteration in individual risk factors which changes the impact of subsequent exposures (e.g., dust exposures, air pollution, cigarette smoking, other environmental agents, respiratory infections).

THE RESPIRATORY END POINTS

This review of respiratory tract responses to a variety of different types of environmental agents has shown that a range of respiratory effects is possible in response to each of the different types of irritant exposures. With the exception of effects limited to *upper airway irritation* (non-productive cough, hoarseness), most of these may be grouped together as obstructive lung diseases, a term that includes several different clinical syndromes: *simple chronic bronchitis* (mucus hypersecretion); *chronic obstructive bronchitis* (characterized by mucus hypersecretion and chronic airflow limitation, largely irreversible); *emphysema* (defined in anatomical terms as an increase in the size of the distal airspaces and destruction of their walls); and a variety of airway reactivity conditions including *allergic bronchoconstriction* (acute recurrent episodic reversible airflow limitation with specific airway hypersensitivity), *inflammatory bronchoconstriction* (acquired airway hyperresponsiveness from nonimmunogenic irritant exposures characterized by reversible airflow obstruction and nonspecific airway hypersensitivity), and *pharmacologic bronchoconstriction* (reversible airflow limitation without evidence of a hypersensitive subgroup of the population of exposed). As with any set of clinical syndromes, there is a degree of overlap, and classification changes as understanding the underlying mechanism improves.

Simple Chronic Bronchitis

Basically there are no good population data on this condition as in and of itself it is not considered to be a disabling (therefore relevant) condition.

Chronic Obstructive Bronchitis and Emphysema

The British hypothesis suggests that chronic bronchitis and chronic airflow limitation (by implication, emphysema) are separate parallel disease processes affecting different parts of the respiratory tract. Both were related at least to cigarette smoking and asthma was unrelated to either. The Dutch hypothesis focuses on individual susceptibility, hyperreactive airways, as key in both conditions and independent of cigarette smoking. Follow-up study suggests both are correct. Smokers who develop chronic airflow limitation (a minority) have been shown to have increased airway reactivity while both air pollutants and cigarettes produced accelerated decline in pulmonary function independent of airway reactivity.

Both chronic obstructive bronchitis and emphysema have been shown to be related independently to the irritant effects of inhaled airborne dust but not to each other. Similar results are seen in response to chronic exposures to irritant gases or vapors. As in the case of cigarettes, although dose-response relationships have generally been demonstrated to both the level and the intensity of exposure, not all those exposed to what appear to be comparable levels are affected. This points to a role of individual susceptibility. In turn, the key factor which makes an individual susceptible may well be the capacity of his or her airways to become reactive to inhaled materials.

Allergic Bronchoconstriction

Different patterns of asthmatic reactions have been noted in response to high and low molecular weight agents. The former (proteins, polysaccharides and peptides) produce specific IgE (sometimes IgG) antibodies, generally have a positive immediate skin test to extracts, and produce results in isolated immediate or biphasic (immediate and late) reactions, but generally do not show isolated late reactions. These appear not to differ in mechanism from asthma due to common allergens such as house dust.

The latter, low molecular weight agents appear to be of two types. Some (e.g., the anhydrides or platinum salts) act as haptens and show asthma patterns similar to the high molecular weight agents. Others, best exemplified by the isocyanates, do not produce IgE in most responders. The asthma associated with isocyanates affects 5-10% of the exposed, and is associated predominantly with a late phase (isolated or part of a biphasic reaction) response to inhalation challenge studies. The asthma persists after removal in many and appears to affect atopic and non-atopic individuals equally.

Inflammatory Bronchoconstriction

Immunologically active substances can cause occupational asthma in some exposed workers while exposure to nonimmunogenic substances (i.e., irritants) may cause reactive airway dysfunction (RADS) or irritant induced occupational asthma in a wider population. Asthma and airway hyperresponsiveness typically occur together although they are not synonymous, so irritant-induced asthma is not necessarily caused through an acquired airway hyperreactivity mechanism. Documentation of the mechanism, however, would be strongly suggestive.

Hyperresponsiveness is an amplification of the normal physiologic response to irritant stimulation. It is a characteristic of asthma, but is not always associated with overt asthma or with respiratory symptoms. It can be an inherent characteristic of the person or an acquired one, and it can be either temporary or permanent. Distribution of hyperresponsiveness in population studies is skewed, possibly bimodal. It has been hypothesized that it may lead to, or may be a predisposing factor in, subsequent chronic obstructive lung disease. This is a candidate for a biologically plausible mechanism for an acute irritant exposure resulting in a postponed but long-term effect which is chronic in nature.

Pharmacologic Bronchoconstriction

Although not without controversy, this type of environmentally induced asthma is similar to airway reactivity in response to pharmacologic agonists. In this regard, there is a common dose-response relationship with high enough exposures causing reactions in all exposed subjects. Since these reactions are not associated with eosinophilia or nonspecific bronchial hyperreactivity, some argue they should not be considered asthma. Regardless, the causative agents are clearly associated with airway reactions such as those represented by byssinosis and the related chronic effects of exposures to cotton dust (or endotoxin).

HOST FACTORS

In both the examination of chronic airflow limitation and the examination of allergic asthma, less effort has been spent on objective characterization of the environmental exposures and more on delineating host factors such as the sensitization status of the individual. The preceding review, however, suggests that many respiratory irritants and toxins affect broad portions of the population. Host factors certainly interact with these agents, but it should be recognized that the following commonly discussed individual risk factors or habits play a more variable role in the non-malignant respiratory diseases than is often recognized.

There is little evidence to suggest that gender or race are important risk factors in differentiating the types of respiratory tract responses to the above agents. Age, similarly, is relatively unimportant except for those of increasing age having increasing probability of experiencing a wider variety of irritant exposure events.

Atopic status appears to be a risk factor for asthma due to some of the high molecular weight agents but does not appear important for many if not all of the low molecular weight agents. The other respiratory conditions do not seem to be affected by atopic status.

Although nonspecific bronchial reactivity is often noted in the majority of patients with occupational asthma, it is not known whether this is a result of the exposure or a predisposing factor. Studies in red cedar asthma suggest that the increased bronchial reactivity is reduced or returns to normal after exposure ceases, suggesting that the reactivity change is a result of the exposure.[80] Similar conclusions might be drawn from a recent study comparing subjects responding to cotton dust (with both byssinotic and non-byssinotic symptoms) and asymptomatic workers, which showed nonspecific bronchial reactivity most prevalent among byssinotics and least prevalent in the asymptomatic subjects. However, when reactivity was examined without respect to symptoms, cumulative cotton dust level was a significant predictor of reactivity.[81] Kennedy recently addressed the question of whether nonspecific bronchial reactivity was an acquired or inherent feature of respiratory reactions in persons exposed to non-immunogenic irritant agents.[5] She reluctantly concluded that the question remains unanswered, although there is growing evidence that acquired increased bronchial reactivity is of likely importance.

Smoking (a personal habit rather than a host factor) is well described as a risk factor for chronic obstructive bronchitis and chronic airflow limitation (or emphysema). Agents that cause these conditions most probably result in effects which are additive to those of cigarette smoking.[82] The fact that all smokers do not experience the same level of risk suggests that smoking, itself, must be interacting with some other host factor in regard to these respiratory outcomes. With respect to asthma, the role of smoking is quite unclear. One hypothesis suggests that increased membrane permeability of smokers allows greater penetration of antigens. However, smoking has not been associated with work-related symptoms among those exposed to such agents as detergent enzymes[83] or colophony[84] while it has been noted to be related instudies of workers exposed to phthalic anhydride[85] and of soy bean workers.[86] In contrast, in studies of those exposed to plicatic acid or to isocyanates, asthma was mostly noted in non-smokers.[87,88]

CONCLUSIONS

In this review a variety of materials toxic to the respiratory tract have been examined. The review was not designed to be comprehensive, yet, it was also not unduly selective. In making the selection, agents were included which are commonly considered primarily as acute respiratory irritants, those known to induce extrinsic asthma, those which are related to non-immunogenic bronchial hyperreactivity, those which are most often responsible for slowly developing chronic fibrosis or granulomatous disease, and one which is believed to cause pharmacologic bronchoconstriction. The chronic respiratory effects associated with each of the agents reviewed included several of the general types of chronic respiratory response rather than being limited to only one type of reaction. Now, as a final step in the evaluation of the agents reviewed, an attempt should be made to answer the original three general questions about pulmonary reactions.

The questions and suggested answers are:

1. **Does the occurrence of an acute pulmonary reaction identify an individual at risk for long-term respiratory sequelae?** Ample evidence

has been provided for all of the agents reviewed that a significant portion of individuals who react acutely to short, high exposures (and even some to short, relatively low exposures) go on to develop a variety of long-term respiratory effects. The agents differ in the probability of long-term adverse outcomes, but none appear to be free of this risk. The literature reviewed does not allow identification of a minimum magnitude of acute response necessary in order for there to be long-term sequelae.

2. **If acute pulmonary reactions can identify individuals at risk for long-term sequelae, can the probability or degree of damage be predicted from the magnitude of the acute response?** The second question has generally not been studied. The fact that an association of dose is found with the acute as well as the chronic responses, provides support for the association, but not for a relationship between the *magnitude* of the acute response and the *magnitude* of the chronic response. Only in the case of the non-asthma-like isocyanate effects has the question been directly addressed. Here the acute cross-shift change in FEV_1 has several times been significantly correlated with accelerated decrement in function. This association, however, is not invariant since the highly significant correlations were still under 0.5. Furthermore, the acute symptomatic response has not been a reliable predictor of accelerated functional losses.

   If the disease model invoked requires the acute exposure to cause acute irreversible damage one might propose that it is likely the magnitude of acute response would predict the magnitude of the chronic. However, if the model invoked is that the acute exposure resulted in or led to an alteration in individual risk factors, then it is quite likely that the magnitude of the acute and chronic responses would be unrelated.

3. **What assurance is there that the absence of an acute pulmonary reaction identifies individuals who will not develop long-term sequelae?** This may be the most important question to ask and, unfortunately, the one for which no direct evidence could be found. The indirect evidence reported, however, would suggest that it would have to be an unusual disease model which would need to be invoked that could exclude the possible mechanism of change in individual risk factors so that the absence of an acute reaction would eliminate the possibility of a chronic effect related to the acute exposure.

LITERATURE CITED

1. Somani SM, Babu SR. Toxicodynamics of sulfur mustard. Int J Clin Pharm Therapy Tox (1989) 27:419-435.
2. World Health Organization. 1984 report of the specialists appointed by the Secre

tary-General to investigate allegations by the Islamic Republic of Iran concerning the use of chemical weapons. Report S/16433, United Nations, New York, March 26, 1986.
3. World Health Organization. 1986 report of the mission dispatched by the SecretaryGeneral to investigate allegations of the use of chemical weapons in the conflict between the Islamic Republic of Iran and Iraq. Report S/17911, United Nations, New York, March 12, 1986.
4. Becklake MR, Bourbeau J, Menzies R, Ernst P. The relationship between acute and chronic responses to occupational exposures. Curr Pulmonol (1988) 9:25-66.
5. Kennedy SM. Acquired airway hyperresponsiveness from nonimmunogenic irritant exposure. Occup Med: State of Art Rev (1992) 7(2):287-300.
6. Weill H, George R, Schwarz M, Ziskind M. Late evaluation of pulmonary function after acute exposure to chlorine gas. Am Rev Respir Dis (1969) 99:374-379.
7. Jones RN, Hughes JH, Glindmeyer H, Weill H. Lung function after acute chlorine exposure. Am Rev Respir Dis (1986) 134:1190-1195.
8. Kowitz TA, Reba RC, Parker RT, Spicer WS. Effects of chlorine gas upon respiratory function. Arch Environ Health (1967) 14:545-558.
9. Kaufman J, Burkons D. Clinical, roentgenologic, and physiologic effects of acute chlorine exposure. Arch Environ Health (1971) 23:29-34.
10. Schwartz DA, Smith DD, Lakshminarayan S. The pulmonary sequelae associated with accidental inhalation of chlorine gas. Chest (1990) 97:820-825.
11. Moore BB, Sherman M. Chronic reactive airway disease following acute chlorine gas exposure in an asymptomatic atopic patient. Chest (1991) 100:855-856.
12. Chester EH, Gillespie DG, Krause FD. The prevalence of chronic obstructive pulmonary disease in chlorine gas workers. Am Rev Respir Dis (1969) 99:365-373.
13. Kennedy SM, Enarson DA, Janssen RG, Chan-Yeung M. Lung health consequences of reported accidental chlorine exposures among pulpmill workers. Am Rev Respir Dis (1991) 143:74-79.
14. Salisbury DA, Enarson DA, Chan-Yeung M, Kennedy SM. First-aid reports of acute chlorine gassing among pulpmill workers as predictors of lung health consequences. Am J Indust Med (1991) 20:71-81.
15. Smith TJ, Peters JM, Reading JC, et al. Pulmonary impairment from chronic exposure to sulfur dioxide in a smelter. Am Rev Respir Dis (1977) 116:31-39.
16. Peters JM, Smith TJ, Bernstein I, et al. Pulmonary effects of exposures in silicon carbide manufacturing. Brit J Indust Med (1984) 41:109-115.
17. Charan NB, Myers CG, Lakshminarayan S, Spencer TM. Pulmonary injuries associated with acute sulfur dioxide inhalation. Am Rev Respir Dis (1979) 119:555-560.
18. Harkonen H, Nordman H, Korhonen O, Winblad I. Long-term effects of exposure to sulfur dioxide. Am Rev Respir Dis (1983) 128:890-893.
19. Rabinovitch S, Greyson ND, Weiser W, Hoffstein V. Clinical and laboratory features of acute sulfur dioxide inhalation poisoning: two-year follow-up. Am Rev Respir Dis (1989) 139:556-558.
20. Guidotti TL, Clough VM. Health concerns of firefighting. Ann Rev Publ Health (1992) 13:151-171.
21. Musk AW, Smith TJ, Peters JM, McLaughlin E. Pulmonary function in firefighters: acute changes in ventilatory capacity and their correlates. Brit J Indust Med (1979) 36:29-34.
22. Sheppard D, Distefano S, Morse L, Becker C. Acute effects of routine firefighting on lung function. Am J Indust Med (1986) 9:333-340.
23. Markowitz JS. Self-reported short- and long-term respiratory effects among PVC exposed firefighters. Arch Environ Health (1989) 44:30-32.
24. Axford AT, McKerrow CB, Jones AP, and LeQuesne PM. Accidental exposure to isocyanate fumes in a group of firemen. Brit J Indust Med (1976) 33:65-71.

25. Bergstrom, Tornling G, Unge G. Acquired progressive asthma in a fire-fighter. Eur Respir J (1988) 1:469-470.
26. Fogarty PW, George PJM, Solomon M, Spiro SG, Armstrong RF. Longterm effects of smoke inhalation in survivors of the King's Cross underground station fire. Thorax (1991) 46:914-918.
27. Brooks SM, Weiss MA, Bernstein IL. Reactive airways dysfunction syndrome (RADS): persistent asthma syndrome after high level irritant exposures. Chest (1985) 88:376-384.
28. Kern DG. Outbreak of the reactive airways dysfunction syndrome after a spill of glacial acetic acid. Am Rev Respir Dis (1991) 144:1058-1064.
29. Boulet, L-B. Increases in airway responsiveness following acute exposure to respiratory irritants: reactive airway dysfunction syndrome or occupational asthma? Chest (1988) 94:476-481.
30. Promisloff RA, Lenchner GS, Cichelli AV. Reactive airway dysfunction syndrome in three police officers following a roadside chemical spill. Chest (1990) 98:928-929.
31. Schilling RSF. Byssinosis in cotton and other textile workers. Lancet (1956) 2:261-265 (August 11).
32. Fox JJ, Tombleson JBL, Watt A, Wilkie AG. A survey of respiratory disease in cotton operatives. Part II. Symptoms, dust estimation and the effect of smoking habit. Brit J Indust Med (1973) 30:48-53.
33. Merchant JA, Kilburn KH, O'Fallon WM, et al. Byssinosis and chronicbronchitis among cotton textile workers . Ann Intern Med (1972) 76:423-433.
34. Haglind P, Lundholm M, Rylander. Prevalence of byssinosis in Swedish cotton mills. Brit J Indust Med (1981) 138-143.
35. Merchant JA, Lumsden JC, Kilburn KH, et al. Dose-response studies in cotton textile workers. J Occup Med (1973) 15:222-230.
36. Fox JJ, Tombleson JBL, Watt A, Wilkie AG. A survey of respiratory disease in cotton operatives. Part I, Symptoms and ventilation test results. Brit J Indust Med (1973) 30:42-48.
37. Schilling RSF. Worldwide problems of byssinosis. Chest (1981) 79(suppl 4):35-65.
38. Berry G, McKerrow CB, Molyneux MKB, et al. A study of the acute and chronic changes in ventilatory capacity of workers in Lancashire cotton mills. Brit J Indust Med (1973) 30:25-36.
39. Imbus HR, Suh MW. Byssinosis. A study of 10,133 textile workers. Arch Environ Health (1973) 26:183-191.
40. Christiana DC, Eisen EA, Wegman DH, Ye TT, Gong ZC, Lu PL and Dai HL. Respiratory disease in cotton textile workers in the People's Republic of China, II: pulmonary function results. Scandinavian Journal of Work Environment Health (1986) 12:46-50.
41. Castellan RM, Olenchock SA, Kinsley KB, Hankinson JL. Inhaled endotoxin and decreased spirometric values. New Eng J Med (1987) 317:605-610.
42. Molyneux MKB, Tombleson JBL. An epidemiological study of respiratory symptoms in Lancashire cotton mills. Brit J Indust Med (1970) 27:225-234.
43. Beck GJ, Schacter EN, Maunder LR, et al. A prospective study of chronic lung disease in cotton textile workers. Am J Epid (1984) 119:33-43.
44. Elwood JH, Elwood PC, Campbell MJ, et al. Respiratory disability in ex-flax workers. Brit J Indust Med (1986) 43:300-306.
45. Chan-Yeung M, Lam S. State of the art: occupational asthma. Am Rev Respir Dis (1986) 133:686-703.
46. Berry G, Molyneux MKB. A mortality study of workers in the Lancashire cotton mills. Chest (1981) 79(suppl 4):11S-15S.

47. Merchant JA, Ortmeyer C. Mortality of employees of two cotton mills in North Carolina. Chest (1981) 79(suppl 4):6S-11S.
48. Jones RN, Diem JE, Glindmeyer H, et al. Mill effect and dose-response relationships in byssinosis. Brit J Indust Med (1979) 36:305-313.
49. Kamat SR, Kamat GR, Salpeker H, et al. Distinguishing byssinosis from chronic obstructive pulmonary disease. Am Rev Respir Dis (1981) 124:31-40.
50. Schacter EN, Maunder LR, Beck GJ. The pattern of lung function abnormalities in cotton textile workers. Am Rev Respir Dis (1984) 129:523-527.
51. Glindmeyer HW, Lefante JJ, Jones RN, et al. Exposure-related declines in the lung function of cotton textile workers: relationship to current workplace standards. Am Rev Respir Dis (1991) 144:675-683.
52. Berry G, Molyneux MKB, Tombleson JBL. Relationships between dust levels and byssinosis and bronchitis in Lancashire cotton mills. Brit J Indust Med (1974) 31:18-27.
53. Christiani D. Unpublished data.
54. Martin CVF, Higgins JE. Byssinosis and other respiratory ailments. (1976) J Occup Med 16:455-462.
55. Musk AW, Peters JM, Wegman DH. Isocyanates and respiratory disease: current status. Am J Indust Med (1988) 13:331-349.
56. Fuchs S, Valade P. Etudes cliniques et experimentales sur quelques cas d'intoxication par le desmodur T (diisocyanate de toluylene) 1-2-4 et 1-2-6. Arch Mal Prof (1951) 12:191-196.
57. Moller DR, McKay RT, Bernstein II, Brooks SM. Persistent airways disease caused by toluene diisocyanate. Am Rev Respir Dis (1986) 134:175-176.
58. Adams WGF. Lung function of men engaged on the manufacture of toluene diisocyanate (TDI). Proc Roy Soc Med (1975) 63:378-379.
59. Porter CV, Higgins RI, Scheel LD. A retrospective study of clinical, physiologic and immunologic changes in workers exposed to toluene diisocyanate. Am Ind Hyg Assoc J (1975) 36:159-168.
60. Smith AB, Brooks SM, Blanchard J, Bernstein IL, Gallagher J. Absence of airway hyperreactivity to methacholine in a worker sensitized to toluene diisocyanate (TDI). J Occup Med (1980) 22:327-331.
61. Mapp CE, Dal Vecchio L, Boschetto P, De Marzo N, Fabbri LM. Toluene diisocyanate-induced asthma without airway hyperresponsiveness. Eur J Respir Dis (1986) 68:89-95.
62. Paggiaro PL, Baci E, Dente F, Talini D, Giutini C. Prognosis of occupational asthma induced by isocyanates. Bull Eur Physiopathol Respir (1988) 23:565-569.
63. Lozewicz S, Assoufi BK, Hawkins R, Newman-Taylor AJ. Outcome of asthma induced by isocyanates. Brit J Dis Chest (1987) 81:14-22.
64. Mapp CE, Corona PC, DeMarzo N, Fabbri L. A follow-up study of subjects with occupational asthma due to toluene diisocyanate (TDI). Am Rev Respir Dis (1988) 137:1326-1329.
65. Axford AT, McKerrow CB, Jones AP, Le Quesne PM. Accidental exposure to isocyanate fumes in a group of firemen. Brit J Indust Med (1976) 33:65-71.
66. Brooks SM, Weiss MA, Bernstein IL. Reactive airways dysfunction syndrome (RADS). Chest (1985) 88:376-384.
67. Peters JM, Murphy RLH, Pagnotto LD, Van Ganse WF. Acute respiratory effects in workers exposed to low levels of toluene diisocyanate (TDI). Arch Environ Health (1968) 16:642-647.
68. Peters JM. Cumulative pulmonary effects in workers exposed to tolylene diisocyanate. Proc Roy Soc Med (1970) 63:14-17.
69. Peters JM, Wegman DH. Epidemiology of toluene diisocyanate (TDI) induced respiratory disease. Environ Health Perspect (1975) 11:97-100.

70. Wegman DH, Pagnotto LD, Fine LJ, Peters JM. A dose-response relationship in TDI workers. J Occup Med (1974) 16:258-260.

71. Wegman DH, Musk AW, Main DM, Pagnotto LD. Accelerated loss of FEV-1 in polyurethane production workers: a four-year prospective study. Am J Ind Med (1982) 3:209-215.

72. Musk AW, Peters JM, Bernstein L. Absence of respiratory effects in subjects exposed to low concentrations of TDI and MDI: a reevaluation. J Occup Med (1985) 27:917-920.

73. Diem JE, Jones RN, Hendrick DJ, Glindmeyer HW, Dharmarajan V, Butcher B, Salvaggio JE, Weill H. Five-year longitudinal study of workers employed in a new toluene diisocyanate manufacturing plant. Am Rev Respir Dis (1982) 126:420-428.

74. Weill H, Butcher B, Dharmarajan V, Glindmeyer H, Jones R, Carr J, O'Neill C, Salvaggio J. Respiratory and immunologic evaluation of isocyanate exposure in a new manufacturing plant. Cincinnati: U.S. Department of Health and Human Services (1981) DHHS (NIOSH) Publication No. 81-125.

75. Beryllium disease and its control (Conference held at the Massachusetts Institute of Technology, Sept 30-Oct 1, 1958). AMA Arch Indust Health (1959) Volume 19.

76. Hall TC, Wood CH, Stoeckle JD, Tepper LB. Case data from the beryllium registry. AMA Arch Indust Health (1959) 19:100-103.

77. Jones RN, Weill H, Ziskind. Pulmonary function in sandblasters' silicosis. Bull Physiopath Resp (1975) 11:589-595.

78. Craighead JE, Vallyathan NV. Cryptic pulmonary lesions in workers occupationally exposed to dust containing silica. J Am Med Assn (1980) 244:1939-1941.

79. Rodriques-Roisin R, Picado C, Roca J, Arrigo S, Agusti-VidalA. Early lung function changes after short heavy exposure to chrysotile asbestos in non-smoking women. Bull Eur Physiopathol Respir (1986) 22:225-229.

80. Lam S, Chan-Yeung M. Nonspecific bronchial reactivity in occupational asthma. J Allergy Clin Immunol (1979) 63:28-34.

81. Fishwick D, Fletcher AM, Anthony C, Pickering C, Niven RM, Faragher EB. Lung function, bronchial reactivity, atopic status, and dust exposure in Lancashire cotton mill operatives. Am Rev Resp Dis (1992) 45:1103-1108.

82. Marine WM, Gurr D, Jacobsen M. Clinically important effects of dust exposure and smoking in British coal miners. Am Rev Respir Dis (1988) 137:106-112.

83. Mitchell CA, Gandevia B. Respiratory symptoms and skin sensitivity in workers exposed to proteolytic enzymes in detergent industry. Am Rev Respir Dis (1971) 104:1-12.

84. Burge PS, Perks WH, O'Brien IM, et al. Occupational asthma in an electronics factory: a case control study to evaluate etiologic factors. Thorax (1979) 300-307.

85. Venables KM, Topping MD, Howe W, Luczynski CM, Hawkins R, NewmanTaylor AJ. Interaction of smoking and atopy in producing specific IgE antibody against a hapten protein conjugate. Brit Med J (1985) 290:201-204.

86. Sunyer J, Anto J, Sabria J, et al. Risk factors of soybean epidemic asthma. Am Rev Respir Dis (1992) 145:1098-1102.

87. Chan-Yeung M, Lam S, Koerner S. Clinical features and natural history of occupational asthma due to western red cedar *(Thuja plicata)* Am Rev Resp Dis (1982) 72:411-415.

88. Paggiaro PL, Loi AM, Rossi O, Ferrante B, Pavdi F, Roselli MG. Follow-up study of patients with respiratory disease due to toluene diisocyanate (TDI). Allergy (1984) 14:463-469.

List of Acronyms and Abbreviations

ADP	Adenine diphosphoribose
ANL	Acute nonlymphocytic leukemia
APA	Aerican Psychiatric Association
ATP	Adenosine triphosphate
BAL	British Anti-Lewisite, an antidote for Lewisite
CAG	Carcinogen Assessment Group of the Environmental Protection Agency
CAS	Chemical Abstracts Service
CBW	Chemical and biological warfare
CDC	Centers for Disease Control
CMR	Committee on Medical Research
COPD	Chronic obstructive pulmonary disease
Ct	Concentration multiplied by time, used to determine cumulative exposure
CTGC	Committee on the Treatment of Gas Casualties
CWS	Chemical Warfare Service
DAV	Disabled American Veterans
DNA	Deoxyribonucleic acid
DoA	Department of the Army
DoD	Department of Defense
EPA	Environmental Protection Agency
FEV1	Forced expiratory volume in 1 second
FOIA	Freedom of Information Act
FVC	Forced vital capacity
HCN	Hydrogen cyanide
HD	Distilled sulfur mustard; bis(2-chloroethyl) sulfide

HN2	Mechlorethamine, a nitrogen mustard
HN3	Tris(β-chloroethyl) methylamine, a nitrogen mustard
hprt	Gene locus coding for hypoxanthine phosphoribosyl transferase
HS	Sulfur mustard (British acronym for agent HD); bis(2-chloroethyl) sulfide
HT	Impure mixture of distilled sulfur mustard, stabilizer compound, and sulfur impurities
IARC	International Agency for Research on Cancer
ICt_{50}	Statistically estimated concentration over a period of time that would cause incapacitation to 50 percent of ex posed test subjects
IOM	Institute of Medicine
IP	Intraperitoneal
IV	Intravenous
LCt_{50}	Statistically estimated concentration over a period of time that would be lethal to 50 percent of test subjects; median lethal count
LD_{50}	Dose level that would be lethal to 50 percent of test subjects; median lethal dose
MEDLARS	Medical Literature Analysis and Retrieval System
MeSH	Medical Subject Headings
MSH	Melanocyte-stimulating hormone
NAD	Nicotinamide adenine dinucleotide
NAS	National Academy of Sciences
NDRC	National Defense Research Committee
NLM	National Library of Medicine
NOEL	No observed effect level
NRC	National Research Council
NRL	Naval Research Laboratory
NTIS	National Technical Information Service
NTP	National Toxicology Program
OSHA	Occupational Safety and Health Administration
OSRD	Office of Scientific Research and Development
PADPRP	Poly-(adenosine diphosphateribose) polymerase
PBB	Polybrominated biphenyl
PF	Protection factor
PTSD	Post-traumatic stress disorder
RADS	Reactive airways disfunction syndrome
RASH	Rapid screening of hazard
RNA	Ribonucleic acid
SCE	Sister chromatid exchange
SIPRI	Stockholm International Peace Research Institute
SM	Sulfur mustard

SMR	Standardized mortality ratio
TD_{50}	Statistically estimated dose that would produce a toxic effect in 50 percent of test subjects; median toxic dose
TDI	Toluene diisocyanate
TMI	Three Mile Island
USUHS	Uniformed Services University of the Health Sciences
UV	Ultraviolet light
UVB	Region in the ultraviolet spectrum
VA	Department of Veterans Affairs
WHO	World Health Organization
WWI	World War I
WWII	World War II

Index

R

S

T

U

V

W